Cosmological Models of the Universe

Cosmological Models of the Universe

Guest Editors

Panayiotis Stavrinos
Emmanuel N. Saridakis

MDPI

Basel • Beijing • Wuhan • Barcelona • Belgrade • Novi Sad • Cluj • Manchester

Guest Editors
Panayiotis Stavrinos
Department of Mathematics
National and Kapodistrian
University of Athens
Athens
Greece

Emmanuel N. Saridakis
National Observatory of Athens
Athens
Greece

Editorial Office
MDPI AG
Grosspeteranlage 5
4052 Basel, Switzerland

This is a reprint of the Special Issue, published open access by the journal *Universe* (ISSN 2218-1997), freely accessible at: https://www.mdpi.com/journal/universe/special_issues/T0P2T501H4.

For citation purposes, cite each article independently as indicated on the article page online and as indicated below:

Lastname, A.A.; Lastname, B.B. Article Title. *Journal Name* **Year**, *Volume Number*, Page Range.

ISBN 978-3-7258-8221-2 (Hbk)
ISBN 978-3-7258-8222-9 (PDF)
https://doi.org/10.3390/books978-3-7258-8222-9

Contents

About the Editors

Panayiotis Stavrinos

Panayiotis Stavrinos, of the National and Kapodistrian University of Athens, is a founding member of the Balkan Society of Geometers and the Hellenic Relativity, Gravitation and Cosmology Society. He has authored over 100 research papers, with significant contributions to Finslerian Gravitation and Cosmology. In particular, he derived the Einstein–Finsler-like field equations on the Lorentz tangent bundle of spacetime and he also obtained generalized Schwarzchild solutions in Finsler–Randers gravity, as well as the Raychaudhuri equation in the Finslerian framework of spacetime. Furthermore, he studied cosmological models and derived generalized Friedmann equations in Finsler and Finsler-like frameworks. An invited speaker at universities and workshops worldwide, he also serves as a referee, editorial board member, and Guest Editor for international journals, as well as an organizer of numerous international conferences.

Emmanuel N. Saridakis

Emmanuel N. Saridakis is an author of 330 publications, with a total of 32,000 citations and an h-index of 98 (according to Google Scholar). His research interests lie in modified theories of gravity, early- and late-Universe cosmology, dark energy phenomenology, inflation, and black holes. He is a representative of Greece and a Working Group Leader in the COST Actions "Addressing observational tensions in cosmology with systematics and fundamental physics" (CosmoVerse), "COSMIC WISPers in the Dark Universe: Theory, astrophysics and experiments", and "Bridging high and low energies in search of quantum gravity (BridgeQG)", and he is a member of the LISA CosWG. Additionally, he is on the review boards of various journals and has performed more than 1500 reviews of scientific works.

Preface

The primary motivation of the subject 'Cosmological Models of the Universe' is to answer fundamental questions regarding the origin, structure, and future of the universe. In recent years, this has specifically involved resolving 'cosmological tensions,' such as discrepancies in the Hubble constant and the nature of dark matter, dark energy, or cosmological anisotropy. This research is also addressed to scholars who examine the existential implications of cosmic theories. Furthermore, investigation into this topic provides the necessary details for peers to evaluate the validity of results and conclusions, allowing researchers to replicate experimental methods and compare devices.

Panayiotis Stavrinos and Emmanuel N. Saridakis
Guest Editors

universe

MDPI

Article

Searching for Gravitational-Wave Bursts from Cosmic String Cusps with the Parkes Pulsar Timing Array's Third Data Release

Yong Xia [1,2,3], Jingbo Wang [3,*], Sachiko Kuroyanagi [4,5], Wenming Yan [1,6,*], Yirong Wen [1,2,3], Agastya Kapur [7], Jing Zou [1,2,3], Yi Feng [8,9], Valentina Di Marco [10,11,12], Saurav Mishra [13], Christopher J. Russell [14], Shuangqiang Wang [1], De Zhao [1] and Xingjiang Zhu [15,16]

1 Xinjiang Astronomical Observatories, Chinese Academy of Sciences, Urumqi 830011, China
2 University of Chinese Academy of Sciences, Beijing 100049, China
3 Institute of Optoelectronic Technology, Lishui University, Lishui 323000, China
4 Department of Physics, Nagoya University, Nagoya 464-8602, Japan
5 Instituto de Fisica Teorica UAM-CSIC, Universidad Autonoma de Madrid, 28049 Madrid, Spain
6 Xinjiang Key Laboratory of Radio Astrophysics, Urumqi 830011, China
7 School of Mathematical and Physical Sciences, Macquarie University, Sydney, NSW 2109, Australia
8 Research Center for Astronomical Computing, Zhejiang Laboratory, Hangzhou 311100, China
9 Institute for Astronomy, School of Physics, Zhejiang University, Hangzhou 310027, China
10 School of Physics and Astronomy, Monash University, Clayton, VIC 3800, Australia
11 OzGrav: The Australian Research Council Centre of Excellence for Gravitational Wave Discovery, Clayton, VIC 3800, Australia
12 Australia Telescope National Facility, CSIRO, Space and Astronomy, P.O. Box 76, Epping, NSW 1710, Australia
13 Centre for Astrophysics and Supercomputing, Swinburne University of Technology, P.O. Box 218, Hawthorn, VIC 3122, Australia
14 CSIRO Scientific Computing, Australian Technology Park, Locked Bag 9013, Alexandria, NSW 1435, Australia
15 Department of Physics, Faculty of Arts and Sciences, Beijing Normal University, Zhuhai 519087, China
16 Institute for Frontier in Astronomy and Astrophysics, Beijing Normal University, Beijing 102206, China
* Correspondence: 1983wangjingbo@163.com (J.W.); yanwm@xao.ac.cn (W.Y.)

Abstract: Pulsar timing arrays (PTAs) are designed to detect nanohertz-frequency gravitational waves (GWs). Since GWs are anticipated from cosmic strings, PTAs offer a viable approach to testing their existence. We present the results of the first Bayesian search for gravitational-wave bursts from cosmic string cusps (GWCSs) using the third PPTA data release for 30 ms pulsars. In this data collection, we find no evidence for GWCS signals. We compare a model with a GWCS signal to one with only noise, including a common spatially uncorrelated red noise (CURN), and find that our data are more consistent with the noise-only model. We then establish upper limits on the strain amplitude of GWCSs at the pulsar term, based on the analysis of 30 ms pulsars, after finding no compelling evidence. We find the addition of a CURN with different spectral indices into the noise model has a negligible impact on the upper limits. And the upper limit range of the amplitude of the pulsar-term GWCSs is concentrated between 10^{-12} and 10^{-11}. Finally, we set upper limits on the amplitude of GWCS events, parametrized by width and event epoch, for a single-pulsar PSR J1857 + 0943. Moreover, we derive the upper limit on the cosmic string tension as a function of burst width and compare it with previous results.

Keywords: pulsars; gravitational waves; Bayesian statistical

1. Introduction

Low-frequency GWs between 10^{-9} and 10^{-7} Hz can be detected with high sensitivity using a pulsar timing array (PTA), which consists of a group of millisecond pulsars with

 https://doi.org/10.3390/universe11030081

extremely stable rotational periods [1]. Owing to their stability, it is anticipated that timing residuals affected by a gravitational wave (GW) can be observed by closely monitoring the times of arrival (TOAs) of radio pulses from these millisecond pulsars [2]. Timing residuals are the discrepancies between the TOAs and the predictions of the pulsar timing model [3]. Factors other than GWs that can lead to timing residuals include clock faults, unmodeled variability in the solar wind, interstellar medium scattering, flaws in the solar system ephemeris (SSE), and offsets caused by the observational instrument [4]. Therefore, there will be a significant difficulty in detecting gravitational waves. The North American Nanohertz Observatory for Gravitational Waves (NANOGrav) [5], the European Pulsar Timing Array (EPTA) [6], the Parkes Pulsar Timing Array (PPTA) [7], and the Indian Pulsar Timing Array (InPTA) [8] are among the PTA collaborations that are currently in operation. These partnerships result in the International Pulsar Timing Array (IPTA) [9]. Furthermore, the first analyses from the Chinese Pulsar Timing Array (CPTA) [10] and the MeerTime Pulsar Timing Array (MPTA) [11] have been published recently.

PTAs are currently the most promising method for detecting nanohertz GWs, primarily generated by the inspiraling of supermassive black hole binaries (SMBHBs) [12]. The incoherent superposition of these binary systems produces a stochastic GW background. Recent studies by NANOGrav [13], PPTA [14], EPTA [15], and CPTA [10] have reported that the significance of a GW origin is between 2σ and 4σ. The observed correlations between pulsar timing residuals closely match the predicted Hellings–Downs pattern, a characteristic signature of such a background. This finding has significant implications for our understanding of supermassive black holes and the large-scale structure of the Universe.

Beyond the stochastic gravitational-wave background, PTA data can also be used to probe a variety of other astrophysical phenomena. For instance, continuous gravitational waves emitted by single supermassive black holes [16], known as continuous waves, can be detected or constrained using PTA observations. Additionally, PTAs are sensitive to gravitational wave memory [17,18], a permanent distortion of spacetime caused by extreme gravitational events like the merger of two black holes. Furthermore, PTAs can provide insights into the nature of dark matter [19]. By searching for characteristic signals from ultralight scalar-field dark matter, PTAs can constrain the properties of this elusive component of the Universe.

One particularly intriguing possibility is the detection of GWCSs, hypothetical defects in the fabric of spacetime that may have formed during the early Universe. Such bursts, if they exist, could be within the detectable frequency range of PTAs. Theoretical models suggest that a spontaneous symmetry-breaking phase transition in the early Universe [20] could have resulted in the formation of macroscopic, one-dimensional, stable, and highly energetic cosmic strings. Quantum field theory and condensed matter models predict these cosmic strings as topological defects, which play a role in various supersymmetric unified field theories, including D-brane models [21]. Initially, cosmic strings were proposed as a potential explanation for the formation of large-scale structure in the early Universe, with symmetry breaking occurring at the grand unification scale [22].

Due to the predicted emission of GWCSs, experiments like the PTAs offer a potential method to verify their existence. The most powerful GW bursts are generated at cusps, which are highly Lorentz-boosted singularities on string loops [20]. A cusp is a region in the loop that is highly Lorentz-boosted and produces a powerful beam of GWs. In general, the gravitational wave background is primarily the result of the overlap of multiple bursts. However, in specific regions of parameter space, a limited number of distinct GW bursts can contribute significantly to the overall background. A sufficiently large GW amplitude from such an event could render it detectable as a single burst. The Laser

Interferometer Gravitational-Wave Observatory (LIGO) [23–25] has conducted searches for individual GWs from cosmic strings in the direction of fast radio burst sources. Leveraging the second data release from the PPTA project, Yonemaru et al. [21] performed the first search for GWCSs, employing a frequentist-based approach. In this paper, we will search and place limits on GWCSs based on the third PPTA data release (PPTA-DR3) using a Bayesian approach.

This paper is organized as follows. In Section 2, we provide a short description of PPTA-DR3. In Section 3, we examine the impact of GWCSs on TOAs from a PTA. We provide a concise overview of the Bayesian method and software employed in this search in Section 4. In Section 5, we demonstrate the implementation of the proposed algorithm on the PPTA data and discuss the resulting outcomes. Finally, the conclusion is presented in Section 6.

2. Data

This analysis is based on the PPTA-DR3 data and the noise analyses of individual pulsars. We will provide a concise summary of the most significant aspects of this dataset, with additional details available in the work of Zic et al. [26] and Reardon et al. [27].

The 64-meter Parkes "Murriyang" radio telescope was utilized to observe 32 pulsars, which are included in the release. Observations commenced in 2004, and the data were collected over the course of 18 years (MJD 53,040–59,640) consisting of approximately 48 h of observations made every 2–3 weeks. This release combines an updated version of the second PPTA data release (PPTA-DR2) with approximately three years of more recent data. The data were predominantly obtained using an ultra-wide-bandwidth receiver system that operates between 704 and 4032 MHz. One of the pulsars that was previously included in PPTA-DR2, PSR J1732 − 5049, was excluded due to the significant uncertainties in its TOAs, which resulted in periodic observations ceasing in 2011. Six additional binary pulsars (PSRs J0125 − 2327, J0614 − 3329, J0900 − 3144, J1741 + 1351, J1902 − 5105, and J1933 − 6211) and one solitary pulsar (PSR J0030 + 0451) are included in this release, in contrast to previous data releases. After the UWL receiver was commissioned, these pulsars were incorporated into PPTA observations, despite not having been previously included. In large part, the enhanced observation efficiency of the UWL receiver enabled the inclusion of these pulsars in the PPTA observations.

The dataset utilized for this analysis, which encompasses the pulsar ephemerides and TOAs, is detailed in Zic et al.'s work [26]. Commencing with the timing analyses of the preceding data releases [28,29], the initial timing models of each pulsar were fitted using TEMPO2 [3]. For the pulsars newly included in the PPTA following the release of PPTA-DR2, we adopt the initial timing models presented by Curyło et al. [30].

Despite the inclusion of observations from 32 pulsars in PPTA-DR3, we will analyze only 30 pulsars in this study. We excluded the globular cluster MSP PSR J1824 − 2452A because its steep-spectrum red noise is too intense for a common process to affect it. Although globular-cluster dynamics may also be involved, the disturbance is likely intrinsic to the pulsar [31]. This analysis, as well as the subsequent GWCS search, excludes PSR J1741 + 1351 because the dataset contains only 16 unique observations of this pulsar, which is insufficient for noise process simulation.

3. Signal of the GWCSs

We will examine the effect of GWCSs on TOA in pulsar timing data in this section and provide an extensive overview of the signal and noise models incorporated. A GW passing through a pulsar will induce a shift in the observed rotational frequency, which may result in either an increase or decrease in the frequency. When a GW passes through the Earth, the rotational frequencies observed for all pulsars in the PTA will be affected, leading to either an increase or decrease. The timing residuals induced by a GWCS will display distinct shapes depending on the epoch and duration of the burst, yet these shapes adhere to a deterministic pattern.

The timing residuals will be introduced as a result of the discrepancy between the observed rotational frequency and the pulsar's timing model-fitted rotational frequency, regardless of the situation. Detweiler [32] provided the timing residuals induced by GWs:

$$r(t) = \sum_{a=+,\times} F^a(\hat{\Omega}, \hat{p}) \int^t \Delta h_a(\hat{\Omega}, t')dt', \tag{1}$$

where $\hat{p}$ and $\hat{\Omega}$ are the directions of the pulsar and the GW propagation, respectively. $F^a(\hat{\Omega}, \hat{p})$ is the antenna pattern defined by Anholm et al. [33]:

$$F^a(\hat{\Omega}, \hat{p}) = \frac{1}{2}\frac{\hat{p}^i\hat{p}^j}{1+\hat{\Omega}\cdot\hat{p}}e^a_{ij}(\hat{\Omega}). \tag{2}$$

The GW polarization tensors, denoted as $e^a_{ij}(a = +, \times)$, are defined as follows:

$$e^+_{ij}(\hat{\Omega}) = \hat{m}_i\hat{m}_j - \hat{n}_i\hat{n}_j \tag{3}$$

$$e^\times_{ij}(\hat{\Omega}) = \hat{m}_i\hat{n}_j + \hat{n}_i\hat{m}_j, \tag{4}$$

where $\hat{m}$ and $\hat{n}$ are the polarization basis vectors. For ease of computation, we define

$$\hat{m} = sin\delta\hat{x} - cos\delta\hat{y} \tag{5}$$

$$\hat{n} = -cos\alpha cos\delta\hat{x} - cos\alpha sin\delta\hat{y} + sin\alpha\hat{z}, \tag{6}$$

where $\hat{x}, \hat{y}$, and $\hat{z}$ are unit vectors in the Cartesian coordinate system, and (α, δ) is the gravitational wave source position. Therefore, we can obtain

$$\hat{\Omega} = -sin\alpha cos\delta\hat{x} - sin\alpha sin\delta\hat{y} - cos\alpha\hat{z}, \tag{7}$$

where $\Delta h_a(\hat{\Omega}, t')$ represents the metric perturbation difference between the Earth and the pulsar, as determined by the linearly polarized GWCSs. The time-domain waveform of a plus-polarized GWCS event is detailed by Yonemaru et al. [21]:

$$h_+(t) = \begin{cases} A_{\rm fit}\left[|t-t_0|^{1/3} - (\frac{1}{2}W)^{1/3}\right] & (t_0 - \frac{1}{2}W \le t < t_0 + \frac{1}{2}W) \\ 0 & (\text{otherwise}) \end{cases} \tag{8}$$

$$h_\times(t) = 0, \tag{9}$$

where $A_{\rm fit}$ is the amplitude, t_0 is the epoch at which the burst peak arrives the Earth, and W is the duration of the burst. Strongly beamed in the direction of the cusp velocity is the GW burst, with cosmic string loops thought to generate cusps with high efficiency ($\mathcal{O}(1)$ per oscillation period). The orientation is assumed to be random and depend on the string configuration [34]. It is important to mention that $A_{\rm fit}$ carries dimensions of $s^{-1/3}$. The peak value at $t = t_0$, which is dimensionless, is defined as follows:

$$A_{\rm peak} \equiv h_+(t_0) = \left(\frac{1}{2}W\right)^{1/3} A_{\rm fit}. \tag{10}$$

By substituting Equation (8) into Equation (1), the analytic expressions for the pre-fit timing residual caused by the GWCSs are derived, using $A_{\rm fit}$, as follows:

$$r(t) = F^{+}(\hat{\Omega}, \hat{p}) \times \begin{cases} 0 & (t < t_0 - \frac{W}{2}) \\ A_{\rm fit}\left[\frac{3}{4}\left(\left(\frac{W}{2}\right)^{4/3} \mp |t - t_0|^{4/3}\right) - \left(\frac{W}{2}\right)^{1/3}\left(t - \left(t_0 - \frac{W}{2}\right)\right)\right] & \left(t_0 - \frac{1}{2}W \leq t < t_0 + \frac{1}{2}W\right) \\ -\frac{1}{4}\left(\frac{1}{2}\right)^{1/3} A_{\rm fit} W^{4/3} & (t \geq t_0 + \frac{1}{2}W) \end{cases}. \quad (11)$$

In the second line, the use of $\mp$ signifies that the $-$ sign is to be placed before t_0 and the $+$ after. The effect of GWCSs has been incorporated into the TEMPO2 timing model. This allows us to simulate residuals (or TOAs) of the GWCSs. The updated timing model parameters are $(A_{\rm fit}, t_0, W, \alpha, \delta, \zeta)$. Here, ζ is the principal polarization angle. This parametrization proves advantageous for the simulation of timing residuals induced by GWCSs. Introducing a second parametrization of the GW burst proves advantageous for the search procedure. The GW burst in this parametrization is represented by two orthogonal components, A_1 and A_2, with $A_1 = A_{\rm fit}\cos(2\zeta)$ and $A_2 = A_{\rm fit}\sin(2\zeta)$. These components are associated with the two polarization modes of the GW, and this parametrization allows for the search of all possible GW polarizations. TEMPO2 software package allows the use of the following parameters in its parameter files: GWCS_A1, GWCS_A2, GWCS_POSITION, GWCS_EPOCH, and GWCS_WIDTH, which correspond to the cosmic string parameters A_1, A_2, sky position, burst epoch, and width, respectively.

Examples of the waveform and simulated timing residuals are shown in Figure 1. The PTASIMULATE software package was used to simulate cosmic string burst events as described above. We assume that the position and polarization angle of the gravitational wave source are $(\alpha = 57.3°, \delta = -57.3°)$ and $\zeta = 30°$, respectively. Timing residuals are produced by infusing a GWCS signal with an amplitude of $A_{\rm peak} = 10^{-14}$ at the center of the observation (MJD 55,000) and 1 µs of Gaussian white noise using Equation (11). The burst widths are set to 500 days (panel on the left side) and 2000 days (panel on the right side), corresponding to $A_{\rm fit} = 1.59 \times 10^{-15}$ and 1.0×10^{-15}, respectively.

Because of the crucial role of detecting a common spatially uncorrelated red noise process (CURN) identified by NANOGrav, PPTA, EPTA, and CPTA, a CURN with a specified spectral index is thus included in our model. A power-law can be used to model the CURN, with its power spectrum governed by two hyperparameters, (A, γ) [35]:

$$P(f) = A^2\left(\frac{f}{\rm yr^{-1}}\right)^{-\gamma}, \quad (12)$$

Here, f denotes the frequency of the spectral component, A represents the characteristic amplitude of the red noise process at a frequency of $\rm yr^{-1}$, and γ is the spectral index associated with the characteristic amplitude. The spectral index of a stochastic gravitational wave background, arising from an ensemble of supermassive black hole binaries, is expected to be 4.33. However, the spectral index of the CURN was found to have a maximum a posteriori value of 5.5 [36]. In this paper, we present two sets of results for the CURN that utilize both of these fixed spectral indices ($\gamma_{\rm CURN} = 4.33$ and 5.5).

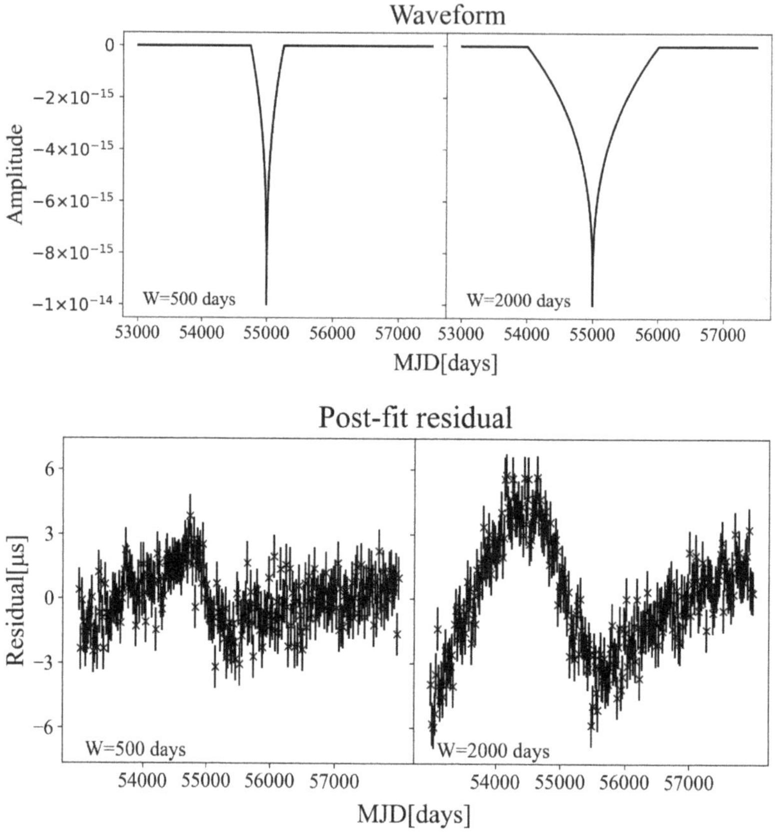

Figure 1. The top panels display the waveforms of GWCSs. The simulated post-fit timing residuals, accompanied by 1 μs of Gaussian white noise, are displayed in the bottom panels. The gravitational wave amplitude is $A_{\text{peak}} = 10^{-14}$, and the peak occurs at MJD 55,000. These post-fit residuals are utilized to fit pulsar parameters like pulse period and spin-down rate, as shown in the bottom panel.

4. Methods

We used the Markov Chain Monte Carlo (MCMC) approach for the search of GWCSs. In this section, we will provide a concise summary of the techniques. The methods employed in this search are detailed in the works of Sun et al. [37] and Aggarwal et al. [38] to search for gravitational wave memory and continuous gravitational wave signals. The TOA residuals for an individual pulsar can be represented as a combination of multiple stochastic deterministic and stochastic processes:

$$\delta t = \delta t_{\text{cs}} + M\epsilon + Fa + F_{\text{gw}} a_{\text{gw}} + n. \tag{13}$$

where δt represents the residual time series for the pulsar. The term δt_{cs} denotes the residuals induced by GWCSs. M is the design matrix representing the linearized timing model, which is responsible for accounting for the uncertainty in the residuals resulting from an imperfect timing model fit ϵ. The design matrix F corresponds to the pulsar's intrinsic Gaussian red noise process, modeled as a Fourier series with coefficients represented by a. Likewise, F_{gw} and a_{gw} are the design matrix and Fourier coefficients for the CURN, respectively. Lastly, the elements of vector n represent uncertainty in the observed TOAs, which follow a Gaussian white noise distribution.

Based on the estimations of the Gaussian process parameters, GWCS signal, and timing model parameters, it is possible to generate residuals n:

$$n = \delta t - \delta t_{\mathrm{cs}} - M\epsilon - Fa - F_{\mathrm{gw}}a_{\mathrm{gw}}. \tag{14}$$

As the terms on the right-hand side are approximations, this constitutes merely an estimation of white noise. Nevertheless, assuming the white noise follows a Gaussian distribution, the likelihood of observing this specific sequence of white noise residuals can be expressed as

$$p(n) = \frac{exp(-\frac{1}{2}n^T N^{-1} n)}{\sqrt{2\pi detN}}. \tag{15}$$

Above, N is a covariance matrix that represents the white noise uncertainties in each observed TOA, and n^T is the transpose of n.

After the removal of deterministic effects, the remaining pulsar timing residuals have the same likelihood of following a Gaussian white noise distribution as the detection of a GWCS signal. This means that

$$p(\delta t \mid \epsilon, a, a_{\mathrm{gw}}, A_{\mathrm{fit}}, t_0, W, \hat{\Omega}, \hat{p}, \zeta) = \frac{exp(-\frac{1}{2}n^T N^{-1} n)}{\sqrt{2\pi detN}}. \tag{16}$$

The parameter space dimensionality can be reduced through the analytical marginalization of the likelihood in Equation (16) with respect to the parameters that describe the Gaussian processes [39–41]. The final marginalized likelihood obtained is

$$p(\delta t \mid A_{\mathrm{fit}}, W, t_0, \hat{\Omega}, \hat{p}, \zeta) = \frac{exp(-\frac{1}{2}q^T C^{-1} q)}{\sqrt{2\pi detC}}, \tag{17}$$

where the following definitions are provided:

$$q = \delta t - \delta t_{\mathrm{cs}}, \tag{18}$$

$$C = N + TDT^T, \tag{19}$$

$$T = \begin{bmatrix} M & F \end{bmatrix}, \tag{20}$$

$$D = \begin{bmatrix} \infty & 0 \\ 0 & \phi \end{bmatrix}, \tag{21}$$

where ∞ is a diagonal matrix of infinities, which effectively provides unconstrained priors on the timing model parameters. ϕ is a covariance matrix that represents the individual red noise and the CURN Fourier coefficients. As a result of employing a CURN, the ϕ matrices are diagonal, with each diagonal element representing the red noise power at the corresponding frequency bin, as specified in Equation (12). C^{-1} is efficiently computed using the Woodbury matrix identity [42]. In this identity, the matrix D is present solely as an inverse. As a result, the diagonal matrix containing infinities is effectively replaced by a matrix of zeros during the likelihood calculation.

We favored ignorance priors for our model parameters and implemented uniform or log-uniform priors for all. We employed the same priors for noise parameters as those used by Reardon et al. [27] and Reardon et al. [14]. Table 1 provides the prior distributions of model parameters in the Bayesian Search for Global GWCSs using the full PTA that we are interested in. The Bayes factor, denoted as $\mathcal{B}_{\mathrm{gw}}$, was employed as the detection statistic to compare the GW model with a noise-only model. Bayes factor was computed using the Savage–Dickey approximation, as described by Dickey [43]:

$$\mathcal{B}_{\rm gw} = \frac{\varepsilon_{\rm gw}}{\varepsilon_{\rm noise}} \approx \lim_{A_{\rm fit}\to 0} \frac{p(A_{\rm fit})}{p(A_{\rm fit}|\delta t)}. \tag{22}$$

The evidence ratio ($\varepsilon_{\rm gw}/\varepsilon_{\rm noise}$) for the GW and noise-only models can be approximated as the ratio of the prior to posterior probability as the GW amplitude approaches zero. This computation is significantly more computationally efficient than a complete evidence integral, as it employs posterior samples that are located near the low-amplitude prior boundary.

Table 1. Prior distributions used in all analyses of this article, including red noise, CURN, and GWCSs.

Parameter	Prior	Description
$\log_{10} A_{\rm rn}$	LinearExp (−17, −11)	Amplitude of intrinsic pulsar red noise
$\gamma_{\rm rn}$	Uniform (0, 7)	Spectral index of intrinsic pulsar red noise
$\log_{10} A_{\rm CURN}$	LinearExp (−17, −11)	Amplitude of GWB
$\gamma_{\rm CURN}$	Uniform (0, 7)	Spectral index of GWB
$\log_{10} A_{\rm GWCS}$	LinearExp (−18, −11)	Amplitude of GWCSs
$W_{\rm GWCS}$	Uniform (0, 5000)	Width of GWCSs
$t_{\rm GWCS}$	Uniform (MJD 53,500, MJD 59,100)	Epoch of GWCSs
$\zeta_{\rm GWCS}$	Uniform (0, π)	Polarization of GWCSs
$\alpha_{\rm GWCS}$	Uniform (0, π)	Polar angle of GWCS source
$\delta_{\rm GWCS}$	Uniform (0, 2π)	Azimuthal angle of GWCS source

Notes: There are a total of six global GWCS parameters. Imposing priors on the logarithm of the amplitude is mathematically equivalent to applying uniform priors directly on the amplitude itself. Red noise and CURN prior parameters satisfy the power-law distribution of Equation (12).

This likelihood calculation and these signal models were implemented in the ENTERPRISE [44] and ENTERPRISE_EXTENSIONS [45]. The PTMCMCSAMPLER [46] package implements the MCMC sampler to extract samples from the posterior distributions.

5. Results

5.1. Earth-Term GWCS Search

By applying MCMC sampling, we initiated a Bayesian search for GWCSs in the Earth term, contrasting two models: (1) a noise-only model and (2) a model that includes both noise and the GWCS signal. The noise-only model included white noise, achromatic red noise (Red), dispersion measure (DM), high fluctuation frequency (HFF), chromatic (Chr), low-frequency band noise (BN), and a CURN process [14]. The signal model incorporated identical noise processes together with GWCS signals. We employed the product-space sampling technique [47] to sample both models simultaneously. This allowed us to determine the posterior probability for the GWCS signal and compute the Bayes factor of $\mathcal{B}_{\rm gw} = 0.7$ for the GWCS signal model. Compared to the noise-only model, this Bayes factor is too small to be considered detected. Figure 2 illustrates the posterior probability distributions for the GWCS signal as well as the global spatially uncorrelated red noise process.

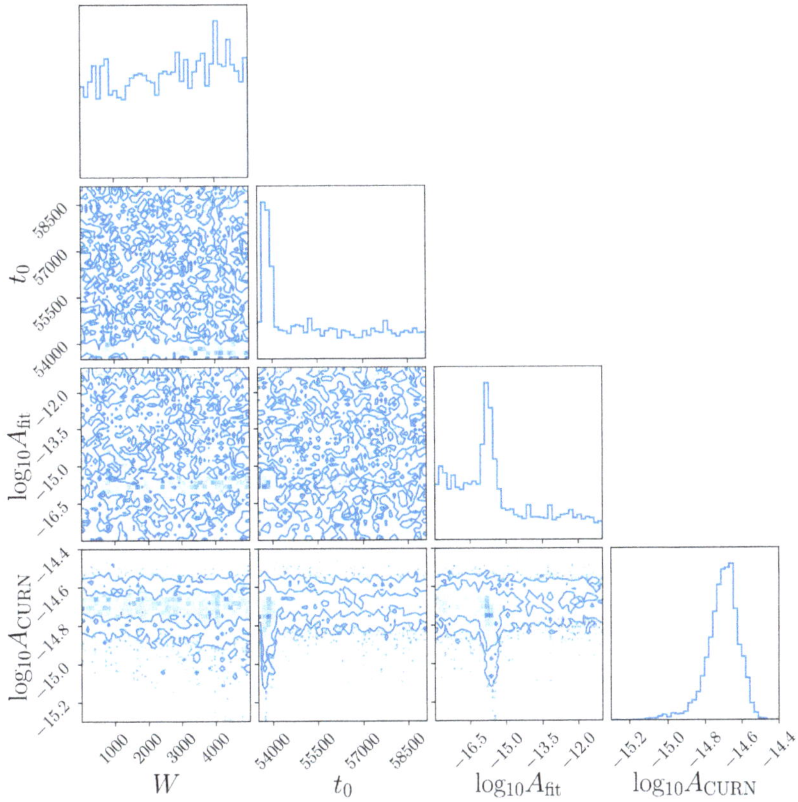

Figure 2. A corner plot displaying the marginalized 1D and 2D posterior distributions for four essential model parameters: burst strain amplitude $\log_{10}A_{\rm fit}$, burst epoch t_0, width W, and CURN amplitude $\log_{10}A_{\rm CURN}$. In the presence of the GWCS model, the favorable localization of $\log_{10}A_{\rm CURN}$ suggests that the CURN remains detectable. Moreover, the tail of $\log_{10}A_{\rm fit}$ stretches to a very small amplitude, suggesting that the model still strongly supports $A_{\rm fit} \sim 0$.

By analyzing the posterior distribution, we can pinpoint geographically significant areas of high activity near MJDs 54,000. The feature near MJD 54,000 is situated approximately in the beginning of our observations, during a period when there were significant gaps in data for multiple pulsars. At the beginning of our dataset, there was a small number of pulsars being observed, and the observations exhibited less regularity. Data sparsity complicates the task of constraining any signal in this time. A heightened degeneracy is observed for events occurring earlier in the dataset when a quadratic pulsar timing model is applied to the pulsar's rotational frequency and its derivative. This implies that the signal model can exhibit consistency with a GWCS event of significant magnitude, which is appropriately eliminated by the marginalization of the timing model.

5.2. Pulsar-Term Upper Limits

It is almost impossible to make a confident detection with a single pulsar term search of GWCSs, but we could still use non-detection in the pulsar term search to set upper limits. We report upper limits on the amplitude of the GWCS strain. The pulsar-term upper limits on GWCSs are illustrated in Figure 3, which employs both fixed spectral indices ($\gamma_{\rm CURN} = 4.33$ and 5.5). As the pulsar-term upper limits are calculated individually for each pulsar, any information about the signal's sky location is effectively lost. This means that it becomes challenging to distinguish between a weak GWCS event and one that originates from a sky location where the antenna pattern is weak relative to the position of the pulsar.

The upper limit $A_{\rm fit}$ is determined by setting the width at $W = 100$ days, and then we calculate the upper limit h for the pulsar term using Equation (10). It is evident that the pulsar-term upper limits are not significantly influenced by the selection of spectral index for the majority of pulsars.

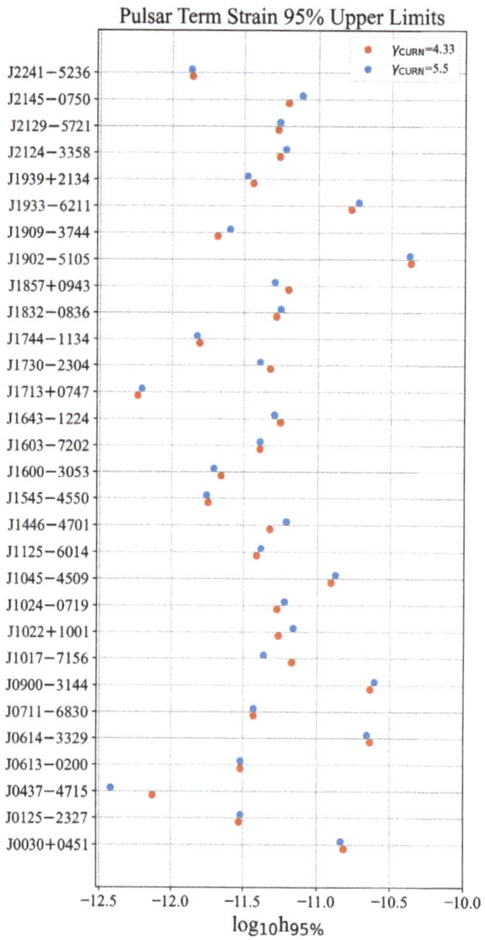

Figure 3. A plot of the upper limits of GWCS strain amplitude in the pulsar-term. In order to obtain the upper limits, we fixed a duration of the burst W = 100 days and obtained the upper limits for the pulsar terms with fixed spectral indices $\gamma_{\rm CURN}$ = 4.33 and 5.5, respectively. We observe minimal variation in the pulsar-term upper limits when different CURN spectral indices are fixed.

The constraints on h can be expressed as a mathematical function that depends on the epoch, width, and sky position. Only the width has a physical significance, which is equivalent to the loop size of the cosmic string. In Figure 4, we obtained the upper limit h as a function of the width for PSR J1857 + 0943. The early- and late-time bursts lack credibility due to insufficient data available prior to or following the occurrence of these bursts, which hinders the ability to accurately determine their amplitude. As the width increases, the constraint on h becomes more stringent since the timing residuals caused by the GWCSs increase as the event duration increases. In Figure 4, the upper limit h is a function of the width. As width increases, the strain amplitude h decreases, then increases, and finally plateaus. The limitation is strongest when the width is about 1200 days and the upper limit is about 10^{-12}.

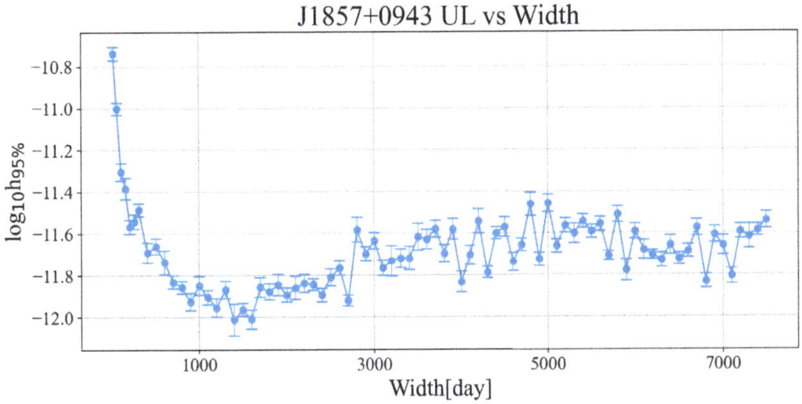

Figure 4. The relationship between the GWCS amplitude and burst width is constrained, assuming a fixed CURN spectral index of $\gamma_{\mathrm{CURN}} = 4.33$.

A term specific to each pulsar is introduced to impose a constraint on the upper time analysis for individual pulsars within the array. The upper bound h is a function of the burst epoch t_0 by fixing $W = 100$ days for PSR $1857 + 0943$ in Figure 5. When epoch t_0 is the median value, the upper limit of the amplitude stabilizes, and the constraint on h is the weakest at the beginning and end. When epoch t_0 is about 59,000, the GWCS strain amplitude upper limit is possibly the strongest because of a new ultra-wideband receiver.

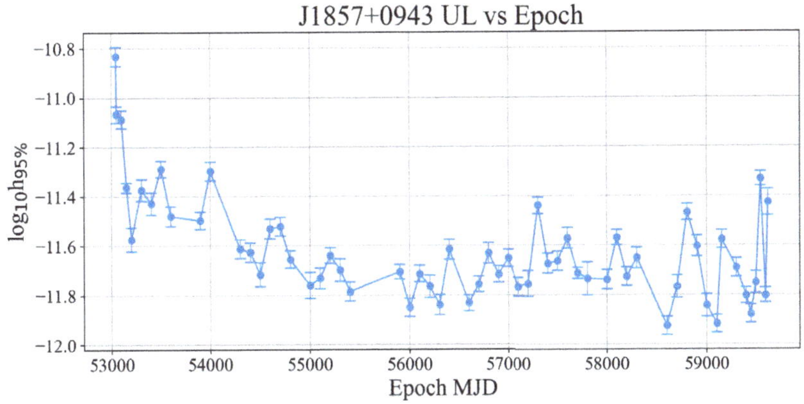

Figure 5. The upper limit on the GWCS strain amplitude as a function of burst epoch t_0 is derived by fixing the burst width at $W = 100$ days and adopting a CURN power-law spectral index of $\gamma_{\mathrm{CURN}} = 4.33$.

Here, we translate the upper bound on the peak amplitude shown in Figure 4 into a constraint on the cosmic string tension. The upper bounds on the time-domain amplitude with different widths shown in Figure 4 can be translated into the Fourier strain amplitude $\tilde{h}_{\mathrm{lim}}$. The expected number of gravitational wave bursts from cosmic string cusps reaching

Earth today with amplitudes exceeding $\tilde{h}_{\rm lim}$ is denoted as $N_{\rm GWCS}$. Given the absence of detected cosmic string GW bursts with amplitudes above $\tilde{h}_{\rm lim}$, a random Poisson process analysis at the 95% confidence level excludes scenarios where $N_{\rm GWCS} > 2.996$. Using this constraint, we establish an upper limit on the cosmic string tension. Details of the translation can be found in [21].

Figure 6 presents the constraint on the cosmic string tension as a function of burst width W, derived from the single-pulsar dataset of PSR 1857 + 0943, alongside the results from [21]. The α in Figure 6 represents the size of the primordial string loop, and here, we compare the cosmic string tension limit specifically for $\alpha = 0.1$. As shown in Figure 6, the upper bound becomes more constraining for larger W due to two factors. First, as shown in Figure 4, the upper bound improves for larger W. Second, larger loops produce stronger GW amplitudes but have lower number densities. While the former enhances detectability, the latter reduces it. Considering both effects, we find that the former effect dominates, and that larger loops, equivalently larger W, are easier to detect, leading to tighter constraints.

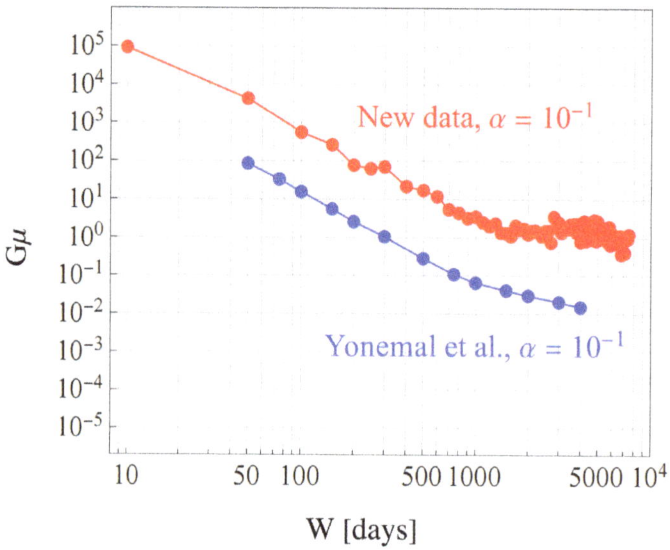

Figure 6. Constraint on $G\mu$ is derived for different widths, with the coupling constant α fixed at 0.1 [21].

However, the upper limit from [21] is more than an order of magnitude tighter than our results. This discrepancy arises because we only used the dataset of a single pulsar (pulsar term), whereas Yonemaru et al. [21] utilized the full PPTA dataset (Earth term). Additionally, MCMC sampling is very computationally intensive and time-consuming. In future work, we plan to extend our analysis to include the Earth-term search.

So far, only Yonemaru et al. [21] and this work have studied individual GW bursts from cosmic string cusps using PTA. In addition, cosmic string parameters can also be constrained by the cosmic microwave background (CMB) [48] and the stochastic GW background. While the stochastic background provides a stronger constraint on string parameters, we emphasize that GW burst searches offer independent constraints. The future SKA telescope will observe a large number of millisecond pulsars with high timing precision. With the improved sensitivity of SKA, the prospects for detecting single GW bursts will be significantly enhanced. A stringent constraint on the cosmic string tension from SKA would be highly valuable for testing various models of cosmic superstrings, such as the KKLMMT model [49].

6. Conclusions

In this paper, we present the first Bayesian search for the GWCSs using PPTA-DR3; there is no significant evidence of GWCSs. Therefore, we place upper limits on the strain amplitude of pulsar-term GWCSs for the 30 ms pulsars. We find that incorporating a CURN with various spectral indices into the noise model has a negligible impact on the upper limits. And the upper limit range of the amplitude of the pulsar-term GWCSs is concentrated between 10^{-12} and 10^{-11}. By analyzing an individual pulsar PSR J1857 + 0943, we place upper limits on the amplitude of GWCS events as a function of width and event epoch. We obtain the upper limit on the cosmic string tension as a function of burst width and will present the results of all-sky Earth-term upper limits in future work. With its improved sensitivity, SKA will place significantly tighter constraints on cosmic string parameters, making it a valuable tool for testing cosmic string models. Moreover, such constraints will provide an independent probe for investigating the clustering of cosmic string loops in our Galaxy, which cannot be tested through CMB analysis or stochastic GW background.

Author Contributions: Software, Y.X. and Y.W.; investigation, J.W., S.K., Y.F., S.M., C.J.R., S.W. and D.Z.; data curation, Y.X. and Y.W.; writing—original draft preparation, Y.X.; writing—review and editing, J.W., W.Y. and A.K.; visualization, Y.X., S.K. and J.Z.; funding acquisition, J.W., W.Y., V.D.M. and X.Z. All authors have read and agreed to the published version of the manuscript.

Funding: This research was funded by the Major Science and Technology Program of Xinjiang Uygur Autonomous Region (No. 2022A03013-4), the Zhejiang Provincial Natural Science Foundation of China (No. LY23A030001), the Natural Science Foundation of Xinjiiang Uygur Autonomous Region (No. 2022D01D85), National Natural Science Foundation of China (No. 12041304); W.M.Y. is supported by the National Natural Science Foundation of China (NSFC) project (No. 12273100, 12041303), the West Light Foundation of Chinese Academy of Sciences (No.WLFC 2021-XBQNXZ-027), the National Key Program for Science and Technology Research and Development and the National SKA Program of China (No. 2022YFC2205201, 2020SKA0120200); V.D.M. is supported by the Australian Research Council (ARC) Centre of Excellence CE170100004 and CE230100016, and receives support from the Australian Government Research Training Program; X.J.Z. is supported by the National Natural Science Foundation of China (Grant No. 12203004) and by the Fundamental Research Funds for the Central Universities.

Data Availability Statement: Observational data used in this paper are quoted from the cited works. Additional data generated from computations can be made available upon reasonable request.

Acknowledgments: The Murriyang Parkes 64 m radio telescope, part of the Australia Telescope National Facility, is funded by the Australian Government and operated as a National Facility under CSIRO's management. We acknowledge the Wiradjuri People as the Traditional Custodians of the land upon which the Observatory is located. We also pay respect to the Wurundjeri People of the Kulin Nation and the Wallumedegal People of the Darug Nation, the Traditional Owners of the land where this research was primarily carried out. Our heartfelt thanks go to the Parkes Observatory staff for their steadfast support over nearly twenty years. We are also thankful for the access provided by the CSIRO Information Management and Technology High-Performance Computing group to the Petrichor cluster and their technical assistance.

Conflicts of Interest: The authors declare no conflicts of interest.

Abbreviations

The following abbreviations are used in this manuscript:

PTA	Pulsar timing array
GWs	Gravitational waves
GWCSs	Gravitational-wave bursts from cosmic string cusps
CURN	Common spatially uncorrelated red noise

References

1. Foster, R.S; Backer, D.S. Constructing a Pulsar Timing Array. *Astrophys. J.* **1990**, *361*, 300. [CrossRef]
2. Hellings, R.; Downs, G. Upper limits on the isotropic gravitational radiation background from pulsar timing analysis. *Astrophys. J.* **1983**, *265*, L39–L42. [CrossRef]
3. Edwards, R.T.; Hobbs, G.; Manchester, R. TEMPO2, a new pulsar timing package–II. The timing model and precision estimates. *Mon. Not. R. Astron. Soc.* **2006**, *372*, 1549–1574. [CrossRef]
4. Tiburzi, C.; Hobbs, G.; Kerr, M.; Coles, W.; Dai, S.; Manchester, R.; Possenti, A.; Shannon, R.; You, X. A study of spatial correlations in pulsar timing array data. *Mon. Not. R. Astron. Soc.* **2016**, *455*, 4339–4350. [CrossRef]
5. Jenet, F.; Finn, L.; Lazio, J.; Lommen, A.; McLaughlin, M.; Stairs, I.; Stinebring, D.; Verbiest, J.; Archibald, A.; Arzoumanian, Z.; et al. The north american nanohertz observatory for gravitational waves. *arXiv* **2009**, arXiv:0909.1058.
6. Kramer, M.; Champion, D.J. The European pulsar timing array and the large European array for pulsars. *Class. Quantum Gravity* **2013**, *30*, 224009. [CrossRef]
7. Manchester, R. The international pulsar timing array. *Class. Quantum Gravity* **2013**, *30*, 224010. [CrossRef]
8. Nobleson, K.; Agarwal, N.; Girgaonkar, R.; Pandian, A.; Joshi, B.C.; Krishnakumar, M.; Susobhanan, A.; Desai, S.; Prabu, T.; Bathula, A.; et al. Low-frequency wideband timing of InPTA pulsars observed with the uGMRT. *Mon. Not. R. Astron. Soc.* **2022**, *512*, 1234–1243. [CrossRef]
9. Verbiest, J.; Lentati, L.; Hobbs, G.; van Haasteren, R.; Demorest, P.B.; Janssen, G.; Wang, J.B.; Desvignes, G.; Caballero, R.; Keith, M.; et al. The international pulsar timing array: First data release. *Mon. Not. R. Astron. Soc.* **2016**, *458*, 1267–1288. [CrossRef]
10. Xu, H.; Chen, S.; Guo, Y.; Jiang, J.; Wang, B.; Xu, J.; Xue, Z.; Caballero, R.N.; Yuan, J.; Xu, Y.; et al. Searching for the nano-Hertz stochastic gravitational wave background with the Chinese Pulsar Timing Array Data Release I. *Res. Astron. Astrophys.* **2023**, *23*, 075024. [CrossRef]
11. Miles, M.T.; Shannon, R.M.; Bailes, M.; Reardon, D.J.; Keith, M.J.; Cameron, A.D.; Parthasarathy, A.; Shamohammadi, M.; Spiewak, R.; van Straten, W.; et al. The MeerKAT pulsar timing array: First data release. *Mon. Not. R. Astron. Soc.* **2023**, *519*, 3976–3991. [CrossRef]
12. Burke-Spolaor, S.; Taylor, S.R.; Charisi, M.; Dolch, T.; Hazboun, J.S.; Holgado, A.M.; Kelley, L.Z.; Lazio, T.J.W.; Madison, D.R.; McMann, N.; et al. The astrophysics of nanohertz gravitational waves. *Astron. Astrophys. Rev.* **2019**, *27*, 1–78. [CrossRef]
13. Agazie, G.; Anumarlapudi, A.; Archibald, A.M.; Arzoumanian, Z.; Baker, P.T.; Bécsy, B.; Blecha, L.; Brazier, A.; Brook, P.R.; Burke-Spolaor, S.; et al. The NANOGrav 15 yr data set: Evidence for a gravitational-wave background. *Astrophys. J. Lett.* **2023**, *951*, L8. [CrossRef]
14. Reardon, D.J.; Zic, A.; Shannon, R.M.; Hobbs, G.B.; Bailes, M.; Di Marco, V.; Kapur, A.; Rogers, A.F.; Thrane, E.; Askew, J.; et al. Search for an isotropic gravitational-wave background with the Parkes Pulsar Timing Array. *Astrophys. J. Lett.* **2023**, *951*, L6. [CrossRef]
15. Antoniadis, J.; Arumugam, P.; Arumugam, S.; Babak, S.; Bagchi, M.; Nielsen, A.S.B.; Bassa, C.; Bathula, A.; Berthereau, A.; Bonetti, M.; et al. The second data release from the European Pulsar Timing Array-III. Search for gravitational wave signals. *Astron. Astrophys.* **2023**, *678*, A50.
16. Zhu, X.J.; Hobbs, G.; Wen, L.; Coles, W.A.; Wang, J.B.; Shannon, R.M.; Manchester, R.N.; Bailes, M.; Bhat, N.; Burke-Spolaor, S.; et al. An all-sky search for continuous gravitational waves in the Parkes Pulsar Timing Array data set. *Mon. Not. R. Astron. Soc.* **2014**, *444*, 3709–3720. [CrossRef]
17. Wang, J.; Hobbs, G.; Coles, W.; Shannon, R.M.; Zhu, X.; Madison, D.; Kerr, M.; Ravi, V.; Keith, M.J.; Manchester, R.N.; et al. Searching for gravitational wave memory bursts with the Parkes Pulsar Timing Array. *Mon. Not. R. Astron. Soc.* **2015**, *446*, 1657–1671. [CrossRef]
18. Agazie, G.; Arzoumanian, Z.; Baker, P.T.; Bécsy, B.; Blecha, L.; Blumer, H.; Brazier, A.; Brook, P.R.; Burke-Spolaor, S.; Burnette, R.; et al. The NANOGrav 12.5 yr Data Set: Search for Gravitational Wave Memory. *Astrophys. J.* **2024**, *963*, 61. [CrossRef]
19. Porayko, N.K.; Zhu, X.; Levin, Y.; Hui, L.; Hobbs, G.; Grudskaya, A.; Postnov, K.; Bailes, M.; Bhat, N.R.; Coles, W.; et al. Parkes Pulsar Timing Array constraints on ultralight scalar-field dark matter. *Phys. Rev. D* **2018**, *98*, 102002. [CrossRef]
20. Damour, T.; Vilenkin, A. Gravitational wave bursts from cosmic strings. *Phys. Rev. Lett.* **2000**, *85*, 3761. [CrossRef]

21. Yonemaru, N.; Kuroyanagi, S.; Hobbs, G.; Takahashi, K.; Zhu, X.; Coles, W.; Dai, S.; Howard, E.; Manchester, R.; Reardon, D.; et al. Searching for gravitational-wave bursts from cosmic string cusps with the Parkes Pulsar Timing Array. *Mon. Not. R. Astron. Soc.* **2021**, *501*, 701–712. [CrossRef]
22. Dvali, G.; Vilenkin, A. Formation and evolution of cosmic D strings. *J. Cosmol. Astropart. Phys.* **2004**, *2004*, 010. [CrossRef]
23. Abbott, B.P.; Abbott, R.; Adhikari, R.; Ajith, P.; Allen, B.; Allen, G.; Amin, R.; Anderson, S.; Anderson, W.; Arain, M.; et al. First LIGO search for gravitational wave bursts from cosmic (super) strings. *Phys. Rev. D—Part. Fields Gravit. Cosmol.* **2009**, *80*, 062002. [CrossRef]
24. Abbott, B.P.; Abbott, R.; Abbott, T.; Abernathy, M.; Acernese, F.; Ackley, K.; Adams, C.; Adams, T.; Addesso, P.; Adhikari, R.; et al. Search for transient gravitational waves in coincidence with short-duration radio transients during 2007–2013. *Phys. Rev. D* **2016**, *93*, 122008.
25. Abbott, B.P.; Abbott, R.; Abbott, T.D.; Acernese, F.; Ackley, K.; Adams, C.; Adams, T.; Addesso, P.; Adhikari, R.X.; Adya, V.B.; et al. Constraints on cosmic strings using data from the first Advanced LIGO observing run. *Phys. Rev. D* **2018**, *97*, 102002. [CrossRef]
26. Zic, A.; Reardon, D.J.; Kapur, A.; Hobbs, G.; Mandow, R.; Curyło, M.; Shannon, R.M.; Askew, J.; Bailes, M.; Bhat, N.R.; et al. The parkes pulsar timing array third data release. *Publ. Astron. Soc. Aust.* **2023**, *40*, e049. [CrossRef]
27. Reardon, D.J.; Zic, A.; Shannon, R.M.; Di Marco, V.; Hobbs, G.B.; Kapur, A.; Lower, M.E.; Mandow, R.; Middleton, H.; Miles, M.T.; et al. The gravitational-wave background null hypothesis: Characterizing noise in millisecond pulsar arrival times with the Parkes Pulsar Timing Array. *Astrophys. J. Lett.* **2023**, *951*, L7. [CrossRef]
28. Reardon, D.; Hobbs, G.; Coles, W.; Levin, Y.; Keith, M.; Bailes, M.; Bhat, N.; Burke-Spolaor, S.; Dai, S.; Kerr, M.; et al. Timing analysis for 20 millisecond pulsars in the Parkes Pulsar Timing Array. *Mon. Not. R. Astron. Soc.* **2016**, *455*, 1751–1769. [CrossRef]
29. Reardon, D.J.; Shannon, R.M.; Cameron, A.D.; Goncharov, B.; Hobbs, G.; Middleton, H.; Shamohammadi, M.; Thyagarajan, N.; Bailes, M.; Bhat, N.; et al. The Parkes pulsar timing array second data release: Timing analysis. *Mon. Not. R. Astron. Soc.* **2021**, *507*, 2137–2153. [CrossRef]
30. Curyło, M.; Pennucci, T.T.; Bailes, M.; Bhat, N.R.; Cameron, A.D.; Dai, S.; Hobbs, G.; Kapur, A.; Manchester, R.N.; Mandow, R.; et al. Wide-band timing of the Parkes Pulsar Timing Array UWL data. *Astrophys. J.* **2023**, *944*, 128. [CrossRef]
31. Shannon, R.M.; Cordes, J.M. Assessing the role of spin noise in the precision timing of millisecond pulsars. *Astrophys. J.* **2010**, *725*, 1607. [CrossRef]
32. Detweiler, S. Pulsar timing measurements and the search for gravitational waves. *Astrophys. J.* **1979**, *234*, 1100–1104. [CrossRef]
33. Anholm, M.; Ballmer, S.; Creighton, J.D.; Price, L.R.; Siemens, X. Optimal strategies for gravitational wave stochastic background searches in pulsar timing data. *Phys. Rev. D-Part. Fields Gravit. Cosmol.* **2009**, *79*, 084030. [CrossRef]
34. Blanco-Pillado, J.; Olum, K.D. Form of cosmic string cusps. *Phys. Rev. D* **1999**, *59*, 063508. [CrossRef]
35. Phinney, E. A practical theorem on gravitational wave backgrounds. *arXiv* **2001**, arXiv:astro-ph/0108028.
36. Arzoumanian, Z.; Baker, P.T.; Blumer, H.; Bécsy, B.; Brazier, A.; Brook, P.R.; Burke-Spolaor, S.; Chatterjee, S.; Chen, S.; Cordes, J.M.; et al. The NANOGrav 12.5 yr data set: Search for an isotropic stochastic gravitational-wave background. *Astrophys. J. Lett.* **2020**, *905*, L34. [CrossRef]
37. Sun, J.; Baker, P.T.; Johnson, A.D.; Madison, D.R.; Siemens, X. Implementation of an efficient bayesian search for gravitational-wave bursts with memory in pulsar timing array data. *Astrophys. J.* **2023**, *951*, 121. [CrossRef]
38. Aggarwal, K.; Arzoumanian, Z.; Baker, P.; Brazier, A.; Brook, P.; Burke-Spolaor, S.; Chatterjee, S.; Cordes, J.; Cornish, N.; Crawford, F.; et al. The NANOGrav 11 yr data set: Limits on gravitational wave memory. *Astrophys. J.* **2020**, *889*, 38. [CrossRef]
39. Lentati, L.; Alexander, P.; Hobson, M.P.; Taylor, S.; Gair, J.; Balan, S.T.; van Haasteren, R. Hyper-efficient model-independent Bayesian method for the analysis of pulsar timing data. *Phys. Rev. D-Part. Fields Gravit. Cosmol.* **2013**, *87*, 104021. [CrossRef]
40. van Haasteren, R.; Vallisneri, M. New advances in the Gaussian-process approach to pulsar-timing data analysis. *Phys. Rev. D* **2014**, *90*, 104012. [CrossRef]
41. van Haasteren, R.; Vallisneri, M. Low-rank approximations for large stationary covariance matrices, as used in the Bayesian and generalized-least-squares analysis of pulsar-timing data. *Mon. Not. R. Astron. Soc.* **2015**, *446*, 1170–1174. [CrossRef]
42. Woodbury, M.A. *Inverting Modified Matrices*; Statistical Research Group, Memorandum Report No. 42; Princeton University: Princeton, NJ, USA, 1950.
43. Dickey, J.M. The weighted likelihood ratio, linear hypotheses on normal location parameters. *Ann. Math. Stat.* **1971**, *42*, 204–223. [CrossRef]
44. Ellis, J.A.; Vallisneri, M.; Taylor, S.R.; Baker, P.T. *ENTERPRISE: Enhanced Numerical Toolbox Enabling a Robust PulsaR Inference SuitE*; version 3.0.0; Zenodo: Geneva, Switzerland, 2020. [CrossRef]
45. Taylor, S.R.; Baker, P.T.; Hazboun, J.S.; Simon, J.; Vigeland, S.J. *enterprise_extensions*, versio 2.4.3; GitHub, Inc.: San Francisc, CA, USA, 2021. Available online: https://github.com/nanograv/enterprise_extensions (accessed on 22 December 2024).
46. Ellis, J.; Van Haasteren, R. *Jellis18/Ptmcmcsampler: Official Release*, version 1.0.0; Zenodo: Geneva, Switzerland, 2017. Available online: https://ui.adsabs.harvard.edu/abs/2017zndo...1037579E (accessed on 22 December 2024).

47. Carlin, B.P.; Chib, S. Bayesian Model Choice Via Markov Chain Monte Carlo Methods. *J. R. Stat. Soc. Ser. Methodol.* **2018**, *57*, 473–484. [CrossRef]
48. Planck, C.; Ade, P.A.R.; Aghanim, N.; Armitage-Caplan, C.; Arnaud, M.; Ashdown, M.; Atrio-Barandela, F.; Aumont, J.; Baccigalupi, C.; Banday, A.J.; et al. Planck 2013 results. XXV. Searches for cosmic strings and other topological defects. *Astron. Astrophys.* **2014**, *571*, A25. [CrossRef]
49. Kachru, S.; Kallosh, R.; Linde, A.; Maldacena, J.; McAllister, L.; Trivedi, S.P. Towards inflation in string theory. *J. Cosmol. Astropart. Phys.* **2003**, *2003*, 013. [CrossRef]

universe MDPI

Article

Electromagnetic Waves in Cosmological Space–Time II. Luminosity Distance

Denitsa Staicova † and Michail Stoilov *,†

Institute for Nuclear Research and Nuclear Energy, Bulgarian Academy of Sciences, 72 Tszarigradsko Chaussee, 1784 Sofia, Bulgaria; dstaicova@inrne.bas.bg
* Correspondence: mstoilov@inrne.bas.bg
† These authors contributed equally to this work.

Abstract: In this article, we continue our investigation on how the electromagnetic waves propagate in the Friedman–Lemaître–Robertson–Walker spacetime. Unlike the standard approach, which relies on null geodesics and geometric optics approximation, we derive explicit solutions for electromagnetic waves in expanding spacetime and examine their implications for cosmological observations. In particular, our analysis reveals potential modifications to the standard luminosity distance formula. Its effect on other cosmological parameters, e.g., the amount of cold dust matter in the Universe, is considered and estimated from Type Ia supernovae data. We see that this alternative model is able to fit the supernova data, but it gives a qualitatively different Universe without a cosmological constant but with stiff or ultra-stiff matter.

Keywords: electromagnetic waves; Friedman–Lemaître–Robertson–Walker spacetime; luminosity distance; supernovae data fit

1. Introduction

The cosmos is probably the first thing that really puzzled humans in the beginning of history. Its mystery was the origin of the first sciences, and it will probably be one of the last mysteries solved. For centuries, millions of people gazed into the sky. For centuries, all our knowledge about the cosmos was obtained from the light of stars (and planets) that reached our eyes (and later—telescopes) after a long journey through space and time [1–4]. Now, we are at the dawn of multi-messenger observations [2] with gravitational [3] and neutrino [4,5] facilities finally providing qualitatively new information. Nevertheless, our current understanding of the Universe and its evolution is based on experiments like Planck [6], Wilkinson Microwave Anisotropy Probe [7], WiggleZ Dark Energy Survey [8], Background Imaging of Cosmic Extragalactic Polarization [9], All Sky Automated Survey for Super Novae [10], Sloan Digital Sky Survey [11], Hubble Space Telescope [12,13], James Webb Space Telescope [14], etc., all of them detecting electromagnetic signals.

It is obvious that recording celestial electromagnetic signals is not enough. We need additional knowledge in order to correctly interpret the observed data—we have to find a way to determine cosmic distances across all scales. Large cosmic distances cannot be measured directly, so we have to rely on theory, which means that distances in cosmology are model-dependent. In astronomy, distances are determined through the so-called cosmic distance ladder. This is a succession of methods which allows the estimation of distances, starting from close objects and extending to ones which are arbitrarily far away. For cosmological distances, the basic approach is to use certain sources as "standard candles". Therefore, knowledge of how light propagates through spacetime is of primordial importance. The problem is not new and it has been addressed, e.g., in [15–19] and many others. One of the possibilities is to treat the electromagnetic waves according to the geometric optics approximation. Using this approach and combining the experimental data with Einstein's theory of gravity, we find that at the moment, the Universe is accelerating its expansion [20,21]; that the flat Friedman–Lemaître–Robertson–Walker (FLRW) metric [22] is most likely the large-scale metric of the Universe beyond, say, 100 Mpc; and that the Universe contains approximately 28% dust matter and 72% dark energy (cosmological constant). This is the so-called ΛCDM model.

However, in an expanding Universe, the complete treatment of the electromagnetic wave propagation according to the Einstein–Maxwell equations may lead to a modification of the results from the geometric optics approximation. The estimation of distance (especially gravitational distance) is an open question in alternative gravitational theories as well. For example, in Ref. [23], a quadratic modification of gravity leads to an exponential growth in the ratio of gravity to electromagnetic luminosity distances. Other examples of the problem treatment can be found, for instance, in Refs. [24–36] and others in connection to gravitational astronomy. See also Ref. [37] and the references therein. The reason for such interest is that in the modified gravity theories, the luminosity distance for gravitational waves can differ substantially from that of electromagnetic signals (up to an order of magnitude), suggesting that this effect needs to be examined in detail. Therefore, a simultaneous measurement of gravitational and electromagnetic signals from a single cosmic event can shed light on the subject.

In the present work, we try to obtain a formula for the luminosity distance as a result of the explicit functional form of the propagation of electromagnetic waves in the flat FLRW metric. Our approach is very close to the one followed in Ref. [38] but without any assumptions on the form of the electromagnetic potential. We wish to recall that the electromagnetic field in curved spacetime is governed not by the simple wave equation $\Box A^\mu = 0$ assumed in the geometric optics approximation, but by an equation with an additional term proportional to the product of the Ricci tensor and the electromagnetic potential. The extra term is quite interesting, because for a not-Ricci-flat spacetime, its effect cannot be nullified even in an inertial coordinate system (and thus, it resembles a tidal force). For instance, in the case of an exponentially expanding Universe (which happened during the inflation epoch at the beginning of the Universe and is happening again today), this term effectively renders mass to the electromagnetic potential [1] (each of the electromagnetic potential component satisfies a Klein–Gordon equation). This, in turn, suggests a non-zero-length world line of the light ray, which surely affects the wave propagation and, as a consequence, the resulting luminosity distance. Therefore, in order to clarify the problem, we construct an exact solution for the electromagnetic potential suitable for describing light propagation (for example, light emitted by a distant star and observed locally) and we use it to determine the luminosity distance. The luminosity distance obtained in the described way differs from the standard one. We test it against the supernovae of type Ia (SN Ia) data. The fit is successful, but instead of a cosmological constant term, it predicts the presence of another type of exotic matter. We discuss the possible origin of such a term and the consequences of such a model.

This paper is organized as follows: First, we sketch the derivation of distance luminosity using the properties of 0-world lines. Second, we construct a solution of the electromagnetic equation describing the light emitted by a star. Third, we construct the corresponding stress–energy tensor, we find the energy density and energy flow as functions of the distance between observer and emitter, and we determine the corresponding luminosity distance. Fourth, because the newly obtained formula for the luminosity distance differs from the standard one, we consider how it affects the matter content of the Universe, mainly using data inferred from SN Ia.

2. Redshift, Co-Moving Distance, Luminosity Distance—The Null Distance Approach

Hereafter, we shall use the Cartesian form of the FLRW metric. In this case, the invariant length is

$$ds^2 = -dt^2 + a(t)^2(dx^2 + dy^2 + dz^2), \tag{1}$$

where $a(t)$ is the scale factor, and we work in units where the speed of light $c = 1$.

We begin with a derivation of the redshift and luminosity distance based simply on the 0-world-line assumption. Consider a signal, propagating in a straight direction with speed 1 (i.e., c), which is emitted at time t_e and is observed at time t_o. Without any loss of generality, we can place the origin of the spatial coordinate system at the emission point, and we can consider that the signal is along the x-axis. Therefore, the 0-length world line of the signal is parametrically determined by the equation

$$\{t, x(t), y(t), z(t)\} = \{t, \int_{t_e}^{t} \frac{d\tau}{a(\tau)}, 0, 0\}. \tag{2}$$

Let us denote by d the co-moving distance between observation and emission. Consider another signal emitted from the origin, again in the x direction, but at time $t_e + dt_e$. It will arrive at $x = d$ at $t_o + dt_o$, where

$$\int_{t_e+dt_e}^{t_o+dt_o} \frac{d\tau}{a(\tau)} = d = \int_{t_e}^{t_o} \frac{d\tau}{a(\tau)}, \tag{3}$$

so that

$$\frac{dt_o}{a(t_o)} = \frac{dt_e}{a(t_e)}. \tag{4}$$

If the signal is periodic, then using Equation (4), we obtain that the redshift at t_o is

$$z = \frac{\lambda_o - \lambda_e}{\lambda_e} = \frac{a(t_o)}{a(t_e)} - 1. \tag{5}$$

If we normalize $a(t)$, so that $a(t_o) = 1$, then the equation can be put in the form

$$a = \frac{1}{1+z}. \tag{6}$$

Taking the derivative of the above equation, we obtain $\dot{z} = -\dot{a}/a^2$, where we denote by dot over the function its time derivative ($\dot{f} \equiv df/dt, \forall f$). This formula can be used to change the variables in the definition of the co-moving distance passed by the signal, so that

$$d = \int_0^z \frac{d\zeta}{(\dot{a}/a)} = d_H \int_0^z \frac{d\zeta}{\sqrt{\Omega_m(1+\zeta)^3 + \Omega_\Lambda}}, \tag{7}$$

where we used the Friedman equation $\dot{a}/a \sim \sqrt{\rho}$ to obtain the final equation (for simplicity, we stuck to ΛCDM model). In the above equation, d_H is Hubble distance ($d_H = 1/H_0$; recall that here the speed of light is 1), and Ω_m and Ω_Λ are, respectively, the normalized values of dust matter and dark energy densities at t_o. Note that the integral on the right hand side can be evaluated in a closed form (see the paragraph after Equation (28)).

A basic quantity called *the luminosity distance* connects the observed energy flux at a co-moving distance of d to the intrinsic luminosity of the isotropic emitter.

$$S = \frac{(dn/dt_o)h\nu_o}{4\pi d^2} = \frac{(dn/dt_e)h\nu_e}{4\pi d_L^2}. \tag{8}$$

Here, dn/dt is the total photon flux, $h\nu$ is the energy of a photon with frequency ν, and for simplicity, we consider only monochromatic photons. It follows from Equations (4), (7), and (8) that

$$d_L = (1+z)d. \tag{9}$$

3. Electromagnetic Waves in FLRW Metric

In this section, we wish to obtain a formula for the luminosity distance using exact solutions for electromagnetic wave propagation in FLRW spacetime, thus providing a more fundamental approach.

The Lagrangian of the free electromagnetic field is

$$\mathfrak{L} = -\frac{1}{4}F_{\mu\nu}F^{\mu\nu}. \tag{10}$$

Here, F is the electromagnetic field tensor, $F_{\mu\nu} = \nabla_\mu A_\nu - \nabla_\nu A_\mu = \partial_\mu A_\nu - \partial_\nu A_\mu$, and A_μ is the electromagnetic potential. Note that because of the antisymmetric structure of the electromagnetic field tensor, both F and $\mathfrak{L}$ are exactly the same as in the usual Minkowski spacetime. The corresponding equation of motion is $\partial_\nu(\sqrt{-g}F^{\mu\nu}) = 0$, where g is the determinant of the metric tensor $g_{\mu\nu}$. This equation can be recast into an explicitly covariant form, namely

$$\nabla_\nu F^{\mu\nu} = 0. \tag{11}$$

Here, ∇^μ is the covariant derivative in the spacetime.

The electromagnetic field tensor is invariant under the usual gauge symmetry $A_\mu(x) \to A_\mu(x) + \partial_\mu f(x)$, where $f(x)$ is an arbitrary function, and therefore, a gauge fixing is required. In this article, we use the generalized Lorenz gauge

$$\nabla_\nu A^\nu = 0 \tag{12}$$

in which Equation (11) takes the form

$$\Box A^\mu - R^\mu{}_\nu A^\nu = 0. \tag{13}$$

Here, $\Box = \nabla^\nu\nabla_\nu$ is the d'Alembert operator, and R^μ_ν is the Ricci tensor $R_{\mu\nu} = \partial_\alpha\Gamma^\alpha_{\mu\nu} - \partial_\mu\Gamma^\alpha_{\alpha\nu} + \Gamma^\alpha_{\mu\nu}\Gamma^\beta_{\alpha\beta} - \Gamma^\beta_{\alpha\mu}\Gamma^\alpha_{\beta\nu}$.

Obviously, Equation (13) is not the wave equation $\Box A^\mu = 0$. Note that if the spacetime is not Ricci-flat, then the equation satisfied by the electromagnetic waves will differ from the one in the Minkowski spacetime even in the locally inertial coordinate frame. The plane wave solution of Equation (13) can be found in Ref. [1]. There, we have constructed a global solution in radial coordinate system as well. At present, we consider this solution as describing dipole radiation, and therefore, it will not be used here.

At this point, we wish to make a short detour to the geometric optics approach [39,40] to the light propagation in curved spacetime. This approach relies on the approximation that the tail term in Equation (14) is much smaller than the other one. This is supposed to be true for small spacetime curvature (which, in turn, is determined by the matter fields dominating the Universe). The electromagnetic waves are treated in this approximation as rays following null geodesics. According to Ref. [38], in the geometric optics approach, the solution for the electromagnetic potential is supposed to be in the form:

$$A_a = g(\psi)\alpha_a + \text{small tail terms}, \tag{14}$$

where $g(\psi)$ is an *arbitrary* function of the phase ψ, and α_a is an amplitude which is supposed to vary slowly compared to g. Further, based on the arbitrariness of function g, it is assumed that all coefficients in front of the derivatives of g of different orders in Equations (12) and (13) are *separately* equal to zero, thus giving rise to the following system:

$$2k^b\nabla_b\alpha_a + \alpha_a\nabla_b k^b = 0 \tag{15}$$

$$\nabla_b\nabla^b\alpha_a + R_{ab}\alpha^b = 0 \tag{16}$$

$$\alpha^a k_a = 0, \tag{17}$$

$$\nabla_a\alpha^a = 0, \tag{18}$$

$$k_a k^a = 0, \tag{19}$$

where $k_a = \psi_{,a}$ is the propagation vector.

Under the geometric optics approximation, the observed flux is $\mathfrak{F} = T_{ab}u^a u^b \approx \alpha^2(k_a u^a)^2 g(\psi)$, where α^2 varies inversely with the cross-sectional area δS of the ray bundle. Accounting for the redshift relation $(k_a u^a)_o = (k_a u^a)_e/(1+z)$, this yields $\mathfrak{F}\delta S = C\frac{d\Omega}{(1+z)^2}$, where C is constant along null geodesics. Since the source luminosity $L = \int_S (1+z)^2\mathfrak{F}dS$ gives $C = \frac{L}{4\pi}$, we obtain the fundamental relation $\mathfrak{F} = \frac{L}{4\pi d_L^2}$ with the $(1+z)^2$ factor emerging naturally from the wave propagation in curved spacetime rather than from the photon energy and count arguments.

We consider the assumptions of the geometric optics as too restrictive, and in what follows, we shall construct an appropriate solution to Equation (13) without relying on them. We can further use such solutions to determine the luminosity of an isotropic light source, following the procedure outlined in Ref. [38].

We are looking for a solution in the form $A^\mu = \mathcal{A}^\mu(t,r)/r$, where $r = \sqrt{x^2+y^2+z^2}$. This ansatz naturally incorporates the $1/r$ falloff expected for radiative solutions; it maintains consistency with the spherical symmetry of FLRW spacetime and allows for the explicit tracking of the cosmological expansion effect.

There is a topological obstacle (hairy ball theorem) in the construction a global solution describing the electromagnetic potential of spherical waves. This is the reason why we construct only a local one, but this is good enough for our purposes, since in cosmology, we can perform only local observations. We consider a specially orientated coordinate system, where the observer is on the x-axis, and in its vicinity, we have the following equation for the physical (transverse) components of $\mathcal{A}$:

$$-\ddot{\mathcal{A}}^i + \frac{1}{a^2}\mathcal{A}^{i\prime\prime} - 5\frac{\dot{a}}{a}\dot{\mathcal{A}}^i - 2\left(2\frac{\dot{a}^2}{a^2} + \frac{\ddot{a}}{a}\right)\mathcal{A}^i = 0, \quad i = 2,3. \tag{20}$$

Here (again), $\dot{\mathcal{A}}^i = \partial_t\mathcal{A}^i$ and $\mathcal{A}^{i\prime} = \partial_x\mathcal{A}^i$. Surprisingly, Equation (20) is exactly the equation for a plane wave we found in Ref. [1]. Therefore, on the world line $\{t,x,y,z\} = \{t,x,0,0\}$, we have the following solution for the transverse components of the electromagnetic potential:

$$A^i(t,x) = \frac{A^i_\pm}{x\,a(t)^2}\cos\left(k\int^t\frac{d\tau}{a(\tau)} \pm kx + \phi_\pm\right). \tag{21}$$

Here, $A^i_\pm$ and $\phi_\pm$ are some constants, and k is another constant indicating the wave number. In what follows, we shall consider only the outgoing solution ("-" sign). Mathematically, our solution satisfies the outgoing boundary condition at infinity. Due to the singularity at the spatial coordinate origin of the solutions given by Equation (21), we need some boundary condition at $x > 0$ to specify the amplitudes A^i_-. We normalize the solution to match the luminosity of SN Ia at 10 pc.

Using Equation (21), we can show that the redshift and co-moving distance are given by Equation (6) and Equations (3) and (7), respectively.

4. Luminosity Distance Based on Equation (21)

The energy flux of an electromagnetic wave can be found from its energy density or Poynting vector at the observation point. For this purpose, we plug the solution found in the previous section into the electromagnetic stress–energy tensor. In order to work with invariant quantities, we shall need two additional normalized 4-vectors u and v, the first one being the 4-velocity of the observer and the second one determining the spatial orientation of the surface through which the electromagnetic energy passes. Taking into account that $g^{\mu\nu} = \mathrm{diag}\{-1, 1/a(t)^2, 1/a(t)^2, 1/a(t)^2\}$, we define the vectors u and v as follows:

$$u_\mu = \{1,0,0,0\}, \quad |u|^2 = -1, \tag{22}$$

$$v_\mu = \{0,a(t),0,0\}, \quad |v|^2 = 1. \tag{23}$$

The stress–energy tensor of the electromagnetic field is

$$T_{\mu\nu} = F_{\mu\alpha} F_\nu{}^\alpha - \frac{1}{4} g_{\mu\nu} F_{\alpha\beta} F^{\alpha\beta}. \tag{24}$$

Therefore, the averaged observed energy density at point $\{t, x, 0, 0\}$ for the solution given by Equation (21) is

$$\overline{u \cdot T \cdot u} = \frac{k^2}{2x^2 a(t)^4}((A^2_-)^2 + (A^3_-)^2), \tag{25}$$

where we have neglected terms $\sim x^{-3}$ and $\sim x^{-4}$.

The energy flux in the v direction in the observer frame is

$$\overline{u \cdot T \cdot v} = \frac{k^2}{2x^2 a(t)^4}((A^2_-)^2 + (A^3_-)^2), \tag{26}$$

where, again, we have neglected terms $\sim x^{-3}$ and $\sim x^{-4}$. Both quantities are the same, as expected; see, e.g., [38].

Comparing the flux determined by Equations (25) and (26) at different distances, we obtain

$$d_L = (1+z)^2 d. \tag{27}$$

5. Numerical Fits

The data for SN Ia have been used in the last decades to determine the geometry of FLRW. Now, there are really good compilations of SN Ia observations which provide the necessary information. For example, the SCP "Union2.1" SN Ia compilation [41] contains the data for 580 SN Ia[1]. For each supernova, it contains the corresponding distance modulus and redshift. The distance modulus is (by definition) $d_m = 5\log_{10}(d_L/d_0)$ where $d_0 = 10$ pc, and d_L is the luminosity distance, given either by Equation (9) or by Equation (27).

In the ΛCDM case, the SN Ia fit is based on the function

$$f_{\alpha,\beta}(z) = \alpha + 5\log_{10}\left((1+z)\int_1^{1+z}\frac{d\zeta}{\sqrt{\zeta^3+\beta}}\right), \tag{28}$$

where α and β are the fitting parameters. Their physical interpretation is $\beta = \Omega_\Lambda/\Omega_m$ and $\alpha = 5\log_{10}(d_H/d_o\sqrt{\Omega_m})$. The integral in the above equation represents the co-moving distance d (see Equation (7)) without the multiplicative constant $d_H/\sqrt{\Omega_m}$ which is moved in α. The integral can be taken in a closed form by means of an integral representation of the ordinary hypergeometric function ${}_2F_1[\frac{1}{6},\frac{1}{2};\frac{7}{6};-\beta/y^3] = \int dy/2\sqrt{y^3+\beta}$. The integral also has a closed functional expression for other combinations of a two-component Universe (including curvature as well), but in the general case, it can only be calculated numerically. Taking into account that $\Omega_\Lambda + \Omega_m = 1$, the physical quantities can be extracted from the fitting parameters as follows: $\Omega_m = 1/(1-\beta)$, $\Omega_\Lambda = 1-\Omega_m$, and the dimensionless Hubble constant is $h = 3\times 10^{(40-\alpha)/5}/\sqrt{\Omega_m}$.

Equation (27) differs from Equation (9), and correspondingly, the fitting function for the SN Ia data should be different and will lead to different preferable matter content in the Universe. Hereafter, we consider a Universe made of two components—a cold dust matter with normalized density fraction Ω_m and additional term with density fraction Ω_{ex} and equation of state w_{ex}, which we consider as a free parameter. As usual, the densities satisfy the critical density condition $\Omega_m + \Omega_{ex} = 1$. Let us recall that the equation of state in the perfect fluid analogy for the stress tensor relates the pressure p to the energy density ϵ (matter density ρ for cold matter). Usually, the dependence is $p = w\epsilon$ (or $p = w\rho$). For example, the equation of state for radiation is $w = 1/3$; for pressureless (cold) matter, it is $w = 0$; and for the cosmological constant, $w = -1$. In any case, the (energy) density scales as $a(t)^{-3(1+w)}$. Therefore, for the model in consideration, the co-moving distance is as follows:

$$d = \int_{t_e}^{t_o}\frac{dt}{a(t)} = \int_0^z\frac{d\zeta}{(\dot{a}/a)} = d_H\int_0^z\frac{d\zeta}{\sqrt{\Omega_m(1+\zeta)^3+\Omega_{ex}(1+\zeta)^{3(w_{ex}+1)}}}.$$

Having in mind these considerations and Equation (27), we choose the fitting function for the SN Ia data to be

$$\hat{f}_{\alpha,\beta}(z) = \alpha + 5\log_{10}\left((1+z)^2\int_1^{1+z}\frac{d\zeta}{\sqrt{\zeta^3+\beta\zeta^{3(1+w_{ex})}}}\right). \tag{29}$$

Note that $\alpha = 5\log_{10}(d_H/d_o\sqrt{\Omega_m})$ in both Equations (28) and (29). Also, Ω_m is the same: $\Omega_m = 1/(1-\beta)$. So, Equation (29) generalizes Equation (28), replacing the cosmological constant with a matter characterized by a free parameter w_{ex}. This matter can, in principle, represent the cosmological constant as well.

In Figure 1, we juxtapose the standard fit and the new one for $w_{ex} = 5/3$. As can be seen, they are almost identical. The fits with other $w_{ex} \geq 1$ differ marginally from them but have different matter content. The values of Ω_m, Ω_{ex}, and h, and the quality of the fit based on Equation (29) as functions of w_{ex} are shown in Figure 2. We wish to stress that this fit is possible only for equations of state $w_{ex} \lesssim 1.7$. It is also clear from Figure 2 that the fit is reasonable for $w_{ex} \gtrsim 0.9$.

Furthermore, we investigate the same fit with the nested sampler Polychord [42]), applied to the data of the binned Pantheon dataset with their covariance, consisting 40 supernovae luminosity measurements in the redshift range $z \in (0.01, 2.3)$ [43]. We do not use the newer Pantheon Plus dataset since this study is a proof of concept for which the

Pantheon dataset is good enough. In this case, we do not use the fitting function, but we directly use the modified luminosity distance, plugged into the Friedman equations with various combinations of Ω_m and higher terms $\sim \Omega_{ex}$—sixth and eighth order on $(1+z)$.

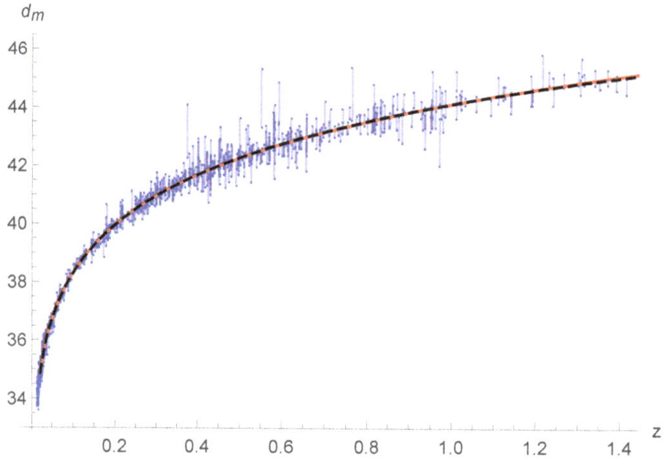

Figure 1. SN Ia fits. Black dashed line—the standard fit based on Equation (28) with 28% dust with equation of state $w = 0$ and 72 % dark energy with equation of state $w = -1$. Red line—the new fit, based on Equation (29) and a mixture of 70% dust with equation of state $w = 0$ and 30% matter with equation of state $w = 5/3$. The fits with other $w_{ex} \geq 1$ differ marginally from it.

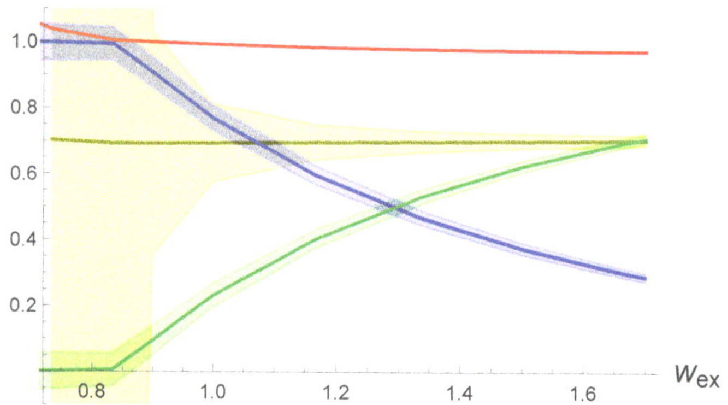

Figure 2. The Universe parameters as functions of the new matter equation of state. In blue—Ω_{ex} with its standard variation as a shaded corridor. In green—Ω_m with its standard variation. Yellow—the dimensionless Hubble constant h and its standard variation. The red line gives the fit's estimated variance ($\chi^2/$(number of degrees of freedom)).

In Figure 3, we show the posteriors of our tests, combining different terms as mentioned in the legend of the plot: $(1+z)^6$, $(1+z)^8$, with and without radiation term $(1+z)^4$ (thus generalizing slightly the model in Equation (29)). Note that we also tested the inclusion of Quasars [44] to the datasets, and the inferred cosmological parameters were practically the same (up to the second decimal digit); thus, we do not include their contour plot in the figure.

We fix $\Omega_r = 0.001$ for simplicity, and because there are numerical problems with the likelihood with a wide uniform prior on Ω_r.

We obtain that some of the combinations are not convergent, while others are very slowly convergent. On the plot, we show the normally convergent combinations. One can see the contour plots in Figure 3. Notably, the expansions $\sim (1+z)^6$, ($w_{ex} = 1$) are able to fit the expected value of Ω_m for the standard wide prior ($\Omega \in [0,1]$, while the $(1+z)^8$, ($w_{ex} = 5/3$) ones predict a much larger matter density. Both expansions, however, are able to fit the expected H_0 in the uniform prior we used, $H_0 \in [50, 100]$. Note that the $(1+z)^8$ expansions predict a slightly higher value for H_0. Throughout, we have assumed that $\sum(\Omega_i) = 1$. Also, the tight constraint on H_0, despite our wide prior, comes from fixing $M_B = -19.25$ in the distance modulus, which corresponds to a $SH0ES$ prior on H_0. We do not test other models (for example, adding curvature Ω_k, since the luminosity distance was derived under the explicit assumption of a flat FRWL).

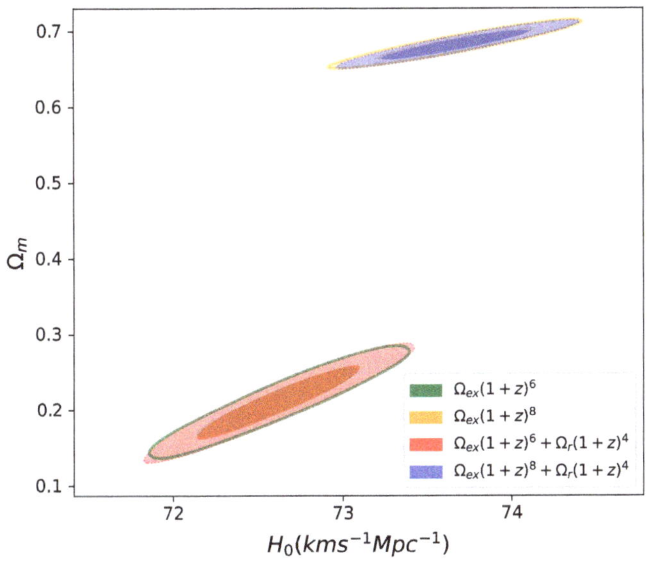

Figure 3. On the plots, the posteriors of the different expansion for the Pantheon dataset. We note two different mean values for Ω_m corresponding to the 6th and 8th powers terms. Adding radiation does not change the posteriors significantly.

6. Discussions

The proposed modification of the luminosity distance formula is without a doubt quite significant. The resulting matter content of the Universe is changed significantly compared to the ΛCDM model. Now, the Universe requires a matter with equation of state $w_{ex} \gtrsim 1$. This seems quite strange compared to the ΛCDM model. However, the ΛCDM model has its peculiarities as well. It is called the "concordance" model, since it fits the observations very well, but at the cost of pronouncing about 95% of the energy content as "dark", i.e., invisible, non-interacting with light, even though the model itself is based on light observations. For now, there are no well-justified candidates for dark matter.

Besides this obvious inconsistency, there are known tensions within the model, such as the Hubble tension, which leads to the search for a proper dark energy theory. All this makes the study of the model and its alternatives even more important. In this paper, we propose, as an alternative, a cosmological model without cosmological constant, which can still fit the SN Ia observations.

Though our fit is possible for $w_{ex} \sim 0.9$, it really prefers matter with $w_{ex} \in [1,\ 5/3]$. Matter with $w = 1$ is known as stiff matter, and we can trace its use in cosmology back to Zeldovich's seminal paper [45]. Subsequently, many other variants of the ideas in this paper have been proposed. We shall mention here Ref. [46], where a non-linear equation of state of the form $p = A\rho - B\sqrt{\rho},\ A,\ B \in \mathbb{R}$ has been considered. It is shown there that a matter with such an equation of state can be interpreted as a mixture of several components with linear equations of state, and one of them has $w = A$. Due to the arbitrariness of A, the model can describe the matter needed in our fit. More examples can be found, e.g., in Ref. [47] and the references therein.

The stiff matter physical realization can be quite simple. For instance, a scalar field without self interaction potential has $w = 1$. For $w > 1$ (ultra-stiff matter), the realization we know is within the $k-$essence model. The $k-$essence models have a non-standard kinetic term in the form of $L_{kin} = f_1(\varphi)f_2(X)$, where $X = (\nabla\varphi)^2/2$ and can have any equation of state [48]. An example in an isotropic and homogeneous Universe is a scalar field without potential for which $f_2(X) = X^a$. In this case, the equation of state will be

$$w = \frac{f_1(\varphi)f_2(X)}{f_1(\varphi)\ 2\ X\ (df_2(X)/dX) - f_1(\varphi)f_2(X)} = \frac{1}{2a-1}, \tag{30}$$

which does not depend on f_1. So, for $a \in \{1/2, 1\}$, we have $w \in \{\infty, 1\}$. A related point to consider comes from Ref. [49], where a mechanism for the "stiffening" of the equation of state is considered. It allows us to obtain matter with $w_{ex} = 1$ and $w_{ex} = 5/3$ from matter with $w = 0$ and $w = 1/3$, respectively.

Models with ultra-stiff matter have the intrinsic capacity to explain the existence of a homogeneous and isotropic Universe without the need for a special inflation epoch. The point is that the density of such matter decreases faster than the density of ordinary matter with the increase in scale factor $a(t)$. This, in turn, means that at the beginning of the Universe, there was an epoch where ultra-stiff matter was dominant. As discussed in our previous paper [1], in such an environment, any $U(1)$ field behaves as tachyons. The tachyons can be considered as mirror reflections of the usual matter with respect to the speed of light—they cannot move slower than the speed of light, and their energy increases with the decrease in their speed v, becoming infinite at $v = c$ and tending to zero at $v \to \infty$. Thus, they are ideal for the large-scale thermalization of the Universe.

7. Conclusions and Outlook

Here, we have presented how solving the Maxwell equations in an FLRW spacetime background can change the basic premise of cosmology. By deriving the analytical solutions without relying on the geometric optics approximation and using it to find the corresponding luminosity distance, we see it differs from the standard one applied to electromagnetic observations. Our key finding is the possibility of a matter with equation of state $w_{ex} \geqq 1$ together with a cold dust matter to describe the data from the SN Ia.

An important result we obtained is that when utilizing the modified luminosity distance, the fits reject the cosmological constant. Therefore, there is no accelerated expansion of the Universe in this case—see Figure 4, where we compare the deceleration parameter and the Hubble parameter for ΛCDM and the expansions considered in this paper. We see that for the stiff-matter models, the Universe is now in a stage of deceleration. While this possibility is interesting, it needs to be confirmed by adding more observational data (such as Baryonic Acoustic Oscillations and CMB). Its connection to modified gravity theories predicting an ultra-stiff term, however, makes it a promising new direction for theoretical work.

In conclusion, we wish to stress again the importance of thorough investigation of the propagation of light in curved spacetime. Here, a small change in the assumptions results in a significant change in the whole picture.

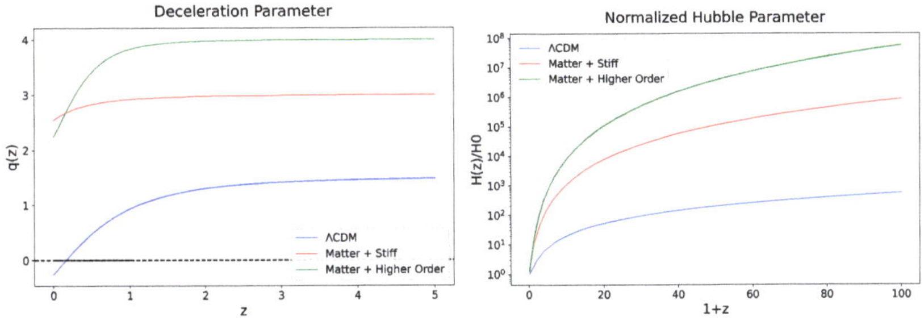

Figure 4. The deceleration parameter and the Hubble parameter for the parameters we obtained from fitting the data, where $H(z) = H_0 E(z)$ and $q(z) = -\frac{\ddot{a}a}{\dot{a}^2}$.

Author Contributions: Conceptualization, M.S.; methodology, D.S. and M.S.; software, D.S. and M.S.; validation, D.S. and M.S.; formal analysis, D.S. and M.S.; investigation, D.S. and M.S.; resources, D.S. and M.S.; data curation, D.S. and M.S.; writing—original draft preparation, M.S.; writing—review and editing, D.S. and M.S.; funding acquisition, M.S. All authors have read and agreed to the published version of the manuscript.

Funding: This research was funded by Bulgarian National Science Fund grant number KP-06-N58/5.

Institutional Review Board Statement: Not applicable.

Informed Consent Statement: Not applicable.

Data Availability Statement: The original contributions presented in this study are included in the article. Further inquiries can be directed to the corresponding author.

Conflicts of Interest: The authors declare no conflicts of interest.

Abbreviations

The following abbreviations are used in this manuscript:

FLRW	Friedman–Lemaître–Robertson–Walker
ΛCDM	Cosmological constant—Cold Dust Matter
SH0ES	Supernova, H0, for the Equation of State of Dark Energy
SN Ia	Supernovae of Type Ia

Note

1 Available online: https://supernova.lbl.gov/Union/figures/SCPUnion2.1_mu_vs_z.txt (accessed on 27 January 2025).

References

1. Staicova, D.; Stoilov, M. Electromagnetic Waves in Cosmological Spacetime. *Universe* **2023**, *9*, 292. [CrossRef]
2. Addazi, A.; Alvarez-Muniz, J.; Batista, R.A.; Amelino-Camelia, G.; Antonelli, V.; Arzano, M.; Asorey, M.; Atteia, J.L.; Bahamonde, S.; Bajardi, F.; et al. Quantum Gravity Phenomenology at the Dawn of the Multi-messenger Era—A Review. *Prog. Part. Nucl. Phys.* **2022**, *125*, 103948.
3. LIGO Scientific Collaboration; Virgo Collaboration; Fermi GBM; INTEGRAL; IceCube Collaboration; AstroSat Cadmium Zinc Telluride Imager Team; IPN Collaboration; The Insight-Hxmt Collaboration; ANTARES Collaboration; The Swift Collaboration; et al. Multi-messenger Observations of a Binary Neutron Star Merger. *Astrophys. J. Lett.* **2017**, *848*, L12.
4. Aartsen, M. et al. [IceCube Collaboration] Observation of High-Energy Astrophysical Neutrinos in Three Years of IceCube Data. *Phys. Rev. Lett.* **2014**, *113*, 101101. [CrossRef]
5. Huang, T.; Li, Z. Neutrino Observations of LHAASO Sources: Present Constraints and Future Prospects. *Mon. Not. Roy. Astron. Soc.* **2022**, *514*, 852–862. [CrossRef]
6. Aghanim, N. et al. [Planck Collaboration] Planck 2018 Results. VI. Cosmological Parameters. *Astron. Astrophys.* **2020**, *641*, A6. Erratum in *Astron. Astrophys.* **2021**, *652*, C4.

7. Hinshaw, G.; Larson, D.; Komatsu, E.; Spergel, D.N.; Bennett, C.L.; Dunkley, J.; Nolta, M.R.; Halpern, M.; Hill, R.S.; Odegard, N.; et al. [WMAP Collaboration] Nine-Year Wilkinson Microwave Anisotropy Probe (WMAP) Observations: Cosmological Parameter Results. *Astrophys. J. Suppl.* **2013**, *208*, 19. [CrossRef]
8. Blake, C.; Brough, S.; Colless, M.; Contreras, C.; Couch, W.; Croom, S.; Croton, D.; Davis, T.M.; Drinkwater, M.J.; Forster, K.; et al. The WiggleZ Dark Energy Survey: Joint measurements of the expansion and growth history at z < 1. *Mon. Not. Roy. Astron. Soc.* **2012**, *425*, 405–414.
9. Ade, P.; Ahmed, Z.; Amiri, M.; Barkats, D.; Thakur, R.B.; Bischoff, C.; Beck, D.; Bock, J.; Boenish, H.; Bullock, E.; et al. [BICEP/Keck Collaboration] Improved Constraints on Primordial Gravitational Waves using Planck, WMAP, and BICEP/Keck Observations through the 2018 Observing Season. *Phys. Rev. Lett.* **2021**, *127*, 151301. [CrossRef]
10. Neumann, K.D.; Holoien, T.W.; Kochanek, C.S.; Stanek, K.Z.; Vallely, P.J.; Shappee, B.J.; Prieto, J.L.; Pessi, T.; Jayasinghe, T.; Brimacombe, J.; et al. The ASAS-SN Bright Supernova Catalog—V. 2018–2020. *Mon. Not. Roy. Astron. Soc.* **2023**, *520*, 4356–4369. [CrossRef]
11. Zhao, C.; Variu, A.; He, M.; Forero-Sánchez, D.; Tamone, A.; Chuang, C.; Kitaura, F.; Tao, C.; Yu, J.; Kneib, J.; et al. [eBOSS Collaboration] The Completed SDSS-IV Extended Baryon Oscillation Spectroscopic Survey: Cosmological Implications from Multitracer BAO Analysis with Galaxies and Voids. *Mon. Not. Roy. Astron. Soc.* **2022**, *511*, 5492–5524. [CrossRef]
12. Riess, A.G.; Strolger, L.G.; Tonry, J.; Casertano, S.; Ferguson, H.C.; Mobasher, B.; Challis, P.; Filippenko, A.V.; Jha, S.; Li, W.; et al. Type Ia Supernova Discoveries at z > 1 from the Hubble Space Telescope: Evidence for Past Deceleration and Constraints on Dark Energy Evolution. *Astrophys. J.* **2004**, *607*, 665–687. [CrossRef]
13. Freedman, W.L.; Madore, B.F.; Gibson, B.K.; Ferrarese, L.; Kelson, D.D.; Sakai, S.; Mould, J.R.; Kennicutt, R.C., Jr.; Ford, H.C.; Graham, J.A.; et al. Final Results from the Hubble Space Telescope Key Project to Measure the Hubble Constant. *Astrophys. J.* **2001**, *553*, 47–72. [CrossRef]
14. Yuan, W.; Riess, A.; Casertano, S.; Macri, L. A First Look at Cepheids in a Type Ia Supernova Host with JWST. *Astrophys. J. Lett.* **2022**, *940*, L17. [CrossRef]
15. Mashhoon, B. Electromagnetic Waves in an Expanding Universe. *Phys. Rev. D* **1973**, *8*, 4297. [CrossRef]
16. Cohen, J.; Kegeles, L. Electromagnetic Fields in Curved Spaces - A Constructive Procedure. *Phys. Rev. D* **1974**, *10*, 1070–1084. [CrossRef]
17. Cabral, F.; Lobo, F. Electrodynamics and Spacetime Geometry: Foundations. *Found. Phys.* **2017**, *47*, 208–228. [CrossRef]
18. Cabral, F.; Lobo, F. Electrodynamics and Spacetime Geometry: Astrophysical Applications. *Eur. Phys. J. Plus* **2017**, *132*, 304. [CrossRef]
19. Asenjo, F.; Hojman, S. Birefringent Light Propagation on Anisotropic Cosmological Backgrounds. *Phys. Rev. D* **2017**, *96*, 044021. [CrossRef]
20. Riess, A.G.; Filippenko, A.V.; Challis, P.; Clocchiatti, A.; Diercks, A.; Garnavich, P.M.; Gilliland, R.L.; Hogan, C.J.; Jha, S.; Kirshner, R.P.; et al. Observational Evidence from Supernovae for an Accelerating Universe and a Cosmological Constant. *Astron. J.* **1998**, *116*, 1009–1038. [CrossRef]
21. Perlmutter, S.; Aldering, G.; Goldhaber, G.; Knop, R.A.; Nugent, P.; Castro, P.G.; Deustua, S.; Fabbro, S.; Goobar, A.; Groom, D.E.; et al. Measurements of Ω and Λ from 42 High-redshift Supernovae. *Astrophys. J.* **1999**, *517*, 565. [CrossRef]
22. Peacock, J. *Cosmological Physics*; Cambridge University Press: Cambridge, UK, 1999.
23. Fanizza, G.; Franchini, G.; Gasperini, M.; Tedesco, L. Comparing the Luminosity Distance for Gravitational Waves and Electromagnetic Signals in a Simple Model of Quadratic Gravity. *Gen Relativ Gravit* **2020**, *52*, 111. [CrossRef]
24. Tomita, K. Angular Distances in Clumpy Friedman Universes. *Prog. Theor. Phys.* **1998**, *100*, 79–90. [CrossRef]
25. Sereno, M.; Covone, G.; Piedipalumbo, E.; de Ritis, R. Distances in Inhomogeneous Quintessence Cosmology. *Mon. Not. Roy. Astron. Soc.* **2001**, *327*, 517. [CrossRef]
26. Demianski, M.; de Ritis, R.; Marino, A.A.; Piedipalumbo, E. Approximate Angular Diameter Distance in a Locally Inhomogeneous Universe with Nonzero Cosmological Constant. *Astron. Astrophys.* **2003**, *411*, 33–40. [CrossRef]
27. Fleury, P.; Dupuy, H.; Uzan, J.P. Light Propagation in Inhomogeneous and Anisotropic Cosmologies. *Phys. Rev. D* **2015**, *91*, 043511. [CrossRef]
28. Nikolaev, A.V.; Chervon, S.V. The Effect of Universe Inhomogeneities on Cosmological Distance Measurements. *Grav. Cosmol.* **2016**, *22*, 208–211. [CrossRef]
29. Belgacem, E.; Dirian, Y.; Foffa, S.; Maggiore, M. Gravitational-Wave Luminosity Distance in Modified Gravity Theories. *Phys. Rev. D* **2018**, *97*, 104066. [CrossRef]
30. Belgacem, E.; Calcagni, G.; Crisostomi, M.; Dalang, C.; Dirian, Y.; Ezquiaga, J.M.; Fasiello, M.; Foffa, S.; Ganz, A.; García-Bellido, J.; et al. Testing Modified Gravity at Cosmological Distances With Lisa Standard Sirens. *JCAP* **2019**, *07*, 024. [CrossRef]
31. Mukherjee, S.; Wandelt, B.D.; Silk, J. Testing The General Theory of Relativity Using Gravitational Wave Propagation From Dark Standard Sirens. *Mon. Not. Roy. Astron. Soc.* **2021**, *502*, 1136–1144. [CrossRef]

32. Garoffolo, A.; Raveri, M.; Silvestri, A.; Tasinato, G.; Carbone, C.; Bertacca, D.; Matarrese, S. Detecting Dark Energy Fluctuations with Gravitational Waves. *Phys. Rev. D* **2021**, *103*, 083506. [CrossRef]
33. Fonseca, J.; Zazzera, S.; Baker, T.; Clarkson, C. The Observed Number Counts in Luminosity Distance Space. *JCAP* **2023**, *08*, 050. [CrossRef]
34. Romano, A.E.; Sakellariadou, M. Mirage of Luminal Modified Gravitational-Wave Propagation. *Phys. Rev. Lett.* **2023**, *130*, 231401. [CrossRef] [PubMed]
35. Gupta, R.P. JWST Early Universe Observations and ΛCDM Cosmology. *Mon. Not. Roy. Astron. Soc.* **2023**, *524*, 3385–3395. [CrossRef]
36. Cuzinatto, R.R.; de Melo, C.A.M.; Neves, J.C.S. Shadows of Black Holes at Cosmological Distances in the co-varying Physical Couplings Framework. *Mon. Not. Roy. Astron. Soc.* **2023**, *526*, 3987–3993. [CrossRef]
37. Antoniou, I.; Papadopoulos, D.; Perivolaropoulos, L. Propagation of Gravitational Waves in an Expanding Background in the Presence of a Point Pass. *Phys. Rev. D* **2016**, *94*, 084018. [CrossRef]
38. Ellis, G.; Maartens, R.; MacCallum, M. *Relativistic Cosmology*; Cambridge University Press: Cambridge, UK, 2012. [CrossRef]
39. Sachs, R.K. Gravitational Waves in General Relativity. 6. The Outgoing Radiation Condition. *Proc. Roy. Soc. Lond. A* **1961**, *264*, 309–338. [CrossRef]
40. Ehlers, J.; Schneider, P. Gravitational Lensing. In Proceedings of the 13th Conference on General Relativity and Gravitation (GR-13), Cordoba, Argentina, 28 June–4 July 1993; pp. 21–40.
41. Suzuki, N.; Rubin, D.; Lidman, C.; Aldering, G.; Amanullah, R.; Barbary, K.; Barrientos, L.F.; Botyanszki, J.; Brodwin, M.; Connolly, N.; et al. The Hubble Space Telescope Cluster Supernova Survey. V. Improving The Dark-Energy Constraints Above z > 1 And Building An Early-Type-Hosted Supernova Sample. *Astrophys. J.* **2012**, *746*, 85. [CrossRef]
42. Handley, W.J.; Hobson, M.P.; Lasenby, A.N. PolyChord: Nested Sampling for Cosmology. *Mon. Not. Roy. Astron. Soc.* **2015**, *450*, L61–L65. [CrossRef]
43. Scolnic, D.M.; Jones, D.O.; Rest, A.; Pan, Y.C.; Chornock, R.; Foley, R.J.; Huber, M.E.; Kessler, R.; Narayan, G.; Riess, A.G.; et al. The Complete Light-curve Sample of Spectroscopically Confirmed SNe Ia from Pan-STARRS1 and Cosmological Constraints from the Combined Pantheon Sample. *Astrophys. J.* **2018**, *859*, 101. [CrossRef]
44. Risaliti, G.; Lusso, E. A Hubble Diagram for Quasars. *Astrophys. J.* **2015**, *815*, 33. [CrossRef]
45. Zeldovich, Y.B. A Hypothesis Unifing the Stucture and the Entropy of the Universe. *Mon. Not. R. astr. Soc* **1972**, *160*, 1. [CrossRef]
46. Mukherjee, S.; Paul, B.C.; Dadhich, N.K.; Maharaj, S.D.; Beesham, A. Emergent Universe with Exotic Matter. *Class. Quantum Grav.* **2006**, *23*, 6927. [CrossRef]
47. Chavanis, P.H. Cosmology with a Stiff Matter Era. *Phys. Rev. D* **2015**, *92*, 103004. [CrossRef]
48. Armendariz-Picon, C.; Mukhanov, V.; Steinhardt, P.J. Essentials of k-essence. *Phys. Rev. D* **2001**, *63*, 103510. [CrossRef]
49. Maartens, R.; Koyama, K. Brane-World Gravity. *Living Rev. Rel.* **2010**, *13*, 5. [CrossRef]

universe

MDPI

Article

A Scale Invariant Fully Conformal Cosmological Model and Its Support by Astrophysical Data †

Richard Dvorsky

Faculty of Materials Science and Technology, Centre for Advanced Innovation Technologies, VSB—Technical University of Ostrava, 17. Listopadu 15/2172, 708 00 Ostrava, Czech Republic; richard.dvorsky@vsb.cz

† This work is a new modified and extended version of a preprint: Dvorsky, Richard, A Simple Fully Conformal Solution of Einstein's Gravitational Equations and the Agreement of its Implications with Astrophysical Data (1 March 2023) on the preprint server SSRN. Available at SSRN: https://ssrn.com/abstract=4373727 or http://dx.doi.org/10.2139/ssrn.4373727 from (1 March 2023).

Abstract: According to general relativity, the cosmological redshift may be caused by other mechanisms than the source moving away from the observer. It can occur on a global scale, similar to the gravitational redshift near massive stars. In principle, these are differences in the time-dependent global metric field between the source in the past and the observer in the present. In this paper we attempt a new interpretation of the simple solution of Einstein's equations within a fully conformal metric for the case of a time-independent energy-momentum tensor. The scaling factor here acts identically on all four space-time coordinates and the speed of light is independent of the conformal time. The fully conformal metric is interpreted here as a universal geometric background which is scale invariant and acts universally on all objects, including gauges and clocks, regardless of their dimensions and internal interactions. The associated scale invariant exponential expansion is thus only relative and all observers at different times are completely equal. The model introduces the concept of the appearent age of the universe, which is the limiting consequence of time dilation into the past, and corresponds to the present value of the age of the universe H^{-1} according to the standard model. This appearent age is the same for all observers, and the Hubble constant is thus a true universal constant, invariant to time translations. The motivation of this work was to test the possibility of the above cosmological redshift mechanism in confrontation with astrophysical data. Probably the most important consequence is the generalized formulation and interpretation of the Hubble-Lemaître law $z(r) = (e^{Hr/c} - 1)$, which shows good agreement with astrophysical data even for the most distant supernovae. Confronting the conformal metric model with some astrophysical data shows an interesting agreement with the observed spatial distribution of astrophysical sources such as γ-ray bursts and quasars. On a cosmological scale, the above fully conformal metric naturally determines the global energy density, spatial flatness, and solves the horizon problem and Olbers' paradox in infinite spacetime.

Keywords: conformal metric; scale invariancy; cosmologhical redshift; gamma burst; quasars; spatial flatness; critical density; Olbers' paradox

1. Introduction

According to the standard model, the currently accepted explanation for the cosmological redshift is a real spatial expansion of the universe with an origin in the Big Bang. However, could the cosmological redshift also be interpreted as the result of a difference between the global metric field around a radiating astrophysical source in the past and the global metric field around an observer in the present?

Astrophysical data for the cosmological redshift are in most cases already published in the form of a dependence of the source receding rate on the distance from the observer. This transformation of real experimental data into terms of the preferred standard cosmological model may be misleading for the further development of the field. The above question about an alternative interpretation of the cosmological redshift was raised in the past by Fritz Zwicky [1]. However, his concept of "tired light" corresponded to a gravitational redshift based on local curvatures of spacetime by the masses of galaxies, and the redshift effect depended on local spatial distances from galactic centres. This created a problem with observational data showing a global dependence of redshift on distance from any observer.

In this paper we attempt to interpret the cosmological redshift based on the aforementioned differences in the global metric field between the two time levels in spacetime. This interpretation should correspond to a global solution of Einstein's gravitational equations whose metric field varies with time $g_{\mu\nu} = g_{\mu\nu}(t)$, while at the same time ensuring full covariance of the laws of nature at every time of the space-time continuum. One of the simplest solutions that satisfies the above requirements is a flat, fully conformal Minkowski metric with time scaling $g_{\mu\nu}(t) = \psi(t) \cdot \eta_{\mu\nu}$, $g_{\mu\nu}(0) = \eta_{\mu\nu}$, where the time distance of the source t is measured into the past from the zero point $r = 0$ of the origin of the observer's coordinates. When the source radiation is observed at a distance $r = ct$ from the observer, we obtain redshift generally defined as $z = \sqrt{g_{44}(r)/g_{44}(0)} - 1 = \sqrt{g_{44}(t)/g_{44}(0)} - 1$. In this article, the aforementioned metric is derived on the basis of the a priori presumption of time-independent energy-momentum tensor and requirements of full conformity of all space-time coordinates due to acceptance of "perfect cosmological principle" [2,3]. Its physical meaning is then analysed in confrontation not only with cosmological redshift data, but also with some other astrophysical data.

The approach used in this work rests in the finding of the solution of Einstein gravity equations in the form of the fully conformal metric, where natural physical time t has a conformal character and the scaling factor acts equivalently on all space-time coordinates. In the case of such solutions, the covariant form of Maxwell's equations is preserved and the speed of light does not change over time (c = const.), which can be considered a positive property and a simplification within the respective models. However, the primary aim of this paper is not to derive a specific form of the fully conformal Minkowski metric per se, but a deeper analysis of its physical content assuming time-independent energy-momentum tensor and the analysis of its possibility to describe astrophysical observational data.

For the various variants of energy-momentum tensors in current cosmological models, most solutions of Einstein's gravitational equations are found in the assumed form of the FLRW metric with linear physical parameter of cosmological time t and scaling factor $a(t)$ acting only on space coordinates. By the establishment of FLRW-conformal time $d\eta = dt/a(t)$ it is possible these solutions transform into the formally identical mathematical form of fully conformal metric.

In the field of conformal solutions of Einstein's gravitational equations, a number of mostly theoretical papers have been published [4–10] including conformal forms of the well-known Lamaitre-Friedman-Robertson-Walker (FLRW) metric [11–14]. In these FLRW metrics, the Weyl conformal tensor disappears [15], which in most of these works allowed to achieve full space-time conformality by a suitable transformation of the space-time coordinates. A comprehensive survey of these secondary conformal forms of the primary FLRW metric has been summarised by Ibison [16].

It may seem that the whole message of this work can be "solved by naming and tagging" and completely identifying the fully conformal metric, with "FLRW-conformal metric", which was derived from FLRW by a transformation—therefore "nothing new". However, is this derived "FLRW-conformal metric" physically identical with above mentioned "the fully conformal metric", which was analyzed in this article? A general calculation shows, that both mentioned conformal metrics are not completely identical. The corresponding Friedmann equations for a flat space without a cosmological term show a different structure and the corresponding equations of state will also differ.

The fully conformal global metric with a time scaling is derived based on four premises:

PREMISE 1: Extrapolation of Einstein's gravitational equations [17] to a global cosmological scale is the correct description of the geometry of the physical universe.

PREMISE 2: The space-time structure of the universe is homogeneous and its spatial part is isotropic ("perfect cosmological principle" applies [2,3]), and the average spatial distribution of astrophysical objects does not vary on global cosmological scales.

PREMISE 3: The local reference observer system has a Minkowski metric: $g_{\mu\nu} \rightarrow \eta_{\mu\nu}$.

PREMISE 4: The global metric field acts universally on all units of length and time scales, independent of how they are dimensioned and materialised.

Although references [2,3], refer to the steady-state theory of the expanding universe, this paper differs significantly from that concept. Unlike these papers, the solution of Einstein's gravitational equations is sought here in the form of a fully conformal metric and not the Robertson-Walker metric. The fully conformal metric is interpreted in this paper as a universal geometric background, that affects all objects, including gauges and clocks, regardless of their dimensions and internal coupling interactions. As a consequence of this interpretation, the associated expansion is only relative and all observers at different times are completely equal. In this paper, the implications of the fully conformal metric are confronted with astrophysical data such as cosmological redshift and the observable spatial distribution of quasars and γ-ray bursts. On a cosmological scale, the fully conformal metric naturally determines energy density and spatial flatness, and solves the horizon problem and Olbers' paradox in infinite spacetime.

2. Fully Conformal Spacetime with Time Scaling

Let the coordinates' origin of the local inertial system (observer location), in which the microwave background of the universe is isotropic, be our "central observer point" (in next text COP). It can be located at any place and time (for simplicity let us accept the condition of local absence of gravitational bodies). Taking into account the experimentally confirmed spatial flatness in the vicinity of the COP, we establish Minkowski coordinates $\mathrm{x_o} = (\mathrm{x_{o1}}, \mathrm{x_{o2}}, \mathrm{x_{o3}}, \mathrm{x_{o4}} = \mathrm{ct_o})$ with own time t_o for this local inertial system with the origin just at the COP. Since the real universe allows for an infinite number of alternative COPs, for the following analysis we restrict ourselves to a single COP at the origin of the above local Minkowski coordinate system $\boldsymbol{x_o} = (x_{o1}, x_{o2}, x_{o3}, x_{o4} = ct_o)$. This local Minkowski coordinate system near the origin merges with the general curvilinear global coordinate systém $\boldsymbol{x} = (x_1, x_2, x_3, x_4 = ct)$, which is used to locate points on a whole variety of the corresponding spacetime, and which parametrizes its global metric field $g_{\mu\nu}(\boldsymbol{x})$. All further considerations are based on PREMISE 1, PREMISE 2 and PREMISE 3 above.

Let us find solutions of Einstein's gravitational equations without a cosmological constant in the form of a fully conformal Minkowski metric $\eta_{\mu\nu}$ with purely temporal global scaling $\psi(x_4)$

$$R_{\mu\nu} - \frac{1}{2}Rg_{\mu\nu} = \kappa T_{\mu\nu} \cdots \left\{ \begin{array}{l} g_{\mu\nu}(x_4) \overset{\text{def}}{=} \psi(x_4)\eta_{\mu\nu} = \psi(x_4)\begin{pmatrix} 1 & 0 & 0 & 0 \\ 0 & 1 & 0 & 0 \\ 0 & 0 & 1 & 0 \\ 0 & 0 & 0 & -1 \end{pmatrix} \\ ds^2 = g_{\mu\nu}(x_4)dx_\mu dx_\nu = \psi(x_4)\left(dx_1^2 + dx_2^2 + dx_3^2 - dx_4^2\right) \end{array} \right. \quad (1)$$

From the corresponding Christoffel symbols $\Gamma^{\mu}{}_{\nu\lambda}$ as functions of single time variable x_4, only the following functions are nonzero for the conformal metric tensor (1) Appendix A:

$$\Gamma^{\mu}{}_{\nu\lambda}(x_4) = \tfrac{1}{2}g^{\mu\alpha}(x_4)(\partial_\nu g_{\alpha\lambda}(x_4) + \partial_\lambda g_{\alpha\nu}(x_4) - \partial_\alpha g_{\nu\lambda}(x_4))$$

$$\left. \begin{array}{r} \Gamma^{1}{}_{14}(x_4) = \Gamma^{1}{}_{41}(x_4) \\ \Gamma^{2}{}_{24}(x_4) = \Gamma^{2}{}_{42}(x_4) \\ \Gamma^{3}{}_{34}(x_4) = \Gamma^{3}{}_{43}(x_4) \\ \Gamma^{4}{}_{11}(x_4) = \Gamma^{4}{}_{22}(x_4) = \Gamma^{4}{}_{33}(x_4) = \Gamma^{4}{}_{44}(x_4) \end{array} \right\} = \tfrac{1}{2}\partial_4 \ln\psi(x_4) \quad . \quad (2)$$

The following Ricci tensor is non-zero only in the diagonal components, and the scalar curvature R is also a non-zero function of single time variable x_4 Appendix C.

$$\begin{array}{l} R_{11} = R_{22} = R_{33} = \tfrac{1}{2}\partial_4^2 \ln\psi(x_4) + \tfrac{1}{2}(\partial_4 \ln\psi(x_4))^2 \\ R_{44} = -\tfrac{3}{2}\partial_4^2 \ln\psi(x_4) \\ R = \left(-3\partial_4^2 \ln\psi(x_4) - \tfrac{3}{2}(\partial_4 \ln\psi(x_4))^2\right)\psi^{-1}(x_4) \end{array} \quad . \quad (3)$$

While the purely spatial part of the 3D has the character of a flat space E3 [18], the total global curvature R of the spacetime according to (1) is generally non-zero.

In a uniform (homogeneous and spacial isotropic) universe, the energy-momentum tensor $T_{\mu\nu}$ on a global cosmological scale can be approximated by a model of inert incoherent dust (particulate matter) with energy density ε_p uniformly dispersed in an ideal static space-time fluid with energy density ε_s and global pressure p. With respect to PREMISE 2, all the parameters mentioned are independent of time.

$$T_{\mu\nu} = \underbrace{\begin{pmatrix} 0 & 0 & 0 & 0 \\ 0 & 0 & 0 & 0 \\ 0 & 0 & 0 & 0 \\ 0 & 0 & 0 & \varepsilon_p \end{pmatrix}}_{T^{par}{}_{\mu\nu}} + \underbrace{\begin{pmatrix} p & 0 & 0 & 0 \\ 0 & p & 0 & 0 \\ 0 & 0 & p & 0 \\ 0 & 0 & 0 & \varepsilon_s \end{pmatrix}}_{T^{spa}{}_{\mu\nu}} \rightarrow T_{\mu\nu} = \begin{pmatrix} p & 0 & 0 & 0 \\ 0 & p & 0 & 0 \\ 0 & 0 & p & 0 \\ 0 & 0 & 0 & (\varepsilon_p + \varepsilon_s) \end{pmatrix}. \quad (4)$$

While the $T^{par}{}_{\mu\nu}$ tensor corresponds to localised particulate forms of energy, such as gases, dust and stars, the $T^{spa}{}_{\mu\nu}$ tensor corresponds to the global energy structure of a continuous metric space-time field. After substituting the energy-momentum tensor (4) into (1), we obtain conditions for a currently unknown scaling function $\psi(x_4)$ Appendix D.

$$R_{\mu\nu} - \frac{1}{2}Rg_{\mu\nu} = \kappa T_{\mu\nu} \rightarrow \left\{ \begin{array}{l} \left. \begin{array}{l} R_{11} - \tfrac{1}{2}R\psi(x_4)\eta_{11} = \kappa p \\ R_{22} - \tfrac{1}{2}R\psi(x_4)\eta_{22} = \kappa p \\ R_{33} - \tfrac{1}{2}R\psi(x_4)\eta_{33} = \kappa p \end{array} \right\} \rightarrow -\partial_4^2[\ln\psi(x_4)] - \tfrac{1}{4}(\partial_4[\ln\psi(x_4)])^2 = \kappa p \text{ (a)} \\ R_{44} - \tfrac{1}{2}R\psi(x_4)\eta_{44} = \kappa\varepsilon \ \} \rightarrow \tfrac{3}{4}(\partial_4[\ln\psi(x_4)])^2 = \kappa(\varepsilon_p + \varepsilon_s) \quad \text{(b)} \end{array} \right. . \quad (5)$$

The solution of Equation (5b) gives a general form of the scaling function $\psi(x_4)$

$$\psi_{\pm}(x_4) = \psi(0)e^{\pm x_4\sqrt{\frac{4}{3}\kappa(\varepsilon_p+\varepsilon_s)}}. \tag{6}$$

The problem of alternatives $\psi_{-}(x_4)$ and $\psi_{+}(x_4)$ of the scaling function (6) will be discussed below. With respect to PREMISE 3, the scaling function (6) becomes:

$$\psi(0) = 1 \rightarrow \psi_{\pm}(x_4) = e^{\pm x_4\sqrt{\frac{4}{3}\kappa(\varepsilon_p+\varepsilon_s)}}. \tag{7}$$

At this stage, the exponent contains an as yet unspecified total density of homogeneously distributed energy (ε_p + ε_s). The formulation of the scaling function (7) will be confronted with the present cosmological redshift data in the next section to provide a concrete form of the conformal metric tensor (1).

3. Cosmological Redshift

Let us analyse the relative time flow at different radial distances from the COP. The spacetime interval $ds(x)$ is a function of the four-vector $x(x_1, x_2, x_3, x_4)$ due to the generally varying metric tensor $g_{\mu\nu}(x)$. Similarly, the differentials of the proper Minkowski coordinates ($x_{o1}(x)$, $x_{o2}(x)$, $x_{o3}(x)$, $x_{o4}(x)$) of the local inertial systems around any point x of spacetime, are a function of the four-vector x

$$\begin{gathered} g_{\mu\nu}(x)dx_\mu dx_\nu = ds(x)^2 = \eta_{\mu\nu}dx_{o\mu}(x)dx_{o\nu}(x) \\ ds(x)^2 = \underbrace{g_{11}(x)dx_1^2}_{\eta_{11}dx_{o1}(x)^2} + \underbrace{g_{22}(x)dx_2^2}_{\eta_{22}dx_{o2}(x)^2} + \underbrace{g_{33}(x)dx_3^2}_{\eta_{33}dx_{o3}(x)^2} + \underbrace{g_{44}(x)dx_4^2}_{\eta_{44}dx_{o4}(x)^2}. \end{gathered} \tag{8}$$

If the chosen position of the COP in spacetime is denoted by $\boldsymbol{x}(0, 0, 0, 0) \equiv \boldsymbol{0}$, then the Minkowski coordinates ($x_{o1}(\boldsymbol{0})$, $x_{o2}(\boldsymbol{0})$, $x_{o3}(\boldsymbol{0})$, $x_{o4}(\boldsymbol{0})$) of the corresponding local inertial system in the conformal metric (1) according to (7), coincide with the geodetic coordinates (x_1, x_2, x_3, x_4) in the vicinity of $\boldsymbol{0}$ (which is consistent with the Minkowski metric of the surrounding spacetime, confirmed by experimental experience-3D-spatial flatness and a large region with negligible cosmological redshift)

$$ds(\boldsymbol{0})^2 = \underbrace{g_{11}(\boldsymbol{0})dx_1^2}_{\eta_{11}dx_{o1}(\boldsymbol{0})^2} + \underbrace{g_{22}(\boldsymbol{0})dx_2^2}_{\eta_{22}dx_{o2}(\boldsymbol{0})^2} + \underbrace{g_{33}(\boldsymbol{0})dx_3^2}_{\eta_{33}dx_{o3}(\boldsymbol{0})^2} + \underbrace{g_{44}(\boldsymbol{0})dx_4^2}_{\eta_{44}dx_{o4}(\boldsymbol{0})^2}. \tag{9}$$

There is a well-known formula for the own time flow τ in a point event (dx_i = 0, i = 1,2,3) at the COP $\boldsymbol{0}$ at any point $\boldsymbol{x}$ in spacetime [15,19]

$$\left.\begin{aligned} \eta_{44}dx_{o4}(x)^2 = g_{44}(x)dx_4^2 \\ \eta_{44}dx_{o4}(\boldsymbol{0})^2 = g_{44}(\boldsymbol{0})dx_4^2 \end{aligned}\right\} \rightarrow dx_{o4}(x) = dx_{o4}(\boldsymbol{0})\sqrt{\frac{g_{44}(x)}{g_{44}(\boldsymbol{0})}} \rightarrow dt_o(x) = dt_o(\boldsymbol{0})\sqrt{\frac{g_{44}(x)}{g_{44}(\boldsymbol{0})}}. \tag{10}$$

Within the conformal metric (1), the (10) takes the form of

$$dt_o(x) = dt_o(\boldsymbol{0})\sqrt{\frac{\eta_{44}\psi(x_4)}{\eta_{44}}} \rightarrow dt_o(x_4) = dt_o(0)\sqrt{\psi(x_4)}. \tag{11}$$

As a result of the assumption of the conformal metric (1) and relation (7), the Equation (11) takes the form

$$dt_o(x_4) = dt_o(0)\sqrt{\psi_{\pm}(x_4)} = dt_o(0)e^{\pm x_4\sqrt{\frac{\kappa}{3}(\varepsilon_p+\varepsilon_s)}}. \tag{12}$$

The negative time coordinate $x_4 \leq 0$ of the past defines a radial "observation sphere with positive radius" r (13)with respect to the COP

$$\left.\begin{array}{r} dr^2 = dx_1^2 + dx_2^2 + dx_3^2 \\ ds(x_4)^2 = 0 \end{array}\right\} \to \psi(x_4)dr^2 - \psi(x_4)dx_4^2 = 0 \to dr = -dx_4 \to r = -x_4. \quad (13)$$

If two different points emit light from a source with the same frequency $\nu(0)$, then we measure an identical value of $\nu(0) \sim 1/dt_o(0)$ for a source at $r = 0$ near the COP, while for a source at a large distance r the measured frequency $\nu(r) \sim 1/dt_o(r)$ will theoretically be higher or lower depending on the polarity of the exponent in (12)

$$\nu(r) = \nu(0)e^{\mp r\sqrt{\frac{\kappa}{3}(\varepsilon_p+\varepsilon_s)}}. \quad (14)$$

At this stage, the polarity of the exponent (14) can be decided based on experimental experience. Since a positive sign would represent a blue cosmological shift, it is appropriate to see it as "non-physical" for now, but this does not exclude its possible meaning in the future. On the contrary, a red cosmological shift ($\nu(r) < \nu(0)$) has been clearly confirmed experimentally, and therefore we will currently accept only the negative sign

$$\nu(r) = \nu(0)e^{-r\sqrt{\frac{\kappa}{3}(\varepsilon_p+\varepsilon_s)}}. \quad (15)$$

Let's reformulate the (15) to the "redshift" Equation (16)

$$z \stackrel{\text{def}}{=} \frac{\nu(0) - \nu(r)}{\nu(r)} \to z = e^{r\sqrt{\frac{\kappa}{3}(\varepsilon_p+\varepsilon_s)}} - 1, \quad (16)$$

which converges into the familiar linear form for small values of the exponent (small distances r from the COP)

$$z = \left(e^{r\sqrt{\frac{\kappa}{3}(\varepsilon_p+\varepsilon_s)}} \approx r\sqrt{\frac{\kappa}{3}(\varepsilon_p + \varepsilon_s)} + 1\right) - 1 \to z = \left[\sqrt{\frac{\kappa}{3}(\varepsilon_p + \varepsilon_s)}\right] r. \quad (17)$$

Equation (17) is formally equivalent to a linear formulation of the standard Hubble–Lemaître law for small cosmological distances

$$z = \frac{H}{c} r. \quad (18)$$

Combining (17) and (18), we obtain a final form that allows us to directly calculate the equivalent inertial mass density of the universe ($\rho_p + \rho_s$). Its value corresponds to the critical density ρ_{crit} [20] for the flat space solution of the Friedmann model

$$\left.\begin{array}{l} \frac{H}{c} = \sqrt{\frac{\kappa}{3}(\varepsilon_p + \varepsilon_s)} \\ \kappa = \frac{8\pi G}{c^4} \end{array}\right\} \to \rho_p + \rho_s = \frac{3H^2}{8\pi G} = \rho_{\text{crit}} = 8 \times 10^{-27}\ \text{kg} \cdot \text{m}^{-3} \to \boxed{\nu(r) = \nu(0)e^{-\frac{H}{c}r}}. \quad (19)$$

Based on this match (19), the Equation (16) can be expressed as the final form of the generalized Hubble–Lemaître law

$$\boxed{z = e^{\frac{H}{c}r} - 1}. \quad (20)$$

Its nonlinear character is clearly visible only at large cosmological distances r (see Figure 1).

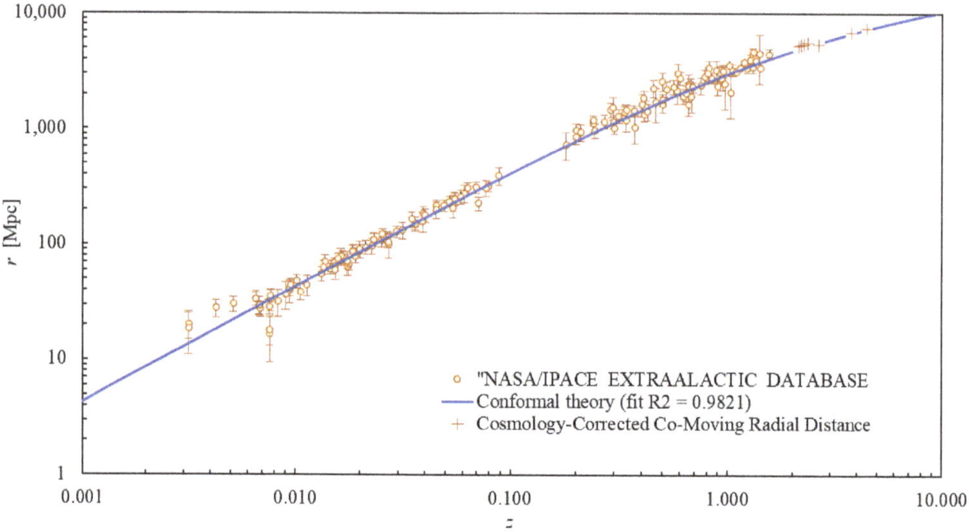

Figure 1. Hubble–Lemaître diagram: experimental points are taken from NASA/IPACE EXTRAGALACTIC DATABASE [21] and the blue curve—represents the fit of the generalized Hubble–Lemaître law (20) for a Hubble's parameter of 69.7 [km s^{-1}/Mpc] [22].

The coefficient of determination R^2 = 0.9821 in Figure 1 documents approximately a 98% fit of the model regression function (20) to the observational data and confirms the realistic nature of the mathematical formula of the generalized Hubble–Lemaître law (20). In this situation, it is worth pointing out that its nonlinearity is a direct consequence of the conformal solution of Einstein's gravitational Equation (1) without the cosmological term and it does not require the hypothetical concept of dark energy. Based on relations (17) and (18), the scaling function (7) and hence the metric tensor $g_{\mu\nu}$ takes a concrete form

$$\left.\begin{array}{l} \psi(x_4) = e^{x_4\sqrt{\frac{4}{3}\kappa(\varepsilon_p+\varepsilon_s)}} \\ \frac{H}{c} = \sqrt{\frac{\kappa}{3}(\varepsilon_p+\varepsilon_s)} \end{array}\right\} \rightarrow \psi(x_4) = e^{2\frac{H}{c}x_4} \rightarrow \boxed{g_{\mu\nu}(x_4) = e^{2\frac{H}{c}x_4}\eta_{\mu\nu}}, \left\{\begin{array}{c} \text{past}: x_4 < 0 \\ \text{COP}: x_4 = 0 \\ \text{future}: x_4 > 0 \end{array}\right\} \quad (21)$$

The global spacetime determined by the metric tensor (21) is the fundamental geometry background of the universe on a cosmological scale. Due to the non-linear nature of Einstein's equations, the "composition" of local gravitational fields is not simply additive with the above cosmological background. For example, the flat Minkowski spacetime $\eta_{\mu\nu}$ can be replaced by a conformal metric (21) in the asymptotic limit of the Schwarzschild field for $r \rightarrow \infty$, but in the vicinity of a source of mass it is problematic to separate those two metric factors. This non-trivial situation is closely related to the still debated problem of whether cosmological expansion also occurs on a local scale in gravitationally bound systems.

4. Relativity of Observer Time and the Apparent Age of the Universe

Despite the formal similarity of the conformal metric factor to the e^{2Ht} expansion factor of the de Sitter solution [23], the factor (21) corresponds to different physical conditions than the de Sitter solution without particulate matter. With a non-zero fraction of particulate matter, it cannot be derived by simply introducing the standard conformal time into the FLRW metric for solving Einstein's equations (*the derived Friedmann equations are significantly different for the two cases*). The energy-momentum tensor contains a non-zero fraction of particulate matter and the equations, on the other hand, do not contain the cosmological term that is the basis of de Sitter's solution.

Assuming global spacetime homogeneity and spatial isotropy of the universe (PREMISE 2), all COPs of such a spacetime must be completely equal, which corresponds to the "perfect cosmological principle" as formulated by Bondi, Gold and Hoyle in [2,3]. Here, however, two fundamental differences should be emphasized: in a scale invariant and fully conformal universe, expansion has only a relative observational character, similar to mass in special relativity, and the time-invariant global energy-momentum tensor does not require the continuous creation of new matter, unlike the familiar steady-state model.

The metric allows "linear extrapolation" of coordinates (x_1, x_2, x_3, x_4) from the vicinity of any COP to global Minkowski coordinates ($x_{o1}, x_{o2}, x_{o3}, x_{o4}$), so in the absence of gravitational field sources, the observer's local inertial frame in the COP is extended over the entire region of infinite spacetime

$$\left[\lim_{x_4 \to 0} g_{\mu\nu}(x_4) dx_\mu dx_\nu = \eta_{\mu\nu} dx_\mu dx_\nu\right] = ds^2 = \left[\eta_{\mu\nu} dx_{o\mu} dx_{o\nu}\right] \tag{22}$$

and corresponds to the linear metric background used by Schwartschild as an asymptotic condition for the metric description of the central gravitational field. The 3D subspace ($x_{o1}, x_{o2}, x_{o3}, x_{o4} = 0$) then represents a "time slice of presence" starting at COP. Although the extrapolated coordinates ($x_{o1}, x_{o2}, x_{o3}, x_{o4}$) continuously cover the abstract Minkowski spacetime, they are not directly measurable from the COP due to the finite speed of light.

For directly measurable relative coordinates (x_1, x_2, x_3, x_4) with origin at any COP, we can set a term "observation coordinates". The transformation of differentials between the two types of coordinates is explicitly determined by the conformal metric and the extrapolation condition (22)

$$dx_{o\mu} = e^{\frac{H}{c} x_4} dx_\mu \tag{23}$$

which gives the quantitative ratio of spatiotemporal scales for two different time levels $x_4 = 0$ and $x_4 < 0$. Especially for the time coordinate x_4, we can derive the transformation formula ($x_4 \to x_{o4}$)$\leftrightarrow$($t \to t_o$) by integration

$$x_{o4} = \int_0^{x_4} e^{\frac{H}{c} x_4} dx_4 \to \begin{cases} x_{o4} = \frac{c}{H}\left(e^{\frac{H}{c} x_4} - 1\right) \to \boxed{t_o = \frac{1}{H}\left(e^{Ht} - 1\right)} \\ \min t_o = \frac{1}{H} \lim_{t \to -\infty} \left(e^{Ht} - 1\right) = -\frac{1}{H} \approx -14 \times 10^9 \text{ yr} \end{cases} . \tag{24}$$

The relationship between local and observation time is shown graphically in Figure 2 and their mutual deviation is at most 3.5% up to $t \approx -10^9$ years, but below values of $t \approx -14 \times 10^9$ yr (into the past), the deviation becomes larger, and the extreme dilation of the observational time t results in a limiting boundary of the apparent origin of the local time min t_o, which appears to be the "age of the universe".

As we go deeper into the past, the increasingly significant dilation of observational time causes real physical effects such as the redshift, which around this threshold time level gradually grows beyond all limits.

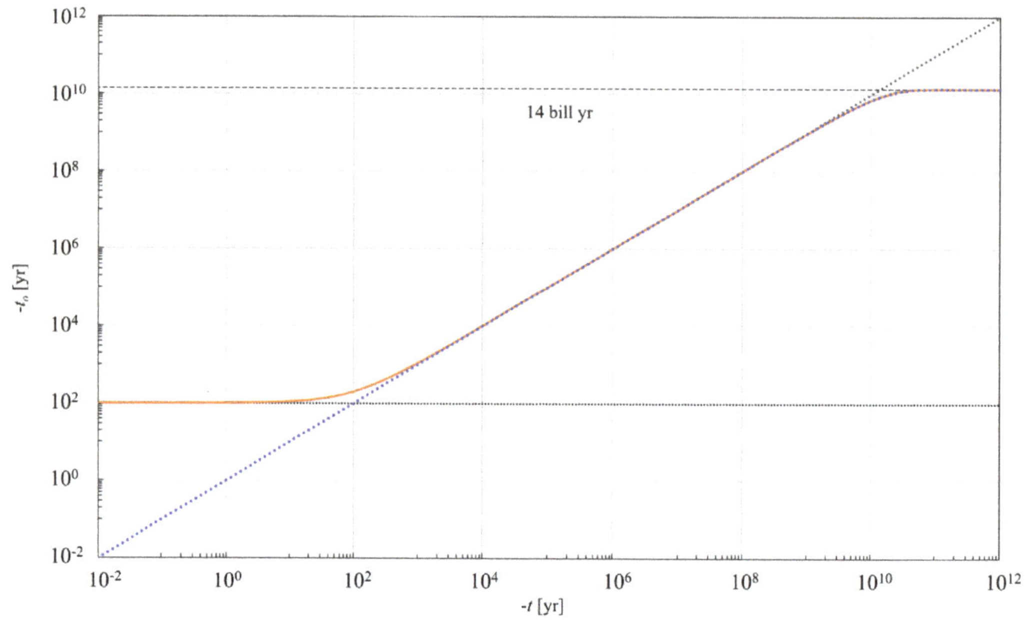

Figure 2. The dependence of the local time t_o on the observation time t of the observer in COP_0 according to (24) (blue dotted curve ·····) shows a significant increase in the dilation of the observation time around 14 billion years (black dashed line -----) into the past, where it forms a kind of boundary that appears in the local time t_o as its beginning. For comparison, the orange curve — shows an observing time against the observer, whose COP_1 is shifted into the past by Δt = 100 yr relative to the first COP (black dotted line ·····).

But what is this limiting value of the apparent "origin" for the observer in COP_1, which is shifted by Δt of observational time into the past relative to the original COP_0 (see Δt = 100 yr in Figure 2)? Due to the conformal metric (21), both minimal time limits t_{00} a t_{10} obey condition

$$\left.\begin{array}{l} COP_0 \rightarrow \min t_{00} = \int\limits_{0}^{-\infty} e^{Ht}dt = -\dfrac{1}{H} \\ COP_1 \rightarrow \min t_{10} = \int\limits_{\Delta t<0}^{-\infty} e^{Ht}dt = -\dfrac{e^{H\Delta t}}{H} \end{array}\right\} \rightarrow \boxed{|\min t_{10}| < |\min t_{00}|}. \tag{25}$$

It seems that an observer from COP_1 in the past is closer to this border resembling "age of the universe", than an observer from COP_0. With respect to PREMISE 4, let us denote the equivalent time intervals (transferred local units) by $[\tau_o]$ for COP_o a $[\tau_1]$ for COP_1. In cosmological scales, they can be naturally represented by differentials, so that (23) gives a relationship between the two local units $[\tau_o] \leftrightarrow [\tau_1]$

$$\left.\begin{array}{r} dt_o = e^{Ht}dt \\ dt_1 = e^{H(t+\Delta t)}dt \end{array}\right\} \rightarrow dt_1 = e^{H\Delta t}dt_o \rightarrow [\tau_1] = e^{H\Delta t}[\tau_o]. \tag{26}$$

While the value of min t_o is naturally expressed in time units $[\tau_o]$ of COP_0, the similarly calculated value of min t_{1o} must be converted into its own time units $[\tau_1]$ COP_1. Let us compare the dimensionless numbers n_{00} and n_{10} of time units $[\tau_o]$ and $[\tau_1]$ at the "age of the universe" values min t_{00} and min t_{10} from (25)

$$[\tau_1] = e^{H\Delta t}[\tau_0] \rightarrow \left\{ \begin{array}{l} n_{00} = \frac{\min t_{00}}{[\tau_0]} = \left(\frac{1}{H} \cdot \frac{1}{[\tau_0]}\right) \\ n_{10} = \frac{\min t_{10}}{[\tau_1]} = \left(\frac{e^{H\Delta t}}{H} \cdot \frac{1}{e^{H\Delta t}[\tau_0]}\right) = \left(\frac{1}{H} \cdot \frac{1}{[\tau_0]}\right) \end{array} \right\} \rightarrow n_{10} = n_{00} \quad (27)$$

It turns out that for both observers the apparent "age of universe" is the same, so there is no real "beginning of time" in global conformal spacetime. This limit value is the same for all observers (min $t_o \approx -14 \times 10^9$ yr) and the Hubble parameter H is a true universal constant in this model, invariant to time translations.

5. Spatial Distribution of γ-Ray Bursts

Another cosmological test of the conformal Minkowski metric hypothesis can be an experimental verification of the natural homogeneous distribution of matter on a global cosmological scale. Such a test can be performed by observing highly specific astrophysical objects and events such as black holes, quasars, and ultimately γ-ray sources that can be understood as standards (the distribution of galaxies is inappropriate for such a test due to their highly non-standard diversity).

We will discuss the spatial distribution of γ-ray bursts in the following chapter. In [24,25] they published a methodology for applying the $V/V_{\max}$ test to γ-ray sources and subsequently performed a constrained analysis of the spatial distribution for 140 sources [26]. The result shows an increasing deficit in the observed number of low intensity sources. Assuming an identical nature of the sources and the standard Euclidean flat-space metric E3, these results can be interpreted as a decrease in source density towards the past within a fully conformal spacetime. Such an interpretation leads to the conclusion that the γ-ray density changes over time on a global scale.

Let us now perform a similar analysis under a hypothetical conformal metric. The time-independent nature of the energy-momentum tensor (4) requires a global density of γ-ray sources invariant over time. The number of sources N within in a sphere of radius $r_{\max}$ around the COP is given by the time-invariant global source density $n_{\gamma o}$:

$$N(r_{\max}) = n_{\gamma o}\frac{4}{3}\pi r_{\max}^3. \quad (28)$$

In the mentioned $V/V_{\max}$ test [24], the ratio of the intensity $C_{\max} = C(r_{\max})$ of the farthest detected source at distance $r_{\max}$ to the threshold intensity $C_{\min} = \min C$ at the detection limit is chosen as the independently variable distribution parameter. The intensity of the source $C_{\max}$ decreases not only with the inverse square of the distance $r_{\max}$ but also due to the conformal metric (21). As a result, the intensity on a model source surface of radius R decreases with distance $r_{\max}$ from the observer.

Now we analyse the measurement of the "observed intensity" C [J m^{-2} s^{-1}] at distant sources, taking into account the effect of the conformal metric (21). For simplicity, consider a spherical radiation source of radius R. Let us denote by $C_o(R)$ the surface intensity of the radiation in its local inertial system $(x_{o1}, x_{o2}, x_{o3}, x_{o4})$ (8) and then denote by $C(R)$ the same relative surface intensity of the radiation in the global system (x_1, x_2, x_3, x_4) (8) of the COP at distance r. While the intensity $C(r)$ observed in the COP decreases according to the standard $C(r) \sim 1/r^2$ dependence, the transition between the $C_o(R) \rightarrow C(R)$ intensities is due to the effect of the conformal metric (21) on the intrinsic surface intensity $C_o(R)$. With a suitable orientation of $x_1 = r$, the remaining two space coordinates x_2 and x_3 define a plane perpendicular to the line between source and the COP. The two surface differentials

dS_o and dS of the source surface then lie in this plane, radiating towards the COP, and the relation $C_o(R) \rightarrow C(R)$ can be written as

$$\left.\begin{array}{r} C_0(R) \stackrel{\text{def}}{=} \frac{d^2E}{dS_o dt_o} \\ dS_o = dx_{o2}dx_{o3} \\ C(R) \stackrel{\text{def}}{=} \frac{d^2E}{dSdt} \\ dS = dx_2 dx_3 \end{array}\right\} \rightarrow C(R) = C_0(R)\frac{dS_o dt_o}{dSdt}. \tag{29}$$

As a consequence of the (8), (13) and the conformal metric (21) we can write

$$\left.\begin{array}{r} dx_{o\mu}(x_4) = e^{\frac{H}{c}x_4}dx_\mu \\ x_4 = -r \end{array}\right\} \rightarrow \left\{\begin{array}{r} dx_{o4} = e^{-\frac{H}{c}r}dx_4 \rightarrow dt_o = e^{-\frac{H}{c}r}dt \\ dS_o = dx_{o1}dx_{o2} = e^{-2\frac{H}{c}r}dx_1dx_2 = e^{-2\frac{H}{c}r}dS \end{array}\right\} \rightarrow \frac{dS_o dt_o}{dSdt} = e^{-3\frac{H}{c}r}, \tag{30}$$

and for the relative radiation intensity (29) on a model source surface of radius R it is

$$C(R) = C_o(R)e^{-3\frac{H}{c}r_{\max}}. \tag{31}$$

As a result of the simultaneous decrease in intensity with the square of the distance, Equation (31) takes its final form:

$$\underbrace{C(r_{\max})}_{C_{\max}} \cdot \left(4\pi r_{\max}^2\right) = C(R) \cdot \left(4\pi R^2\right) \rightarrow C_{\max} = C_o(R)\frac{R^2}{r_{\max}^2}e^{-3\frac{H}{c}r_{\max}}. \tag{32}$$

The NSSTC, BATSE4B astrophysical database [27] contains data as dependence of the number of sources on the ratio of the minimum intensity to the threshold intensity $N(C_{\max}/C_{\min})$. To test the conformal metric hypothesis (1), (21), we express Equation (32) in terms of an intensity ratio

$$\frac{C_{\max}}{C_{\min}} = \left(C_o(R)\frac{R^2}{C_{\min}}\right)e^{-3\frac{H}{c}r_{\max}}(r_{\max})^{-2}. \tag{33}$$

The mutual dependence of the two parameters then comes from the simultaneous validity of (28) and (33):

$$\left\{\begin{array}{l} A = \left(\frac{C_o(R)}{C_{\min}}R^2\right)\left(\frac{3}{4\pi n_{\gamma o}}\right)^{-\frac{2}{3}}[-] \\ B = 3\frac{H}{c}\left(\frac{3}{4\pi n_{\gamma o}}\right)^{\frac{1}{3}}[-] \end{array}\right\} \rightarrow \boxed{\frac{C_{\max}}{C_{\min}} = AN^{-\frac{2}{3}}e^{-BN^{\frac{1}{3}}}} \tag{34}$$

The fit of the theoretical dependence (34) to the mentioned experimental data of NSSTC, BATSE4B [27] is shown in Figure 3.

The first requirement of the fit is the limit transition $B \rightarrow 0$ of Formula (34) to the standard version of the Minkowski metric for small distances from the COP, which shows asymptotic behaviour for well-detectable sources with large $C_{\max}$ corresponding to smaller $r_{\max}$. The coefficient of determination $R^2 \approx 0.888$ demonstrates approximately 90% fit of the regression model function to the observational data in the region $C_{\max}/C_{\min} \geq 10$. The above fit for $B = 0.12$ gives a time-invariant global γ-ray burst density of $n_{\gamma o} \approx 1.47 \times 10^{-75}$ m^{-3} ≈ 43 Gpc^{-3} according to (34).

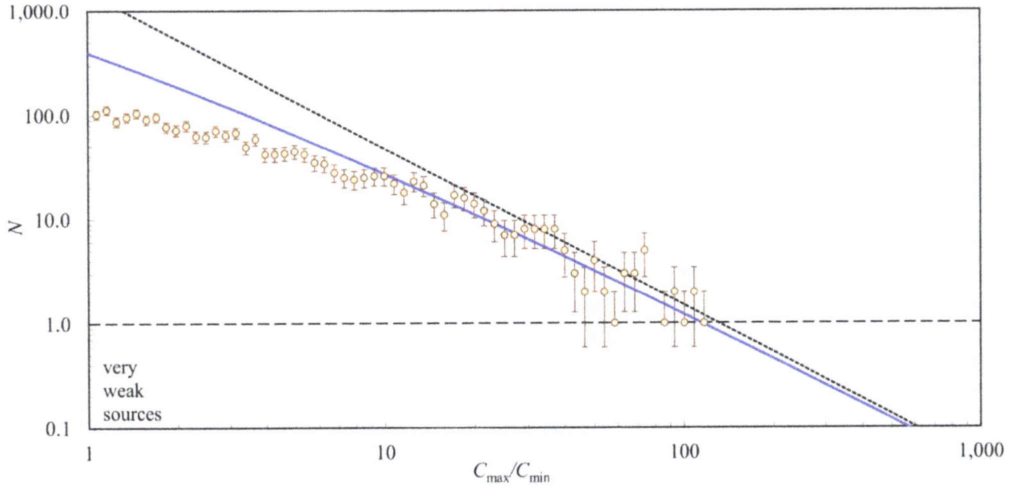

Figure 3. Theoretical fit — distribution of the total number N of γ-ray burst sources as a function of the C_{max}/C_{min} ratio for 2372 measured sources (A = 130, B = 0.12)—orange points. The black dashed line would correspond to a uniform distribution of sources in flat space with the standard Euclidean metric E3, as indicated by a formal extrapolation of the plot below the single-source detection level of $N = 1$.

The detection of the vast majority of strong sources in the nearby region ($C_{max}/C_{min} \geq 10$) is highly probable both in terms of their high intensity and their frequency of occurrence in a relatively small volume of the vicinity of the COP. However, in more distant regions, the detection probability is negatively affected both by the decreasing intensity of the sources and by their more difficult localization in an extremely increasing volume. In the Gaussian angular distribution of the intensity of narrow γ-ray bursts [28], the minimum measurable intensity C_{min} determines the edge of the Gaussian profile, so the effective solid angle detection of the γ-ray burst decreases with distance. This naturally reduces the probability of passage through the ground-based observation point and hence the frequency of detected sources. Detection of "missing" sources could therefore serve as one way to verify or falsify this hypothesis of a global conformal metric of the universe.

6. Spatial Distribution of Quasars

D. Sciama and M. Rees published a paper in 1966 [29] with the intention of falsifying the steady state theory of the universe. The paper showed that the frequency of observed quasars depends on the magnitude of the redshift, and from this fact, they deduced a time evolution of the source density that contradicts the ideas of the steady-state model of the universe. This raises the question of how to correctly interpret observations of sources with different redshift values. Although the estimated time evolution of the source density was contrary to steady-state concepts, it has not been explained in principle even within the standard ΛCMD cosmological model. It is associated with the general assumption of the genesis and evolution of astrophysical objects within any cosmological theory with a time origin. Of course, the time evolution cannot be excluded even in models without said beginning, because with limited observational experience, we may miss slow global processes ("Too brief glimpse into the past from the beginning of summer may lead one to believe that the world was created in winter").

Thus, we examine whether the hypothesis of the conformal metric is consistent with the assumption that the distribution of quasars seems to be homogeneous in space from the

COP. For a time-invariant global source density n_{qo}, the number of sources dN contained in a differential spherical shell of radius r around the COP is given by

$$dN = n_{qo} 4\pi r^2 dr. \tag{35}$$

Now we derive the distribution of dN/dz as a function of redshift from all observed quasars. From the generalized Hubble–Lemaître law (20), the radial parameters r and dr on the right-hand side of Equation (35) can be expressed as a function of the redshift z

$$z = e^{\frac{H}{c}r} - 1 \rightarrow \begin{cases} r = \frac{c}{H}\ln(z+1) \\ \frac{dz}{dr} = \frac{H}{c}e^{\frac{H}{c}r} \rightarrow dr = \frac{c}{H}e^{-\frac{H}{c}r}dz \rightarrow dr = \frac{c}{H}\frac{dz}{z+1} \end{cases}. \tag{36}$$

Substitution into (35) then gives the final formula for comparison with experimental data

$$A = 4\pi n_{qo}\left(\frac{c}{H}\right)^3 \rightarrow \boxed{\frac{dN}{dz} = A\frac{(\ln(z+1))^2}{z+1}}. \tag{37}$$

The relevant data were used from The Million Quasars (Milliquas) catalog, v7.2aa, 18 October 2021 [30], which contains experimental data obtained by observing about 977,000 sources. The fit of the theoretical dependence (37) to these experimental data is shown in Figure 4.

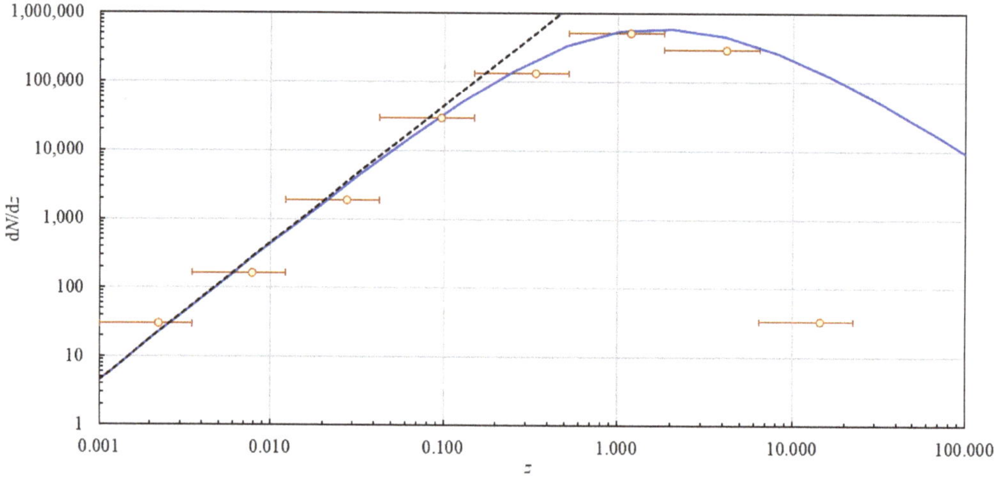

Figure 4. Theoretical fit — of the dependence of the number of quasars with the increase in redshift dN/dz on the magnitude of their redshift z (for parameter 4.5×10^6) [30]. Hollow orange points correspond to the average of all measured z values within the respective interval, and the horizontal line defines its width. The black dashed line --- corresponds to a uniform distribution of sources in flat space with Euclidean metric E3 and the standard linear form of Hubble–Lemaître law for redshift.

The requirement of the fit is again the limit transition $z \rightarrow 0$ of Equation (37) to the standard version of the Minkowski metric at small distances from the COP. This requirement of asymptotic behaviour is satisfied by the transition of the function (37) to the standard form z^2. The theoretical shape of the fit function is confronted in good agreement with the number of nearer quasars in the first six statistical intervals, which are shown in Figure 4 as horizontal lines.

The calculated coefficient of determination $R^2 \approx 0.9511$ demonstrates approximately 95% fit of the regression model function to the observational data for the first six points.

Using the regression parameter $A \approx 4.5 \times 10^6$ in Equation (37) we can estimate the quasar number density in the universe to $n_{qo} \approx 1.55 \times 10^{-73}$ m^{-3} $\approx$ 4559 Gpc^{-3}.

Similarly to the previous section, we can also justify the deficit in the number of sources in the last interval for high values of z. And again, additional measurements could serve to partially verify, or refute the hypothesis of a global conformal metric of the universe.

All the model predictions with observational data in Sections 3, 5 and 6 documents approximately a 90% fit or better. To consider such a degree of coincidence as the work of chance seems very problematic. What conclusions can be drawn from this? The analysis of the "selected data sets" seems to confirm with a high probability the validity of the "perfect cosmological principle", which was the initial assumption for the theoretical derivation of the respective formulae (20), (34) and (37). At the same time, all three formulae provide predictions for high redshift z values above the "selected data sets". These outlying regions are currently characterised by a deficit of observed objects, so model predictions can be used to subsequently verify the homogeneous space-time distribution that was the basic and simplest assumption of the steady state. This hypothesis of a homogeneous spatio-temporal distribution of astrophysical objects can be considered much more robust than current ideas of evolution, which are often problematised by recent observational data from the James Webb Space Telescope and others.

7. Olbers' Paradox and the Observable Universe

The assumption of a luminous sky in the visible light region does not match experimental experience (Olbers' paradox), and in a universe with a fully conformal metric (1) this paradox can be explained via redshift. Assume that on sufficiently large cosmological scales the number density of model astrophysical light sources n can be approximated by a constant. We introduce the intensity contribution dI_Ω of the average model sources from the differential volume $dV = d\Omega r^2 dr$ at distance r from the COP, oriented towards the COP, which is given by the emission of radiative energy ε from the source surface to the differential solid angle $d\Omega$

$$dI_\Omega(r) = \left(\frac{d^2\varepsilon(r)}{dt(r)dS(r)}\right) n d\Omega r^2 dr. \tag{38}$$

As a result of the metric transformation of the derivative variables (30), it decreases exponentially with distance r from the COP

$$\left\{\frac{dt(0)dS(0)}{dt(r)dS(r)} = e^{-3\frac{H}{c}r}\right\} \rightarrow dI_\Omega(r) = \left(\frac{d^2\varepsilon(r)}{dt(0)dS(0)}\right) e^{-3\frac{H}{c}r} n d\Omega r^2 dr. \tag{39}$$

For simplicity of the model derivation, assume purely thermal emission from an average model source with surface temperature $T(0)$. After including all the mentioned effects, the intensity contribution dI_Ω (39) takes the form

$$\left.\begin{array}{r} \text{inverse square law} \rightarrow \frac{d^2\varepsilon(r)}{dt(0)dS(0)} = \left(\frac{\sigma}{\pi}T(r)^4\right)\frac{1}{r^2} \\ \text{redshift} + \text{Wien's law} \rightarrow \frac{T(r)}{T(0)} = \frac{\nu_{\max}(r)}{\nu_{\max}(0)} = e^{-\frac{H}{c}r} \end{array}\right\} \rightarrow dI_\Omega(r) = \left(\frac{\sigma}{\pi}T(0)^4\right) e^{-7\frac{H}{c}r} n d\Omega dr. \tag{40}$$

The experimentally measured intensity I_Ω (R) [Wm^{-2}sr^{-1}] in the direction of the unit angle axis is the sum of all contributions from the opposite half-space 2π to the distance R of the observation horizon

$$I_{\Omega}(R) = \iiint\limits_{V_R^{2\pi}} dI_{\Omega} = \int_0^{2\pi}\int_0^R \left(\frac{\sigma}{\pi}T(0)^4\right)e^{-7\frac{H}{c}r}\, n d\Omega dr = 2\pi\left(\frac{\sigma}{\pi}T(0)^4\right)n\left(\int_0^R e^{-7\frac{H}{c}r}dr\right) \quad (41)$$

Without knowledge of the model parameters n and $T(0)$, the dependence of the intensity I_{Ω} (R) on the distance R from the COP can be characterized by the function Φ (R), which is defined by the parametric integral in (41)

$$\Phi(R) = \int_0^R e^{-7\frac{H}{c}r}dr = \frac{c}{7H}\left(1 - e^{-7\frac{H}{c}R}\right), \quad (42)$$

and its graph is shown in Figure 5.

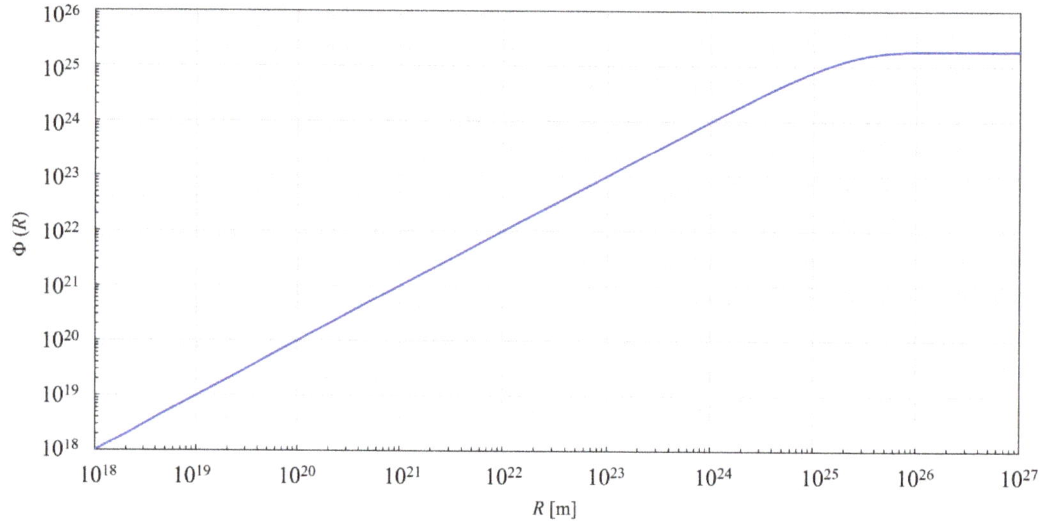

Figure 5. Dependence of the function $\Phi(R)$ (42) on the distance R of the observation sphere from the COP. The asymptotic plateau of the total intensity corresponds to a slightly smaller radius than predicted $R_{obs} \approx 4.3 \times 10^{26}$ m by the calculations of the standard model [31].

The asymptotic behaviour of the dependence (42) leads to the final value of the total intensity max I_{Ω}

$$\max I_{\Omega}(R) = \lim_{R\to\infty} I_{\Omega}(R) = 2n\sigma T(0)^4 \cdot \lim_{R\to\infty}\Phi(R) \to \boxed{\max I_{\Omega}(R) = 2n\sigma T(0)^4 \frac{c}{7H} \ll \infty}, \quad (43)$$

which is approximately reached already at distances of the order of $R \approx 10^{26}$ m, and the outer region contributes negligibly. This is how the problem of the Olbers' paradox is solved within the fully conformal metric.

While in a space with a standard Minkowski metric the radiation of all differential spheres would contribute equally to the total intensity at the COP, under the conditions of a conformal metric there is a kind of limit at which the differential contributions to the total intensity drop significantly to negligible values. This decrease can be expressed on the basis of relation (40) in the form of function

$$\Psi(r) = \frac{dI_{\Omega}(r)/dr}{dI_{\Omega}(0)/dr} = e^{-7\frac{H}{c}r} \tag{44}$$

The dependence (44) is plotted in Figure 6. The radial distance $r = 4 \times 10^{26}$ m from the COP; at which the intensity contribution drops by nine orders of magnitude to 1 ppb of the original value; agrees very well with the $R_{obs} \approx 4.3 \times 10^{26}$ m of the observable universe according to the standard model [31].

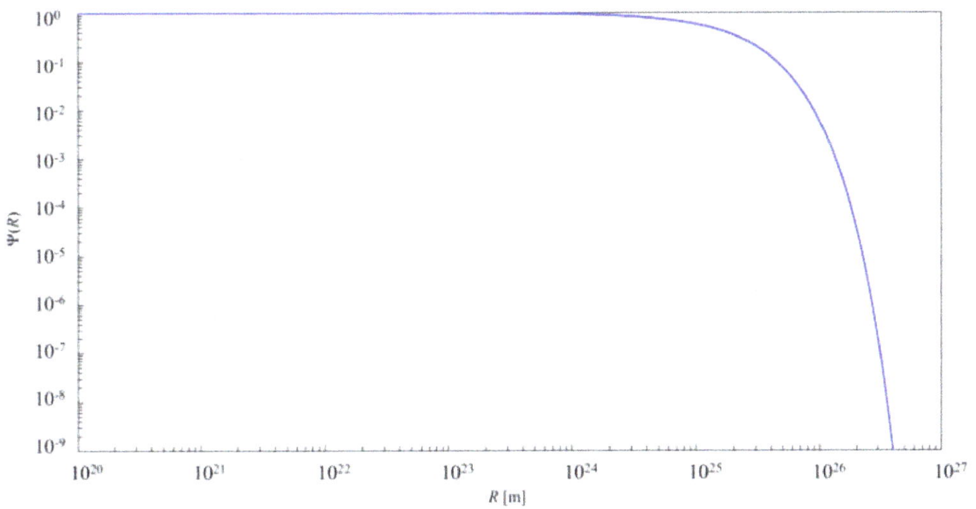

Figure 6. Dependence of the relative intensity decrease function $\Psi(R)$ (44) on the distance R from the COP.

The fully conformal metric thus defines a region of the observable universe where the contributions from the outer region are not zero in principle as in the case of causal cosmological horizon of the standard model, but they are completely negligible from an energy perspective. According to one of the most recent observations from the James Webb Telescope in 2022 [32], we can consider detecting objects with redshifts up to $z = 12$, which according to (20), corresponds to a distance of 3.5×10^{26} m located the steep intensity drop of eight orders of magnitude in Figure 6.

8. Conclusions

This paper presents a scale invariant fully conformal cosmological model based on a new physical interpretation of the relatively simple global solution of Einstein's gravitational equations with a time-independent energy-momentum tensor. This solution takes the form of a fully conformal Minkowski metric that is globally scaled by the same time function $\psi(x_4)$ in all four space-time coordinates. The time dependences of the metric tensor $g_{\mu\nu}(x_4)$ in the components of the Einstein tensor on the left-hand side of the gravity equations are fully compensated in accordance with the time-independent components of the energy-momentum tensor on the right-hand side. The above conformal metric preserves the time independence of the speed of light, and similar to the rest mass, other local physical parameters such as the electromagnetic spectrum etc. are conserved around an arbitrary COP. Clocks and gauges in the vicinity of all COPs must also be viewed as fully equivalent, and the universe on a global scale as stationary spacetime is scale invariant, and exponential expansion is only relative observational in nature. The changes in scales and time flow observed over large distances (in the past) are only relative to a given COP, like the relativistic change in mass of particles in motion. Given the universal nature of the local quantum definition of the time unit 1s and the possibility of naturally choosing the origin of the COP ($x_4 = 0$) in the local presence of an arbitrary observer, the exponential form of the scaling function $\psi(x_4)$ is covariant. Instead of a real time origin, the model introduces the "appearent age of the universe", which, due to dilation, is the limiting value of the time parameter towards the past and corresponds to the present value of the "age of the universe" H^{-1} according to the standard model. This "appearent age" is the same for all observers at different time levels, and the Hubble parameter is thus a true universal constant, invariant to time translations. The experimental part of the paper then discusses some observational implications of the presented model. Without the need for a cosmological term in Einstein's gravity equations, the model provides a physical interpretation for the formulation of the generalized Hubble-Lemaître law $z(r) = (e^{Hr/c} - 1)$, which agrees very well with current astrophysical data even for very distant supernovae Ia. The whole concept was motivated by considerations of whether cosmological redshifts can arise in the time evolution of the global metric field due to its differences at different time levels, similar to the case of gravitational redshifts in the radiation emission of very massive objects. The question of an alternative interpretation of the cosmological redshift has been raised in the past by Fritz Zwicky. However, his concept of "tired light" corresponded to a gravitational redshift based on the local curvature of spacetime by the masses of galaxies, and the redshift depended on the local spatial distances from the galactic centers. This does not agree with the observations that showed a global dependence of the redshift on the distance from any observer, regardless of the distribution of masses. On the other hand, the global conformal solution shows interesting agreement with experimental observations of the spatial distribution of real astrophysical sources such as γ-ray bursts and quasars. The conformal metric field model also solves in principle the problem of critical global density, flatness of 3D space and Olbers' paradox. The strictly sharp cosmological horizon in the standard model is here replaced by a much softer formulation a region of the practically observable universe, that is not bounded by a sharp causal boundary. The follow-up work will focus on the generalization of the Schwartschild solution based on the background of the above fully conformal metric, the derivation of the corresponding equation of state, and a description of the origin of the CMB.

Funding: For Region Sustainability and High-tech Industries project number CZ.10.03.01/00/22_003/0000048 via the Operational Programme Just Transition. I thank, to my colleague Jiri Bednar for his careful review and preparation of the English text, and first and foremost I am very grateful to Jeyninka for great support in my work.

Data Availability Statement: NASA (1996) Gamma-Ray Astrophysics: NSSTC, BATSE4B, CMAXMIN TABLE. Available at: https://gammaray.nsstc.nasa.gov/batse/grb/catalog/4b/4br_cmax_cmin.html. NASA (2022) NASA/IPAC EXTRAGALACTIC DATABASE (2022): Date and Time of the Query: Sun Nov 12 09:25:30 2017PDT. Available at: https://ned.ipac.caltech.edu/. Planck 2013 (2013) Planck 2013 results.: I. Overview of products and scientific results, Astronomy & Astrophysics manuscript no. Planck Mission. Available at: https://arxiv.org/pdf/1303.5062.pdf.

Acknowledgments: The work was supported by the project No. CZ.02.01.01/00/22_008/0004631. Materials and technologies for sustainable development within the Jan Amos Komensky Operational Program financed by the European Union and from the state budget of the Czech Republic and the European Union under the REFRESH—Research Excellence For Region Sustainability and High-tech Industries project number CZ.10.03.01/00/22_003/0000048 via the Operational Programme Just Transition. I thank, to my colleague Jiri Bednar for his careful review and preparation of the English text, and first and foremost I am very grateful to Jeyninka for great support in my work.

Conflicts of Interest: The author declares no conflict of interest.

Appendix A

M4 - space

$$\Gamma^{\alpha}_{\nu\lambda} = \tfrac{1}{2} g^{\alpha\alpha}\left(\partial_{\nu} g_{\alpha\lambda} + \partial_{\lambda} g_{\alpha\nu} - \partial_{\alpha} g_{\nu\lambda}\right)$$

$$\boxed{\Gamma_{\nu\lambda} = \frac{1}{2} g^{11}\left(\partial_{\nu} g_{1\lambda} + \partial_{\lambda} g_{1\nu}\right) + \frac{1}{2} g^{22}\left(\partial_{\nu} g_{2\lambda} + \partial_{\lambda} g_{2\nu}\right) + \frac{1}{2} g^{33}\left(\partial_{\nu} g_{3\lambda} + \partial_{\lambda} g_{3\nu}\right) + \frac{1}{2} g^{44}\left(\partial_{\nu} g_{4\lambda} + \partial_{\lambda} g_{4\nu} - \partial_{4} g_{\nu\lambda}\right)}$$

$$\boxed{\begin{aligned} g_{11}(x_4) &= \psi(x_4)\eta_{11} = \psi(x_4) \\ g_{22}(x_4) &= \psi(x_4)\eta_{22} = \psi(x_4) \\ g_{33}(x_4) &= \psi(x_4)\eta_{33} = \psi(x_4) \\ g_{44}(x_4) &= \psi(x_4)\eta_{44} = -\psi(x_4) \end{aligned}}$$

$$\Gamma^{1}_{\nu\lambda} = \frac{1}{2} g^{11}\left(\partial_{\nu} g_{1\lambda} + \partial_{\lambda} g_{1\nu}\right) \rightarrow \begin{cases} \Gamma^{1}_{1\lambda} = \frac{1}{2} g^{11}\left(\underbrace{\partial_{1} g_{1\lambda}}_{0} + \partial_{\lambda} g_{11}\right) = \frac{1}{2} g^{11}\partial_{\lambda} g_{11} \rightarrow \boxed{\Gamma^{1}_{14} = \frac{1}{2} g^{11}\partial_{4} g_{11} \rightarrow \Gamma^{1}_{14} = \frac{1}{2}\psi(x_4)^{-1}\partial_4\psi(x_4)} \\ \Gamma^{1}_{2\lambda} = \frac{1}{2} g^{11}\left(\underbrace{\partial_{2} g_{1\lambda}}_{0} + \partial_{\lambda} \underbrace{g_{12}}_{0}\right) = 0 \\ \Gamma^{1}_{3\lambda} = \frac{1}{2} g^{11}\left(\underbrace{\partial_{3} g_{1\lambda}}_{0} + \partial_{\lambda} \underbrace{g_{13}}_{0}\right) = 0 \\ \Gamma^{1}_{4\lambda} = \frac{1}{2} g^{11}\left(\partial_{4} g_{1\lambda} + \partial_{\lambda} \underbrace{g_{14}}_{0}\right) = \frac{1}{2} g^{11}\partial_{4} g_{1\lambda} \rightarrow \boxed{\Gamma^{1}_{41} = \frac{1}{2} g^{11}\partial_{4} g_{11} \rightarrow \Gamma^{1}_{41} = \frac{1}{2}\psi(x_4)^{-1}\partial_4\psi(x_4)} \end{cases}$$

$$\Gamma^{2}_{\nu\lambda} = \frac{1}{2} g^{22}\left(\partial_{\nu} g_{2\lambda} + \partial_{\lambda} g_{2\nu}\right) \rightarrow \begin{cases} \Gamma^{2}_{1\lambda} = \frac{1}{2} g^{22}\left(\underbrace{\partial_{1} g_{2\lambda}}_{0} + \partial_{\lambda} \underbrace{g_{21}}_{0}\right) = 0 \\ \Gamma^{2}_{2\lambda} = \frac{1}{2} g^{22}\left(\underbrace{\partial_{2} g_{2\lambda}}_{0} + \partial_{\lambda} g_{22}\right) = \frac{1}{2} g^{22}\partial_{\lambda} g_{22} \rightarrow \boxed{\Gamma^{2}_{24} = \frac{1}{2} g^{22}\partial_{4} g_{22} \rightarrow \Gamma^{2}_{24} = \frac{1}{2}\psi(x_4)^{-1}\partial_4\psi(x_4)} \\ \Gamma^{2}_{3\lambda} = \frac{1}{2} g^{22}\left(\underbrace{\partial_{3} g_{2\lambda}}_{0} + \partial_{\lambda} \underbrace{g_{23}}_{0}\right) = 0 \\ \Gamma^{2}_{4\lambda} = \frac{1}{2} g^{22}\left(\partial_{4} g_{2\lambda} + \partial_{\lambda} \underbrace{g_{24}}_{0}\right) = \frac{1}{2} g^{22}\partial_{4} g_{2\lambda} \rightarrow \boxed{\Gamma^{2}_{42} = \frac{1}{2} g^{22}\partial_{4} g_{22} \rightarrow \Gamma^{2}_{42} = \frac{1}{2}\psi(x_4)^{-1}\partial_4\psi(x_4)} \end{cases}$$

$$\Gamma^{3}_{\nu\lambda} = \frac{1}{2} g^{33}\left(\partial_{\nu} g_{3\lambda} + \partial_{\lambda} g_{3\nu}\right) \rightarrow \begin{cases} \Gamma^{3}_{1\lambda} = \frac{1}{2} g^{33}\left(\underbrace{\partial_{1} g_{3\lambda}}_{0} + \partial_{\lambda} \underbrace{g_{31}}_{0}\right) = 0 \\ \Gamma^{3}_{2\lambda} = \frac{1}{2} g^{33}\left(\underbrace{\partial_{2} g_{3\lambda}}_{0} + \partial_{\lambda} \underbrace{g_{32}}_{0}\right) = 0 \\ \Gamma^{3}_{3\lambda} = \frac{1}{2} g^{33}\left(\underbrace{\partial_{3} g_{3\lambda}}_{0} + \partial_{\lambda} g_{33}\right) = \frac{1}{2} g^{33}\partial_{\lambda} g_{33} \rightarrow \boxed{\Gamma^{3}_{34} = \frac{1}{2} g^{33}\partial_{4} g_{33} \rightarrow \Gamma^{3}_{34} = \frac{1}{2}\psi(x_4)^{-1}\partial_4\psi(x_4)} \\ \Gamma^{3}_{4\lambda} = \frac{1}{2} g^{33}\left(\partial_{4} g_{3\lambda} + \partial_{\lambda} \underbrace{g_{34}}_{0}\right) = \frac{1}{2} g^{33}\partial_{4} g_{3\lambda} \rightarrow \boxed{\Gamma^{3}_{43} = \frac{1}{2} g^{33}\partial_{4} g_{33} \rightarrow \Gamma^{3}_{43} = \frac{1}{2}\psi(x_4)^{-1}\partial_4\psi(x_4)} \end{cases}$$

$$\Gamma^{4}_{\nu\lambda} = \frac{1}{2} g^{44}\left(\partial_{\nu} g_{4\lambda} + \partial_{\lambda} g_{4\nu} - \partial_{4} g_{\nu\lambda}\right) \rightarrow \begin{cases} \Gamma^{4}_{1\lambda} = \frac{1}{2} g^{44}\left(\underbrace{\partial_{1} g_{4\lambda}}_{0} + \partial_{\lambda} \underbrace{g_{41}}_{0} - \partial_{4} g_{1\lambda}\right) = -\frac{1}{2} g^{44}\partial_{4} g_{1\lambda} \rightarrow \boxed{\begin{aligned}\Gamma^{4}_{11} &= -\frac{1}{2} g^{44}\partial_{4} g_{11} \rightarrow \\ \Gamma^{4}_{11} &= \frac{1}{2}\psi(x_4)^{-1}\partial_4\psi(x_4)\end{aligned}} \\ \Gamma^{4}_{2\lambda} = \frac{1}{2} g^{44}\left(\underbrace{\partial_{2} g_{4\lambda}}_{0} + \partial_{\lambda} \underbrace{g_{42}}_{0} - \partial_{4} g_{2\lambda}\right) = -\frac{1}{2} g^{44}\partial_{4} g_{2\lambda} \rightarrow \boxed{\begin{aligned}\Gamma^{4}_{22} &= -\frac{1}{2} g^{44}\partial_{4} g_{22} \rightarrow \\ \Gamma^{4}_{22} &= \frac{1}{2}\psi(x_4)^{-1}\partial_4\psi(x_4)\end{aligned}} \\ \Gamma^{4}_{3\lambda} = \frac{1}{2} g^{44}\left(\underbrace{\partial_{3} g_{4\lambda}}_{0} + \partial_{\lambda} \underbrace{g_{43}}_{0} - \partial_{4} g_{3\lambda}\right) = -\frac{1}{2} g^{44}\partial_{4} g_{3\lambda} \rightarrow \boxed{\begin{aligned}\Gamma^{4}_{33} &= -\frac{1}{2} g^{44}\partial_{4} g_{33} \rightarrow \\ \Gamma^{4}_{33} &= \frac{1}{2}\psi(x_4)^{-1}\partial_4\psi(x_4)\end{aligned}} \\ \Gamma^{4}_{4\lambda} = \frac{1}{2} g^{44}\left(\partial_{4} g_{4\lambda} + \partial_{\lambda} g_{44} - \partial_{4} g_{4\lambda}\right) \rightarrow \boxed{\Gamma^{4}_{44} = \frac{1}{2} g^{44}\partial_{4} g_{44} \rightarrow \Gamma^{4}_{44} = \frac{1}{2}\psi(x_4)^{-1}\partial_4\psi(x_4)} \end{cases}$$

Appendix B

E3 - space

$$\Gamma^{\alpha}_{\nu\lambda} = \tfrac{1}{2} g^{\alpha\alpha}\left(\partial_{\nu} g_{\alpha\lambda} + \partial_{\lambda} g_{\alpha\nu} - \partial_{\alpha} g_{\nu\lambda}\right)$$

$$\boxed{\Gamma_{\nu\lambda} = \frac{1}{2} g^{11}\left(\partial_{\nu} g_{1\lambda} + \partial_{\lambda} g_{1\nu}\right) + \frac{1}{2} g^{22}\left(\partial_{\nu} g_{2\lambda} + \partial_{\lambda} g_{2\nu}\right) + \frac{1}{2} g^{33}\left(\partial_{\nu} g_{3\lambda} + \partial_{\lambda} g_{3\nu}\right)}$$

$$\Gamma^{1}_{\nu\lambda} = \frac{1}{2} g^{11}\left(\partial_{\nu} g_{1\lambda} + \partial_{\lambda} g_{1\nu}\right) \rightarrow \begin{cases} \Gamma^{1}_{1\lambda} = \frac{1}{2} g^{11}\left(\underbrace{\partial_{1} g_{1\lambda}}_{0} + \partial_{\lambda} g_{11}\right) = \frac{1}{2} g^{11}\partial_{\lambda} g_{11} = 0 \\ \Gamma^{1}_{2\lambda} = \frac{1}{2} g^{11}\left(\underbrace{\partial_{2} g_{1\lambda}}_{0} + \partial_{\lambda} \underbrace{g_{12}}_{0}\right) = 0 \\ \Gamma^{1}_{3\lambda} = \frac{1}{2} g^{11}\left(\underbrace{\partial_{3} g_{1\lambda}}_{0} + \partial_{\lambda} \underbrace{g_{13}}_{0}\right) = 0 \end{cases}$$

$$\Gamma^{2}_{\nu\lambda} = \frac{1}{2} g^{22}\left(\partial_{\nu} g_{2\lambda} + \partial_{\lambda} g_{2\nu}\right) \rightarrow \begin{cases} \Gamma^{2}_{1\lambda} = \frac{1}{2} g^{22}\left(\underbrace{\partial_{1} g_{2\lambda}}_{0} + \partial_{\lambda} \underbrace{g_{21}}_{0}\right) = 0 \\ \Gamma^{2}_{2\lambda} = \frac{1}{2} g^{22}\left(\underbrace{\partial_{2} g_{2\lambda}}_{0} + \partial_{\lambda} g_{22}\right) = \frac{1}{2} g^{22}\partial_{\lambda} g_{22} = 0 \\ \Gamma^{2}_{3\lambda} = \frac{1}{2} g^{22}\left(\underbrace{\partial_{3} g_{2\lambda}}_{0} + \partial_{\lambda} \underbrace{g_{23}}_{0}\right) = 0 \end{cases}$$

$$\Gamma^{3}_{\nu\lambda} = \frac{1}{2} g^{33}\left(\partial_{\nu} g_{3\lambda} + \partial_{\lambda} g_{3\nu}\right) \rightarrow \begin{cases} \Gamma^{3}_{1\lambda} = \frac{1}{2} g^{33}\left(\underbrace{\partial_{1} g_{3\lambda}}_{0} + \partial_{\lambda} \underbrace{g_{31}}_{0}\right) = 0 \\ \Gamma^{3}_{2\lambda} = \frac{1}{2} g^{33}\left(\underbrace{\partial_{2} g_{3\lambda}}_{0} + \partial_{\lambda} \underbrace{g_{32}}_{0}\right) = 0 \\ \Gamma^{3}_{3\lambda} = \frac{1}{2} g^{33}\left(\underbrace{\partial_{3} g_{3\lambda}}_{0} + \partial_{\lambda} g_{33}\right) = \frac{1}{2} g^{33}\partial_{\lambda} g_{33} = 0 \end{cases}$$

Appendix C

M4 - space

$$R_{\lambda\nu}=\partial_{\mu}\Gamma^{\mu}{}_{\nu\lambda}-\partial_{\nu}\Gamma^{\mu}{}_{\mu\lambda}+\Gamma^{\eta}{}_{\nu\lambda}\Gamma^{\mu}{}_{\mu\eta}-\Gamma^{\eta}{}_{\mu\lambda}\Gamma^{\mu}{}_{\nu\eta}\rightarrow\begin{cases}R_{11}=\partial_{4}\Gamma^{4}{}_{11}+\Gamma^{\eta}{}_{11}\Gamma^{\mu}{}_{\mu\eta}-\Gamma^{\eta}{}_{\mu1}\Gamma^{\mu}{}_{1\eta}\\ R_{22}=\partial_{4}\Gamma^{4}{}_{22}+\Gamma^{\eta}{}_{22}\Gamma^{\mu}{}_{\mu\eta}-\Gamma^{\eta}{}_{\mu2}\Gamma^{\mu}{}_{2\eta}\\ R_{33}=\partial_{4}\Gamma^{4}{}_{33}+\Gamma^{\eta}{}_{33}\Gamma^{\mu}{}_{\mu\eta}-\Gamma^{\eta}{}_{\mu3}\Gamma^{\mu}{}_{3\eta}\\ R_{44}=\partial_{4}\Gamma^{4}{}_{44}-\partial_{4}\Gamma^{\mu}{}_{\mu4}+\Gamma^{\eta}{}_{44}\Gamma^{\mu}{}_{\mu\eta}-\Gamma^{\eta}{}_{\mu4}\Gamma^{\mu}{}_{4\eta}\end{cases}$$

$$\boxed{R_{11}=\boxed{\partial_{1}\Gamma^{1}{}_{11}+\partial_{2}\Gamma^{2}{}_{11}+\partial_{3}\Gamma^{3}{}_{11}}+\partial_{4}\Gamma^{4}{}_{11}+\Gamma^{\eta}{}_{11}\Gamma^{\mu}{}_{\mu\eta}-\Gamma^{\eta}{}_{\mu1}\Gamma^{\mu}{}_{1\eta}}$$

$$R_{11}=\partial_{4}\Gamma^{4}{}_{11}+\begin{bmatrix}\left(\Gamma^{1}{}_{11}\Gamma^{1}{}_{11}-\Gamma^{1}{}_{11}\Gamma^{1}{}_{11}\right)+\left(\Gamma^{1}{}_{11}\Gamma^{2}{}_{21}-\Gamma^{1}{}_{21}\Gamma^{2}{}_{11}\right)+\left(\Gamma^{1}{}_{11}\Gamma^{3}{}_{31}-\Gamma^{1}{}_{31}\Gamma^{3}{}_{11}\right)+\left(\Gamma^{1}{}_{11}\Gamma^{4}{}_{41}-\Gamma^{1}{}_{41}\Gamma^{4}{}_{11}\right)\\+\left(\Gamma^{2}{}_{11}\Gamma^{1}{}_{12}-\Gamma^{2}{}_{11}\Gamma^{1}{}_{12}\right)+\left(\Gamma^{2}{}_{11}\Gamma^{2}{}_{22}-\Gamma^{2}{}_{21}\Gamma^{2}{}_{12}\right)+\left(\Gamma^{2}{}_{11}\Gamma^{3}{}_{32}-\Gamma^{2}{}_{31}\Gamma^{3}{}_{12}\right)+\left(\Gamma^{2}{}_{11}\Gamma^{4}{}_{42}-\Gamma^{2}{}_{41}\Gamma^{4}{}_{12}\right)\\+\left(\Gamma^{3}{}_{11}\Gamma^{1}{}_{13}-\Gamma^{3}{}_{11}\Gamma^{1}{}_{13}\right)+\left(\Gamma^{3}{}_{11}\Gamma^{2}{}_{23}-\Gamma^{3}{}_{21}\Gamma^{2}{}_{13}\right)+\left(\Gamma^{3}{}_{11}\Gamma^{3}{}_{33}-\Gamma^{3}{}_{31}\Gamma^{3}{}_{13}\right)+\left(\Gamma^{3}{}_{11}\Gamma^{4}{}_{43}-\Gamma^{3}{}_{41}\Gamma^{4}{}_{13}\right)\\+\left(\Gamma^{4}{}_{11}\Gamma^{1}{}_{14}-\Gamma^{4}{}_{11}\Gamma^{1}{}_{14}\right)+\left(\Gamma^{4}{}_{11}\Gamma^{2}{}_{24}-\Gamma^{4}{}_{21}\Gamma^{2}{}_{14}\right)+\left(\Gamma^{4}{}_{11}\Gamma^{3}{}_{34}-\Gamma^{4}{}_{31}\Gamma^{3}{}_{14}\right)+\left(\Gamma^{4}{}_{11}\Gamma^{4}{}_{44}-\Gamma^{4}{}_{41}\Gamma^{4}{}_{14}\right)\end{bmatrix}$$

$$R_{11}=\partial_{4}\Gamma^{4}{}_{11}+\begin{bmatrix}\left(\boxed{\Gamma^{1}{}_{11}\Gamma^{1}{}_{11}-\Gamma^{1}{}_{11}\Gamma^{1}{}_{11}}\right)+\left(\boxed{\Gamma^{1}{}_{11}}\Gamma^{2}{}_{21}-\boxed{\Gamma^{1}{}_{21}}\Gamma^{2}{}_{11}\right)+\left(\boxed{\Gamma^{1}{}_{11}}\Gamma^{3}{}_{31}-\boxed{\Gamma^{1}{}_{31}}\Gamma^{3}{}_{11}\right)+\left(\boxed{\Gamma^{1}{}_{11}}\Gamma^{4}{}_{41}-\Gamma^{1}{}_{41}\Gamma^{4}{}_{11}\right)\\+\left(\boxed{\Gamma^{2}{}_{11}\Gamma^{1}{}_{12}-\Gamma^{2}{}_{11}\Gamma^{1}{}_{12}}\right)+\left(\boxed{\Gamma^{2}{}_{11}}\Gamma^{2}{}_{22}-\boxed{\Gamma^{2}{}_{21}}\Gamma^{2}{}_{12}\right)+\left(\boxed{\Gamma^{2}{}_{11}}\Gamma^{3}{}_{32}-\boxed{\Gamma^{2}{}_{31}}\Gamma^{3}{}_{12}\right)+\left(\boxed{\Gamma^{2}{}_{11}}\Gamma^{4}{}_{42}-\boxed{\Gamma^{2}{}_{41}}\Gamma^{4}{}_{12}\right)\\+\left(\boxed{\Gamma^{3}{}_{11}\Gamma^{1}{}_{13}-\Gamma^{3}{}_{11}\Gamma^{1}{}_{13}}\right)+\left(\boxed{\Gamma^{3}{}_{11}}\Gamma^{2}{}_{23}-\boxed{\Gamma^{3}{}_{21}}\Gamma^{2}{}_{13}\right)+\left(\boxed{\Gamma^{3}{}_{11}}\Gamma^{3}{}_{33}-\boxed{\Gamma^{3}{}_{31}}\Gamma^{3}{}_{13}\right)+\left(\boxed{\Gamma^{3}{}_{11}}\Gamma^{4}{}_{43}-\boxed{\Gamma^{3}{}_{41}}\Gamma^{4}{}_{13}\right)\\+\left(\boxed{\Gamma^{4}{}_{11}\Gamma^{1}{}_{14}-\Gamma^{4}{}_{11}\Gamma^{1}{}_{14}}\right)+\left(\Gamma^{4}{}_{11}\Gamma^{2}{}_{24}-\boxed{\Gamma^{4}{}_{21}}\Gamma^{2}{}_{14}\right)+\left(\Gamma^{4}{}_{11}\Gamma^{3}{}_{34}-\boxed{\Gamma^{4}{}_{31}}\Gamma^{3}{}_{14}\right)+\left(\Gamma^{4}{}_{11}\Gamma^{4}{}_{44}-\boxed{\Gamma^{4}{}_{41}}\Gamma^{4}{}_{14}\right)\end{bmatrix}$$

$$R_{11}=\partial_{4}\Gamma^{4}{}_{11}+\left[\left(-\Gamma^{1}{}_{41}\Gamma^{4}{}_{11}\right)+\left(\Gamma^{4}{}_{11}\Gamma^{2}{}_{24}\right)+\left(\Gamma^{4}{}_{11}\Gamma^{3}{}_{34}\right)+\left(\Gamma^{4}{}_{11}\Gamma^{4}{}_{44}\right)\right]$$

$$R_{11}=\partial_{4}\left(\frac{1}{2}\partial_4\ln\psi(x_4)\right)+\left[-\left(\frac{1}{2}\partial_4\ln\psi(x_4)\right)^2+\left(\frac{1}{2}\partial_4\ln\psi(x_4)\right)^2+\left(\frac{1}{2}\partial_4\ln\psi(x_4)\right)^2+\left(\frac{1}{2}\partial_4\ln\psi(x_4)\right)^2\right]$$

$$\boxed{R_{11}=\frac{1}{2}\partial_4{}^2\ln\psi(x_4)+\frac{1}{2}\left(\partial_4\ln\psi(x_4)\right)^2}$$

$$\boxed{R_{22}=\partial_{4}\Gamma^{4}{}_{22}+\Gamma^{\eta}{}_{22}\Gamma^{\mu}{}_{\mu\eta}-\Gamma^{\eta}{}_{\mu2}\Gamma^{\mu}{}_{2\eta}}$$

$$R_{22}=\partial_{4}\Gamma^{4}{}_{22}+\Gamma^{\eta}{}_{22}\Gamma^{\mu}{}_{\mu\eta}-\Gamma^{\eta}{}_{\mu2}\Gamma^{\mu}{}_{2\eta}$$

$$\boxed{R_{22}=\frac{1}{2}\partial_4{}^2\ln\psi(x_4)+\frac{1}{2}\left(\partial_4\ln\psi(x_4)\right)^2}$$

$$\boxed{R_{33}=\partial_{4}\Gamma^{4}{}_{33}+\Gamma^{\eta}{}_{33}\Gamma^{\mu}{}_{\mu\eta}-\Gamma^{\eta}{}_{\mu3}\Gamma^{\mu}{}_{3\eta}}$$

$$R_{33}=\partial_{4}\Gamma^{4}{}_{33}+\Gamma^{\eta}{}_{33}\Gamma^{\mu}{}_{\mu\eta}-\Gamma^{\eta}{}_{\mu3}\Gamma^{\mu}{}_{3\eta}$$

$$\boxed{R_{33}=\frac{1}{2}\partial_4{}^2\ln\psi(x_4)+\frac{1}{2}\left(\partial_4\ln\psi(x_4)\right)^2}$$

$$\left(\Gamma^{1}{}_{14}=\Gamma^{1}{}_{41}\right)=\left(\Gamma^{2}{}_{24}=\Gamma^{2}{}_{42}\right)=\left(\Gamma^{3}{}_{34}=\Gamma^{3}{}_{43}\right)=\left(\Gamma^{4}{}_{11}=\Gamma^{4}{}_{22}=\Gamma^{4}{}_{33}=\Gamma^{4}{}_{44}\right)=\frac{1}{2}\partial_4\ln\psi(x_4)$$

$$\boxed{R_{44}=\boxed{\partial_{1}\Gamma^{1}{}_{11}+\partial_{2}\Gamma^{2}{}_{11}+\partial_{3}\Gamma^{3}{}_{11}}+\partial_{4}\Gamma^{4}{}_{11}-\partial_{4}\Gamma^{\mu}{}_{\mu4}+\Gamma^{\eta}{}_{44}\Gamma^{\mu}{}_{\mu\eta}-\Gamma^{\eta}{}_{\mu4}\Gamma^{\mu}{}_{4\eta}}$$

$$R_{44}=\partial_{4}\Gamma^{4}{}_{44}-\left[\partial_{4}\Gamma^{1}{}_{14}+\partial_{4}\Gamma^{2}{}_{24}+\partial_{4}\Gamma^{3}{}_{34}+\partial_{4}\Gamma^{4}{}_{44}\right]$$
$$+\begin{bmatrix}\left(\Gamma^{1}{}_{44}\Gamma^{1}{}_{11}-\Gamma^{1}{}_{14}\Gamma^{1}{}_{41}\right)+\left(\Gamma^{1}{}_{44}\Gamma^{2}{}_{21}-\Gamma^{1}{}_{24}\Gamma^{2}{}_{41}\right)+\left(\Gamma^{1}{}_{44}\Gamma^{3}{}_{31}-\Gamma^{1}{}_{34}\Gamma^{3}{}_{41}\right)+\left(\Gamma^{1}{}_{44}\Gamma^{4}{}_{41}-\Gamma^{1}{}_{44}\Gamma^{4}{}_{41}\right)\\+\left(\Gamma^{2}{}_{44}\Gamma^{1}{}_{12}-\Gamma^{2}{}_{14}\Gamma^{1}{}_{42}\right)+\left(\Gamma^{2}{}_{44}\Gamma^{2}{}_{22}-\Gamma^{2}{}_{24}\Gamma^{2}{}_{42}\right)+\left(\Gamma^{2}{}_{44}\Gamma^{3}{}_{32}-\Gamma^{2}{}_{34}\Gamma^{3}{}_{42}\right)+\left(\Gamma^{2}{}_{44}\Gamma^{4}{}_{42}-\Gamma^{2}{}_{44}\Gamma^{4}{}_{42}\right)\\+\left(\Gamma^{3}{}_{44}\Gamma^{1}{}_{13}-\Gamma^{3}{}_{14}\Gamma^{1}{}_{43}\right)+\left(\Gamma^{3}{}_{44}\Gamma^{2}{}_{23}-\Gamma^{3}{}_{24}\Gamma^{2}{}_{43}\right)+\left(\Gamma^{3}{}_{44}\Gamma^{3}{}_{33}-\Gamma^{3}{}_{34}\Gamma^{3}{}_{43}\right)+\left(\Gamma^{3}{}_{44}\Gamma^{4}{}_{43}-\Gamma^{3}{}_{44}\Gamma^{4}{}_{43}\right)\\+\left(\Gamma^{4}{}_{44}\Gamma^{1}{}_{14}-\Gamma^{4}{}_{14}\Gamma^{1}{}_{44}\right)+\left(\Gamma^{4}{}_{44}\Gamma^{2}{}_{24}-\Gamma^{4}{}_{24}\Gamma^{2}{}_{44}\right)+\left(\Gamma^{4}{}_{44}\Gamma^{3}{}_{34}-\Gamma^{4}{}_{34}\Gamma^{3}{}_{44}\right)+\left(\Gamma^{4}{}_{44}\Gamma^{4}{}_{44}-\Gamma^{4}{}_{44}\Gamma^{4}{}_{44}\right)\end{bmatrix}$$

$$R_{44}=\partial_{4}\Gamma^{4}{}_{44}-\left[\partial_{4}\Gamma^{1}{}_{14}+\partial_{4}\Gamma^{2}{}_{24}+\partial_{4}\Gamma^{3}{}_{34}+\partial_{4}\Gamma^{4}{}_{44}\right]$$
$$+\begin{bmatrix}\left(\boxed{\Gamma^{1}{}_{44}}\Gamma^{1}{}_{11}-\Gamma^{1}{}_{14}\Gamma^{1}{}_{41}\right)+\left(\boxed{\Gamma^{1}{}_{44}}\Gamma^{2}{}_{21}-\boxed{\Gamma^{1}{}_{24}}\Gamma^{2}{}_{41}\right)+\left(\boxed{\Gamma^{1}{}_{44}}\Gamma^{3}{}_{31}-\boxed{\Gamma^{1}{}_{34}}\Gamma^{3}{}_{41}\right)+\left(\boxed{\Gamma^{1}{}_{44}}\Gamma^{4}{}_{41}-\boxed{\Gamma^{1}{}_{44}}\Gamma^{4}{}_{41}\right)\\+\left(\boxed{\Gamma^{2}{}_{44}}\Gamma^{1}{}_{12}-\boxed{\Gamma^{2}{}_{14}}\Gamma^{1}{}_{42}\right)+\left(\boxed{\Gamma^{2}{}_{44}}\Gamma^{2}{}_{22}-\Gamma^{2}{}_{24}\Gamma^{2}{}_{42}\right)+\left(\boxed{\Gamma^{2}{}_{44}}\Gamma^{3}{}_{32}-\boxed{\Gamma^{2}{}_{34}}\Gamma^{3}{}_{42}\right)+\left(\boxed{\Gamma^{2}{}_{44}}\Gamma^{4}{}_{42}-\boxed{\Gamma^{2}{}_{44}}\Gamma^{4}{}_{42}\right)\\+\left(\boxed{\Gamma^{3}{}_{44}}\Gamma^{1}{}_{13}-\boxed{\Gamma^{3}{}_{14}}\Gamma^{1}{}_{43}\right)+\left(\boxed{\Gamma^{3}{}_{44}}\Gamma^{2}{}_{23}-\boxed{\Gamma^{3}{}_{24}}\Gamma^{2}{}_{43}\right)+\left(\boxed{\Gamma^{3}{}_{44}}\Gamma^{3}{}_{33}-\Gamma^{3}{}_{34}\Gamma^{3}{}_{43}\right)+\left(\boxed{\Gamma^{3}{}_{14}}\Gamma^{4}{}_{43}-\boxed{\Gamma^{3}{}_{44}}\Gamma^{4}{}_{43}\right)\\+\left(\Gamma^{4}{}_{44}\Gamma^{1}{}_{14}-\boxed{\Gamma^{4}{}_{14}}\Gamma^{1}{}_{44}\right)+\left(\Gamma^{4}{}_{44}\Gamma^{2}{}_{24}-\boxed{\Gamma^{4}{}_{24}}\Gamma^{2}{}_{44}\right)+\left(\Gamma^{4}{}_{44}\Gamma^{3}{}_{34}-\boxed{\Gamma^{4}{}_{34}}\Gamma^{3}{}_{44}\right)+\left(\boxed{\Gamma^{4}{}_{44}\Gamma^{4}{}_{44}-\Gamma^{4}{}_{44}\Gamma^{4}{}_{44}}\right)\end{bmatrix}$$

$$R_{44}=-\left[\partial_{4}\Gamma^{1}{}_{14}+\partial_{4}\Gamma^{2}{}_{24}+\partial_{4}\Gamma^{3}{}_{34}\right]+\left[\left(-\Gamma^{1}{}_{14}\Gamma^{1}{}_{41}\right)+\left(-\Gamma^{2}{}_{24}\Gamma^{2}{}_{42}\right)+\left(-\Gamma^{3}{}_{34}\Gamma^{3}{}_{43}\right)+\left(\Gamma^{4}{}_{44}\Gamma^{1}{}_{14}\right)+\left(\Gamma^{4}{}_{44}\Gamma^{2}{}_{24}\right)+\left(\Gamma^{4}{}_{44}\Gamma^{3}{}_{34}\right)\right]$$

$$R_{44}=-\frac{3}{2}\partial_4{}^2\ln\psi(x_4)+\begin{bmatrix}-\left(\frac{1}{2}\partial_4\ln\psi(x_4)\right)^2-\left(\frac{1}{2}\partial_4\ln\psi(x_4)\right)^2-\left(\frac{1}{2}\partial_4\ln\psi(x_4)\right)^2\\+\left(\frac{1}{2}\partial_4\ln\psi(x_4)\right)^2+\left(\frac{1}{2}\partial_4\ln\psi(x_4)\right)^2+\left(\frac{1}{2}\partial_4\ln\psi(x_4)\right)^2\end{bmatrix}\rightarrow\boxed{R_{44}=-\frac{3}{2}\partial_4{}^2\ln\psi(x_4)}$$

$$\boxed{R_{11}=R_{22}=R_{33}=\frac{1}{2}\partial_4{}^2\ln\psi(x_4)+\frac{1}{2}\left(\partial_4\ln\psi(x_4)\right)^2\ ,\ R_{44}=-\frac{3}{2}\partial_4{}^2\ln\psi(x_4)}$$

Appendix D

$$
\begin{aligned}
&R_{11}-\frac{1}{2}Rg_{11}=\kappa p\\
&R_{11}-\frac{1}{2}R\psi(x_4)\eta_{11}=\kappa p\\
&\frac{1}{2}\partial_4{}^2\ln\psi(x_4)+\frac{1}{2}\left(\partial_4\ln\psi(x_4)\right)^2-\frac{1}{2}\left(3\partial_4{}^2\ln\psi(x_4)+\frac{3}{2}\left(\partial_4\ln\psi(x_4)\right)^2\right)\psi^{-1}(x_4)\psi(x_4)\eta_{11}=\kappa p\\
&\boxed{-\partial_4{}^2\ln\psi(x_4)-\frac{1}{4}\left(\partial_4\ln\psi(x_4)\right)^2=\kappa p}
\end{aligned}
$$

$$
\begin{aligned}
&R_{44}-\frac{1}{2}Rg_{44}=\kappa\left(\varepsilon_p+\varepsilon_s\right)\\
&R_{44}-\frac{1}{2}R\psi(x_4)\eta_{44}=\kappa\left(\varepsilon_p+\varepsilon_s\right)\\
&-\frac{3}{2}\partial_4{}^2\ln\psi(x_4)-\frac{1}{2}\left(3\partial_4{}^2\ln\psi(x_4)+\frac{3}{2}\left(\partial_4\ln\psi(x_4)\right)^2\right)\psi^{-1}(x_4)\psi(x_4)\eta_{44}=\kappa\left(\varepsilon_p+\varepsilon_s\right)\\
&\boxed{\frac{3}{4}\left(\partial_4\ln\psi(x_4)\right)^2=\kappa\left(\varepsilon_p+\varepsilon_s\right)}
\end{aligned}
$$

Appendix E

E3 - space

$$
\left.\begin{aligned}
\partial_1 g_{\mu\nu}\left(x_4\right)=0\\
\partial_2 g_{\mu\nu}\left(x_4\right)=0\\
\partial_3 g_{\mu\nu}\left(x_4\right)=0
\end{aligned}\right\}\rightarrow R_{\lambda\nu}=\partial_\mu\Gamma^\mu{}_{\nu\lambda}-\partial_\nu\Gamma^\mu{}_{\mu\lambda}+\Gamma^\eta{}_{\nu\lambda}\Gamma^\mu{}_{\mu\eta}-\Gamma^\eta{}_{\mu\lambda}\Gamma^\mu{}_{\nu\eta}=0\rightarrow\boxed{R=0}
$$

Appendix F

$$
\begin{aligned}
g_{\mu\nu}&=\psi(x_4)\eta_{\mu\nu}\\
g^{\mu\nu}&=\psi^{-1}(x_4)\eta^{\mu\nu}\\
R_{11}&=R_{22}=R_{33}=\frac{1}{2}\partial_4{}^2\ln\psi(x_4)+\frac{1}{2}\left(\partial_4\ln\psi(x_4)\right)^2\ ,\\
R_{44}&=-\frac{3}{2}\partial_4{}^2\ln\psi(x_4)\\
R&=R_{\mu\nu}g^{\mu\nu}=R_{\mu\nu}\eta^{\mu\nu}\psi^{-1}(x_4)=\left(R_{11}+R_{22}+R_{33}-R_{44}\right)\psi^{-1}(x_4)=\left(3\partial_4{}^2\ln\psi(x_4)+\frac{3}{2}\left(\partial_4\ln\psi(x_4)\right)^2\right)\psi^{-1}(x_4)
\end{aligned}
$$

$$
\left.\begin{aligned}
R=\left(3\partial_4{}^2\ln\psi(x_4)+\frac{3}{2}\left(\partial_4\ln\psi(x_4)\right)^2\right)\psi^{-1}(x_4)\\
\psi_\pm(x_4)=e^{\left|\pm x_4\sqrt{\frac{4}{3}\kappa\left(\varepsilon_p+\varepsilon_s\right)}\right|}\geq 0
\end{aligned}\right\}\left\{\begin{aligned}
&R=\left(\underbrace{3\partial_4{}^2\ln\left(e^{\left|\pm x_4\sqrt{\frac{4}{3}\kappa\left(\varepsilon_p+\varepsilon_s\right)}\right|}\right)}_{0}+\frac{3}{2}\left(\partial_4\ln\left(e^{\left|\pm x_4\sqrt{\frac{4}{3}\kappa\left(\varepsilon_p+\varepsilon_s\right)}\right|}\right)\right)^2\right)\psi_\pm{}^{-1}(x_4)\\
&\boxed{R\left(\psi_\pm(x_4)\right)=2\kappa\left(\varepsilon_p+\varepsilon_s\right)\psi_\pm{}^{-1}(x_4)\geq 0}
\end{aligned}\right.
$$

References

1. Zwicky, F. On the Redshift of Spectral Lines Through Interstellar Space. *Proc. Natl. Acad. Sci. USA* **1929**, *15*, 773–779. [CrossRef] [PubMed]
2. Bondi, H.; Gold, T. The Steady-State Theory of the Expanding Universe. *Mon. Not. R. Astron. Soc.* **1948**, *108*, 252–270. [CrossRef]
3. Hoyle, F. A New Model for the Expanding Universe. *Mon. Not. R. Astron. Soc.* **1948**, *108*, 372–382. [CrossRef]
4. Bekenstein, J.D. Exact solutions of Einstein-conformal scalar equations. *Ann. Phys. N. Y.* **1974**, *82*, 535–547. [CrossRef]
5. Lester, J.A. Conformal Minkowski space-time—I—Relative infinity and proper time. *Nuovo Cim. B* **1982**, *72*, 261–272. [CrossRef]
6. Lester, J.A. Conformal Minkowski spacetime—II—A Cosmological model. *Nuovo Cim. B* **1983**, *73*, 139–149. [CrossRef]
7. Varieschi, G.U. A kinematical approach to conformal cosmology. *Gen. Relativ. Gravit.* **2010**, *42*, 929–974. [CrossRef]
8. Behnke, D.; Blaschke, D.B.; Pervushin, V.N.; Proskurin, D. Description of supernova data in conformal cosmology without cosmological constant. *Phys. Lett. Sect. B Nucl. Elem. Part High-Energy Phys.* **2002**, *530*, 20–26. [CrossRef]
9. Libanov, M.; Rubakov, V.; Rubtsov, G. Towards conformal cosmology. *JETP Lett.* **2015**, *102*, 561–570. [CrossRef]
10. Kumar, M.M. Some significant conformal metrics. *Nuovo Cim. A Ser.* **1969**, *63*, 559–573. [CrossRef]
11. Lematre, A.G. A Homogeneous Universe of Constant Mass and Increasing Radius accounting for the Radial Velocity of Extra-galactic Nebulae. *Mon. Not. R. Astron. Soc.* **1931**, *91*, 483–490. [CrossRef]

12. Friedmann, A. On the Possibility of a World with Constant Negative Curvature of Space. *Gen. Relativ. Gravit.* **1999**, *31*, 2001–2008. [CrossRef]
13. Robertson, H.P. Kinematics and World-Structure. *Astrophys. J.* **1935**, *82*, 284. [CrossRef]
14. Walker, A.G. On Milne's Theory of World-Structure. *Proc. Lond. Math. Soc.* **1937**, *S2–S42*, 90–127. [CrossRef]
15. Weinberg, S. *Relativity, Gravitation and Cosmology*; John Wiley & Sons Inc.: Hoboken, NJ, USA, 1972.
16. Ibison, M. On the conformal forms of the Robertson-Walker metric. *J. Math. Phys.* **2007**, *48*, 122501. [CrossRef]
17. Misner, C.W.; Thorne, K.S.; Wheeler, J.A. MTW-Gravitation. In *Gravitation*; W. H. Freeman and Company: New York, NY, USA, 2018.
18. Peebles, P.J. *Principles of Physical Cosmology*; Princeton University Press: Princeton, NJ, USA, 2020. [CrossRef]
19. Kuchař, K. *Fundamentals of General Relativity*; ACADEMIA: Prahy, Czech Republic, 1968.
20. Weinberg, S. *Cosmology*; Oxford Press: Oxford, UK, 2008.
21. NASA. NASA/IPAC Extragalactic Database (2022): Date and Time of the Query: Sun Nov 12 09:25:30 2017PDT. Available online: https://ned.ipac.caltech.edu/ (accessed on 1 March 2023).
22. Planck 2013. Planck 2013 Results: I. Overview of Products and Scientific Results, Astronomy & Astrophysics Manuscript no. Planck Mission. Available online: https://arxiv.org/pdf/1303.5062.pdf (accessed on 1 March 2023).
23. Mitra, A. Interpretational conflicts between the static and non-static forms of the de Sitter metric. *Sci. Rep.* **2012**, *2*, 923. [CrossRef] [PubMed]
24. Schmidt, M.; Higdon, J.C.; Hueter, G. Application of the V/V (max) test to gamma-ray bursts. *Astrophys. J.* **1988**, *329*, L85–L87. [CrossRef]
25. Band, D.L. The effect of repeating burst sources on ⟨V/V max⟩. *AIP Conf. Proc.* **1994**, *307*, 39–43. [CrossRef]
26. Meegan, C.A.; Fishman, G.J.; Wilson, R.B.; Paciesas, W.S.; Pendleton, G.N.; Horack, J.M.; Brock, M.N.; Kouveliotou, C. Spatial distribution of γ-ray bursts observed by BATSE. *Nature* **1992**, *355*, 143–145. [CrossRef]
27. NASA. Gamma-Ray Astrophysics: NSSTC, BATSE4B, CMAXMIN Table. Available online: https://gammaray.nsstc.nasa.gov/batse/grb/catalog/4b/4br_cmax_cmin.html (accessed on 1 March 2023).
28. Salafia, O.S.; Ghirlanda, G. The Structure of Gamma Ray Burst Jets. *Galaxies* **2022**, *10*, 93. [CrossRef]
29. Sciama, D.W.; Rees, M.J. Cosmological significance of the relation between red-shift and flux density for quasars. *Nature* **1966**, *211*, 1283. [CrossRef]
30. MILLIQUAS v7.9. MILLIQUAS v7.9: Quasars, Redshifts and the 'Accelerating-Expansion' Universe. Data, Sky Fields, & Some Thoughts on the Right Cosmological Model. Available online: https://quasars.org/milliquas.htm (accessed on 6 January 2025).
31. Halpern, P.; Tomasello, N. Size of the Observable Universe. *Adv. Astrophys.* **2016**, *1*, 135–137. [CrossRef]
32. Finkelstein, S.L.; Bagley, M.B.; Haro, P.A.; Dickinson, M.; Ferguson, H.C.; Kartaltepe, J.S.; Papovich, C.; Burgarella, D.; Kocevski, D.D.; Huertas-Company, M.; et al. A Long Time Ago in a Galaxy Far, Far Away: A Candidate z∼12 Galaxy in Early JWST CEERS Imaging. *Astrophys. J. Lett.* **2022**, *940*, L55. [CrossRef]

universe

MDPI

Article

A Model for Cosmic Magnetic Field Amplification: Effects of Pressure Anisotropy Perturbations

Ji-Hoon Ha

Korea AeroSpace Administration, Sacheon-si 52535, Republic of Korea; jhha223@korea.kr; Tel.: +82-64-797-7043

Abstract: Magnetic field amplification in the early universe is a long-standing problem that has been extensively studied through theoretical and numerical approaches, focusing on turbulent dynamos and the growth of collisionless plasma instabilities. In the post-recombination era, pressure anisotropy drives plasma instabilities, and magnetic field amplification through these instabilities can be faster than that driven by turbulent dynamos. By considering the balance between isotropization through magnetic field amplification and anisotropy generation by external sources such as turbulence and anisotropic cosmic-ray scattering, it is reasonable to assume that the system evolves around an equilibrium anisotropy value. To improve the theoretical modeling of magnetic field amplification in such systems, this study specifically examines pressure anisotropy perturbations near the equilibrium anisotropy value, which may destabilize the system. By analyzing the effects of pressure anisotropy perturbations and their damping rates on the time evolution of cosmic magnetic fields, we highlight the importance of these perturbations in driving plasma instabilities and boosting cosmic magnetic field amplification.

Keywords: early universe; intergalactic medium; magnetic field; plasma physics

1. Introduction

It has been shown that the large-scale structure of the universe could be magnetized, and the presence of magnetic fields in clusters of galaxies is particularly suggested by observations [1–7]. Magnetic fields play a significant role in particle acceleration and transport in the intergalactic and intracluster media [8–12]. Additionally, they have a significant impact on the anisotropies of the cosmic microwave background [13–15] as well as on the physics of Big Bang nucleosynthesis [16]. However, the origin and/or amplification of cosmic magnetic fields are not yet fully understood, and these remain long-standing problems.

Theoretically, weak seed magnetic fields can be generated in the early universe through several mechanisms, each of which plays a crucial role in the subsequent evolution of cosmic magnetic fields. These mechanisms include: (1) large-scale magnetohydrodynamic (MHD) processes, such as the Biermann battery effect, which generates magnetic fields from the gradients of pressure and density [17–20]; (2) currents induced by the electronic scattering of anisotropic radiation fields, a process that can create localized magnetic fields by transferring momentum from the radiation to charged particles [21,22]; (3) the evolution of turbulent fields, where the amplification of magnetic fields occurs through the dynamo action within the turbulent plasma, which could be particularly significant during the post-recombination era [23,24]; and (4) Weibel instabilities excited by shocks produced during structure formation, which can generate magnetic fields through the interaction of relativistic particles with the surrounding plasma [12,25,26]. These processes contribute to the generation of cosmic magnetic fields in the early universe. It has been shown that cosmic magnetic fields undergo damping from before the epoch of neutrino decoupling through to recombination, limiting their growth during this period [27]. This suggests that while the early universe experienced significant damping of magnetic fields, the amplification processes became more efficient after recombination, leading to the formation of the large-scale magnetic fields observed in galaxy clusters today.

Magnetic field amplification through turbulent dynamo has been extensively studied using both analytical and numerical methods [23,24,28–32]. The turbulent dynamo is a physical process in which the kinetic energy of turbulent motions within a conducting fluid or plasma is converted into magnetic energy. Turbulence stretches and folds the magnetic field lines, a process that leads to the amplification of magnetic field strength over time. This mechanism plays a crucial role in generating and enhancing magnetic fields in astrophysical environments, particularly in regions like galaxy clusters and the intergalactic medium. Here, and throughout this paper, we assume that intergalactic plasma is a highly conducting medium in the sense that its resistivity (η) is much smaller than its viscosity (μ), i.e., the Prandtl number $Pm \equiv \mu/\eta \gg 1$. In this case, the small-scale dynamo theory predicts exponential growth of magnetic energy up to equipartition at the viscous scale. Specifically, magnetic field amplification through small-scale dynamo is enhanced during the radiation-dominated epoch due to turbulence generated by first-order phase transitions or primordial density perturbations [32]. In principle, magnetic field amplification through small-scale dynamo is efficient; however, it has been shown that pressure anisotropies generated by turbulence can drive plasma instabilities in locally unstable regions. These instabilities enhance effective collisionality, which could affect the exponential growth through turbulent dynamo [33]. In this regard, another mechanism may also contribute to the amplification of cosmic magnetic fields alongside small-scale dynamo.

As Grasso and Rubinstein [34] have highlighted, several processes, such as those occurring during cosmic inflation, phase transitions, and vorticity-driven processes, have been proposed as key contributors to the generation of primordial magnetic fields. In particular, magnetogenesis during inflation is often associated with the amplification of seed magnetic fields due to the rapid expansion of the universe, with quantum fluctuations in the electromagnetic field being stretched to cosmological scales. Similarly, phase transitions such as those occurring during the electroweak or QCD transitions can produce magnetic fields through the breaking of symmetries and the subsequent generation of currents. Vorticity-driven processes related to the motion of fluids or plasmas in the early universe can also lead to the generation of magnetic fields, as rotating fluids can create local currents that contribute to magnetic field generation. Within this broader framework, it has been highlighted that pressure-anisotropy-driven instabilities can be seen as an additional mechanism that could amplify or alter magnetic fields [35–38]. These instabilities, which arise from the anisotropic distribution of plasma particles, could act on already-existing magnetic fields, possibly leading to nonlinear amplification. The role of pressure anisotropy in magnetogenesis should, therefore, be considered alongside other mechanisms, such as vorticity-driven processes, to provide a more complete picture of the processes responsible for the generation and evolution of cosmic magnetic fields.

To enhance the understanding of magnetogenesis mechanisms, this study specifically focuses on the amplification of cosmic magnetic fields through pressure-anisotropy-driven plasma instabilities. In previous work, the factor describing pressure anisotropy is often assumed to be constant, which is justified by the fact that any process amplifying the pressure anisotropy is precisely counterbalanced by processes leading to isotropization [35]. Even in such a balanced system, small perturbations around an equilibrium anisotropy factor could be generated by turbulence. In this work, we additionally consider perturbations to the pressure anisotropy around a constant value. These perturbations could lead to a marginally unstable system, driving plasma instabilities in very weakly magnetized plasmas. Such effects could modify the evolution of the magnetic field, including the saturation timescales of magnetic field amplification.

2. Statement of the Problem

In this section, we introduce the problem setup based on pressure anisotropy and the perturbations that modify it. Depending on the characteristic frequencies for isotropizing pressure anisotropy, we distinguish between two different regimes for examining magnetic field amplification.

2.1. Pressure Anisotropy

Throughout the paper, we define the pressure anisotropy factor with respect to the magnetic field direction as follows:

$$\Delta \equiv \frac{p_\perp - p_\parallel}{p_\perp}. \tag{1}$$

To consider the time evolution of pressure anisotropy, we apply the time derivative to Equation (1):

$$\frac{d\Delta}{dt} = \frac{1}{p_\perp}\left[(1-\Delta)\frac{dp_\perp}{dt} - \frac{dp_\parallel}{dt}\right]. \tag{2}$$

Here, the first term on the right-hand side of Equation (2) accounts for the time evolution of magnetic field intensity. Considering the conservation of the first adiabatic invariant $\mu = m_i v_\perp^2/2B$, the emergence of pressure anisotropy is a natural consequence of changes in magnetic field strength. By summing up the first adiabatic invariants of all particles following a Maxwellian distribution with thermal velocity $v_{\rm th}$, we obtain $p_\perp/B = \text{const}$. Including the pressure isotropization due to collisions, parameterized by the Braginskii collision frequency ν_{ii}, Equation (2) can be rewritten as follows:

$$\frac{d\Delta}{dt} \approx (1-\Delta)\frac{1}{B}\frac{dB}{dt} - \frac{1}{p_\perp}\frac{dp_\parallel}{dt} - \nu_{ii}|\Delta|. \tag{3}$$

The term $p_\perp^{-1}\left(dp_\parallel/dt\right)$ describes the interchange of parallel and perpendicular pressures, mediated by magnetic field fluctuations, which represents the net magnetic scattering of the particle distributions. The relevant scattering frequency is defined as follows:

$$\nu_{\rm scatt} \sim \frac{1}{p_\perp}\frac{dp_\parallel}{dt}\frac{1}{\Delta}. \tag{4}$$

Based on the characteristic frequencies for isotropization of pressure anisotropy, this work considers two different regimes, summarized as follows:

- Collision-dominated regime, for $\nu_{ii} \gg \nu_{\rm scatt}$;
- Intermediate regime, for $\nu_{ii} \sim \nu_{\rm scatt}$;
- Scattering-dominated regime, for $\nu_{ii} \ll \nu_{\rm scatt}$.

As the magnetic field evolves in the system, the relative significance of ν_{ii} and $\nu_{\rm scatt}$ also changes. Initially, the pressure anisotropy is primarily isotropized by collisions. As the magnetic fields increase, scattering processes mediated by plasma instabilities become more significant. In this context, the paper describes the evolution of three regimes, as outlined below.

2.2. Magnetic Field Amplification in the Collision-Dominated Regime and Perturbations Enhancing Pressure Anisotropy

The analytic approach with a constant Δ in regions involving pressure, anisotropy-driven instabilities can be conducted by assuming that any process amplifying the pressure anisotropy, such as turbulence or anisotropic cosmic-ray scattering, is precisely counterbalanced by processes leading to isotropization. Under this assumption, the evolution of the system can be formulated using the equilibrium anisotropy value $\Delta(t) \approx \Delta_0$ [35]. We additionally consider perturbations to the pressure anisotropy around Δ_0 that marginally destabilize the system near equilibrium ($\delta\Delta \ll \Delta_0$). Depending on the damping rate (Γ_d) of the oscillation frequency ($\omega_{\rm pert}$) of the perturbations, $\Delta(t)$ can be described as follows:

$$\Delta(t) \approx \Delta_0 + \delta\Delta \exp(-\Gamma_d t)\cos(\omega_{\rm pert} t). \tag{5}$$

Assuming that oscillation is negligible compared to the damping (i.e., $\Gamma_d \gg \omega_{\rm pert}$), $\Delta(t)$ is simplified as follows:

$$\Delta(t) \approx \Delta_0 + \delta\Delta \exp(-\Gamma_d t). \tag{6}$$

The generation of $\delta\Delta$ through turbulence is further examined in the regime of collision-dominated isotropization, which could be relevant to the early stages of the evolution of a cosmological seed field after the recombination era, especially given that $\beta = 2v_{\rm th}^2/v_{\rm A}^2 \gg 1$ and $\nu_{ii} \gg \Omega_i$. Ignoring the isotropization governed by magnetic scattering, the equation describing the time evolution of the anisotropy factor can be reduced as follows:

$$\frac{d\Delta}{dt} \approx (1-\Delta)\frac{1}{B}\frac{dB}{dt} - \nu_{ii}|\Delta|. \tag{7}$$

In this regime, the magnetic field is primarily amplified by the dynamics of the plasma (i.e., the turbulence dynamo in the kinetic phase) rather than by instabilities driven by the pressure anisotropy.

Throughout this paper, we assume the intergalactic plasma to be a highly conductive medium (i.e., the resistivity is much smaller than the viscosity). In such a system, the motion of the fluid elements is coupled to the magnetic field (i.e., magnetic field lines are "frozen" into the fluid). Applying the "frozen-in" condition to the plasma, we describe the magnetic field amplification through turbulent dynamos in the collision-dominated regime [37]:

$$\frac{1}{B}\frac{dB}{dt} \approx \hat{\boldsymbol{b}}\hat{\boldsymbol{b}} : \nabla u - \nabla \cdot \boldsymbol{u} \sim \frac{\delta v_L}{L}{\rm Re}^{1/2} = \left(\frac{\delta v_L}{v_{\rm th}}\right)^{3/2}\left(\frac{\lambda_{\rm mfp}}{L}\right)^{1/2}\nu_{ii}, \tag{8}$$

where $\lambda_{\rm mfp} = v_{\rm th}/\nu_{ii}$ is the mean free path with the thermal velocity $v_{\rm th}$, $\boldsymbol{u}$ is the fluid velocity, $\hat{\boldsymbol{b}} = \boldsymbol{B}/B_0$ is the normalized magnetic field with respect to the seed field intensity B_0, and δv_L is the velocity fluctuations at the outer scale of turbulence L. The term $\hat{\boldsymbol{b}}\hat{\boldsymbol{b}} : \nabla u$ represents the projection of the gradient of the fluid velocity ∇u onto the magnetic field direction, represented by the unit vector $\hat{\boldsymbol{b}}$. The Reynolds number (Re) is defined as ${\rm Re} = \delta v_L L/\nu$, where ν is the fluid viscosity. The Kolmogorov scaling relation is employed to scale the turbulent velocity with respect to the viscous scale $\delta v_{l_\nu} \propto l_\nu^{1/3}$. In the case without the perturbation term, we obtain the value of Δ_0 from Equation (8):

$$|\Delta_0| \sim \left(\frac{v_{\rm th}}{\delta v_L}\right)^{3/2}\left(\frac{L}{\lambda_{\rm mfp}}\right)^{1/2} - 1. \tag{9}$$

Equation (9) indicates that the turbulence during the early evolutionary phase generates pressure anisotropy by itself [35], as the rapid growth of the magnetic field is not counterbalanced by collision-mediated isotropization.

Next, we take into account the perturbation term in Equation (7) to examine the perturbation produced by turbulence during the early evolutionary phase. Assuming that the time variation of the perturbation term can be neglected compared to the damping rate (i.e., we consider the damping-dominated regime, $\Gamma_d \gg \omega_{\rm pert}$), we obtain the following expression from Equation (7):

$$\frac{d\Delta}{dt} \approx -\Gamma_d \delta\Delta \exp(-\Gamma_d t), \tag{10}$$

$$\left|\frac{\delta\Delta}{\Delta_0}\right| \exp(-\Gamma_d t) \approx \left|\frac{\left(1+|\Delta_0|^{-1}\right)\left(\frac{1}{B}\frac{dB}{dt} - \nu_{ii}\right)}{\Gamma_d + \nu_{ii} - \frac{1}{B}\frac{dB}{dt}}\right| = \frac{1}{\frac{\Gamma_d}{\nu_{ii}} + \frac{1}{1+|\Delta_0|^{-1}}}. \tag{11}$$

Equation (11) shows that the time evolution of the perturbation term depends on the ratio between the damping and collision rates, as well as the anisotropy factor around the equilibrium state, Δ_0. For systems with sufficiently small Δ_0 satisfying $\beta \gg |\Delta_0|^{-1} \gg 1$, the perturbation term at the saturation timescale $\tau_c^{(turb)}$, which satisfies $\tau_c^{(turb)} \ll \Gamma_d^{-1}$, is given as follows:

$$\left|\frac{\delta\Delta}{\Delta_0}\right| \sim \left(\frac{\Gamma_d}{\nu_{ii}}\right)^{-1} \sim \left(\frac{\Gamma_d}{\Omega_i}\right)^{-1}. \tag{12}$$

This is because the strength of the saturated magnetic field through turbulent dynamo is roughly $\Omega_i \sim \nu_{ii}$, which is relevant to the condition for a gyrotropic collisionless plasma (i.e., the Larmor radius of the ions ($r_{L,i} \sim v_{\rm th}/\Omega_i$) is smaller than the mean free path ($\lambda_{\rm mfp}$)). Figure 1 shows the calculated perturbation terms induced by the turbulent dynamo. It is evident that Equation (11) for the case with sufficiently small Δ_0 provides an anti-correlation between the perturbation term and Γ_d/ν_{ii}.

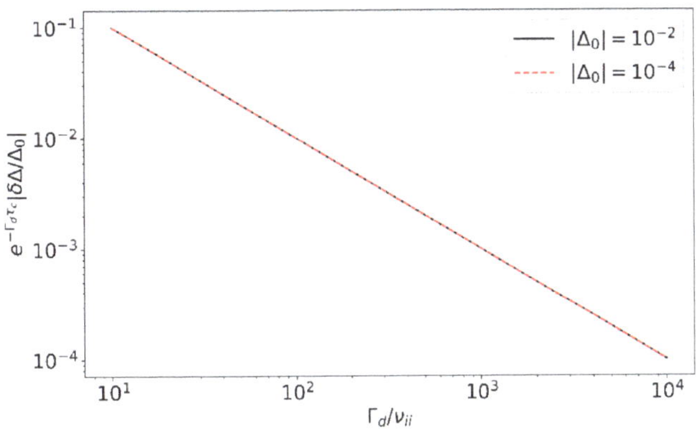

Figure 1. The perturbation term at $t \sim \tau_c^{(turb)}$ as a function of the ratio between the damping rate and collision frequency (Γ_d/ν_{ii}) for different values of $|\Delta_0|$.

3. A Model for Amplifying the Cosmic Magnetic Field Mediated by Plasma Instabilities

In this section, we investigate the impact of perturbations on the evolution of the magnetic field in a regime where isotropization is affected by scattering. Once the magnetic field reaches a sufficient strength for the plasma to become gyrotropic, magnetic field amplification will be primarily driven by instabilities. The condition for a plasma to be gyrotropic is given by $B_{\rm gyro}(\rm G) > \sim 10^{-20} n(\rm cm^{-3}) T(\rm eV)^{-1/2}$. This leads to the conclusion that for magnetic fields stronger than $\sim 10^{-14}$G, the condition $r_{L,i} < \lambda_{\rm mfp}$ holds, which is relevant to the parameters of standard Λ CDM cosmology (with baryon density $n_{z1000} \sim 10^6$ cm^{-3} and temperature $T_{z1000} \sim 1$ eV at $z \sim 1000$). For the gyrotropic collisionless plasma, we adopt a magnetic field amplification model based on plasma instabilities that enable fast interchange of internal, kinetic, and magnetic energies. The characteristics of pressure-anisotropy-driven instabilities are discussed in Section 3.1, while the model for magnetic field amplification is presented in Sections 3.2 and 3.3. In particular, the effects of perturbations in pressure anisotropy on the magnetic field amplification process are examined. A discussion of the nonlinear saturation stage of magnetic field amplification is provided briefly in Section 3.4.

3.1. Instabilities Induced by Pressure Anisotropy

Pressure-anisotropy-driven instabilities have been explored as possible mechanisms for magnetic field amplification [35–38]. In this paper, we repeat the formulation of pressure-driven instabilities introduced in earlier studies, including the linear wave analysis [36], to ensure the paper remains self-contained. Additionally, we investigate the impact of perturbation terms on the growth of these instabilities.

For magnetized ions satisfying $\Omega_i \gg \nu_{ii}$, the following equations for the fluid velocity $\boldsymbol{u}$ and magnetic field $\boldsymbol{B}$ are valid, where $u = |\boldsymbol{u}|$ and $B = |\boldsymbol{B}|$ represent the moduli of the respective vectors:

$$\rho \frac{d\boldsymbol{u}}{dt} = -\nabla\left(p_\perp + \frac{B^2}{8\pi}\right) + \nabla\cdot\left[\hat{\boldsymbol{b}}\hat{\boldsymbol{b}}\left(p_\perp - p_\parallel\right)\right] + \frac{\boldsymbol{B}\cdot\nabla B}{4\pi}\hat{\boldsymbol{b}}, \tag{13}$$

$$\frac{d\boldsymbol{B}}{dt} = \boldsymbol{B}\cdot\nabla u - \boldsymbol{B}\nabla\cdot\boldsymbol{u}. \tag{14}$$

Equation (13) represents the equation of motion for the fluid velocity, where the term $\rho d\boldsymbol{u}/dt$ accounts for the inertial term and the right-hand side includes forces due to pressure gradients, magnetic pressure, and the interaction between the magnetic field and the velocity gradient. Equation (14) governs the evolution of the magnetic field by accounting for the advection of the field by the plasma flow and the stretching of the magnetic field lines due to velocity gradients. Additionally, the kinetic equation for the magnetized ions, following an anisotropic ion distribution function $f\left(v_\parallel, v_\perp\right)$, is written as follows:

$$\left(\frac{df}{dt} + v_\parallel \hat{\boldsymbol{b}}\cdot\nabla f\right) + \frac{1}{2}v_\perp\left(\hat{\boldsymbol{b}}\hat{\boldsymbol{b}} : \nabla u - \nabla\cdot\boldsymbol{u} + v_\parallel\frac{\hat{\boldsymbol{b}}\cdot\nabla B}{B}\right)\frac{\partial f}{\partial v_\perp} \\ - \left[\hat{\boldsymbol{b}}\cdot\left(\frac{d\boldsymbol{u}}{dt} + v_\parallel\hat{\boldsymbol{b}}\cdot\nabla u\right) + \frac{1}{2}v_\perp^2\frac{\hat{\boldsymbol{b}}\cdot\nabla B}{B}\right]\frac{\partial f}{\partial v_\parallel} = C[f_0, f], \tag{15}$$

$$C[f_0, f] = \frac{1}{2}\nu_{ii}\left(\frac{v_{th}}{v}\right)^3\frac{\partial}{\partial\xi}\left(1-\xi^2\right)\frac{\partial f}{\partial\xi}. \tag{16}$$

here, $C[f_0, f]$ represents the collision operator with $\xi = v_\parallel/v$, and $v = \sqrt{v_\parallel^2 + v_\perp^2}$, where f_0 is the zeroth-order Maxwellian distribution function. Equation (15) describes how the distribution function changes due to the interactions between the plasma velocity and magnetic field. The first term accounts for the convective transport of ions along the magnetic field direction and their movement in the velocity space. The second term describes the effect of the plasma velocity and the magnetic field gradient on the distribution function, incorporating the anisotropic velocity and field interactions. Particularly, the term

$\hat{b}\hat{b} : \nabla u$ represents how the velocity of the plasma varies in the direction of the magnetic field. The third term addresses the magnetic field's effect on the ion dynamics, including energy exchange between the field and the plasma.

Under the assumption that the fluid velocity is much smaller than the thermal speed and varies over scales much longer than the mean free path, we can construct a perturbation theory to derive the dispersion relation for pressure-anisotropy-driven instabilities. Assuming Kolmogorov scaling, the velocity at the viscous scale (u) and the normalized wavenumber associated with the viscous scale relative to the mean free path ($\epsilon = k_\nu \lambda_{\rm mfp}$) are given as follows:

$$u \sim \mathrm{Re}^{-1/4} v_{\rm th}, \tag{17}$$

$$\epsilon = k_\nu \lambda_{\rm mfp} \sim \mathrm{Re}^{-1/4}. \tag{18}$$

To perform the ordering of Equation (15) with respect to ϵ, we assume the following relations:

$$k_\parallel \lambda_{\rm mfp} \sim \epsilon, \tag{19}$$

$$\frac{d}{dt} \sim \nabla u \sim \epsilon^2 \frac{v_{\rm th}}{\lambda_{\rm mfp}}. \tag{20}$$

The anisotropic ion distribution function, up to order ϵ^2, is then calculated as follows:

$$f\left(v_\parallel, v_\perp\right) \sim \frac{n_0}{\left(2\pi v_{\rm th}^2\right)^{3/2}} \exp\left(-\frac{v^2}{2v_{\rm th}^2}\right)\left[1 + \frac{\hat{b}\hat{b} : \nabla u}{\nu_{ii}} \frac{v^3\left(v^2 - 3v_\parallel^2\right)}{3v_{\rm th}^5}\right], \tag{21}$$

where n_0 is the plasma number density.

Note that the kinetic model in this study assumes small velocity fluctuations and long spatial scales relative to the mean free path, which are standard approximations in low-collision and weakly magnetized plasmas. However, these assumptions may be sensitive to variations in collisional frequency (ν_{ii}) and magnetic field dynamics ($\hat{b}\hat{b} : \nabla u$, ∇B), especially under highly dynamic conditions such as those in the early universe. A sensitivity analysis reveals that decreasing ν_{ii} enhances the anisotropic ion distribution's deviation from Maxwellian behavior, increasing susceptibility to pressure-anisotropy-driven instabilities. Conversely, increasing ν_{ii} suppresses these deviations, stabilizing the plasma. Furthermore, changes in magnetic field strength and spatial gradients (∇B) significantly modify the anisotropic distribution by amplifying the contribution of terms proportional to $\hat{b}\hat{b} : \nabla u$ in the distribution function. Strong magnetic fields or steep gradients exacerbate anisotropy, potentially triggering instabilities such as mirror or firehose instabilities.

We proceed with the stability analysis of the velocity and magnetic fields that vary on time scales larger than $\sim (\nabla u)^{-1}$ and spatial scales larger than $\sim k_\nu^{-1}$. By considering the linear perturbations $\delta \boldsymbol{u}$ and $\delta \boldsymbol{B}$, with frequencies $\omega \gg \nabla u$ and wavenumbers $k \gg k_\nu$, we linearize Equations (13) and (14), neglecting temporal and spatial derivatives of the unperturbed quantities. The resulting equations are as follows:

$$\begin{aligned} -i\omega\rho\delta\boldsymbol{u} = {} & -i\boldsymbol{k}\delta\left(p_\perp + \frac{B^2}{8\pi}\right) + ik_\parallel \delta\hat{\boldsymbol{b}}\left(p_\perp - p_\parallel + \frac{B^2}{4\pi}\right) \\ & + \hat{\boldsymbol{b}}\delta\left[\hat{\boldsymbol{b}}\cdot\nabla\left(p_\perp - p_\parallel\right) - \left(p_\perp - p_\parallel + \frac{B^2}{4\pi}\right)\frac{\hat{\boldsymbol{b}}\cdot\nabla B}{B}\right], \end{aligned} \tag{22}$$

$$\delta\hat{\boldsymbol{b}} = -\frac{k_\parallel}{\omega}\delta\boldsymbol{u}_\perp, \tag{23}$$

$$\frac{\delta B}{B} = \frac{1}{\omega} \boldsymbol{k}_\perp \cdot \delta \boldsymbol{u}_\perp. \tag{24}$$

The dispersion relation for modes with shear-Alfvén-wave polarization ($\delta \boldsymbol{u} \propto \boldsymbol{k}_\perp \times \hat{\boldsymbol{b}}$) is derived as follows:

$$\frac{\omega^2}{k_\parallel^2 v_{\mathrm{th}}^2} = \Delta + 2\beta^{-1}. \tag{25}$$

In the limit, $\beta \gg |\Delta|^{-1}$, and for $\Delta < 0$ (i.e., $p_\parallel > p_\perp$), an instability appears with the growth rate $\Gamma_g \approx k_\parallel v_{\mathrm{th}} |\Delta|^{1/2}$, which is known as the firehose instability. The instability stabilizes when $|\Delta| \sim 2\beta^{-1}$, with the maximum instability occurring around the ion-gyration scale (i.e., $k_\parallel r_{L,i} \sim 1$) [37,39]:

$$\Gamma_{g,max} \sim \left(|\Delta| - 2\beta^{-1} \right)^{1/2} \Omega_i \tag{26}$$

To analyze the stability of perturbations with other polarizations, it is necessary to compute the pressure perturbations $\delta p_\perp$ and $\delta p_\parallel$. To obtain these, we first linearize the kinetic equation for magnetized ions around the distribution function in Equation (21) and calculate $\delta p_\perp$ and $\delta p_\parallel$ using the perturbed distribution function. The perturbations in the distribution function are assumed to have frequency $\omega \gg \nabla u$ and wavenumber $k \gg \lambda_{\mathrm{mfp}}^{-1}$. We then linearize Equation (13) with the perturbations $\delta \boldsymbol{u}$, $\delta \boldsymbol{B}$, $\delta p_\perp$, and $\delta p_\parallel$, while neglecting the terms of the order $\left(\omega / k_\parallel v_{\mathrm{th}} \right)^3$ and higher, as high-frequency perturbations are subject to strong collisionless damping and cannot become unstable with small anisotropies. Under these assumptions, we obtain the following equation:

$$\begin{aligned} \omega^2 \delta \boldsymbol{u} = {} & k_\parallel^2 v_{\mathrm{th}}^2 (\Delta + 2\beta^{-1}) \delta \boldsymbol{u}_\perp - \hat{\boldsymbol{b}} \frac{\omega^2}{k_\parallel} \boldsymbol{k}_\perp \cdot \delta \boldsymbol{u}_\perp \\ & + \boldsymbol{k}_\perp (\boldsymbol{k}_\perp \cdot \delta \boldsymbol{u}_\perp) \left[2(-\Delta + \beta^{-1}) v_{\mathrm{th}}^2 - i\omega \sqrt{2\pi} \frac{v_{\mathrm{th}}}{|k_\parallel|} + 2 \frac{\omega^2}{k_\parallel^2} \right]. \end{aligned} \tag{27}$$

The first term on the right-hand side represents modes with shear-Alfvén-wave polarization. To analyze the characteristics of $\delta u_\parallel$, we take the dot product using both sides of Equation (27) with $\hat{b}$. This gives the following:

$$\delta u_\parallel \propto -\boldsymbol{k}_\perp \cdot \frac{\delta \boldsymbol{u}_\perp}{k_\parallel}. \tag{28}$$

Perturbations with $\delta u_\parallel \neq 0$ are incompressible and slow-wave-polarized. The dispersion relation for these perturbations, with the condition $\delta \boldsymbol{u}_\perp \propto \boldsymbol{k}_\perp$, is as follows:

$$\left(1 - \frac{2k_\perp^2}{k_\parallel^2} \right) \omega^2 + i\omega \sqrt{2\pi} \frac{k_\perp^2 v_{\mathrm{th}}}{|k_\parallel|} - \left[\left(k_\parallel^2 - 2k_\perp^2 \right) \Delta + 2k^2 \beta^{-1} \right] v_{\mathrm{th}}^2 = 0. \tag{29}$$

For parallel propagating modes where $k_\perp \ll k_\parallel$, the second term is negligible. For $\Delta < 0$, the growth rate is $\Gamma_g \approx k_\parallel v_{\mathrm{th}} |\Delta|^{1/2}$. For modes propagating at an oblique angle to the magnetic field (i.e., $k_\perp$ is not small), the second term in Equation (29) dominates over the first term. In this case, the growth rate is as follows:

$$\frac{\Gamma_g}{k_\parallel v_{\mathrm{th}}} \approx \left(\frac{2}{\pi} \right)^{\frac{1}{2}} \left[\Delta \left(1 - \frac{k_\parallel^2}{2k_\perp^2} \right) - \beta^{-1} \left(1 + \frac{k_\parallel^2}{k_\perp^2} \right) \right]. \tag{30}$$

Slow-wave-polarized firehose instability can be triggered when $k_\parallel > \sqrt{2}k_\perp$ for $\Delta < 0$, while mirror instability can develop if $k_\perp > k_\parallel/\sqrt{2}$ for $\Delta > 0$. In both cases, the growth rate is approximately $\Gamma_g \sim k_\parallel v_{\rm th}|\Delta|$. For the regime where $|\Delta| \ll 1$, the growth of oblique modes occurs much more slowly than the growth of parallel modes or shear-Alfvén-wave polarized modes.

The dispersion relations derived above do not account for second-order perturbation terms, such as $\delta p_\perp \delta \boldsymbol{u}_\perp$, $\delta p_\parallel \delta \boldsymbol{u}_\perp$, $\delta p_\perp \delta \boldsymbol{u}_\parallel$, and $\delta p_\parallel \delta \boldsymbol{u}_\parallel$, which include the effects of perturbations in the pressure anisotropy factor ($\delta\Delta$). These terms account for more complex interactions between the plasma pressure anisotropy and the plasma velocity fields, which allow for a more accurate description of the system's response to pressure anisotropy, especially near the critical stability condition, where plasma instabilities are triggered. In this context, we specifically investigate the impact of pressure anisotropy perturbations on the saturation timescale of magnetic field amplification due to plasma instabilities, denoted by $\tau_c^{(inst)} = 1/\Gamma_c^{(inst)}$. Near the stability condition for the system where $|\Delta| \sim 2\beta^{-1}$, the instability driven by pressure anisotropy perturbations may influence the system. In the case of rapid damping ($\Gamma_d \gg \Gamma_c^{(inst)}$), the effects of pressure anisotropy are minimal, and the system tends toward equilibrium. In contrast, when $\Gamma_d \ll \Gamma_c^{(inst)}$, the perturbation can trigger plasma instabilities in a marginally unstable system, depending on the anisotropy factor $|\delta\Delta|$.

Considering the perturbations $\delta p_\perp \delta \boldsymbol{u}_\perp$ and $\delta p_\parallel \delta \boldsymbol{u}_\perp$, the dispersion relation for the modes with shear-Alfvén-wave polarization is modified to include the pressure anisotropy perturbation factor ($\delta\Delta$) as follows:

$$-i\omega\rho\delta\boldsymbol{u}_\perp = -i\frac{k_\parallel^2}{\omega}\delta\boldsymbol{u}_\perp\left(p_\perp + \delta p_\perp - p_\parallel - \delta p_\parallel + \frac{B^2}{4\pi}\right), \quad (31)$$

$$\frac{\omega^2}{k_\parallel^2 v_{\rm th}^2} = \Delta + \delta\Delta\exp\left(-\frac{\Gamma_d}{\Gamma_c^{(inst)}}\right) + 2\beta^{-1}. \quad (32)$$

In the limit, $|\Delta| \sim 2\beta^{-1}$, and if $\Delta < 0$ (i.e., $p_\parallel > p_\perp$), the growth rate of the instability is given as follows:

$$\frac{\tilde{\Gamma}_{g,c}}{k_\parallel v_{\rm th}} \approx \left|\delta\Delta\exp\left(-\frac{\Gamma_d}{\Gamma_c^{(inst)}}\right)\right|^{\frac{1}{2}}. \quad (33)$$

Figure 2 illustrates the growth rate due to the perturbed anisotropy factor at $t \sim \tau_c^{(inst)}$, with $|\Delta| = 10^{-2}$ and $|\delta\Delta/\Delta| = 10^{-1}$. This demonstrates how the perturbed anisotropy affects the growth of plasma instabilities. In the regime where $\Gamma_d < \Gamma_c^{(inst)}$, the growth rate depends weakly on the value of Γ_d, whereas in the regime $\Gamma_d > \Gamma_c^{(inst)}$, the growth rate rapidly decreases as Γ_d increases. The growth of plasma instabilities and its dependence on the damping of the perturbed anisotropy can also accelerate or decelerate the growth of the magnetic field through plasma instabilities, which is discussed in the following sections.

Considering the perturbations $\delta p_\perp \delta \boldsymbol{u}_\parallel$ and $\delta p_\parallel \delta \boldsymbol{u}_\parallel$, the dispersion relation for the parallel modes with slow-wave-polarization around the saturation timescale ($|\Delta| \sim 2\beta^{-1}$) becomes the following:

$$\left(1 - \frac{2k_\perp^2}{k_\parallel^2}\right)\omega^2 - \left[\left(k_\parallel^2 - 2k_\perp^2\right)\left(\Delta + \delta\Delta\exp\left(-\frac{\Gamma_d}{\Gamma_c^{(inst)}}\right)\right) + 2k^2\beta^{-1}\right]v_{\rm th}^2 \approx 0. \quad (34)$$

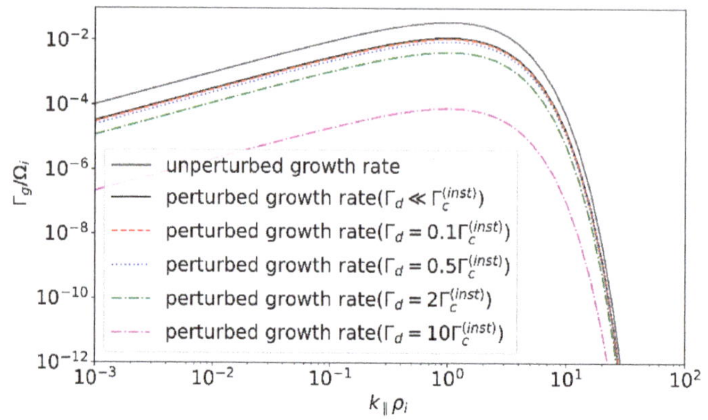

Figure 2. Growth rate of instabilities driven by the perturbed anisotropy factor around $t \sim \tau_c^{(inst)}$ for different values of $\Gamma_d/\Gamma_c^{(inst)}$. For comparison, the growth rate of the instability, without the effect of perturbed anisotropy (as described in Equation (26)), is shown as a gray solid line for $|\Delta| \gg 2\beta^{-1}$.

The growth rate for the instability is as follows:

$$\frac{\tilde{\Gamma}_{g,c}}{k_\parallel v_{\rm th}} \approx \left|\delta\Delta \exp\left(-\frac{\Gamma_d}{\Gamma_c^{(inst)}}\right)\right|^{\frac{1}{2}}. \tag{35}$$

For the oblique modes with slow-wave-polarization around the saturation timescale ($|\Delta| \sim 2\beta^{-1}$), the dispersion relation becomes:

$$i\omega\sqrt{2\pi}\frac{k_\perp^2 v_{\rm th}}{|k_\parallel|} - \left[\left(k_\parallel^2 - 2k_\perp^2\right)\left(\Delta + \delta\Delta \exp\left(-\frac{\Gamma_d}{\Gamma_c^{(inst)}}\right)\right) + 2k^2\beta^{-1}\right]v_{\rm th}^2 \approx 0. \tag{36}$$

The growth rate is then as follows:

$$\frac{\tilde{\Gamma}_{g,c}}{k_\parallel v_{\rm th}} \approx \left(\frac{2}{\pi}\right)^{\frac{1}{2}}\left[\delta\Delta \exp\left(-\frac{\Gamma_d}{\Gamma_c^{(inst)}}\right)\left(1 - \frac{k_\parallel^2}{2k_\perp^2}\right) - 3\beta^{-1}\right]. \tag{37}$$

Assuming $|\delta\Delta| \ll \beta^{-1}$, the mirror modes are stable regardless of the amplitude of $\delta\Delta$. Firehose modes could grow for $\Gamma_d \ll \Gamma_c^{(inst)}$, and the minimum condition for unstable modes is as follows:

$$\frac{k_\parallel}{k_\perp} > \left[2\left(\frac{3\beta^{-1}}{|\delta\Delta|}\exp\left(\frac{\Gamma_d}{\Gamma_c^{(inst)}}\right) - 1\right)\right]^{\frac{1}{2}}. \tag{38}$$

Additionally, the maximum condition for $k_\parallel/k_\perp$ depends on whether the contribution of $k_\perp$ is negligible. This condition implies that the second term in Equation (29) must be larger than the first term:

$$\left(1 - \frac{2k_\perp^2}{k_\parallel^2}\right)\omega^2 \le i\omega\sqrt{2\pi}\frac{k_\perp^2 v_{\rm th}}{|k_\parallel|}. \tag{39}$$

The growth rate is then constrained by the following:

$$\frac{\tilde{\Gamma}_{g,c}}{k_\parallel v_{\rm th}} \le (2\pi)^{\frac{1}{2}}\frac{k_\perp^2}{k_\parallel^2 - 2k_\perp^2}. \tag{40}$$

This inequality provides the upper limit for $k_{\parallel}/k_{\perp}$, which is given by the following:

$$\frac{k_{\parallel}}{k_{\perp}} \leq \left\{1+\frac{3\beta^{-1}}{|\delta\Delta|}+\left[\left(1+\frac{3\beta^{-1}}{|\delta\Delta|}\right)^2-\frac{12\beta^{-1}-2\pi}{|\delta\Delta|}\right]^{\frac{1}{2}}\right\}^{\frac{1}{2}}. \tag{41}$$

Note that the conditions for $k_{\parallel}/k_{\perp}$, described in Equations (38) and (41), represent the obliquity of the oblique mode with respect to the background magnetic field. Figure 3 shows the possible range of $k_{\parallel}/k_{\perp}$ with $|\delta\Delta/\Delta| = 0.1$–$0.3$ as a function of $\Gamma_d/\Gamma_c^{(inst)}$. This analysis demonstrates how the properties of oblique modes depend on Γ_d and $\Gamma_c^{(inst)}$. The minimum condition is saturated below $\Gamma_d/\Gamma_c^{(inst)} \sim 10^{-1}$. The properties of waves can be modified with respect to $|\delta\Delta/\Delta|$, where the parallel mode is favored (i.e., $k_{\parallel}/k_{\perp}$ becomes larger) as $|\delta\Delta/\Delta|$ decreases.

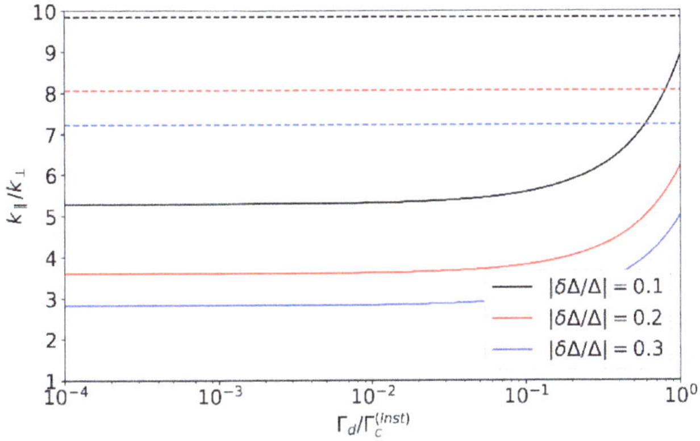

Figure 3. The range of $k_{\parallel}/k_{\perp}$ for the unstable modes with $|\delta\Delta/\Delta| = 0.1 - 0.3$ as a function of $\Gamma_d/\Gamma_c^{(inst)}$. The lower and upper limits are shown as solid and dashed lines, respectively.

3.2. *Amplification of the Cosmic Magnetic Field in the Intermediate Regime*

Based on the maximum growth rate of the shear-Alfvén-wave polarized firehose instability, $\Gamma_{g,max} \sim (|\Delta| - 2\beta^{-1})^{1/2}\Omega_i$, the saturated magnetic fluctuation is estimated as follows:

$$\left.\frac{\delta B^2}{B^2}\right|_{sat} \sim \left(|\Delta| - 2\beta^{-1}\right)^{\frac{1}{2}}. \tag{42}$$

We further assume that the effect of the saturated instabilities is to scatter particles at a rate given as follows:

$$\nu_{scatt} \sim \left.\frac{\delta B^2}{B^2}\right|_{sat} \Gamma_{g,max} \sim \left(|\Delta| - 2\beta^{-1}\right)^{\frac{3}{2}}\Omega_i. \tag{43}$$

In the intermediate regime, the equation for the evolution of Δ is written as follows:

$$\frac{d\Delta}{dt} \approx (1-\Delta)\frac{1}{B}\frac{dB}{dt} - \nu_{ii}|\Delta| + \left(|\Delta| - 2\beta^{-1}\right)^{\frac{3}{2}}\Omega_i. \tag{44}$$

Using the time derivative of Δ described in Equation (10), the time variation of the magnetic field is obtained from Equation (44):

$$\frac{dB}{dt} \approx -\Omega_{i,0}\frac{1}{1+|\Delta|}\frac{\nu_{ii}}{\Omega_i}\frac{B}{B_0} + \Omega_{i,0}\frac{\left(|\Delta|-2\beta^{-1}\right)^{\frac{3}{2}}}{1+|\Delta|}\left(\frac{B}{B_0}\right)^2 + \Omega_{i,0}\frac{2\exp\left(-\Gamma_d t - \frac{1}{2}\frac{\Gamma_d}{\Gamma_c^{(inst)}}\right)|\delta\Delta|^{\frac{3}{2}}}{1+|\Delta|}\left(\frac{B}{B_0}\right)^2. \tag{45}$$

Here, $\Omega_{i,0}$ represents the cyclotron frequency for the seed magnetic field B_0. The first term represents the collisional damping term, while the second and third terms are responsible for the magnetic field growth through pressure anisotropy effects. Particularly, the third term describes the magnetic field amplification due to the anisotropy perturbation alone. Figure 4 shows that the magnetic field growth is suppressed at higher collisional rates.

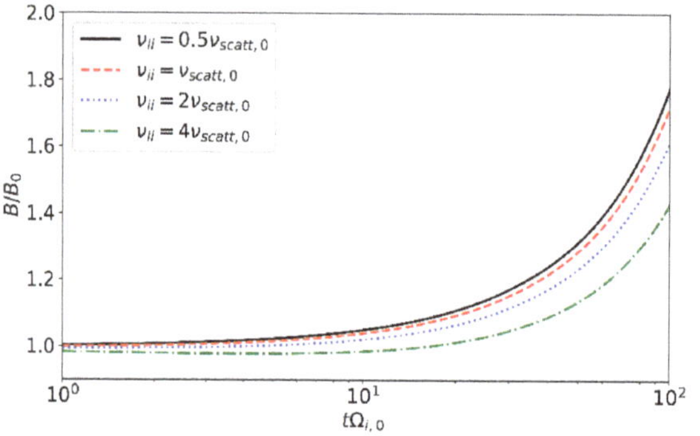

Figure 4. Time evolution of the magnetic field with different collisional frequency (ν_{ii}). For all cases, $|\delta\Delta/\Delta_0| = 10^{-1}$ is used, and $\nu_{scatt,0}$ indicates ν_{scatt} measured at $t\Omega_{i,0} = 1$.

In the very early stages of magnetic field amplification through plasma instabilities, collisional effects are significant. However, However, as the magnetic field strengthens (i.e., Ω_i increases), the collisional effects become negligible (i.e., $\nu_{ii}/\Omega_i \ll 1$). In this context, magnetic field growth through plasma instabilities is primarily driven by the scattering-dominated regime.

3.3. *Amplification of the Cosmic Magnetic Field in the Scattering-Dominated Regime*

In the scattering-dominated regime, the equation for the evolution of Δ is written as follows without the collisional term:

$$\frac{d\Delta}{dt} \approx (1-\Delta)\frac{1}{B}\frac{dB}{dt} + \left(|\Delta| - 2\beta^{-1}\right)^{\frac{3}{2}}\Omega_i. \tag{46}$$

The time variation of the magnetic field is obtained from Equation (46):

$$\frac{dB}{dt} \approx \Omega_{i,0}\frac{\left(|\Delta|-2\beta^{-1}\right)^{\frac{3}{2}}}{1+|\Delta|}\left(\frac{B}{B_0}\right)^2 + \Omega_{i,0}\frac{2\exp\left(-\Gamma_d t - \frac{1}{2}\frac{\Gamma_d}{\Gamma_c^{(inst)}}\right)|\delta\Delta|^{\frac{3}{2}}}{1+|\Delta|}\left(\frac{B}{B_0}\right)^2, \tag{47}$$

In the limit $|\delta\Delta| \to 0$, the result is consistent with that obtained for a constant $\Delta = \Delta_0$ [35].

It is possible to obtain an analytical solution for the exponential growth of the magnetic field before saturation (i.e., $t \ll 1/\Gamma_d \ll 1/\Gamma_c^{(inst)}$ and $|\Delta| \ll \beta^{-1}$). In this limit, the system reduces to a first-order nonlinear ordinary differential equation with an analytical solution:

$$B(t) \sim B_0 \left[1 - \left(\frac{|\Delta|^{\frac{3}{2}} + 2\exp\left(-\frac{1}{2}\frac{\Gamma_d}{\Gamma_c^{(inst)}}\right)|\delta\Delta|^{\frac{3}{2}}}{1+|\Delta|} \right) \Omega_{i,0} t \right]^{-1}. \quad (48)$$

From the quasi-stability condition, given by $\beta \sim 2|\Delta|^{-1}$, the saturation timescale for $|\Delta| \ll 1$ is obtained as follows:

$$\tau_c^{(inst)} \sim \left(|\Delta|^{\frac{3}{2}} + 2\exp\left(-\frac{1}{2}\frac{\Gamma_d}{\Gamma_c^{(inst)}}\right)|\delta\Delta|^{\frac{3}{2}} \right)^{-1} \Omega_{i,0}^{-1}. \quad (49)$$

Equations (48) and (49) indicate that the time evolution of the magnetic field is influenced by the characteristic damping rate and the amplitude of anisotropy perturbations. We examine these effects on magnetic field evolution, as shown in Figure 5. While the contribution of anisotropy perturbations to amplifying the magnetic field strength is negligible, they significantly modify the saturation timescale. When considering a strong damping rate (i.e., $\Gamma_d > \Gamma_c^{(inst)}$) or very weak perturbations (i.e., $|\delta\Delta/\Delta_0| < \sim 10^{-3}$), the solution converges to the result without anisotropy perturbation. Figure 6 illustrates the saturation timescales for varying damping rates, ranging from weak ($\Gamma_d \ll \Gamma_c^{(inst)}$) to strong ($\Gamma_d \gg \Gamma_c^{(inst)}$) damping. For instance, in the weak damping regime, the saturation timescale can vary by up to $\sim 10\%$ when $|\delta\Delta/\Delta_0| \sim 10^{-1}$. This highlights that perturbations driving pressure anisotropy around an equilibrium value can substantially affect the evolution of cosmic magnetic fields, although the field amplitude remains independent of such effects.

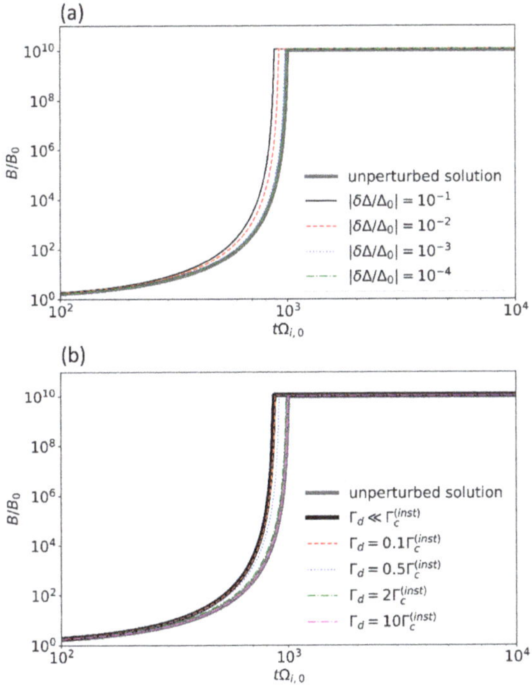

Figure 5. (**a**) Time evolution of the magnetic field with different perturbation levels ($|\delta\Delta/\Delta_0|$) ranging from 10^{-4} to 10^{-1}. (**b**) Time evolution of the magnetic field with different damping rates of anisotropy perturbations for $|\delta\Delta/\Delta_0| = 10^{-1}$. For all cases, $|\Delta_0| = 10^{-2}$ is used. The gray solid lines represent the time evolution of the magnetic field without the anisotropy perturbation $\delta\Delta$.

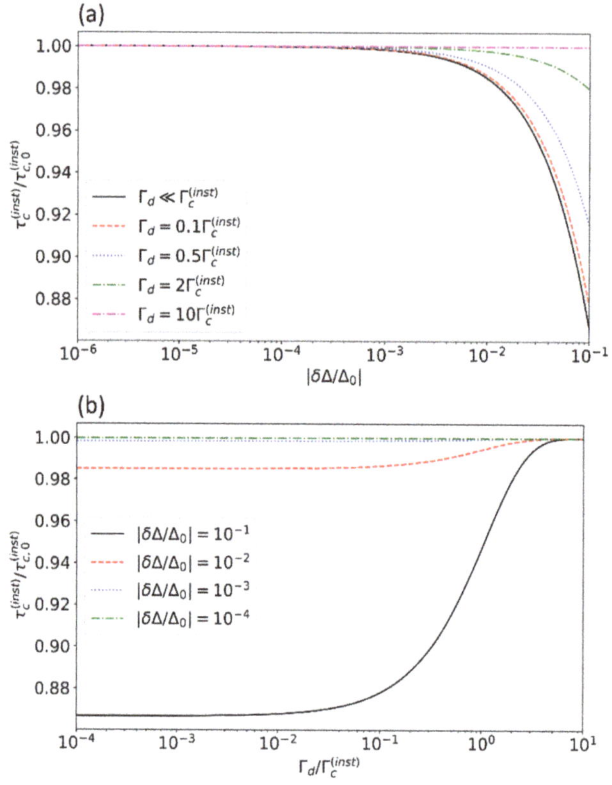

Figure 6. The ratio of the saturation timescale with and without anisotropy perturbation ($\tau_c^{(inst)}$ and $\tau_{c,0}^{(inst)}$, respectively) as a function of $|\delta\Delta/\Delta_0|$ (panel (**a**)) and $\Gamma_d/\Gamma_c^{(inst)}$ (panel (**b**)). The saturated timescales are measured by the point at which the magnetic field strength reaches the quasi-stability condition, $\beta \sim 2|\Delta|^{-1}$.

3.4. Saturation of Plasma Instabilities Driven by Pressure Anisotropy

In this section, we discuss the saturation of plasma instability in turbulent media. Assuming the outer scale of the turbulence, L, and the rms velocity, U, at this scale, the effective Reynolds number ($\mathrm{Re_{eff}}$) is defined as follows:

$$\mathrm{Re_{eff}} \sim \frac{UL}{\mu_{\|,\mathrm{eff}}} \sim \frac{UL}{v_{th}^2}\nu_{\mathrm{eff}}. \tag{50}$$

Here, $\nu_{\mathrm{eff}} \equiv \nu_{ii} + \nu_{scatt}$ is the effective collision rate, and $\mu_{\|,\mathrm{eff}} \sim v_{th}\lambda_{\mathrm{mfp,eff}}$ is the effective viscosity of the turbulent system, with the effective mean free path of the particles $\lambda_{\mathrm{mfp,eff}} \sim v_{th}/\nu_{\mathrm{eff}}$. $\mathrm{Re_{eff}}$ takes into account the combined effects of turbulent motions and collisional damping in the plasma, and it plays a crucial role in determining how efficiently magnetic fields can grow through turbulence and plasma instabilities. In the saturation stage ($d\Delta/dt \sim 0$), Equation (3) is rewritten as follows:

$$\nu_{\mathrm{eff}}\Delta \sim \frac{1}{B}\frac{dB}{dt} \sim \mu_{\|,\mathrm{eff}}\hat{\boldsymbol{b}}\hat{\boldsymbol{b}} : \nabla u. \tag{51}$$

Using $|\nabla u| \sim (U/L)\mathrm{Re}_{\mathrm{eff}}^{1/2}$, we obtain the saturated $|\Delta|$ as follows:

$$|\Delta| \sim \left(\frac{U}{v_{th}}\right)^2 \mathrm{Re}_{\mathrm{eff}}^{-\frac{1}{2}}. \tag{52}$$

Considering that the instability is suppressed when the magnetic field is sufficiently amplified (i.e., $|\Delta| \sim 2\beta^{-1}$), we obtain the critical effective Reynolds number ($\mathrm{Re}_{\mathrm{eff,c}}$) for the magnetic field amplification through plasma instabilities:

$$\mathrm{Re}_{\mathrm{eff,c}} \sim \frac{\beta^2}{4}\left(\frac{U}{v_{th}}\right)^4. \tag{53}$$

This indicates that the magnetic field has grown enough to suppress further instability-driven growth when $\mathrm{Re}_{\mathrm{eff}} < \mathrm{Re}_{\mathrm{eff,c}}$.

While this paper focuses on the exponential growth of the magnetic field before the saturation stage, as the magnetic field grows, nonlinear effects become increasingly important, particularly the saturation of the amplification process. Wagstaff et al. [32], for instance, emphasize the critical role of nonlinear feedback mechanisms that arise as the magnetic field interacts with the plasma. As the magnetic field strength increases, it induces back-reactions on the plasma, altering the dynamics of the turbulence and affecting plasma transport properties, such as viscosity and conductivity. These changes can significantly influence the efficiency of the dynamo process and ultimately lead to the saturation of magnetic field growth. In particular, as the field strength increases, the magnetic back-reaction acts to dampen further growth, leading to a state where the energy injected into the system by turbulence is balanced by the magnetic field energy. This nonlinear feedback loop causes the magnetic field to approach an equilibrium, limiting its growth. The timescale for saturation is dependent on various factors, including the plasma's resistivity, the intensity of turbulence, and the rate of energy transfer between the kinetic and magnetic components of the system. Moreover, the feedback between magnetic field growth and plasma properties can lead to modifications in the plasma's viscosity and conductivity, further influencing the saturation process. As the magnetic field amplifies, it can change the turbulence dynamics, leading to a decrease in the effective viscosity and a potential reduction in the efficiency of magnetic field amplification. These effects are particularly significant in the context of the early universe, where pressure anisotropies and instabilities could accelerate or slow down the saturation process, depending on the specific plasma conditions at the time.

4. Summary and Discussion

In an unstable system driven by pressure anisotropy, the possible scenario for amplifying the cosmic magnetic field is summarized as follows: (1) pressure anisotropies could naturally arise in the intergalactic medium during the early universe, at timescales comparable to the eddy turnover time at viscous dissipation scales (i.e., the post-recombination era); (2) these pressure anisotropies drive instabilities, leading to the exponential amplification of the magnetic field via the isotropization of the pressure anisotropy. In such a system, this study aims to examine the effects of pressure anisotropy perturbations on the evolution of cosmic magnetic fields.

The magnetic field amplification in three different regimes is summarized as follows. In the collision-dominated regime, the amplitude of pressure anisotropy can be calculated based on the evolution of the magnetic field through turbulent dynamo processes. Specifically, the amplitude of anisotropy perturbations around the equilibrium anisotropy value can be determined by the ratio of the perturbation damping rate to the collision frequency. It is possible to expect the presence of pressure anisotropy perturbations near the equilibrium value even after the magnetic field evolution has saturated. In the intermediate region, we examine the influence of collisional effects on magnetic field growth through plasma instabilities during the very early stages of magnetic field amplification. In the scattering-dominated regime, we examine the time evolution of the magnetic field driven by pressure anisotropy instabilities, as well as the effects of pressure anisotropy perturbations on the saturation of the magnetic field evolution. The significance of these effects is measured in terms of the amplitude and damping rate of the anisotropy perturbations. While the strength of the saturated magnetic field is not significantly affected by small anisotropy perturbations near the equilibrium value (denoted as Δ_0), the saturation timescale of magnetic field amplification through instabilities is notably influenced by the pressure anisotropy perturbations. For perturbations with amplitude $|\delta\Delta/\Delta_0| > \sim 10^{-3}$ in the weak damping regime ($\Gamma_d \ll \Gamma_c^{(inst)}$), the system reaches saturation faster compared to a system without anisotropy perturbation. However, the effect of perturbations becomes negligible in the strong damping regime ($\Gamma_d \gg \Gamma_c^{(inst)}$).

Before concluding, we further discuss the following issues: (1) Scaling of magnetic fields in an expanding universe; (2) Observational implications of pressure-anisotropy-driven instabilities.

The scaling relation for magnetic fields in an expanding universe is derived under the assumption of conformal invariance, which assumes that the plasma behaves ideally with no resistivity or anisotropic effects. This scaling is typically valid in the early universe, when ideal magnetohydrodynamics (MHD) conditions hold, and magnetic field strength decreases as the universe expands. However, deviations from this ideal scaling law may occur under non-ideal MHD conditions [34]. Specifically, in the presence of finite resistivity or anisotropy-driven effects such as pressure anisotropy, these ideal scaling relations may no longer apply. Resistive effects introduce dissipation into the magnetic field evolution, leading to a slower decay rate or possible amplification of magnetic fields in certain regions. Similarly, anisotropy-driven instabilities can lead to nonlinear evolution of the magnetic field structure, altering its scaling behavior. Therefore, a more comprehensive treatment of magnetic field evolution should account for these non-ideal effects, especially in later stages of the universe's expansion or in regions where plasma anisotropy is significant.

Magnetogenesis scenarios driven by such instabilities can have significant effects on observational signatures, particularly in the cosmic microwave background (CMB) anisotropies and the intergalactic magnetic field [40]. These effects could manifest as subtle distortions in CMB anisotropies, where the pressure anisotropy in the early universe could affect the polarization patterns and the temperature fluctuations of the CMB. Such distortions could serve as indirect evidence for anisotropic instabilities in the early universe, providing a testable prediction of this model. Additionally, gamma-ray halos produced by high-energy particles interacting with cosmic microwave background photons could be used to probe the nature of pressure-anisotropy instabilities. Specifically, energetic particles accelerated by anisotropic instabilities may generate gamma-ray emission, which could be detected through high-resolution gamma-ray telescopes. Furthermore, the presence of a large-scale intergalactic magnetic field, potentially shaped by pressure-anisotropy-driven processes, could leave its imprint on the polarization and intensity of gamma rays from distant astrophysical sources. Therefore, building the theoretical framework for magnetic field amplification through plasma instabilities would enhance the understanding of these observational signatures.

Funding: This research received no external funding.

Data Availability Statement: No datasets were generated or analyzed during the current study.

Acknowledgments: We thank the referees for their comments and suggestions, which have improved the quality of this manuscript.

Conflicts of Interest: The author declares no conflicts of interest.

References

1. Ryu, D.; Kang, H.; Cho, J.; Das, S. Turbulence and Magnetic Fields in the Large Scale Structure of the Universe. *Science* **2008**, *320*, 909–912. [CrossRef] [PubMed]
2. Kronberg, P.P. Extragalactic magnetic fields. *Rep. Prog. Phys.* **1994**, *57*, 325. [CrossRef]
3. Zweibel, E.G.; Heiles, C. Magnetic fields in galaxies and beyond. *Nature* **1997**, *385*, 131–136. [CrossRef]
4. Widrow, L.M. Origin of Galactic and Extragalactic Magnetic Fields. *Rev. Mod. Phys.* **2002**, *74*, 775–823. [CrossRef]
5. Govoni, F.; Feretti, L. Magnetic Field in Clusters of Galaxies. *Int. J. Mod. Phys. D* **2004**, *D13*, 1549–1594. [CrossRef]
6. Ryu, D.; Schleicher, D.R.G.; Treumann, R.A.; Tsagas, C.G.; Widrow, L.M. Magnetic Fields in Large-Scale Structure of the Universe. *Space Sci. Rev.* **2012**, *166*, 1–35. [CrossRef]
7. Di Gennaro, G.; van Weeren, R.J.; Brunetti, G.; Cassano, R.; Brüggen, M.; Hoeft, M.; Shimwell, T.W.; Röttgering, H.J.A.; Bonafede, A.; Botteon, A.; et al. Fast magnetic field amplification in distant galaxy clusters. *Nat. Astron.* **2021**, *5*, 268–275. [CrossRef]
8. Bagchi, J.; Ensslin, T.A.; Miniati, F.; Stalin, C.S.; Singh, M.; Raychaudhury, S.; Humeshkar, N.B. Evidence for shock acceleration and intergalactic magnetic fields in a large-scale filament of galaxies ZwCl 2341.1+0000. *New Astron.* **2002**, *7*, 249–277. [CrossRef]
9. Strong, A.W.; Moskalenko, I.V.; Ptuskin, V.S. Cosmic-ray propagation and interactions in the Galaxy. *Annu. Rev. Nucl. Part. Sci.* **2007**, *57*, 285–327. [CrossRef]
10. Kim, J.; Ryu, D.; Kang, H.; Kim, S.; Rey, S.-C. Filaments of galaxies as a clue to the origin of ultra-high-energy cosmic rays. *Sci. Adv.* **2019**, *5*, eaau8227. [CrossRef] [PubMed]
11. Wittor, D. Cosmic-Ray Acceleration and Magnetic Fields in Galaxy Clusters and Beyond: Insights from Radio Observations. *Universe* **2023**, *9*, 319. [CrossRef]
12. Ha, J.-H.; Ryu, D.; Kang, H. Cosmic-Ray Acceleration and Nonthermal Radiation at Accretion Shocks in the Outer Regions of Galaxy Clusters. *Astrophys. J.* **2023**, *943*, 119. [CrossRef]
13. Barrow, J.D.; Ferreira, P.G.; Silk, J. Constraints on a Primordial Magnetic Field. *Phys. Rev. Lett.* **1997**, *78*, 3610–3613. [CrossRef]
14. Subramanian, K. Primordial magnetic fields and CMB anisotropies. *Astron. Nachrichten* **2006**, *327*, 403–409. [CrossRef]
15. Minoda, T.; Ichiki, K.; Tashiro, H. Small-scale CMB anisotropies induced by the primordial magnetic fields. *J. Cosmol. Astropart. Phys.* **2021**, *3*, 093. [CrossRef]
16. Grasso, D.; Rubinstein, H.R. Revisiting nucleosynthesis constraints on primordial magnetic fields. *Phys. Lett. B* **1996**, *379*, 73–79. [CrossRef]
17. Biermann, L. Uber den Ursprung der Magnetfelder auf Sternen und im interstellaren Raum (miteinem Anhang von A. *Schluter). Zs. Naturforschung* **1950**, *5*, 65–71.
18. Ryu, D.; Kang, H.; Biermann, P.L. Cosmic magnetic fields in large scale filaments and sheets. *Astron. Astrophys.* **1998**, *335*, 19–25.
19. Naoz, S.; Narayan, R. Generation of Primordial Magnetic Fields on Linear Overdensity Scales. *Phys. Rev. Lett.* **2013**, *111*, 051303. [CrossRef]
20. Langer, M.; Durrive, J.-B. Magnetizing the Cosmic Web during Reionization. *Galaxies* **2018**, *6*, 124. [CrossRef]
21. Langer, M.; Aghanim, N.; Puget, J.-L. Magnetic fields from reionisation. *Astron. Astrophys.* **2005**, *443*, 367–372. [CrossRef]
22. Durrive, J.-B.; Langer, M. Intergalactic magnetogenesis at Cosmic Dawn by photoionization. *Mon. Not. R. Astron. Soc.* **2015**, *453*, 345–356. [CrossRef]
23. Subramanian, K.; Shukurov, A.; Haugen, N.E.L. Evolving turbulence and magnetic fields in galaxy clusters. *Mon. Not. R. Astron. Soc.* **2006**, *366*, 1437–1454. [CrossRef]
24. Schekochihin, A.A.; Iskakov, A.B.; Cowley, S.C.; McWilliams, J.C.; Proctor, M.R.E.; Yousef, T.A. Fluctuation dynamo and turbulent induction at low magnetic Prandtl numbers. *New J. Phys.* **2007**, *9*, 300. [CrossRef]
25. Schlickeiser, R.; Shukla, P.K. Cosmological Magnetic Field Generation by the Weibel Instability. *Astrophys. J. Lett.* **2003**, *599*, L57. [CrossRef]
26. Schlickeiser, R. On the origin of cosmological magnetic fields by plasma instabilities. *Plasma Phys. Control. Fusion* **2005**, *47*, A205. [CrossRef]
27. Jedamzik, K.; Katalinić, V.; Olinto, A.V. Damping of cosmic magnetic fields. *Phys. Rev. D* **1998**, *57*, 3264–3284. [CrossRef]
28. Brandenburg, A.; Sokoloff, D.; Subramanian, K. Current Status of Turbulent Dynamo Theory. *Space Sci. Rev.* **2012**, *169*, 123–157. [CrossRef]
29. Beresnyak, A. Universal Nonlinear Small-Scale Dynamo. *Phys. Rev. Lett.* **2012**, *108*, 035002. [CrossRef] [PubMed]
30. Cho, J. Origin of Magnetic Field in the Intracluster Medium: Primordial or Astrophysical? *Astrophys. J.* **2014**, *797*, 133. [CrossRef]

31. Martin-Alvarez, S.; Devriendt, J.; Slyz, A.; Sijacki, D.; Richardson, M.L.A.; Katz, H. Towards convergence of turbulent dynamo amplification in cosmological simulations of galaxies. *Mon. Not. R. Astron. Soc.* **2022**, *513*, 3326–3344. [CrossRef]
32. Wagstaff, J.M.; Banerjee, R.; Schleicher, D.; Sigl, G. Magnetic field amplification by the small-scale dynamo in the early Universe. *Phys. Rev. D* **2014**, *89*, 103001. [CrossRef]
33. Mogavero, F.; Schekochihin, A.A. Models of magnetic field evolution and effective viscosity in weakly collisional extragalactic plasmas. *Mon. Not. R. Astron. Soc.* **2014**, *440*, 3226–3242. [CrossRef]
34. Grasso, D.; Rubinstein, H.R. Magnetic Fields in the Early Universe. *Phys. Rep.* **2001**, *348*, 163–266. [CrossRef]
35. Falceta-Goncalves, D.; Kowal, G. Fast Magnetic Field Amplification in The Early Universe: Growth of Collisionless Plasma Instabilities in Turbulent Media. *Astrophys. J.* **2015**, *808*, 65. [CrossRef]
36. Schekochihin, A.A.; Cowley, S.C.; Kulsrud, R.M.; Hammett, G.W.; Sharma, P. Plasma Instabilities and Magnetic Field Growth in Clusters of Galaxies. *Astrophys. J.* **2005**, *629*, 139. [CrossRef]
37. Schekochihin, A.A.; Cowley, S.C. Fast growth of magnetic fields in galaxy clusters: A self-accelerating dynamo. *Astron. Nachrichten* **2006**, *327*, 599–604. [CrossRef]
38. Schekochihin, A.A.; Cowley, S.C. Turbulence, magnetic fields and plasma physics in clusters of galaxies. *Phys. Plasmas* **2006**, *13*, 056501. [CrossRef]
39. Gary, S.P.; Li, H.; O'Rourke, S.; Winske, D. Proton resonant firehose instability: Temperature anisotropy and fluctuating field constraints. *J. Geophys. Res. Space Phys.* **1998**, *103*, 14567–14574. [CrossRef]
40. Durrer, R.; Neronov, A. Cosmololgical Magnetic Fields: Their Generation, Evolution, and Observation. *Astron. Astrophys. Rev.* **2013**, *21*, 62. [CrossRef]

universe

MDPI

Article

Universes Emerging from Nothing and Disappearing into Nothing as an Endless Cosmological Process

Leonid Marochnik

Department of Physics, University of Maryland, College Park, MD 20742, USA; lmarochnik@gmail.com

Abstract: The equation of state of quantum fluctuations of the gravitational field of the universe depends on H^4, where H is the Hubble constant. This means that it is invariant with respect to the Wick rotation, i.e., the transition from Lorentzian space-time to Euclidean space-time and vice versa. It is shown that the quantum birth of universes from Euclidean space-time, i.e., from nothing, and their quantum disappearance to nothing (return to Euclidean space-time) by the time the density of the matter filling the universe becomes negligible could be a likely cosmological scenario. On an infinite time axis, this is an endless process of birth and death of universes appearing and disappearing and replacing each other. Within this scenario, our current universe is going to disappear into nothing at $z \leq -0.68$, i.e., after 18.37 billion years, and the lifetime of our universe and similar universes is about 32 billion years.

Keywords: universes; lifetime; birth; death; endless cosmological process

1. Introduction

A cosmological scenario in which the universe appeared from nothing and finally disappeared into nothing was proposed in [1]. In this paper, we summarize all the known arguments in favor of this scenario, which are based on quantum fluctuations of the gravitational field of the universe (virtual gravitons) [2,3]. In Section 2, which can be considered a preamble to the main body of this paper, we show that, within the framework of general relativity (GR), classical stochastic gravitational waves (linear as well as nonlinear ones) are unable to produce an acceleration of the expansion of the universe if antigravity is not included in the scenario (although there are other points of view; see Section 2).

The main body of this paper (Sections 3–8) is dedicated to the quantum nature of cosmological acceleration and the cosmological scenario that follows. In Section 3, we show that, due to the expansion of the universe, the balance between the creation and annihilation of virtual gravitons is disrupted, in turn leading to the macroscopic quantum effect of cosmological acceleration with an equation of state, which is invariant with respect to the transition from Lorentzian to Euclidean space-time and vice versa. In Section 4, the same equation of state is obtained independently, directly from equations of one-loop quantum gravity. In Section 5, gravitational instantons are used to calculate the probabilities of transition from Lorentzian to Euclidean space-time and vice versa. In Sections 6 and 7, the recent appearance of dark energy and "Hubble tension" is briefly discussed as a consequence of the macroscopic quantum effect of virtual gravitons. Section 8 outlines the cosmological scenario.

2. Classical Gravitational Waves

As is known, without going beyond Einstein's theory, it is impossible to obtain an acceleration of the expansion of the universe. This is the reason why the dark energy (DE) effect (the contemporary exponential acceleration in the expansion of the universe detected observationally [4,5]) and the hypothetical exponential acceleration of the universe at the start of its evolution (inflation) [6] remain unsolved problems of cosmology. In [7], we used Vereshkov's equations for classical stochastic nonlinear gravitational waves (SGW) to try to produce the acceleration of the expansion of the universe. The idea was that Vereshkov's theory (VT), as a bimetric theory (see [8–10]), could go beyond pure Einsteinian theory, with the hope being to derive acceleration from nonlinear gravitational waves. VT uses two metric tensors: standard g_k^i and stochastic $\widetilde{g_k^i}$ (ψ_i^k in the notation of [7]), where the average value of $\widetilde{g_k^i}$ across the statistical ensemble of SGWs is zero. It turns out that the de Sitter accelerating expansion is a solution to the VT equations, which are expressed as Equations (20)–(21) in [7]:

$$a(t) = exp(Ht) = exp(H_E\tau); \kappa\rho_{GW} = 3H^2; \kappa p_{GW} = -3H^2$$

where ρ_{GW}, p_{GW} are the energy density and the pressure of classical gravitational waves, $\kappa = 8\pi G$, and the speed of light, $c = 1, G$, is the gravitational constant. The scale factor is invariant with respect to the Wick rotation, where H and t are the Hubble constant and time in the Lorentzian space-time (LST), respectively, and $H_E = iH$ and $\tau = -it$ are the Hubble constant and time in Euclidean space-time (EST), respectively. According to Equation (8) in [7], this solution of the VT equations leads to $H^2 < 0$, which leads in turn to $\rho_{GW} < 0$. This means that this solution corresponds to an antigravity effect, which seems to have no physical meaning in the framework of classical general relativity.

The same finding applies to classical gravitational waves of small amplitude (linear waves) [11] because they are just a particular case of nonlinear waves. Here, there also exists a de Sitter expansion which is invariant with respect to the Wick rotation. And again, its energy density is negative. So, classical gravitational waves of small and/or finite amplitude cannot produce the acceleration of expansion if antigravity does not exist, although there are other points of view[1] (see [12–14] and the references therein). It can be said another way: classical stochastic gravitational waves are capable of providing the de Sitter accelerated expansion of the universe if they are of antigravity nature.

Of course, one can achieve an acceleration of expansion by making some arbitrary hypotheses, for example, assuming that the universe is filled with a hypothetical scalar field (see the quintessence [15–18]) or assuming that Einstein's equations are incomplete and require generalization (see [19]). In this work, we believe that Einstein's equations are valid. However, GR is a classical theory. If cosmological acceleration is a quantum effect, then GR is unable to describe it.

In the following two sections, using two independent approaches, we show that the equation of the state of virtual gravitons in the universe depends on H^4 so that it is invariant with respect to the Wick rotation $t = i\tau$.

3. Equation of State of Virtual Gravitons: Uncertainty Relation[2]

The creation and annihilation of virtual gravitons in the graviton vacuum is in a balance in standard conditions (as many as created as many are annihilated). However, the universe is rapidly expanding. During the lifetime of virtual particles, the universe manages to expand so quickly that virtual gravitons of the horizon wavelengths do not have time to annihilate back into the vacuum. This leads to a non-zero energy balance, which is spent on the acceleration of the expansion of the universe. That is, some of the gravitons that did not have time to annihilate are "stuck," and their energy is spent on accelerating the expansion of the universe. For the first time, this process was considered in [2], where it was shown that the probability of annihilation and creation processes are

$$w_{cr} = \frac{\alpha}{3\pi^2} H^3 (\overline{N}_{\mathbf{k}} + 1)(\overline{N}_{-\mathbf{k}} + 1), \; w_{ann} = \frac{\alpha}{3\pi^2} H^3 \overline{N}_{\mathbf{k}} \overline{N}_{-\mathbf{k}} \tag{1}$$

where $\alpha = O(1)$, $N_{\pm\mathbf{k}} \sim N_g/2$, and N_g is the average number of gravitons with wavelengths of the order of horizon H^{-1}. Thus, the balance equation reads [2]

$$\rho_g = \hbar\omega(w_{cr} - w_{ann}) = \frac{\alpha}{3\pi^2} \hbar H^4 (\overline{N}_{\mathbf{k}} + \overline{N}_{-\mathbf{k}} + 1) \simeq \frac{\alpha}{3\pi^2} \hbar N_g H^4 \tag{2}$$

where ρ_g is the energy density of gravitons and $\hbar\omega \sim \hbar H$. According to (2), the quantum spontaneous process of graviton creation leads to the non-zero effect. With an accuracy of the numerical factor α of the order of unity, this estimate coincides with the exact calculation of the energy density of virtual gravitons in the framework of one-loop quantum gravity (Section 4). The important fact is that the energy density of quantum gravitational waves (gravitons) depends on H^4: this is of crucial importance for our scenario (Section 8). The energy–momentum conservation law reads

$$\dot{\rho}_g + 3H(\rho_g + p_g) = 0 \tag{3}$$

where the superscript dot means the time derivative, and p_g is the pressure of the gravitons. According to (2), $\varrho_g = const$, so we obtain from (2) and (3)

$$\rho_g = -p_g = \frac{\alpha}{3\pi^2} \hbar N_g H^4 \tag{4}$$

From the Friedmann equation, we obtain

$$3H^2 = 8\pi G\rho \tag{5}$$

Combining (2) and (5), we obtain

$$H^2 = \frac{\pi}{G\hbar N_g} \tag{6}$$

(As we show in Section 4, in the framework of one-loop quantum gravity, $\alpha = \frac{9}{8}$). According to (6), there is no passage to the classical limit $\hbar \to 0$. This means that we meet a situation typical for macroscopic quantum effects such as superfluidity and superconductivity. This conclusion was confirmed by exact calculations [3] in the framework of one-loop quantum gravity, which is finite in the absence of matter and the presence of a ghost sector [20]. In the second loop, it is non-renormalizable [21]. There is a huge number of measurements of the Hubble constant obtained by different methods (see, e.g., Wikipedia's paper, "Hubble Law") that range from $H_{0,z\sim1100}$ = 67.4 ± 0.5 km s^{-1} Mpc^{-1} [22] to H_0 = 73.30 ± 1.04 km s^{-1} Mpc^{-1} [23]. For this range of Hubble constants, (6) demonstrates that the number of virtual gravitons under the horizon is $N_g \sim 10^{122}$. This is a familiar number, of course. The main self-contradiction of the ΛCDM model is the unexplainable ratio of a vacuum energy Λ to its observational value, which is of the order of 10^{122}. In quantum theory, this is the number of virtual gravitons under the universe's horizon. Note that the latest (at the present time) observational data show that the cosmological constant is not perhaps a constant at all[3] [24]. This is not yet a proven fact (the accuracy is only 2 sigma). However, the other contradictions already noted above (an order of 122 of magnitude contradiction between the observational and theoretical value of the cosmological constant, plus the inability to answer the long-standing question "Why Now?"[4]) testify not in favor of an ΛCDM model where Lambda = const by definition. Despite the contradictions noted above, a seemingly strong argument in favor of the Lambda CDM model is its seeming consistency with the available observational data. In reality, however, the GCDM model (where G stands for gravitons) is consistent with the SNe Ia observational data and has the same accuracy as the ΛCDM model [25]. Note also that all observational data (with no exceptions) are based on measurements of the equation of state parameter $w = p/\rho$, where p and ρ are the pressure and energy density of matter filling the universe. For a cosmological constant $w_\Lambda = -1$, all observational data show that w = -1. This fact is widely considered as a confirmation of the validity of the ΛCDM model; it is necessary, however, to note that virtual gravitons also produce a de Sitter expansion with $w_G = -1$. It is possible that there are also other models with w = -1. So, observational data tell us only that w = -1, and nothing more. Thus, the difference between the models can be found only by measuring w(z).

4. Equation of State of Virtual Gravitons from One-Loop Quantum Gravity

The equation of state of virtual gravitons in the framework of one-loop quantum gravity was considered in [2,3,26]. Taking this opportunity in the present paper to correct an inaccuracy in [1][5], we will present in this section a calculation of the equation of state of virtual gravitons from one-loop quantum gravity. For the convenience of the reader, we present here the general scheme of reasoning set out in detail in [3,26]. One-loop quantum gravity cosmology can be represented as a theory of gravitons in a macroscopic space-time with self-consistent geometry if one is dealing with a curvature that is much less than the Planck curvature. The interaction of gravitons with the macroscopic field determines the quantum state of gravitons, and the macroscopic geometry depends on the graviton state. The graviton operator and the background metric are taken from the unified gravitational field, which is described by exact equations of quantum gravity. The classical component of the unified field is a function of coordinates and time, but the quantum component of the same unified field is a tensor operator function, which also depends on coordinates and time. Thus, the original exact equations should be the operator equations of the quantum theory of gravity in the Heisenberg representation. In [3,26], the rigorous mathematical derivation of these equations can be found. It is important to note the unavoidable appearance of the auxiliary ghost fields introduced by Feynman [27] and known as Faddeev–Popov ghosts [28][6]. Namely, ghosts ensure the one-loop finiteness of quantum gravity, making the theory mathematically consistent [2,3]. In the self-consistent theory of gravitons, the background metric is described by regular vacuum Einstein equations [2]

$$R_i^k - \frac{1}{2}\delta_i^k R = \kappa\Big(\Big\langle \Psi_g \Big| \hat{T}^k_{i(grav)} \Big| \Psi_g \Big\rangle + \Big\langle \Psi_{gh} \Big| \hat{T}^k_{i(ghost)} \Big| \Psi_{gh} \Big\rangle\Big) \tag{7}$$

Here, Ψ_g and Ψ_{gh} are the quantum state vectors of gravitons and ghosts, respectively. The explicit form of the energy–momentum tensors (EMT) of gravitons $\hat{T}^k_{i(grav)}$ and ghosts $\hat{T}^k_{i(ghost)}$ is presented in [3].

From [3], we obtain the following system equations for the background (where we use the conformal time $\eta = \int dt/a$):

$$3\frac{a'^2}{a^4} = \kappa\rho_g = \frac{1}{16\pi^2}\int_0^\infty \frac{k^2}{a^2} dk \Big(\sum_\sigma \langle \Psi_g | \hat{\psi}'^{+}_{\mathbf{k}\sigma} \hat{\psi}'_{\mathbf{k}\sigma} + k^2 \hat{\psi}^{+}_{\mathbf{k}\sigma} \hat{\psi}_{\mathbf{k}\sigma} | \Psi_g \rangle \\ -2\Big\langle \Psi_{gh} \Big| \bar{\theta}'_{\mathbf{k}} \theta'_{\mathbf{k}} + k^2 \bar{\theta}_{\mathbf{k}} \theta_{\mathbf{k}} \Big| \Psi_{gh} \Big\rangle \Big) \tag{8}$$

$$2\frac{a''}{a^3}-\frac{a'^2}{a^4}=-\kappa p_g=-\frac{1}{16\pi^2}\int_0^\infty\frac{k^2}{a^2}dk\Big(\sum_\sigma\Big\langle\Psi_g\Big|\hat{\psi}'^{+}_{\mathbf{k}\sigma}\hat{\psi}'_{\mathbf{k}\sigma}-\frac{k^2}{3}\hat{\psi}^{+}_{\mathbf{k}\sigma}\hat{\psi}_{\mathbf{k}\sigma}\Big|\Psi_g\Big\rangle-2\Big\langle\Psi_{gh}\Big|\bar{\theta}'_{\mathbf{k}}\theta'_{\mathbf{k}}-\frac{k^2}{3}\bar{\theta}_{\mathbf{k}}\theta_{\mathbf{k}}\Big|\Psi_{gh}\Big\rangle\Big) \tag{9}$$

And for fluctuations, we obtain

$$\hat{\phi}''_{\mathrm{k},\sigma}+(k^2-\frac{a''}{a})\hat{\phi}_{\mathrm{k},\sigma}=0 \tag{10}$$

$$\hat{\vartheta}''_{\mathrm{k}}+(k^2-\frac{a''}{a})\hat{\vartheta}_{\mathrm{k}}=0 \tag{11}$$

where $\hat{\psi}_{\mathrm{k}\sigma}=\frac{1}{a}\hat{\phi}_{\mathrm{k}\sigma}$ and $\hat{\theta}_{\mathrm{k}}=\frac{1}{a}\hat{\vartheta}_{\mathrm{k}}$; σ is the polarization index; $a(\eta)$ is the FLRW scale factor; and superscript "+" denotes a complex conjugation. Primes denote derivatives over conformal time η. The de Sitter expansion in terms of conformal time is

$$a_S=-(H\eta)^{-1},-\infty<\eta\leq 0 \tag{12}$$

The exact solutions to (10) and (11) over the de Sitter background (12) read [2]

$$\begin{aligned}\hat{\psi}_{\mathrm{k}\sigma}&=\frac{1}{a_S}\sqrt{\frac{2\kappa\hbar}{k}}[\hat{c}_{\mathrm{k}\sigma}f(x)+\hat{c}^{+}_{-\mathrm{k}-\sigma}f^{+}(x)],\\ \hat{\theta}_{\mathrm{k}}&=\frac{1}{a_s}\sqrt{\frac{2\kappa\hbar}{k}}[\hat{\alpha}_{\mathrm{k}}f(x)+\hat{\bar{\beta}}_{-\mathrm{k}}f^{+}(x)],\end{aligned} \tag{13}$$

$$f(x)=(1-\frac{i}{x})e^{-ix} \tag{14}$$

where $x=k\eta$.

The quantum metric fluctuations (13) are unable to form a self-consistent de Sitter solution in real time because of the non-computability of the integrals

$$\int_0^\infty x^n e^{\pm 2ix}dx$$

arising in the right-hand side of Equations (8) and (9). However, they can do this if these incomputable integrals are re-defined in the following way [2]:

$$\int_0^\infty x^n e^{\pm 2ix}dx=\lim_{\delta\to 0}\int_0^\infty x^n e^{\pm 2ix-\delta x}dx$$

In accordance with [2], we finally obtain the following energy density of virtual gravitons:

$$\rho_g=\frac{3\hbar N}{8\pi^2}H^4,\ p_g=-\rho_g \tag{15}$$

where N is a generalized parameter of the graviton vacuum state. By order of magnitude, it is N_g. Note that the energy density of the virtual gravitons depends again on H^4. Another way to calculate the incomputable integrals mentioned above is a passage to EST (Euclidean space-time), where divergences can be removed by a choice of integration constants [1] and then reverting to LST. If one repeats the above calculations in imaginary time $\tau=-it$, i.e., in EST, we obtain [1]

$$\rho_g=\frac{\hbar H_E^4}{2\pi^2}\int_{-\infty}^{0}N_k[\zeta^2-(1-\zeta)^2]exp(2\zeta)\zeta d\zeta \tag{16}$$

$$N_k = \sum_{\sigma}\langle \Psi_g | \hat{Q}^{+}_{\mathbf{k}\sigma}\hat{Q}_{\mathbf{k}\sigma}|\Psi_g\rangle - 2\left\langle \Psi_{gh}\middle|\hat{q}^{+}_{\mathbf{k}}\hat{q}_{\mathbf{k}}\middle|\Psi_{gh}\right\rangle \tag{17}$$

where $\hat{Q}_{\mathbf{k}\sigma}$ and $\hat{q}_k$ are the operator constants of the integration of Equations (10) and (11) for gravitons and ghosts, respectively, in imaginary time. Note that we have to have a flat spectrum $N_k = const$ to obtain a constant in the LHS of (16). In accordance with [1], under reasonable assumptions about the occupation numbers of gravitons and ghosts, we obtain

$$N_k = \left(\langle n_g\rangle - \left\langle n_{gh}\right\rangle\right) \tag{18}$$

In the RHS of (18), we have a difference between the average numbers of gravitons and ghosts under the horizon of the universe. So, from Equations (3) and (16)–(18), we obtain the following for EST:

$$\rho_g = \frac{3\hbar H_E^4}{2\pi^2}\left(\left\langle n_{gh}\right\rangle - \langle n_g\rangle\right),\ p_g = -\rho_g \tag{19}$$

In accordance with (19), for LST, we obtain the same form with the only change of $H_E^4 = H^4$.

$$\rho_g = \frac{3\hbar H^4}{2\pi^2}\left(\left\langle n_{gh}\right\rangle - \langle n_g\rangle\right),\ p_g = -\rho_g \tag{20}$$

Here,

$$\left|\langle n_g\rangle - \left\langle n_{gh}\right\rangle\right| \approx N_g \tag{21}$$

Considering (20) and (21), we see that the α parameter in (4) is in fact

$$\alpha = 9/8 \tag{22}$$

One can see that ghosts must prevail over gravitons to make the energy density positive. Figuratively speaking, ghosts must be "materialized" to produce acceleration.

This is an expected result because the ghost system is an antigravitating one[7]. As was shown in [3], the physical nature of this solution is a macroscopic quantum condensate. All the equations of state listed above in Equations (4), (15), (19), and (20) have a common feature: the energy density depends on H^4, which means that it is invariant with respect to the transfer from Lorentzian to Euclidean space-times and vice versa.

These equations of state are similar to what comes from conformal anomalies, as was shown by Starobinsky [29] and Zeldovich [30].

It is worth noting that all known attempts to consider quantum corrections to Einstein's theory have led to a dependence of energy density on H^4, i.e., to its invariance with respect to a transition from LST to EST and vice versa. It is this fact that underlies the idea of the cosmological scenario proposed in [1,7] and here.

5. Transition from Euclidean Space-Time to Lorentzian Space-Time and Vice Versa: Gravitational Instantons

As a matter of fact, the idea of the appearance of the universe from EST, i.e., from "nothing", has a long history (see, e.g., [31–39] and the references therein). To calculate the probability of a transition from EST to LST, one needs to consider a gravitational instanton which (if it exists) can provide the tunneling of a de Sitter expansion regime from EST to LST. The detailed consideration of this problem is given in [3], Section 7.3. In general, instantons are intermediaries between two vacua. In this particular case, it is about a transition from EST to LST. The probability of such a transition is $\exp(-S)$, where S represents action in EST, i.e., in imaginary time. A full theory of gravitational instantons applicable to the graviton-ghost system was presented in [3] (Section 7). As was shown in [3], in the framework of one-loop quantum gravity, S = 0 in imaginary time; thus, the equation of the state and energy density of gravitons (19) obtained in EST can be tunneled into LST with the probability being equal to 1, which is almost obvious because of the invariance of (19, 20) with respect to Wick's rotation. This type of tunneling could explain the origin of inflation, and this was a reason why an avalanche of publications on gravitational instantons appeared in the 80s (see, e.g., [39] and the references therein).

As was mentioned in [7], the existing inflation theories applicable to the existing universe usually start with approximately $10^8 t_p$ and last about $(10^{11} - 10^{12})t_p$, where the Planck time is $t_p \approx 5.4 \cdot 10^{-44}$ sec. Therefore, a one-loop approximation of quantum gravity is probably applicable to such a case. Recall that the one-loop approximation of quantum gravity is applicable only to empty space (with no matter in it) and the presence of a ghost sector [20]. Obviously, there was no matter at $10^8 t_p$.

To calculate the tunneling probability from LST to EST, i.e., a possible end of our universe, we have to calculate the probability of tunneling from our Lorentzian space-time to Euclidean space-time, i.e., to nothing. In distinction on the previous case, we have to use action in LST, i.e., in real time. This was obtained by [3] (Section 3), and it reads

$$\begin{gathered} S = \int dt\left(-\frac{3\dot{a}^2 a}{N} + L_{\text{grav}} + L_{\text{ghost}}\right), \\ L_{\text{grav}} + L_{\text{ghost}} = \frac{1}{8}\sum_{\mathbf{k}\sigma}\left(\frac{a^3}{N}\dot{\hat{\psi}}^+_{\mathbf{k}\sigma}\dot{\hat{\psi}}_{\mathbf{k}\sigma} - Nak^2\hat{\psi}^+_{\mathbf{k}\sigma}\hat{\psi}_{\mathbf{k}\sigma}\right) - \frac{1}{4}\sum_{\mathbf{k}}\left(\frac{a^3}{N}\dot{\bar{\theta}}_{\mathbf{k}}\dot{\theta}_{\mathbf{k}} - Nak^2\bar{\theta}_{\mathbf{k}}\theta_{\mathbf{k}}\right). \end{gathered} \tag{23}$$

where L_{grav}, L_{ghost} are graviton and ghost Lagrangians, respectively. N is the Lagrange multiplier, which is a variation variable (the choice of N can be made after a variation of action, e.g., $N = 1$).

In (23), the background metric is taken to be in the form

$$ds^2 = dt^2 - a\left(t\right)^2\gamma_{\alpha\beta}dx^\alpha dx^\beta = dt^2 - a\left(t\right)^2\left(dx^2 + dy^2 + dz^2\right)$$

A variation of action (23) over N leads to (24)

$$\delta S = \delta N\int a^3 dt\left\{\left(\frac{3\dot{a}^2}{N^2a^2}\right) + \frac{1}{8}\sum_{k,\sigma}\left(-\frac{1}{N^2}\dot{\hat{\psi}}_{k\sigma}\widehat{\psi_{k\sigma}}^+_{k\sigma} - \frac{k^2}{a^2}\hat{\psi}_{k\sigma}\widehat{\psi_{k\sigma}}^+\right) - \frac{1}{4}\sum_{k}\left(-\frac{1}{N^2}\dot{\theta}^+_k\dot{\theta}_k - \frac{k^2}{a^2}\theta_k\theta^+_k\right)\right\} \tag{24}$$

If we choose $N = 1$ in (24), average over 3-volume and come back from conformal time η to real time t in (8), we can see that the integrand in curly brackets is the left-hand side minus the right-hand side of (8), i.e., $\delta S = 0$. In accordance with the Principle of Least Action $S = const = 0$, exp(−S) = 1. So, the probability of a transition back to EST is again unity, which is also almost obvious for the same reason, namely due to the invariance of (19, 20) with respect to the transfer from LST to EST and vice versa.

6. Why Now?

One-loop quantum gravity also provides an answer to the question: "Why Now?", i.e., why did cosmological acceleration come into play now, i.e., very recently? The reason is a conformal non-invariance of the gravitational field, which manifests itself as the third term in Equations (9) and (10). Note that over the de Sitter background (12), Equation (10) reads

$$\hat{\phi}_{\dot{\mathbf{k}},\sigma} + (k^2 - \frac{2}{\eta^2})\hat{\phi}_{\mathbf{k},\sigma} = 0 \tag{25}$$

Note also that (25) does not change its form when gravitons are propagating inside matter with the equation of state $p = 0$ [25][8]. Thus, when the equation of state of the matter filling the universe changes from $p = \rho/3$ to $\rho = 0$, i.e., around z = 1100–1200 [25,40], gravitons feel as though they are in a vacuum and can produce the acceleration considered above, i.e., start the dark energy effect any time after the matter changes its equation of state (for more details see [25]).

7. "Hubble Tension" and Early Dark Energy

Obviously, the initiation of DE after a change in the equation of state of the universe matter from $p = \rho/3$ to $\rho = 0$ would resemble the initial DE (so-called early dark energy—EDE), which may be needed, possibly, for the explanation of the origin of "Hubble tension" [41]. In the frame of the GCDM model, EDE is simply a choice of initial conditions [25]. However, this topic is beyond the scope of this paper.

8. Conclusion: The Cosmological Scenario

In conclusion, we have arrived at the following cosmological scenario ([1,7] and the present paper). A flat inflationary Universe could have been formed by tunneling from EST to LST, i.e., from "nothing". After that it should evolve in accordance with inflation scenarios that are beyond the scope of this paper (see, e.g., [6,42–44]). Then the standard Big Bang cosmology starts and lasts until the matter of the universe changes its equation of state from $\rho = \rho/3$ to $\rho = 0$. After that quanta of the gravitational field (virtual gravitons) start to form the DE effect. After that, to the extent that space continues to empty as $\rho_M \sim a^{-3}$, the universe gets the opportunity again to form a gravitational instanton (as in the beginning) but to tunnel back from LST to EST, i.e., disappear to nothing. After that, the entire scenario can be repeated indefinitely[9]. In accordance with the "Planck Collaboration" [45], the energy density of ordinary and dark matter today is about $\Omega_m(0) = 0.315 \pm 0.007$. After the Universe expands, e.g., 3.16 times, i.e.,

$$a_{end}/a_{today} = 3.16 = 1/1 + z_{end}$$

the matter content will be close to 1%, i.e., negligible. So, in this scenario, our current universe can disappear into nothing at $z_{end} \leq -0.68$. To recalculate this result in terms of time, one can use the Friedmann equation, which leads to the following:

$$\int_{t_{today}}^{t_{end}} dt = -\frac{1}{H_0} \int_0^{-0.68} \frac{dz}{(1+z)((\Omega_m(0) \cdot (1+z)^3) + (1-\Omega_m(0)))^{\frac{1}{2}}} \quad (26)$$

Given $H_0 = 67.4 \frac{km}{s \cdot Mpc}$ [45], (26) indicates that $t_{end} - t_{today} = 18.37$ billion years. Considering today's age of the universe, $t_{today} = 13.78$ billion years [45], one can say that the lifetime of a universe similar to ours is about $T_{lifetime} = 32.15$ billion years.

Funding: This research received no external funding.

Data Availability Statement: All data listed in this paper is noted in the References section and are available to everyone.

Acknowledgments: I am grateful to John Payne for his invaluable help and to Liliya Payne for providing moral support. I am grateful to Daniel Usikov for his important comments and technical assistance with the formatting of mathematical expressions.

Conflicts of Interest: The author declares no conflicts of interest.

Notes

1 Although negative energy density of gravitational waves can arise in quantum theories (see, e.g., [46] or antigravitation subsystem of Faddeev-Popov ghosts in Section 4), for classical gravitational waves such a solution, apparently, makes no physical sense,

despite the fact that static gravitational energy is negative. We did not mention the antigravity nature of these solutions in our works [7,11] and we are doing it now.

2 An anonymous referee raised the question of how virtual gravitons might be detectable. This question is beyond the scope of this paper, but readers might consult Dyson Freeman (8 October 2013). "Is a Graviton Detectable?". International Journal of Modern Physics A. 28 (25): 1330041–1." See also Jiehui Liang, Ziyu Liu, Zihao Yang et al., 2024, Nature, 628, 78–83.

3 Let me remind you of the well-known fact that George Gamow remarked upon in his book [47]: "Much later, when I was discussing cosmological problems with Einstein, he remarked that the introduction of the cosmological term was the biggest blunder of his life".

4 Why does dark energy appear exactly in our epoch of cosmic evolution, i.e., why now?

5 In [1], instead of (14), we used a complex conjugated solution with $[0, \infty]$ integration limits in (16). It does not change the final result (19, 20), but it was decided to fix this error here. The correct argumentation follows in the main text.

6 For completeness, note the following quotation from Cécile DeWitt-Morette about Bryce De Witt on "Faddeev-Popov ghosts". "And so it happened that Bryce was one of the first few mavericks ready to tackle the problem. There were just a few others, such as Richard Feynman and Ludwig Faddeev. Was his genius undervalued? Bryce once complained about the name "Faddeev Popov ghosts" for the fictitious particles that appear in the Feynman rules for gauge- and gravity theories. His paper contained the same expressions and had been published earlier. But the difference was only two weeks, and, characteristically, Bryce had buried his result in three extremely lengthy and technical papers, where Faddeev and Popov only needed two pages, which was all that was really needed. If you want completeness, Feynman's name should also have been added; he was the first to notice these ghosts, although he could only handle the one-loop case and had not done it quite correctly" [48]. Thus, the correct name of these ghosts must be "Feynman-de Witt-Faddeev-Popov ghosts".

7 At first glance, "ghost materialization" seems unusual because ghosts are not physical particles. A detailed discussion of this fact can be found in footnotes #1 and #2 in [3]. In short, the universe as a whole is a region of interaction for gravitons, so all the gravitons are virtual in the universe. There are no asymptotic states in the universe where ghosts must disappear as happens in the canonical S-matrix theory, which is inapplicable to the universe as whole. So, the question of the fate of ghosts in the universe will be open until a full (not one-loop) quantum gravity theory appears.

8 For matter with the equation of state $p = 0$, the expansion low is $a(t) = const \cdot t^{\frac{2}{3}}$. In conformal time, it is $a(\eta) = const \cdot \eta^2$, so that $\frac{a''}{a} = \frac{2}{\eta^2}$. There are two only regimes in which conformal non-invariance of the gravitational field produces $\frac{a''}{a} = \frac{2}{\eta^2}$. They are de Sitter expansions (12) and matter with the equation of state $p = 0$.

9 At first sight, this scenario looks similar to the "Cycles of Time" scenario of R. Penrose [49]. The main difference (leaving out the details) is that the lifetime of universes is finite in our scenario, which is distinct from [45], where it is infinite, which leads to the conception of cycles of time.

References

1. Marochnik, L. Dark Energy from Instantons. *Grav. Cosmol.* **2013**, *19*, 178–187. [CrossRef]
2. Marochnik, L.; Usikov, D.; Vereshkov, G. Cosmological Acceleration from Virtual Gravitons. *Found. Phys.* **2008**, *38*, 546–555. [CrossRef]
3. Marochnik, L.; Usikov, D.; Vereshkov, G. Macroscopic Effect of Quantum Gravity: Graviton, Ghosts and Instanton Condensation on Horizon Scale of the Universe. *J. Mod. Phys.* **2013**, *4*, 285. arXiv:1306.6172v1. [CrossRef]
4. Riess, A.G.; Filippenko, A.V.; Challis, P.; Clocchiatti, A.; Diercks, A.; Garnavich, P.M.; Gilliland, R.L.; Hogan, C.J.; Jha, S.; Kirshner, R.P. Observational evidence from supernovae for an accelerating universe and a cosmological constant. *Astron. J.* **1998**, *116*, 1009–1038. [CrossRef]
5. Perlmutter, S.; Aldering, G.; Goldhaber, G.; Knop, R.A.; Nugent, P.; Castro, P.G.; Deustua, S.; Fabbro, S.; Goobar, A.; Groom, D.E.; et al. Measurements of Ω and Λ from 42 high-redshift supernovae. *Astrophys. J.* **1999**, *517*, 565–586. [CrossRef]
6. Guth, A. Inflationary universe: A possible solution to the horizon and flatness problems. *Phys. Rev. D* **1981**, *23*, 347. [CrossRef]
7. Marochnik, L. Nothing to Something and Vice Versa: A Cosmological Scenario. *Universe* **2023**, *9*, 445. [CrossRef]
8. Rosen, N. General Relativity and Flat Space. I. *Phys. Rev.* **1940**, *57*, 147–150. [CrossRef]
9. Rosen, N. A bi-metric Theory of Gravitation. *Gen. Rel. Grav.* **1973**, *4*, 435–447. [CrossRef]
10. Milgrom, M. Bimetric MOND gravity. *Phys. Rev. D.* **2009**, *80*, 123536. arXiv:0912.0790. [CrossRef]
11. Marochnik, L. Dark Energy and Inflation from Gravitational Waves. *Universe* **2017**, *3*, 72. [CrossRef]
12. Bondi, H. Negative mass in general relativity. *Rev. Mod. Phys.* **1957**, *29*, 423–428. [CrossRef]
13. Farnes, J. A unifying theory of dark energy and dark matter: Negative masses and matter creation within a modified ΛCDM framework. *Astron. Astrophys.* **2018**, *620*, A92. [CrossRef]
14. Socas-Navarro, H. Can a negative-mass cosmology explain dark matter and dark energy? *Astron. Astrophys.* **2019**, *626*, A5. [CrossRef]
15. Wetterich, C. Cosmology and the fate of dilatation symmetry. *Nucl. Phys. B* **1988**, *302*, 668. [CrossRef]

16. Caldwell, R.; Dave, R.; Steinhardt, P.J. Cosmological imprint of an energy component with general equation of state. *Phys. Rev. Lett.* **1998**, *80*, 1582–1585. [CrossRef]
17. Caldwell, R.; Linder, E.V. Limits of quintessence. *Phys. Rev. Lett.* **2005**, *95*, 1401. [CrossRef]
18. Weinberg, S. *Cosmology*; Oxford University Press: Oxford, UK, 2008.
19. Heisenberg, L. A systematic approach to generalisations of General Relativity and their cosmological implications. *arXiv* **2018**, arXiv:1807.01725v1. [CrossRef]
20. t'Hooft, G.; Veltman, M. One-Loop Diveregencies in the Theory of Gravitation. *Ann. Inst. Henri Poincare* **1974**, *20*, 69–94.
21. Goroff, M.; Sagnotti, A. Quantum Gravity at Two Loops. *Phys. Lett. B* **1985**, *160*, 81–85. [CrossRef]
22. Jia, X.D.; Hu, J.P.; Wang, F.Y. Evidence of a decreasing trend for the Hubble constant. *Astron. Astrophys.* **2023**, *674*, A45. [CrossRef]
23. Riess, A.G.; Yuan, W.; Macri, L.M.; Scolnic, D.; Brout, D.; Casertano, S.; Jones, D.O.; Murakami, Y.; Anand, G.S.; Breuval, L.; et al. A Comprehensive Measurement of the Local Value of the Hubble Constant with 1 km s^{-1} Mpc^{-1} Uncertainty from the Hubble Space Telescope and the SH0ES Team. *Astrophys. J. Lett.* **2022**, *934*, L7. [CrossRef]
24. Adame, A.G.; Aguilar, J.; Ahlen, S.; Alam, S.; Alexander, D.M.; Alvarez, M.; Alves, O.; Anand, A.; Andrade, U.; Armengaud, E.; et al. DESI 2024 VI: Cosmological Constraints from the Measurements of Baryon Acoustic Oscillations. *arXiv* **2024**, arXiv:2404.03002v2.
25. Marochnik, L.; Usikov, D. Dark Energy from Virtual Gravitons (GCDM Model vs. LambdaCDM Model). *Universe* **2022**, *8*, 464. [CrossRef]
26. Vereshkov, G.; Marochnik, L. Quantum Gravity in Heisenberg Representation and Self-Consistent Theory of Gravitons in Macroscopic Spacetime. *J. Mod. Phys.* **2013**, *4*, 285. arXiv: 1108.4256v2. [CrossRef]
27. Feynman, R. Quantum Theory of Gravitation. *Acta Phys. Pol.* **1963**, *24*, 697–722.
28. Faddeev, L.; Popov, V. Feynman diagrams for the Yang-Mills field. *Phys. Lett. B* **1967**, *25*, 30. [CrossRef]
29. Starobinsky, A. A new type of isotropic cosmological models without singularity. *Phys. Lett. B.* **1980**, *91*, 99–102. [CrossRef]
30. Zeldovich, Y. Vacuum theory: A possible solution to the singularity problem of cosmology. *Soviet Phys. Uspekhi* **1981**, *24*, 216. [CrossRef]
31. Tryon, E. Is the universe a vacuum fluctuation? *Nature* **1973**, *246*, 396. [CrossRef]
32. Zeldovich, Y.B. The birth of a closed universe, and the anthropogenic principle. *Soviet. Astron. Lett.* **1981**, *7*, 322.
33. Atkatz, D.; Pagels, H. Origin of the universe as a quantum tunneling event. *Phys. Rev.* **1982**, *25*, 2065.
34. Hartle, B.J.; Hawking, S.W. Wave function of the universe. *Phys. Rev.* **1983**, *8*, 2960. [CrossRef]
35. Grishchuk, L.P.; Zeldovich, Y.B. *Quantum Structure of Space and Time*; Duff, M., Isham, C., Eds.; Cambridge University Press: Cambridge, UK, 1982; p. 409, see also Preprint IKI (Space Research Institute, USSR Academy of Sci.) #726, 1982.
36. Vilenkin, A. Creation of universes from nothing. *Phys. Lett.* **1982**, *117B*, 25. [CrossRef]
37. Gibbons, G.W.; Hartle, B.J. Real tunneling geometries and the large-scale topology of the universe. *Phys. Rev. D* **1990**, *42*, 2458. [CrossRef]
38. Hawking, S.W.; Moss, I.G. Supercooled phase transitions in the very early universe. *Phys. Lett.* **1982**, *110B*, 35. [CrossRef]
39. Gibbons, G.; Hawking, S. *Euclidean Quantum Gravity*; World Scientific: Singapore; London, UK; Hong Kong, China, 1993; pp. 1–585.
40. Hadzhiyska, B.; Spergel, D. Measuring the duration of last scattering. *Phys. Rev. D* **2019**, *99*, 043537. [CrossRef]
41. Kamionkowski, M.; Riess, A. The Hubble Tension and Early Dark Energy. *arXiv* **1922**, arXiv:2211.04492. [CrossRef]
42. Linde, A. A new inflationary universe scenario: A possible solution of the horizon, flatness, homogeneity, isotropy and primordial monopole problems. *Phys. Lett. B* **1982**, *108*, 389. [CrossRef]
43. Mukhanov, V. & Chibisov, G. Vacuum energy and large-scale structure of the universe. *Sov. Phys. JETP* **1982**, *56*, 258.
44. Albrecht, A.; Steinhardt, P. Cosmology for grand unified theories with radiatively induced symmetry breaking. *Phys. Rev. Lett.* **1982**, *48*, 1220. [CrossRef]
45. Aghanim, N.; Akrami, Y.; Ashdown, M.; Aumont, J.; Baccigalupi, C.; Ballardini, M.; Banday, A.J.; Barreiro, R.B.; Bartolo, N.; Basak, S.; et al. Planck 2018 results. VI. Cosmological parameters. *Astron. Astrophys.* **2020**, *641*, A6.
46. Grumiller, D.; Parekh, P.; Riegler, M. Local Quantum Energy Conditions in Non-Lorentz-Invariant Quantum Field Theories. *Phys. Rev. Lett.* **2019**, *123*, 121602. [CrossRef]
47. Gamow, G. *My World Line*; Viking Press: New York, NY, USA, 1970.
48. Hooft, G. DeWitt-Morette, Cécile, 2012, Memoirs of Bryce DeWitt from 1946 to 2004. *Found. Phys.* **2012**, *42*, 685–687.
49. Penrose, R. *Cycles of Time*; Vintage Books: New York, NY, USA, 2012.

universe

MDPI

Article

Ellipsoidal Universe and Cosmic Shear

Luigi Tedesco [1,2]

[1] Dipartimento di Fisica, Università di Bari, Via G. Amendola 173, 70126 Bari, Italy; luigi.tedesco@ba.infn.it
[2] Istituto Nazionale di Fisica Nucleare, Sezione di Bari, 70126 Bari, Italy

Abstract: We consider a Bianchi I geometry of the universe. We obtain a cosmic shear expression related to the eccentricity of the universe. In particular, we study the connections among cosmic shear, eccentricity, and CMB. The equations are self-contained, with only two parameters.

Keywords: anisotropy; Bianchi I; cosmic shear

1. Introduction

The standard model of the universe shows great compatibility with the observational data. The homogeneity and isotropy of the universe seem to be two acquired aspects in the so-called standard model of the universe. Clearly, much depends on the scales we are examining. When we consider local scales, the universe is neither homogeneous nor isotropic, and, generally, galaxies, galaxy clusters, and large-scale structures such as cosmic filaments and voids create anisotropies. On cosmological scales, these anisotropies must be averaged, and the universe becomes homogeneous and isotropic over about 100 megaparsecs. But, despite all the successes, this model shows several dark spots, the nature of dark matter and a dark energy, the Hubble tension, the coincidence and fine-tuning problem, and so on.

There is often a great deal of confusion about this in the sense that one thing is the anisotropy of the universe at the "local" level and another thing is the anisotropy of the universe in the early universe. We discuss here the second anisotropy.

In this paper, we consider a Bianchi I model of the universe that describes a homogenous and anisotropic universe. There is much evidence regarding the anisotropy of the universe, and the possibility of the anisotropy of our universe is not an academic discussion. The origins of the study in this sector are in 1933 and 1968 [1,2], and the works are very interesting and pioneering. To be fair, it was Einstein who directed Lemaitre to study these anisotropic universes. The first classification is due to Luigi Bianchi in his famous classification (Lie algebra is the first classification) [3].

Anisotropy is an important aspect in relation to the evolution of the universe; for example, the anomalies in the CMB indicate that we have a plane-mirroring symmetry in large scales [4–8], or the large-angle anomalies in the CMB scale [9], or the low-quadrupole moment in the CMB [10–15] in which it is possible to indicate an ellipsoidal expansion of the universe because the low-quadrupole moment of the CMB is suppressed at large scales. In fact, the CMB data put in evidence a suppression of the power when we consider large angular scales, the so-called quadrupole problem. Another aspect of this model is the large-scale asymmetry [16,17] or a very strange cold spot [18] or the alignment of quadrupole and octupole moments [19–22]. A very interesting recent work regards the study of more precise observational bounds on global anisotropy [23]. Among other things, this work studies how the difference between the different expansion rates disappears as the universe evolves, eventually showing a universe in which the differences in anisotropy disappear, demonstrating an isotropic universe in evolution. In any case, many interesting works have been conducted regarding the anisotropic universe, for example, [24–34], and, conversely, in this context regarding an ellipsoidal universe. Indeed, several aspects have been studied, including the parallax effect [35]. The importance of an ellipsoidal universe consists of considering this model as a valid alternative to the FLRW models if we mean small deviations from the perfect isotropy.

A very interesting study consists of connecting the anisotropy in the context of the Finslerian structure of space–time because, when we consider these spaces, we are able to investigate the phenomena of the local anisotropy connected with a gravitational field and the cosmology [36–43].

The mechanisms that enable the universe to become anisotropic are different, such as a uniform cosmological magnetic field [12,13] or topological defects (such as cosmic strings or domain walls) [11]. On the other hand, it is not possible to neglect motions whose origin is due to inhomogeneity and local anisotropy [44]. A very interesting aspect regards the new data of the Pantheon supernova from the late universe; these data show new hints of anisotropy [45–47].

In this paper, we analyze cosmic shear, a powerful concept to explore the distribution of matter in the universe, the understanding of the cosmology, and the evolution of the large-scale structures. We will connect the shear with the eccentricity of the ellipsoidal universe. In this work, we also analyze the deep connections between cosmic shear, eccentricity, and CMB, as evident in the last section.

The importance of considering the CMB is linked to a double factor; the first is the fact that the temperature of the cosmic microwave background does not provide a total isotropy, but, with the precision of the order of 10^{-5}, the second factor is the problem of large-angle anomalies and the quadrupole problem, i.e., the discrepancy between the theoretical prediction for l = 2 and the experimental one, which are in deep disagreement. Therefore, understanding the connection between the anisotropy of the universe and the CMB is not a purely academic exercise but can be a good starting point for understanding and decoding those mechanisms that are still unknown and that hide profound connections.

The paper is organized as follows. In Section 2, we discuss the fundamental properties of an ellipsoidal universe. In Section 3, we analyze the evolution equation for anisotropic energy density. In Section 4, we study the cosmic shear with an ellipsoidal universe. The quadrupole of the CMB, shear, and eccentricity are investigated in Section 5. We summarize and discuss our results in the final section.

2. Model of the Ellipsoidal Universe

The line element, spatially homogeneous and anisotropic Bianchi type-I, is described by the most general plane-symmetric line element, the so-called Taub line element [48]

$$ds^2 = dt^2 - a^2(t)(dx^2 + dy^2) - b^2(t)dz^2 \tag{1}$$

where a and b are two time-dependent different scale factors along two different axes of space, $x\,y$ and z, respectively. Equation (1) describes a space in which the expansion is ellipsoidal.

Bianchi I model may be considered an intriguing alternative to FLRW model of the universe if we consider small deviations from the isotropy.

A very interesting relation between these anisotropic models and quadrupole, octupole, and limit on the shear, vorticity, and Weyl tensor has been considered in [49].

For completeness, we write the non-vanishing Christoffel symbols

$$\Gamma^0_{11} = \Gamma^0_{22} = a\,\dot{a}, \qquad \Gamma^0_{33} = b\,\dot{b}, \qquad \Gamma^1_{01} = \Gamma^2_{02} = \dot{a}/a, \qquad \Gamma^3_{03} = \dot{b}/b \tag{2}$$

where $\dot{} \equiv d/dt$ is the derivative with respect to the cosmic time. The non-vanishing Ricci tensor components are

$$R^0_0 = -2\left(\frac{\ddot{a}}{a} + \frac{\ddot{b}}{b}\right), \qquad R^1_1 = R^2_2 = -\left[\frac{\ddot{a}}{a} + \left(\frac{\dot{a}}{a}\right)^2 + \frac{\dot{a}\dot{b}}{ab}\right], \qquad R^3_3 = -\left(\frac{\ddot{b}}{b} + 2\frac{\dot{a}\,\dot{b}}{a\,b}\right). \tag{3}$$

The energy–momentum tensor, considering planar symmetry, is

$$T^{\mu}_{\ \nu} = \text{diag}\,(\rho, -p_{\parallel}, -p_{\parallel} - p_{\perp}) \tag{4}$$

with ρ energy density, $p_{\parallel}$ longitudinal pressure, and $p_{\perp}$ transversal pressure. This tensor, in an anisotropic universe, is provided by the contributions of the sum of two components, one isotropic and one anisotropic. The anisotropic component is

$$(T_A)^{\mu}_{\ \nu} = \text{diag}\,(\rho^A, -p^A_{\parallel}, -p^A_{\parallel}, -p^A_{\perp}) \tag{5}$$

and an isotropic contribution is provided by

$$(T_I)^{\mu}_{\ \nu} = \text{diag}(\rho^I, -p^I, -p^I, -p^I) \tag{6}$$

where it can be obtained by means of the three components: radiation (r), matter (m), and cosmological constant (Λ):

$$\rho^I = \rho_r + \rho_m + \rho_\Lambda \tag{7}$$

$$p^I = p_r + p_m + p_\Lambda \tag{8}$$

where $p_r = \rho_r/3,\ \ p_m = 0,\ \ p_\Lambda = \rho_\Lambda$. Different authors studied the exact solutions of Einstein's equations in relation to different plane-symmetric and isotropic components [11,50,51].

Let A be the mean expansion parameter

$$A \equiv (a^2 b)^{1/3}, \qquad H_a = \frac{\dot{a}}{a}, \qquad H_b = \frac{\dot{b}}{b} \tag{9}$$

with H_a and H_b the directional Hubble parameters along the directions (x,y) and z, respectively. The average expansion rate or "mean Hubble parameter" is provided by

$$H = \frac{\dot{A}}{A} = \frac{2H_a + H_b}{3} \tag{10}$$

and the spatial volume V is

$$V = \sqrt{-g} = a^2 b\,. \tag{11}$$

Another important element is the so-called "scalar expansion" or "expansion rate", indicated by θ and defined as

$$\theta = 3H = 3\frac{\dot{A}}{A} \tag{12}$$

while the scalar shear is [32,34,52]

$$\sigma^2 = \frac{1}{2}\sigma_{\alpha\beta}\sigma^{\alpha\beta} = \frac{1}{2}\left(2H_a^2 + H_b - \frac{\theta^2}{3}\right) \tag{13}$$

where $\sigma_{\alpha\beta} = \frac{1}{2}(u_{\mu;\nu} + u_{\nu;\mu})h^{\mu}{}_{\alpha}h^{\nu}{}_{\beta} - \frac{1}{3}u^{\mu}_{;\mu}h_{\alpha\beta}$ is called shear tensor and $h_{\mu\nu} = g_{\mu\nu} + u_\mu u_\nu$ is the projection tensor, while u_μ is the four-velocity in the comoving coordinates. The components of the cosmic shear vector $\vec{\Sigma}$ are

$$\Sigma_{x,y,z} \equiv \frac{H_{x,y,z} - H}{H}\,. \tag{14}$$

Therefore,

$$\Sigma_a = \frac{H_a - H}{H}, \qquad \Sigma_b = \frac{H_b - H}{H}, \qquad \text{and} \quad \Sigma_a = \frac{H_a - H_b}{2H_a + H_b}. \tag{15}$$

From a physical point of view, the shear or the scalar expansion indicates how the anisotropic universe is.

Now, we want to connect the eccentricity, 'e', of the universe with Σ. Let us first define the eccentricity as

$$e=\sqrt{1-\frac{b^2}{a^2}} \quad \text{if} \quad a>b \qquad or \qquad e=\sqrt{1-\frac{a^2}{b^2}} \quad \text{if} \quad b>a. \tag{16}$$

We normalize the eccentricity in order to obtain $a(t_0)=b(t_0)=1$.

Let us consider the first case, $(a>b)$; we have

$$b^2=a^2(1-e^2). \tag{17}$$

If we derive both sides by dt and divide for b^2, we obtain

$$2\frac{\dot{b}}{b}=2\,a\,\dot{a}\frac{(1-e^2)}{b^2}+\frac{a^2}{b^2}(-2\,e\,\dot{e}) \tag{18}$$

from which we have the important connection between H and e

$$H_b=H_a-\frac{e\,\dot{e}}{1-e^2} \tag{19}$$

(for $e=\sqrt{1-\frac{b^2}{a^2}}$). Of course, the other case, $e=\sqrt{1-\frac{a^2}{b^2}}$, provides $H_a=H_b-\frac{e\,\dot{e}}{1-e^2}$. For completeness, we write important relations

$$H_a=H\pm\frac{e\,\dot{e}}{3(1-e^2)} \tag{20}$$

and

$$H_b=H\mp\frac{2}{3}\frac{e\,\dot{e}}{(1-e^2)} \tag{21}$$

where the signs "+" and "−" in Equation (20) refer to $e=\sqrt{1-b^2/a^2}$ and $e=\sqrt{1-a^2/b^2}$, respectively.

Combining Equations (15) and (19), we can derive an interesting relation linking Σ and e:

$$\Sigma_a=\frac{e\,\dot{e}}{3H(1-e^2)} \qquad \text{for} \qquad e=\sqrt{1-\frac{b^2}{a^2}} \tag{22}$$

and of course $\Sigma_b=-\frac{e\,\dot{e}}{3H(1-e^2)}$ for $e=\sqrt{1-\frac{a^2}{b^2}}$. Now, we derive the evolution equation for the eccentricity e. Let us start with Einstein's equations, in a Bianchi I universe

$$\begin{cases} \left(\frac{\dot{a}}{a}\right)^2+2\frac{\dot{a}\,\dot{b}}{a\,b}=8\pi G\left(\rho^I+\rho^A\right) \\ \frac{\ddot{a}}{a}+\frac{\ddot{b}}{b}+\frac{\dot{a}\,\dot{b}}{a\,b}=-8\pi G\left(p^I+p^A_{\parallel}\right) \\ 2\frac{\ddot{a}}{a}+\left(\frac{\dot{a}}{a}\right)^2=-8\pi G\left(p^I+p^A_{\perp}\right) \end{cases} \tag{23}$$

Taking into account that $\frac{\ddot{a}}{a}=\dot{H}_a+{H_a}^2$, it is possible to write these equations in terms of $H_{a,b}$:

$$\begin{cases} H_a^2+2H_aH_b=8\pi G\rho \\ \dot{H}_a+H_a^2+\dot{H}_b+H_b^2+H_aH_b=-8\,\pi\,G\,p_{\parallel} \\ 2\dot{H}_a+3H_a=-8\,\pi\,G\,p_{\perp}, \end{cases} \tag{24}$$

with $\rho = \rho^I + \rho^A$, $p_\parallel = p^I + p_\parallel^A$ and $p_\perp = p^I + p_\perp^A$. Subtracting the second from the third equation in Equation (24), we obtain

$$\dot{H}_a + 2H_a^2 - \dot{H}_b - H_b^2 - H_a H_b = 8\pi G(p_\parallel - p_\perp). \tag{25}$$

If we set $E = \frac{e\dot{e}}{3(1-e^2)}$, we have $H_a = H + E$ and $H_b = H - 2E$, which resolve Equation (25) as

$$3\dot{E} + 9HE = 8\pi G(p_\parallel - p_\perp), \tag{26}$$

from which we finally have the most general temporal evolution for the eccentricity:

$$\frac{d}{dt}\left(\frac{e\,\dot{e}}{1-e^2}\right) + 3H\left(\frac{e\,\dot{e}}{1-e^2}\right) = \pm 8\pi G\,(p_\parallel - p_\perp) \tag{27}$$

where "+" refers to the eccentricity $e = \sqrt{1 - b^2/a^2}$ and the sign "−" refers to $e = \sqrt{1 - a^2/b^2}$. Equation (27) is the exact equation, which can be conveniently simplified by neglecting second-order terms. It is important to underline that it is the most general expression of the evolution of the eccentricity over time. Clearly, the resolution of the differential equation must take into account the boundary condition that $e(t_0) = 0$; that is to say, the universe today is isotropic.

3. Evolution Equation for Anisotropic Energy Density

We use the conservation of the anisotropic component of the energy momentum tensor provided by Equation (4)

$$(T^A)^\mu{}_{\nu;\mu} = \frac{\partial T_\nu^\mu}{\partial x^\mu} + \Gamma^\mu_{\mu\lambda} T^\lambda_\nu - \Gamma^\lambda_{\mu\nu} T^\mu_\lambda = 0. \tag{28}$$

Let us consider the only component $\nu = 0$ of the conservation of the anisotropic energy momentum tensor, $(T^A)^\mu{}_{0;\mu}$, and the non-vanishing terms are

$$\mu = 0 \quad \rightarrow \quad \frac{\partial T_0{}^0}{\partial x^0} + \Gamma^0_{0\lambda} T^\lambda_0 - \Gamma^\lambda_{00} T^0_\lambda = \dot{\rho}^A \tag{29}$$

$$\mu = 1 \quad \rightarrow \quad \frac{\partial T_0{}^1}{\partial x^1} + \Gamma^1_{1\lambda} T^\lambda_0 - \Gamma^\lambda_{10} T^1_\lambda \quad \rightarrow \quad \begin{cases} \text{for } \lambda = 0 \text{ we have } \frac{\dot{a}}{a}\rho^A \\ \text{for } \lambda = 1 \text{ we have } \frac{\dot{a}}{a} p_\parallel^A \end{cases} \tag{30}$$

$$\mu = 2 \quad \rightarrow \quad \frac{\partial T_0{}^2}{\partial x^2} + \Gamma^2_{2\lambda} T^\lambda_0 - \Gamma^\lambda_{20} T^2_\lambda \quad \rightarrow \quad \begin{cases} \text{for } \lambda = 0 \text{ we have } \frac{\dot{a}}{a}\rho^A \\ \text{for } \lambda = 2 \text{ we have } \frac{\dot{a}}{a} p_\parallel^A \end{cases} \tag{31}$$

$$\mu = 3 \quad \rightarrow \quad \frac{\partial T_0{}^3}{\partial x^3} + \Gamma^3_{3\lambda} T^\lambda_3 - \Gamma^\lambda_{30} T^3_\lambda \quad \rightarrow \quad \begin{cases} \text{for } \lambda = 0 \text{ we have } \frac{\dot{b}}{b}\rho^A \\ \text{for } \lambda = 3 \text{ we have } \frac{\dot{b}}{b} p_\perp^A. \end{cases} \tag{32}$$

Finally, putting all together, Equation (28) provides the 00 component of the conservation of the energy momentum anisotropic tensor:

$$\dot{\rho}^A + 2\frac{\dot{a}}{a}(\rho^A + p^A_{\parallel}) + \frac{\dot{b}}{b}(\rho^A + p^A_{\perp}) = 0. \quad (33)$$

From a cosmological point of view, it is important to find how density evolves. To this end, we put Equations (20) and (21) in Equation (33)

$$\dot{\rho}^A + 2\left(H \pm \frac{e\,\dot{e}}{3(1-e^2)}\right)(\rho^A + p^A_{\parallel}) + \left(H \mp \frac{2}{3}\frac{e\,\dot{e}}{1-e^2}\right)(\rho^A + p^A_{\perp}) \quad (34)$$

and after a few steps we have

$$\dot{\rho}^A + H\,(3\rho^A + 2p^A_{\parallel} + p^A_{\perp}) \pm \frac{2}{3}\frac{e\,\dot{e}}{1-e^2}(p^A_{\parallel} - p^A_{\perp}) = 0. \quad (35)$$

Now, if we neglect the last term because $p^A_{\parallel} \simeq p^A_{\perp}$ and if we assume a classical barotropic relation between pressure and density, that is to say $p^A_{\parallel} = w^A_{\parallel}\rho_A$ and $p^A_{\perp} = w^A_{\perp}\rho^A$, Equation (35) becomes

$$\dot{\rho}^A + H\rho^A(2 + 2w^A_{\parallel} + w_{\perp}) = 0 \quad (36)$$

which is a first-order differential equation with the initial condition $\rho^A(t=0) = \rho^A_{(0)}$, and we have the following solution

$$\rho^A = \rho^A_{(0)}\frac{1}{A^{3+2w^A_{\parallel}+w^A_{\perp}}}. \quad (37)$$

Equation (37) is interesting because we are expressing the anisotropic energy density (which is an elusive quantity) of the universe by the anisotropic energy density now and the barotropic relations involving $w_{\parallel}$ and $w_{\perp}$ (see for example [33,35]).

4. Cosmic Shear and Ellipsoidal Universe

In this section, we want to connect the cosmic shear to the parameter that most characterizes an ellipsoidal universe, namely the eccentricity e. To accomplish this, we start from the most general eccentricity evolution equation, Equation (27), and introduce the shear, Equation (22). Let us put $L(t) \equiv H(t)\,\Sigma(t)$ and we have

$$\frac{dL}{dt} + 3Hy = \frac{8\pi G}{3}(p^A_{\parallel} - p^A_{\perp}). \quad (38)$$

The general solution is

$$L(t) = \frac{1}{A^3}\left[\int dt'\,\frac{8\pi G}{3}(w^A_{\parallel} - w^A_{\perp})\,\rho^A A^3 + cost\right] \equiv \frac{1}{A^3}[M(A) + const] \quad (39)$$

where *const* is the integration constant. Now, t_0 is the actual time, $L(t_0) = M[A(t_0)/A^3(t_0)] + cost/A^3(t_0)$, but $A(t_0) = 1$; this implies $L(t) = 1/A^3[M(A(t)) + L(t_0) - M(t_0)]$; in other terms, we can write (remembering Equation (22))

$$L(t) = \frac{1}{A^3}\left[H_0\Sigma_a + \Omega^{(0)}_A H_0(w^A_{\parallel} - w^A_{\perp})\int_1^A \frac{d\xi}{\xi^{1+2w^A_{\parallel}+w^A_{\perp}}\,H/H_0}\right] = \frac{e\,\dot{e}}{3(1-e^2)} \quad (40)$$

Let us observe that $1 - e^2 \simeq 1$ and $2\,e\,\dot{e} = de^2/dt$; therefore,

$$\frac{de^2}{dt} = \frac{6H_0}{A^3}\left[\Sigma_a + \Omega^{(0)}_A H_0(w^A_{\parallel} - w^A_{\perp})\int_1^A \frac{d\xi}{\xi^{1+2w^A_{\parallel}+w^A_{\perp}}\,H/H_0}\right]. \quad (41)$$

Now, integrate Equation (41) between t_0 time today and t_{dec} time at decoupling, and let us observe that $\int_{t_0}^{t_{dec}} dt = \int_1^{A_{t_{dec}}} \frac{dA}{AH}$

$$\int_{t_0}^{t_{dec}} \frac{de^2}{dt} = e^2_{dec} = 6H_0 \left[\Sigma_a \int_1^{A_{dec}} \frac{dA}{A^4 H} + \int_1^{A_{dec}} \Omega_A^{(0)} H_0 (w_\parallel^A - w_\perp^A) \int_1^{A_{dec}} \frac{d\xi}{\xi^{1+2w_\parallel^A + w_\perp^A} H^2/H_0} \frac{dA}{A^4} \right], \tag{42}$$

and, dividing both members by e^2, we obtain an equation of this type

$$1 = c_1 \Sigma_a + c_2 \Omega_A^{(0)}. \tag{43}$$

The two unknowns are Σ_a and $\Omega_A^{(0)}$; we have managed to simplify the evolution equations in the simplest way by connecting sigma and omega. Clearly, we have one equation and two unknowns; it is necessary to obtain another equation.

5. Quadrupole of CMB and Cosmic Shear

In this section, we want to obtain the second relation in order to fix one of the two important parameters of the previous section. In order to find this equation, we rewrite the metric Equation (1), taking into account the eccentricity

$$ds^2 = dt^2 - A^2(t)\,(\delta_{ij} + h_{ij})\, dx^i dy^j. \tag{44}$$

The null geodesic equation in a perturbed Friedmann–Lemaitre–Robertson–Walker metric provides the temperature anisotropy (Sachs–Wolfe effect)

$$\frac{\Delta T}{< T >} = -\frac{1}{2} \int_{t_0}^{t_{dec}} dt \frac{\partial h_{ij}}{\partial t} n^i n^j \tag{45}$$

where n^i are the directional cosines. Let us find the value of h_{ij} starting with Equation (1), which can be written as

$$ds^2 = dt^2 - A^2(t) \left[\delta_{ij} + \frac{a^2}{A^2}\delta_{ij} + \frac{a^2 - b^2}{A^2}\delta_{i3}\delta_{j3} - \delta_{ij} \right] dx^i dx^j. \tag{46}$$

From the comparison with the two equations, Equation (44) with Equation (46), we obtain

$$h^{ij} = \left(\frac{a^2}{A^2} - 1 \right) \delta^{ij} + \frac{a^2 - b^2}{A^2} \delta^{i3}\, \delta^{j3}. \tag{47}$$

Since in Equation (45) we have the time derivative of h_{ij}, we derive with respect to time Equation (47)

$$\frac{\partial h^{ij}}{\partial t} = 2\frac{a^2}{A^2}(H_a - H)\,\delta^{ij} + 2\frac{a^2}{A^2}\left[H_a - \frac{b^2}{a^2} H_b - H\left(1 - \frac{b^2}{a^2}\right) \right] \delta^{i3}\delta^{j3}. \tag{48}$$

Now, remembering that $H_a - H = H\,\Sigma_a$, we have

$$\frac{\partial h^{ij}}{\partial t} = 2\,\frac{a^2}{A^2}\, H\,\Sigma_a\, \delta^{ij} + 2\frac{a^2}{A^2}\left[H\,\Sigma_a - \frac{b^2}{a^2}\Sigma_b \right] \delta^{i3}\delta^{j3} \tag{49}$$

but, $\Sigma_b = -2\Sigma_a$; therefore,

$$\frac{\partial h^{ij}}{\partial t} = 2\,\frac{a^2}{A^2}\, H\,\Sigma_a \left[\delta^{ij} + \left(1 + 2\frac{b^2}{a^2} \right) \delta^{i3}\delta^{j3} \right]. \tag{50}$$

Let us go back to the starting equation, Equation (45); substituting Equation (50) into it, we finally have

$$\frac{\Delta T}{< T >} = -\int_{t_0}^{t_{dec}} dt \, \frac{a^2}{A^2} \, H \, \Sigma_a \left[1 + \left(1 + 2\frac{b^2}{a^2} \right) n_3^2 \right] \tag{51}$$

where $\delta^{ij} n^i n^j = 1$ and $\delta^{i3} \, \delta^{j3} \, n^i \, n^j = n_3^2$. The two equations, Equations (43) and (51), identify our two unknowns. An indicative value for Ω_A can be deduced from Equation (2.20) of [13], where

$$e_{dec} \simeq 10^{-2} \sqrt{\Omega_A / 10^{-7}}. \tag{52}$$

Now, imposing that the universe is isotropic $e(t_0) = 0$, from the evolution equation of eccentricity, Equation (27), we have at decoupling ($z = 1088$) $e_{dec} = 10^{-2}$. In other terms, $\Omega_A \simeq 10^{-7}$, which is of course a first order of magnitude of this unknown quantity that is not zero. Another observation is connected to Equation (43) because, if we take into account this value of Ω, we have that $\Sigma_a \simeq k_1 - k_2 \Omega_A$; in other terms, they are fundamental to the determination of the constants that can also be orders of magnitude. We have only two parameters to follow.

A final qualitative analysis regards Equation (51), which provides a useful equation for determining our unknowns. The observed quadrupole anisotropy at CMB provides $(\Delta T_2)_{obs} \simeq 250\ \mu K^2$; therefore, considering the equation that relates the temperature variation in CMB with the eccentricity (see Equation (3.13) of [33]) $\Delta T / < T > = -1/2 \, e_{dec}^2 n_3^2$, it is possible to obtain n_3^2 from the inverse formula $n_3^2 = -2 \Delta T / < T > 1/e_{dec}^2$. We set $\Delta T = \sqrt{250 \times 10^{-6}}$ K, $< T > \simeq 2.7$ K, and $e_{dec} \simeq 10^{-2}$ and obtain $n_3^2 \sim 120$, which is a value that can be utilized in Equation (51).

6. Conclusions

In this paper, we have investigated the connections among the eccentricity of an ellipsoidal universe, the lower quadrupole of the large-scale CMB anisotropy data, and the cosmic shear.

A non-negligible aspect is that, in the literature, either anisotropic or non-homogeneous models of the universe are almost always considered separately. In reality, taking both models (at the same time) into consideration is not an "error" because the cosmological principle model is only valid on a very large scale. There are works that consider models of the universe that simultaneously take into account anisotropy and non-homogeneity, such as [53], in which anisotropic and inhomogeneous models of the universe are studied. In this paper, we study a universe in which inhomogeneities and anisotropies coexist and thus obtain Einstein's equation. This approach will enable further investigations regarding universes that are simultaneously both non-isotropic and non-homogeneous, providing variations with respect to considering only one of the two aspects with related properties. It is probable that considering only one aspect (or only isotropy or only non-homogeneity) makes us lose useful information and probably hides properties that would emerge with the considered model.

Different energy densities, synthesized in ρ^A, can contribute to the Hubble parameter in a different weighted way. Therefore, the anisotropic nature of the early universe provides a very interesting energy content in relation to H. This analysis has been completed in recent works [27,54] that analyze the discrepancy between the existence of distant galaxies that are very large, primordial, and, therefore, early, compared to a ΛCDM model that would provide a star formation too slowly.

Equations (43) and (51) are important because no constraint exists on Σ_a in the literature and an order of magnitude of $\Sigma_a \sim 10^{-5}$ [33].

A very important consideration regards "isotropization" because, generally, when we consider an anisotropic universe, the Einstein gravity and the cosmological constant tend to cancel the anisotropy; in other terms, the universe becomes isotropic asymptotically.

Practically, a way to realize the isotropization consists of considering the limit for directional expansion factor $a = b$ for $t \to \infty$; on the other hand, it is possible to obtain the isotropization with [55–58] the mean expansion parameter $A \to 0$ and the scalar shear $\sigma^2 \to 0$.

This work is a completion of [34]. There are substantial differences between the two works because, in [34], we try to frame the physics using the CS to relate everything to the accelerating and decelerating phases of the universe. This aspect has not been addressed here, just as the correlation between the variation in acceleration of the universe, the so-called jerk parameter, eccentricity, and cosmic shear have not been addressed. In this work, we have instead tried to link the cosmic shear to the eccentricity passing through the cosmic microwave background. Interestingly, from a temporal point of view, these three "indicators" are coexisting: we are at about z = 1080. What is new is to correlate a link between these parameters and the variation in the temperature of the CMB related to the average $\frac{\Delta T}{T}$. We have therefore arrived at Equation (51), which, with Equation (43), represent the final synthesis of all the reasoning and enable it to be a good starting point for new developments.

As regards the comparison with [8], there are profound differences. While in this work the pressure is practically the same in the three different expansion axes, in this work, a pressure is introduced parallel and perpendicular to the expansion axes, which is not only different from a calculation point of view but conceptually profound. The work of [8] focuses the analysis on the isotropization of the universe to make it FLRW, all due to a sort of contraction of the scale factors, a process that would allow us to move from a Bianchi I universe to an FLRW universe. For the moment, in this work, this aspect has been put aside because it will be the subject of a work [59] not linked to any contraction but as an evolution of an external factor (a magnetic field, for example). In [8], the authors find, on the anisotropic model, some constraints that are provided by the upper limits of anisotropy from the Planck data. In addition, in that work, isotropization criteria are introduced. In this work, however, the possible experimental constraints are only those related to the position of the quadrupole (l = 2), leaving the cosmic shear free for the moment, although Equation (43) can determine important constraints on the cosmic shear since $\Omega_A^{(0)}$ will have to be derived from the Friedmann equation. On the other hand, in this work, instead, the eccentricity of the universe is linked to the cosmic microwave background through the parameter of cosmic shear, which considers things from another point of view. A determining role is played by eccentricity and its evolution in time, as can be seen from Equation (27). From this point of view, the approach with respect to the work of [8] is completely different since it sheds light on the close connection between eccentricity and cosmic shear and their temporal evolution.

The limits of this model are strictly connected to the transition between the anisotropic universe and the current isotropic universe. In other words, it is necessary to connect the origin of the anisotropy to the causes that generated this ellipsoidal universe. These causes may later disappear by "washing out" the anisotropy itself. An example is the primordial magnetic field that can be a good generator of anisotropy [51,60].

Following the expansion of the universe, the primordial magnetic fields become less and less intense, so they act less and less. Today, traces of these cosmic magnetic fields remain in the universe, whose origin could be connected to the anisotropy of the universe itself. This is a limit of the model itself because it takes little to eliminate the seed magnetic fields, but, at the same time, it is a potential area of research for future developments. An aspect often overlooked in these models, and which could provide great contributions to the understanding of the universe, is this: if the universe expanded like an ellipsoid, it means that one axis expanded more rapidly than the other. Why was there this disparity in expansion? Think of the shape of an egg, in which one axis is larger than the other, generating the characteristic shape because space was stretched more in one direction than the other. Answering these apparently simple questions means entering the realm of the mysteries of the universe. Another important evolution in these anisotropic models can be connected with the current H_0, in which we have a deep discrepancy between the early time and late time determinations of the Hubble constant. On the other hand, the ellipsoidal model of the universe may solve the so-called S_8 tension [61]. Another aspect, as already mentioned, concerns how to isotropize this ellipsoidal universe. It is not at all trivial. There are new studies that approach this problem in a very promising way. The delicate problem concerns the need to find the cosmological limits that enable testing them with experimental data [59].

It is beyond the scope of this work, but, for completeness, next, we will study the problem of isotropizing the model, which does not simply mean reducing the eccentricity to zero. As already demonstrated in [34], the isotropization must take into account several factors, such as Σ , as well as the jerk parameter, which we have not dealt with in this work, a parameter that must be set to 1 in the ΛCDM model. On the other hand it is essential to connect these reasonings to the CMB polarization data, to have a better connection with the experimental data [62].

In [59], the study combines cosmic shear with eccentricity and two important cosmological parameters: the acceleration parameter (which is essentially connected with the second derivative of the expansion parameter) and the jerk (which is essentially connected with the third derivative of the expansion parameter). Preliminary studies indicate that there is a strong correlation between these parameters that are not simply linked by standard relations but provide interesting indicators on their evolution, strictly connected to eccentricity. The study must consider that, if there is a universe with two expansion parameters, like the one considered in this work, it is necessary to consider two acceleration parameters and two jerk parameters that, in the end, when the universe returns to being isotropic, must provide a single contribution. Last but not least, it is interesting to study the luminosity distance in relation to an ellipsoidal universe, where the value of the eccentricity is not negligible. In other terms, considering an ellipsoidal universe is not a painless process for modern cosmology, which cannot ignore it.

Funding: The work is supported in part by INFN under the program TAsP: "Theoretical Astroparticle Physics" and also this work was partially supported by the research grant number 2022E2J4RK "PANTHEON: Perspectives in Astroparticle and Neutrino THEory with Old and New messengers" under the program PRIN 2022 funded by the Italian Ministero dell'Universitá e della Ricerca (MUR) and by the European Union-Next Generation EU.

Data Availability Statement: No new data were created.

Acknowledgments: I want to acknowledge the hospitality of CERN, CERN Department of Theoretical Physics, where this work has been carried out.

Conflicts of Interest: The author declare no conflicts of interest

References

1. Lemaitre, G. Annales de la Société Scientifique de Bruxelles, A53, 51, 1933 . English translation by M.A.H. MacCallum. *Expand. Universe Gen. Rel. Grav.* **1997**, *29*, 641.
2. Jacobs, K.C. Bianchi Type I Cosmological Models. Ph.D. Thesis, California Institute of Technology, Pasadena, CA, USA , 1968.
3. Bianchi, L. Sugli spazi a tre dimensioni che ammettono un gruppo continuo di movimenti. Memorie di Matematica e di Fisica della Societa Italiana delle Scienze 11 267352 (1898). English translation by R.T. Jantzen. On the three-dimensional spaces which admit a continuous group of motions. *Gen. Rel. Grav.* **2001**, *33*, 2171.
4. Gurzadyan, V.G.; Starobinsky, A.A.; Ghahramanyan, T.; Kashin, A.; Khachatryan, H.G.; Kuloghlian, H.; Vetrugno, D.; Yegorian, A.G. Large Scale Plane-mirroring in the cosmic microwave background WMAP5 maps. *Astron. Astrophys.* **2008**, *490*, 929. [CrossRef]
5. Gurzadyan, V.G.; Ghahramanyan, T.; Kashin, A.L.; Khachatryan, H.G.; Kocharyan, A.A.; Kuloghlian, H.; Vetrugno, D.; Yegorian, G. Plane-mirroring anomaly in the cosmic microwave background maps. *Astron. Astrophys.* **2009**, *498*, L1. [CrossRef]
6. Buchert, T.; Coley, A.A.; Kleinert, H.; Roukema, B.F.; Wiltshire, D.L. Observational Challenges for the Standard FLRW Model. *Int. J. Mod. Phys. D* **2016**, *25*, 1630007. [CrossRef]
7. Schwarz, D.J.; Copi, C.J.; Hureter, D.; Starkman, G.D. CMB Anomalies after *Planck*. *Class. Quant. Grav.* **2016**, *33*, 184001. [CrossRef]
8. Russel, E.; Kilinc, C.B.; Pashev, O.K. Bianchi iI model: An alternative way to model the present-day Universe. *Mon. Not. Roy. Astron. Soc.* **2014**, *442*, 2331. [CrossRef]
9. Bennett, C.L.; Hill, R.S.; Hinshaw, G.; Nolta, M.R.; Odegard, N.; Page, L.; Spergel, D.N.; Weiland, J.L.; Wright, E.L.; Halpern, M. First-Year Wilkinson Microwave Anisotropy Probe (WMAP) Observations: Foreground Emission. *Astrophys. J. Suppl. Ser.* **2003**, *148*, 97. [CrossRef]

10. Barrow, J.D. Cosmological limits on slightly skew stresses. *Phys. Rev. D* **1997**, *55*, 7451. [CrossRef]
11. Berera, A.; Buniy, R.V.; Kephart, T.W. The eccentric universe. *J. Cosmol. Astropart. Phys.* **2004**, *10*, 016. [CrossRef]
12. Campanelli, L.; Cea, P.; Tedesco, L. Ellipsoidal Universe Can Solve the Cosmic Microwave Background Quadrupole Problem. *Phys. Rev. Lett.* **2006**, *97*, 131302; Erratum in *Phys. Rev. Lett.* **2006**, *97*, 209903. [CrossRef] [PubMed]
13. Campanelli, L.; Cea, P.; Tedesco, L. Cosmic microwave background quadrupole and ellipsoidal universe. *Phys. Rev. D* **2007**, *76*, 063007. [CrossRef]
14. Copi, C.J.; Huterer, D.; Schwarz, D.J.; Starkman, G.D. Large-scale alignments from WMAP and Planck. *Mon. Not. R. Astron. Soc.* **2015**, *449*, 3458. [CrossRef]
15. Rodrigues, D.C. Anisotropic cosmological constant and the CMB quadrupole anomaly. *Phys. Rev. D* **2008**, *77*, 023534. [CrossRef]
16. Eriksen, H.K.; Hansen, F.K.; Banday, A.J.; Gorski, K.M.; Lilje, P.B. Asymmetries in the Cosmic microwave background anisotropy field. *Astrophys. J.* **2004**, *605*, 14. Erratum in *Astrophys. J.* **2004**, *609*, 1198. [CrossRef]
17. Hansen, F.K.; Banday, A.J.; Gorski, K.M. Testing the cosmological principle of isotropy: Local power-spectrum estimates of the WMAP data. *Mon. Not. R. Astron. Soc.* **2004**, *354*, 641. [CrossRef]
18. Vielva, P.; Martinez-Gonzalez, E.; Barreiro, R.B.; Cayon, J.L.S.L. Detection of Non-Gaussianity in the Wilkinson Microwave Anisotropy Probe First-Year Data Using Spherical Wavelets. *Astrophys. J.* **2004**, *609*, 22. [CrossRef]
19. Land, K.; Magueijo, J. Examination of Evidence for a Preferred Axis in the Cosmic Radiation Anisotropy. *Phys. Rev. Lett.* **2005**, *95*, 071301. [CrossRef]
20. Ralston, J.P.; Jain, P. The Virgo alignment puzzle in propagation of radiation on cosmological scales. *Int. J. Mod. Phys. D* **2004**, *13*, 1857. [CrossRef]
21. de Oliveira-Costa, A.; Tegmark, M.; Zaldarriaga, M.; Hamilton, A. The significance of the largest scale CMB fluctuations in WMAP. *Phys. Rev. D* **2004**, *69*, 063516. [CrossRef]
22. Copi, C.J.; Huterer, D.; Starkman, G.D. Multipole vectors: A new representation of the CMB sky and evidence for statistical anisotropy or non-Gaussianity at $2 < l < 8$. *Phys. Rev. D* **2004**, *70*, 043515.
23. Herzberg, M.P.; Loeb, A. Constraints on an anisotropic universe. *Phys. Rev. D* **2024**, *109*, 083538. [CrossRef]
24. Quercellini, C.; Cabella, P.; Amendola, L.; Quartin, M.; Balbi, A. Cosmic parallax as a probe of late time anisotropic expansion. *Phys. Rev. D* **2009**, *80*, 063527. [CrossRef]
25. Sarmah, P.; Goswami, U.D. Bianchi Type I model of universe with customized scale factors. *Mod. Phys. Lett. A* **2022**, *37*, 2250134. [CrossRef]
26. Campanelli, L. Helium-4 synthesis in an anisotropic universe. *Phys. Rev. D* **2011**, *84*, 123521. [CrossRef]
27. Deliduman, C.; Kasikci, O.; Tugyanoglu, V.K. f(R) Gravity in an Ellipsoidal Universe. *Phys. Dark Univ.* **2024**, *44*, 101469. [CrossRef]
28. Mukherjee, S.; Aluri, P.K.; Das, S.; Shaikh, S.; Souradeep, T. Direction dependence of cosmological parameters due to cosmic hemispherical asymmetry. *J. Cosmol. Astropart. Phys.* **2016**, *06*, 042. [CrossRef]
29. de Andrade, L.C.G. Kinematic fast cosmic dynamos in non-inflationary phases of ellipsoidal universe. *arXiv* **2007**, arXiv:0707.1734.
30. Aluri, P.K.; Cea, P.; Chingangbam, P.; Chu, M.C.; Clowes, R.G.; Hutsemékers, D.; Kochappan, J.P.; Lopez, A.M.; Liu, L.; Martens, N.C.M.; et al. Is the Observable Universe Consistent with the Cosmological Principle? *Class. Quant. Grav.* **2023**, *40*, 094001. [CrossRef]
31. Akarsu, O.; Kumar, S.; Sharma, S.; Tedesco, L. Constraints on a Bianchi type I spacetime extension of the standard ΛCDM model. *Phys. Rev. D* **2019**, *100*, 023532. [CrossRef]
32. Akarsu, O.; Valentino, E.D.; Kumar, S.; Ozyigit, M.; Sharma, S. Testing spatial curvature and anisotropic expansion on top of the ΛCDM model. *Phys. Dark Univ.* **2023**, *39*, 101162. [CrossRef]
33. Campanelli, L.; Cea, P.; Fogli, G.L.; Tedesco, L. Anisotropic dark energy and ellipsoidal Universe. *Int. J. Mod. Phys.* **2011**, *20*, 1153. [CrossRef]
34. Tedesco, L. Ellipsoidal expansion of the Universe, cosmic shear, acceleration and jerk parameter. *Eur. Phys. J. Plus* **2018**, *133*, 188. [CrossRef]
35. Campanelli, L.; Cea, P.; Fogli, G.L.; Tedesco, L. Cosmic parallax in ellipsoidal universe. *Mod. Phys. Lett. A* **2011**, **26**, 1169. [CrossRef]
36. Stavrinos, P.C.; Triantafyllopoulos, A. Cosmology based on Finsler and Finsler-like metric structure of gravitational field. *arXiv* **2024**, arXiv:2405.19318.
37. Kapsabelis, E.; Saridakis, E.N.; Stavrinos, P.C. Finsler-Randers-Sasaki gravity and cosmology. *Eur. Phys. J. C* **2024**, *84*, 538. [CrossRef]
38. Stavrinos, P.C.; Diakogiannis, F.I. A Geometric anisotropic model of space-time based on Finslerian metric. *arXiv* **2002**, arXiv:0203083.
39. Papagiannopoulos, G.; Basilakos, S.; Paliathanasis, A.; Savvidou, S.; Stavrinos, P.C. Finsler-Randers cosmology: Dynamical analysis and growth of matter perturbations. *Class. Quant. Grav.* **2017**, *34* 225008. [CrossRef]
40. Papagiannopoulos, G.; Basilakos, S.; Paliathanasis, A.; Pan, S.; Stavrinos, P.C. Dynamics in varying vacuum Finsler-Randers cosmology. *Eur. Phys. J. C* **2020**, *80*, 816. [CrossRef]
41. Basilakos, S.; Stavrinos, P.C. Cosmological equivalence between the Finsler-Randers space-time and the DGP gravity model. *Phys. Rev. D* **2013**, *87*, 043506. [CrossRef]
42. Gonner, H.F.; Bogoslovsky, G.Y. A Class of anisotropic (Finsler) space-time geometries. *Gen. Rel. Grav.* **1999**, *31*, 1383. [CrossRef]

43. Stavrinos, P.C.; Kouretsis, A.P.; Stathakopoulos, M. Friedmann Robertson Walker model in generalised metric space-time with weak anisotropy. *Gen. Rel. Grav.* **2008**, *40*, 1403. [CrossRef]
44. Colin, J.; Mohayaee, R.; Rameez, M.; Sarkar, S. Evidence for anisotropy of cosmic acceleration. *Astron. Astrophys.* **2019**, *631*, L13. [CrossRef]
45. Zhao, D.; Zhou, Y.; Chang, Z. Anisotropy of the Universe via the Pantheon supernovae sample revisited. *Mon. Not. R. Astron. Soc.* **2019**, *486*, 5679. [CrossRef]
46. Amirhashch, H.; Amirhashch, S. Constraining Bianchi type I universe with type Ia supernova and H(z) data. *Phys. Dark Univ.* **2020**, *29*, 100557. [CrossRef]
47. Colin, J.; Mohayaee, R.; Sarkar, S.; Shafieloo, A. Probing the anisotropic local Universe and beyond with SNe Ia data. *Mon. Not. R. Astron. Soc.* **2011**, *414*, 264. [CrossRef]
48. Taub, A.H. Empty space-times admitting a three parameter group of motions. *Ann. Math.* **1951**, *53*, 472; Reprinted in *Gen. Rel. Grav.* **2004**, *36*, 2689. [CrossRef]
49. Maartens, R.; Ellis, G.F.R.; Stoeger, W.R. Anisotropy and inhomogeneity of the Universe from deltaT/T. *Astron. Astrophys.* **1996**, *309*, L7.
50. Buniy, R.V.; Berera, A.; Kephart, T. Asymmetric inflation: Exact solutions. *Phys. Rev. D* **2006**, *73*, 063529. [CrossRef]
51. Barrow, J.D.; Ferreira, P.G.; Silk, J. Constraints on a primordial magnetic field. *Phys. Rev. Lett.* **1997**, *78*, 3610. [CrossRef]
52. Maurya, D.C. Anisotropic dark energy transit cosmological models with time-dependent ω(t) and redshift-dependent ω(z) EoS parameter. *Int. J. Geom. Meth. Mod. Phys.* **2017**, *15*, 1850019. [CrossRef]
53. Fanizza, G.; Tedesco, L. Inhomogeneous and anisotropic Universe and apparent acceleration. *Phys. Rev. D* **2015**, *91*, 023006. [CrossRef]
54. Binici, S.S.; Deliduman, C.; Dilsiz, F.S. The ages of the oldest astrophysical objects in an ellipsoidal universe. *Phys. Dark Univ.* **2024**, *46*, 101600. [CrossRef]
55. Jacobs, K.C. Spatially Homogeneous and Euclidean Cosmological Models with Shear. *Astrophys. J.* **1968**, *153*, 661. [CrossRef]
56. Bronnikov, K.A.; Chudayeva, E.N.; Shikin, G.N. Magneto-dilatonic Bianchi-I cosmology: Isotropization and singularity problems. *Class. Quantum Gravity* **2004**, *21*, 3389. [CrossRef]
57. Saha, B. Nonlinear spinor field in Bianchi type-I cosmology: Inflation, isotropization, and late time acceleration. *Phys. Rev. D* **2006**, *74*, 124030. [CrossRef]
58. Saha, B. Early inflation, isotropization, and late time acceleration in a Bianchi type-I universe. *Phys. Part. Nucl.* **2009**, *40*, 656. [CrossRef]
59. Tedesco, L. Different evolution of jerk parameters in ellipsoidal universe. *In Preparartion.*
60. Parker, E.N. *Cosmological Magnetic Fields*; Oxford University Press: Oxford, UK, 1979.
61. Cea, P. The Ellipsoidal Universe and the Hubble tension. *arXiv* **2022**, arXiv:2201.04548.
62. Cea, P. Confronting the Ellipsoidal Universe to the Planck 2018 Data. *Eur. Phys. J. Plus* **2020**, *135*, 150. [CrossRef]

universe

MDPI

Article

From de Sitter to de Sitter: A Thermal Approach to Running Vacuum Cosmology and the Non-Canonical Scalar Field Description

Pedro Eleuterio Mendonça Almeida [1], Rose Clivia Santos [2] and Jose Ademir Sales Lima [1,*]

[1] Departamento de Astronomia, Universidade de São Paulo, Rua do Matão 1226, São Paulo 05508-900, SP, Brazil; pedro.emalmeida@usp.br

[2] Departamento de Física, Universidade Federal de São Paulo (UNIFESP), Diadema 210, São Paulo 05580-090, SP, Brazil; rose.clivia@unifesp.br

* Correspondence: jas.lima@iag.usp.br

Abstract: The entire classical cosmological history between two extreme de Sitter vacuum solutions is discussed based on Einstein's equations and non-equilibrium thermodynamics. The initial non-singular de Sitter state is characterised by a very high energy scale, which is equal or smaller than the reduced Planck mass. It is structurally unstable, and all of the continuous created matter, energy, and entropy of the material component comes from the irreversible flow powered by the primeval vacuum energy density. The analytical expression describing the running vacuum is obtained from the thermal approach. It opens a new perspective to solve the old puzzles and current observational challenges plaguing the cosmic concordance model driven by a rigid vacuum. Such a scenario is also modelled through a non-canonical scalar field. It is demonstrated that the resulting scalar field model is shown to be a step-by-step a faithful analytical representation of the thermal running vacuum cosmology.

Keywords: running vacuum; scalar field; cosmological dark sector

PACS: 98.80.-k; 95.36.+x

1. Introduction

Since long ago, investigations on de Sitter and quasi-de Sitter cosmic solutions have attracted much attention along different research lines. Such studies were somewhat inspired by the possible avoidance of the primeval singular state ordinarily predicted by "big-bang" models, and the decaying vacuum emerging from a plethora of inflationary models driven by scalar fields including or excluding quadratic corrections to Einstein gravity [1–7]. A different approach analysed the possible creation of the universe from "nothing" (in the sense of Vilenkin [8]), with the classical universe emerging as a de Sitter solution from an expected, but still observationally unavailable, quantum gravity regime. This primeval state without ordinary matter may be characterised by a physical energy scale, $H_I \leq M_P$, or equivalently, by a vacuum energy density, $\rho_I = 3M_P^2 H_I^2$, where $M_P = (8\pi G)^{-1/2} \simeq 2.4 \times 10^{18}$ GeV is the reduced Planck mass (in our units, $\hbar = k_B = c = 1$).

It is also widely known that a non-singular de Sitter spacetime is structurally unstable both from quantum and classical viewpoints, at least due to three different effects: quantum corrections on the geometric sector of general relativity [4], the existence of thermal fluctuations [9], and the process of gravitational particle creation in the expanding universe [7,10–14], Henceforth, due to its simplicity and interesting physical consequences for the evolution of the emergent classical initial vacuum state, the presence of thermal instabilities conjoined with an irreversible flow of energy, matter, and entropy forming the thermal bath will be taken for granted. Additional reasons will be outlined ahead.

More recently, independent astronomical observations have shown that the cosmic concordance model (ΛCDM + Inflation) provides a quite reasonable and predictive description of the current universe driven by a rigid vacuum energy density, $\rho_{\Lambda_0} = M_p^2\Lambda_0$. However, the incredible discrepancy with what is expected from quantum field theory (QFT) as compared to current astronomical observations, $\rho_{\Lambda_0}/\rho_{\Lambda_I} \sim 10^{-122}$, gave rise to the cosmological constant problem (CCP) [15]. Furthermore, according to the cosmic concordance model, the dominance of rigid vacuum ($\rho_{\Lambda_0} \simeq \rho_M$, where ρ_M is the matter density) took place at a redshift $z \simeq 0.55$ (see, e.g., [16]) thereby leading to the so-called coincidence problem (CP). Another well-known consequence of the rigid vacuum ΛCDM cosmology is the inexorable evolution of the current universe to a final de Sitter spacetime. Here, it is characterised by an extremely low vacuum energy density, say, $\rho_{\Lambda_F} = 3M_p^2H_F^2$, where the Hubble parameter H_F provides the final de Sitter constant energy scale, which will be attained only for $\rho_M \equiv 0$.

Currently, the standard rigid vacuum cosmology is also being observationally challenged by the H_0 and S_8 tensions. In the case of H_0, local determinations of the Hubble constant (H_0) using Cepheid-calibrated supernovae of the SHOES collaboration [17], measurements at intermediate redshifts from a combination of tests [18,19] and strong lensing time-delays [20], which, in comparison with the independent prediction of Planck collaboration plus ΛCDM, shows a clear-cut tension around 5σ [21]. In addition, estimates based on cosmic shear evaluations from weak lensing are favouring values of the parameter $S_8 = \sigma_8\sqrt{\Omega_M/0.3}$, where σ_8 measures the current mass fluctuation in 8 h^{-1}Mpc, which are lower than the ones provided by early time probes [22–24].

In this context, it seems natural to advocate that the whole classical evolution of the universe would be analytically described between two de Sitter stages defined by a pair of extreme and constant energy scales, H_I and H_F. In this case, for all theoretical purposes, we may also consider that the second de Sitter vacuum state of very low energy density is the true vacuum state. In certain sense, the structural instability of the primeval de Sitter (H_I) is somewhat driving the evolution of the observed universe to its 'ground state', corresponding to the final de Sitter vacuum, $H_F < H_0$, where the sub-index 0 refers to present day quantities. Thus, it is not surprising that a lot of attention has been paid to running vacuum cosmologies in the last decade by several reasons, among them: (i) a simple solution to the singularity problem, as well as the old CCP and CP puzzles; (ii) a complete cosmology (from de Sitter to de Sitter) with two accelerating stages driven by the same actor, described by the running vacuum energy density, $\rho_\Lambda(H_I, H_F, H)$, departing slightly at all phases from the rigid vacuum model plus slow-roll inflation; and (iii) a possible solution to the current H_0 and S_8 observational tensions plaguing the current standard cosmology provided by the ΛCDM model, which at late times model is driven by a rigid vacuum [25–32].

As we shall see, a suitable $\Lambda(H)$-term uniting both de Sitter phases can be deduced using non-equilibrium thermodynamics and the Einstein field equations (EFE). The classical evolution of the universe can also be analytically described, and some of the questions outlined above are answered. Perhaps more interestingly, we demonstrate that a minimally coupled ordinary scalar field cannot describe the irreversible two-fluid approach of the running vacuum scenario. However, the whole evolution can perfectly be mimicked by a single minimally coupled non-canonical field when interpreted as a mixture of two minimally coupled interacting perfect fluids [33–42].

The paper is organised as follows. In Section 2, we set up the Einstein equations for an interacting mixture of running vacuum medium plus a perfect fluid. In Section 3, the irreversible thermodynamic approach for such a mixture is discussed in detail. In particular, by taking into account only the thermal effects of the interacting mixture, a viable expression for $\Lambda(H)$ is phenomenological deduced based on Einstein's equations and non-equilibrium thermodynamics. The equation governing the entire classical evolution of the universe from H_I to H_F is also determined. Section 4 is dedicated to the basic results of the cosmic eras, and the solutions to some cosmological problems are also discussed. In Section 5, a representation of the whole process in terms of a non-canonical scalar field and the associated dynamics is investigated. The article is closed in Section 6 with the final comments and conclusions. In Appendix C, the two basic parameters of the non-canonical scalar field potential are calculated in terms of H_I, H_F, M_P, and the mass scale (M) of the scalar field.

2. Basic Equations

In what follows we consider a class of Friedmann–Lemaître–Roberson–Walker (FLRW) cosmologies described by a flat geometry:

$$ds^2 = dt^2 - a^2(t)\left(dx^2 + dy^2 + dz^2\right), \tag{1}$$

where $a(t)$ is the scale factor. In such a background, the EFE for a perfect fluid plus a running vacuum can be written as

$$\rho + M_P^2\Lambda(H) = 3M_p^2 H^2\,, \tag{2}$$

$$p - M_P^2\Lambda(H) = -M_p^2\left[2\dot{H} + 3H^2\right], \tag{3}$$

where ρ and p are the energy density and pressure of the ordinary fluid component, while a dot means a comoving time derivative. Now, by adopting the "ω-law" equation of state (EoS),

$$p = \omega\rho, \quad 0 \leq \omega \leq 1, \tag{4}$$

it is readily seen that Equations (2)–(4) imply that the cosmic dynamics is driven by the "H-equation of motion"

$$\dot{H} + \frac{3(1+\omega)}{2}H^2[1 - \Omega_\Lambda(H)] = 0\,, \tag{5}$$

where $\Omega_\Lambda(H) = \Lambda(H)/3H^2$ is the vacuum density parameter.

In principle, for $\Omega_\Lambda = 1$, it is possible to have two extreme de Sitter solutions ($\dot{H} = 0$) describing the initial and final vacuum states (say, $H = H_I$ and $H = H_F$) and, as such, two accelerating cosmic states (early and late time inflation) may result from the same actor [26,27,32]. The initial de Sitter spacetime (H_I) works like a "repeller" (unstable solution) at early times. However, due to the cosmic evolution, energy, particles and entropy are continuously being transferred to the fluid component, thereby giving rise to an attractor de Sitter spacetime in the distant future, characterised by a very low energy scale (H_F). Accordingly, $\Omega_\Lambda = 1$ implies that the fluid energy densities for both extreme scales are nullified, i.e., $\rho(H_I) = \rho(H_F) = 0$.

Of course, the complete evolution driven by Equation (5), linked to the subsequent radiation-vacuum and matter-vacuum dominated phases, only comes out if the expression for $\Lambda(H)$ is known. Next, it will be derived by combining the EFE, non-equilibrium thermodynamics, and the limiting 'boundary' constraints provided by the extreme de Sitter solutions.

3. Running Vacuum: A Thermal Approach

Let us now consider the energy flow from the decaying vacuum creating particles and entropy to the fluid component. Following standard lines, the thermodynamic states for the interacting mixture is defined by the energy conservation law ($u_\mu T^{\mu\nu}_{;\nu} = 0$) and the balance equations for the number of particles and entropy fluxes, $N^\mu = nu^\mu$ and $S^\mu = su^\mu$:

$$u_\mu T^{\mu\nu}_{;\nu} = 0 \iff \dot{\rho} + 3H(1+\omega)\rho = -M_P^2\dot{\Lambda}, \tag{6}$$

$$N^\mu_{;\mu} = n\Gamma \iff \dot{n} + 3nH = n\Gamma, \tag{7}$$

$$S^\mu_{;\mu} = s\Gamma \iff \dot{s} + 3sH = s\Gamma, \tag{8}$$

where n and s are the particle concentration and entropy density, and Γ is the particle creation rate, being the same for particle number and entropy. Thus, the entropy of the fluid is also a pure consequence of the decaying vacuum and, as such, the bulk viscosity process and gravitational matter creation have been neglected [43]. In principle, one may argue that the second law of thermodynamics should be applied for both components. However, the vacuum fluid behaves like a condensate carrying no entropy, as happens in the two-fluid description ordinarily employed in superfluid dynamics [44]. The vacuum state is assumed to have energy, but no entropy and real particles.

In the course of the decaying vacuum process, one may see from (7) and (8) that the specific entropy, $\sigma = s/n = S/N$, remains constant ($\dot{\sigma} = 0$). Such an interesting result becomes more clear when rewritten as (see Appendix A):

$$\frac{\dot{S}}{S} = \frac{\dot{N}}{N} = \Gamma \geq 0. \tag{9}$$

This inequality implies that the decaying vacuum (due to its own energy density) can only produce matter in the space-time ($\dot{N} \geq 0$), while the reverse process is thermodynamically forbidden.

The condition $\dot{\sigma} = 0$ has another remarkable consequence. Actually, by combining Gibbs' law, $nTd\sigma = d\rho - (\rho + p)dn/n$, with (4) and (5), we find a key result in this framework:

$$\dot{\rho} - (\rho + p)\frac{\dot{n}}{n} = 0 \leftrightarrow M_P^2\dot{\Lambda} = -(\rho + p)\Gamma. \tag{10}$$

The second equality above is a typical first-order thermodynamic relation uniting the stress $\dot{\Lambda}$ ("flux") to its thermodynamic force $\Gamma(H)$. The above coefficient of Γ is the fluid enthalpy density which, due to the weak energy condition, satisfies the inequality $h \equiv \rho + p = (1+\omega)\rho \geq 0$.

Notice that by adding **EFE** (2) and (3), we find $\rho + p = -2M_P^2\dot{H}$, which means that $\dot{H} < 0$. Thus, since $\Gamma \geq 0$ from Equation (9), it follows that the constraint $\dot{\Lambda} \leq 0$ must be obeyed. In this thermal approach, a decaying vacuum from the primeval de Sitter state is also required by the second law of thermodynamics.

Now, by inserting h into Equation (10), a new and interesting result is derived:

$$\Lambda(H) = 2\int \Gamma(H)dH + B, \tag{11}$$

where B is a constant to be fixed by the de Sitter 'boundary conditions'. To the best of our knowledge, the above integral expression defining $\Lambda(H)$ in terms of the creation rate $\Gamma(H)$ has not been found before. Thus, in order to obtain a complete description (from H_I to H_F), an expression for $\Gamma(H)$ needs to be specified.

At this point, it should be remarked that, until now, only classical macroscopic tools were explored, namely **EFE** and non-equilibrium thermodynamics. It is also well known that particle creation from a perturbed vacuum state is essentially a microscopic phenomenon. Hence, from a more rigorous viewpoint, an expression for $\Gamma(H)$ in this context cannot be macroscopically deduced. Nevertheless, under the proviso that the dimension of $[\Gamma] = [H]$, a simple possibility would be $\Gamma(H) = 3\nu H$, where the factor 3 is introduced for mathematical convenience, and ν is a dimensionless positive free parameter, in principle, restricted on the interval $0 \leq \nu < 1$.

Furthermore, such a linear expression generates a term $\Lambda(H) \propto H^2$. However, it is not enough for describing the unified cosmic history from de Sitter to de Sitter, since it generates a singular initial state [38]. At late times, due to the very low expansion rate, ν is taken as a constant parameter. However, it may be a function of H at early times, thereby implying a more intense transference of energy, particles, and entropy to the material component. This is needed for the formation of the primeval thermal bath, both in the macroscopic approach [26] and for a possible scalar field description [45].

In this connection, a renormalized vacuum energy density proportional to H^4 has been derived long ago [7]. This is also the next term obtained by Shapiro and Solà based on the covariance of the renormalized action [46]. From Equation (11), we also see that a term proportional to H^4 can also be generated by a creation rate $\Gamma(H) \propto H^3$. Therefore, inspired by such studies, and also some ad hoc treatments for $\Lambda(H)$ [25,26,32], a quadratic correction, namely $3\gamma(H/H_I)^2$, where the arbitrary $\gamma = \gamma(\nu)$, is added to the natural term, based on dimensional grounds. In particular, for $\gamma = 2(1-\nu)$ it follows that

$$\Gamma(H) = 3\nu H + 6(1-\nu)H(H/H_I)^2, \tag{12}$$

whereas $\Lambda(H)$, as given by Equation (11), reads:

$$\Lambda(H) = 3(1-\nu)H_F^2 + 3\nu H^2 + 3(1-\nu)H^2\left(\frac{H}{H_I}\right)^2. \tag{13}$$

As a simple check of such expressions, one may compute directly $\Gamma(H) = d\Lambda(H)/2dH$ (see Appendix B for a more detailed calculation, including both de Sitter boundary conditions).

Now, it is easy to see that the expressions for the density parameters take the form

$$\Omega_\Lambda(H) = \nu + (1-\nu)\frac{H_F^2}{H^2} + (1-\nu)\left(\frac{H}{H_I}\right)^2, \tag{14}$$

$$\Omega_M(H) = (1-\nu)\left[1 - \frac{H_F^2}{H^2} - \left(\frac{H}{H_I}\right)^2\right], \tag{15}$$

where Ω_M is the matter density parameter. Now, by inserting Ω_Λ or Ω_M into Equation (5), it is readily seen that the classical evolution of the model between the two extreme de Sitter phases is governed by

$$\dot{H} + \frac{3(1+\omega)(1-\nu)H^2}{2}\left[1 - \frac{H_F^2}{H^2} - \frac{H^2}{H_I^2}\right] = 0. \tag{16}$$

The above equation depends on two pairs of positive parameters (ω, ν) and two energy scales (H_I, H_F). Like in the rigid vacuum model, ω is differently fixed by the description of the cosmic eras $\omega = 1/3$ (radiation) and $\omega = 0$ (dust). So, it cannot rigorously be considered a new free parameter. For the remaining three parameters, we remark that the energy scale H_I has a natural upper bound given by the reduced Planck mass M_P (see Section 1), while the final H_F scale can be determined in order to solve the CCP problem, by assuming, for instance, that $H_I = M_P$. It can also be calculated by the model in terms of H_0.

Figure 1 schematically displays the complete evolution of the whole classical scenario. Observe that Equation (16) remains valid, even for $\nu \equiv 0$, since this minimal model is also defined by the same extreme constant physical scales (H_I, H_F). The only difference is that the late time evolution for $H << H_I$ happens in such a way that CMB photons are not produced any more. Moreover, it is easy to see that the scenario becomes the standard, rigid ΛCDM cosmology, described here in terms of the constant final scale H_F. Thus, only the early inflationary stage is modified, because it started from a pure de Sitter vacuum solution. The details of each transition describing the evolution (from de Sitter to de Sitter) through different cosmic eras will be discussed next.

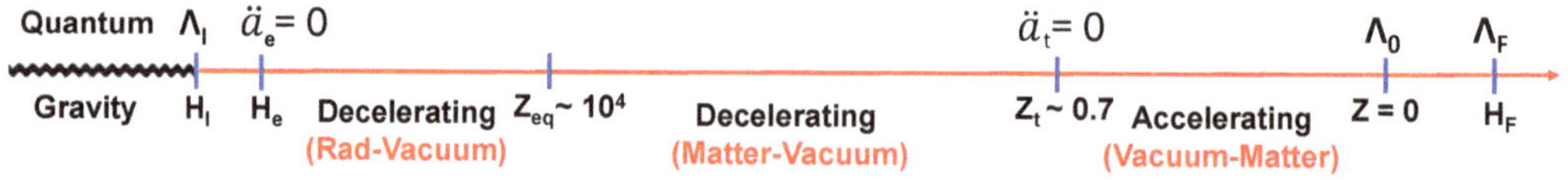

Figure 1. From de Sitter to de Sitter (out of visible scale). Just after the unknown quantum gravity regime, all the energy of the universe is concentrated on the emergent and unstable de Sitter spacetime ($\Lambda_I = 3H_I^2, \rho_{\Lambda_I} = 3M_P^2 H_I^2$, $\rho_M \equiv 0$). The classical evolution shown above is analytically described by the decaying $\Lambda(H)$ model, as given by Equation (13), which is the unique fuel of both inflationary stages. Since $\Lambda(H)$ is a continuous function, the equilibrium redshift z_{eq} is determined in the usual way (sudden transition), with a small contribution by the ν parameter, as discussed in Equation (32). For all allowed values of ν, the ultimate destiny of the model is the pure de Sitter vacuum state fixed by $\Lambda_F = 3H_F^2$. However, it is attained only in the distant future ($a \to \infty$), as shown in Equation (26). For $\nu = 0$, the late time evolution after $\ddot{a}(H_e) = 0$ is just the rigid vacuum ΛCDM model.

4. Cosmic Eras

4.1. From Initial Pure de Sitter Vacuum to Radiation-Vacuum Phase

In the primeval universe, the EoS parameter is $\omega = 1/3$, and the Hubble scale is restricted on the interval $H_I \geq H >> H_F$. In this limit, Equation (16) reduces to

$$\dot{H} + 2(1-\nu)H^2 \left[1 - \frac{H^2}{H_I^2}\right] = 0, \tag{17}$$

whose solution in terms of the scale factor reads

$$H(a) = \frac{H_I}{[1 + Ca^{4(1-\nu)}]^{1/2}}, \tag{18}$$

where $C = a_e^{-4(1-\nu)}/(1-2\nu)$ is an integration constant.

In the same limits for H given above, we see that the vacuum energy density reduces to

$$\rho_\Lambda(H) = 3\nu M_P^2 H^2 + 3(1-\nu)M_P^2 H^2 \left(\frac{H}{H_I}\right)^2, \tag{19}$$

and by inserting solution Equation (18) for $H(a)$ into the above expression, we obtain

$$\rho_\Lambda(a) = \rho_I \frac{1 + \nu C a^{4(1-\nu)}}{\left[1 + Ca^{4(1-\nu)}\right]^2}, \tag{20}$$

where $\rho_I = 3M_P^2 H_I^2 = \rho_\Lambda(H = H_I)$.

The radiation energy density can be obtained by using use the value of Ω_M, or even the Friedmann Equation **EFE** (2), and combining the result with the above $H(a)$ solution. It follows that

$$\rho_R(H) = 3(1-\nu)M_P^2 H^2 \left[1 - \left(\frac{H}{H_I}\right)^2\right], \tag{21}$$

$$\rho_R(a) = \rho_I \frac{(1-\nu)Ca^{4(1-\nu)}}{\left[1 + Ca^{4(1-\nu)}\right]^2}. \tag{22}$$

Note that all of the above equations are in agreement with the physical limits. In particular, for $H = H_I$, $\rho_R(H_I) \equiv 0)$, as expected for a pure de Sitter vacuum state. In this scenario, the classical instability of the initial de Sitter state, which is analytically described by Equation (18), is the unique source of the primeval thermal bath.

Let us now examine, in more detail, when this running $\Lambda(H)$ model performed its transition from the very early inflation dominated by the vacuum state to the radiation-vacuum phase. In principle, since $\omega = 1/3$, this must happen for a value of H_e (see also Figure 1), which is determined by the equality of the energy density of both components, i.e., $\rho_\Lambda(H_e) = \rho_R(H_e)$. From Equations (19) and (21), one finds

$$H_e = \sqrt{\frac{1-2\nu}{2(1-\nu)}}H_I. \tag{23}$$

The meaning of the H_e result can also be understood by taking the deceleration parameter into account. By definition,

$$q(H) = -\frac{\ddot{a}a}{\dot{a}^2} = -1 - \frac{\dot{H}}{H^2} = -1 + 2(1-\nu)\left[1 - \frac{H^2}{H_I^2}\right]. \tag{24}$$

where, in the last equality, we have used the 'H-equation of motion'. For ending inflation ($\ddot{a} = 0$) at $H = H_e$, the above equation yields $H_e = [(1-2\nu)/2(1-\nu)]^{1/2}H_I$, which is exactly the same value determined by the densities equality condition. Therefore, different from many adiabatic scalar field models, the value of H_e provides both the end of inflation and the beginning of the radiation dominated phase.[1]

It should also be remarked that some observations constrain the values of ν to be much smaller than unity [47]. This is an advantage of such models, since the evolution after entering the radiation-vacuum phase is not too drastically different from ΛCDM. The same occurs at low redshifts when $\ddot{a}(z_t) = 0$. In this connection, recent studies have also pointed out that a possible explanation to the current H_0 and σ_8 tensions require mildly deviations of ΛCDM both at early and late times [23,24,30].

4.2. From Radiation-Vacuum to Matter-Vacuum Dominated Phase

In this limit, we assume $\omega = 0$ and $H_F \leq H \ll H_I$, so that Equation (16) now boils down to

$$\dot{H} + \frac{3(1-\nu)H^2}{2}\left[1 - \frac{H_F^2}{H^2}\right] = 0, \tag{25}$$

whose solution is given by

$$H(a) = H_F[1 + Da^{-3(1-\nu)}]^{1/2} = H_F[1 + D(1+z)^{3(1-\nu)}]^{1/2}, \tag{26}$$

where the integration constant D is readily obtained computing the current value of Equation (26). For the sake of simplicity, we fix $a_0 \equiv 1$, and defining $H(a_0) \equiv H_0$, it follows that $D = (H_0/H_F)^2 - 1$. The above expression also shows that the ultimate value of the Hubble parameter (H_F) is attained only in the distant future, $a \to \infty$, or equivalently, $z \to -1$.

Note also that in the same limit, from Equation (13) and the first **EFE**, the vacuum and matter (dust) energy densities are easily obtained:

$$\rho_\Lambda(H) = 3(1-\nu)M_P^2 H_F^2 + 3\nu M_P^2 H^2, \tag{27}$$

$$\rho_M(H) = 3(1-\nu)M_P^2 H^2\left[1 - \left(\frac{H_F}{H}\right)^2\right]. \tag{28}$$

Now, let us determine the transition redshift, which generalises the time of radiation-matter equivalence of the rigid vacuum model. Here, we take it to be $t = t_e$, which is akin to $z(t_{eq}) = z_{eq}$. We will derive this result from the evolution of the temperature law in the presence of a decaying $\Lambda(H)$ which, for $H << H_I$ and $\Gamma = 3\nu H$, is given by [41]

$$T = T_0(1+z)^{1-\nu}. \tag{29}$$

At this time, the initial vacuum (H_I) is almost completely depleted, since the transition radiation-matter should occur at the end of the radiation phase. In addition, since $\dot{\sigma} = 0$ for an 'adiabatic' decaying vacuum, the equilibrium relations of the CMB are preserved, i.e., $\rho_R \propto T^4$ and $n_R \propto T^3$. This means that the radiation energy density and temperature are related by

$$\rho_R = \rho_{R0}\left(\frac{T}{T_0}\right)^4 = \rho_{R0}(1+z)^{4(1-\nu)} \tag{30}$$

where ρ_{R0} is the present day radiation energy density and in the second equality was used the temperature law Equation (29). In addition, by taking the equality between radiation and matter density, we obtain the transition redshift z_{eq},

$$\rho_{R0}(1+z_{eq})^{4(1-\nu)} = \rho_{M0}(1+z_{eq})^{3(1-\nu)}, \tag{31}$$

$$\Rightarrow z_{eq} \simeq \left(\frac{\Omega_{M0}}{\Omega_{R0}}\right)^{1/(1-\nu)}, \tag{32}$$

where, in the approximation, we have used that $z_{eq} >> 1$. For $\nu = 0$, and inserting the observational values, $\Omega_{M0} \approx 0.3$ and $\Omega_{R0} \sim 10^{-5}$, we obtain $z_{eq} \approx 3 \times 10^4$, which is the same result obtained in the case of the rigid vacuum, ΛCDM model. Hence, there is a small correction for $\nu \neq 0$ whose value may be determined by constraining ν from current observations.

At this point, it is interesting to know when the beginning of the second accelerating stage of the universe in the dust-vacuum phase happens, i.e., when $\ddot{a}_t = 0$ (see Figure 1). The deceleration parameter now reads

$$q(H) = -\frac{a\ddot{a}}{\dot{a}^2} = -1 + \frac{3(1-\nu)}{2}\left[1 - \frac{H_F^2}{H^2}\right], \tag{33}$$

so that $q(H) = 0$ leads to

$$H(\ddot{a}_t = 0) = H_F\left[\frac{3(1-\nu)}{1-3\nu}\right]^{1/2}. \tag{34}$$

Note also that the scale H_F can be obtained in terms of H_0 from the expression of Ω_M, as given by Equation (15). For $H << H_I$, we find

$$H_F = H_0\sqrt{\frac{\Omega_{\Lambda_0} - \nu}{1-\nu}}, \tag{35}$$

where the value of H_F in Equation (35) has been used. For small values of ν, it follows that the final de Sitter is nearly characterised by $H_F \simeq \sqrt{\Omega_{\Lambda_0}}H_0 \sim 0.83H_0$.

In the matter-vacuum dominated era, it is also convenient to rewrite the expression of the Hubble parameter Equation (26) in terms of the current observable quantities. One finds

$$H(\Omega_{M0}, z) = H_0\left[\frac{\Omega_{M0}}{1-\nu}(1+z)^{3(1-\nu)} + 1 - \frac{\Omega_{M0}}{1-\nu}\right]^{1/2}. \tag{36}$$

Note that, in the limit $z \to -1$, the same value of Equation (35) obtained by a different way is recovered. The above equation resembles the expression for the Hubble parameter in the rigid Λ_0CDM model, becoming exactly the same for $\nu = 0$. In addition, by defining $\Omega_{M0}^{eff} = \Omega_{M0}/(1-\nu)$, the above equation becomes

$$H(\Omega_{M0}, z) = H_0\left[\Omega_{M0}^{eff}(1+z)^{3(1-\nu)} + 1 - \Omega_{M0}^{eff}\right]^{1/2}. \tag{37}$$

4.3. Solving Some Cosmological Problems

Historically, the 'Big-Bang' singularity predicted by classical general relativity has been considered one of the most challenging cosmological problems. Since long ago, some authors suggested that a natural classical solution in the general relativistic domain comes out whether the universe starts like a de Sitter spacetime, and evolves to a radiation dominated phase (see also, e.g., [3]), similarly to what is proposed here. However, differently from several inflationary solutions, the present scenario does not need a super-cooling process followed by an extreme reheating mechanism in order to solve the horizon and flatness (entropy) problems. In particular, the so-called 'graceful exit' problem is absent (in this connection, see also [45]).

The cosmic evolution in the long run (H_I to H_F) also suggests an interesting perspective to the some cosmological constant problem. Particularly, the ratio between the extreme vacuum energy densities reads

$$\mathcal{R} = \frac{\rho_{\Lambda F}}{\rho_{\Lambda I}} \equiv \frac{\Lambda_F}{\Lambda_I} = \frac{H_F^2}{H_I^2}. \tag{38}$$

Hence, the Λ problem is now reduced to the ratio between two constant scales. From the above calculated result for H_F, and assuming two different values for the initial scale (say $H_I \sim M_P$ and $H_I \sim 10^{16}$ Gev, associated with the Grand Unified Theory (GUT) scale), the values of the above ratio are $\mathcal{R}_{(1)} = 10^{-122}$ and $\mathcal{R}_{(2)} = 10^{-119}$, respectively. Such results are in agreement with naïve expectations from quantum field theory. However, as remarked before, the correct value of the inflationary scale still needs an accurate observational determination from the B-mode power spectrum of CMB [48].

One may also advocate that the coincidence problem is also naturally solved in our scenario. Its naturalness arises because $\rho_v \simeq \rho$ for two different phases. Actually, at the end of the first inflationary phase when $H = H_e$ it is easily checked that $\rho_\Lambda = \rho_R$. Moreover, deep in the matter dominated phase, the second accelerating stage vacuum started when $\ddot{a} = 0$ with $\rho_\Lambda = 2\rho_M$. Hence, such 'coincidences' can also be seen as a kind of *necessity*, i.e., a natural consequence of an evolution between two extreme de Sitter solutions (i.e., the same actor) constrained by two definite scales.

It is also interesting to investigate whether this classical non-singular cosmology (from de Sitter to Sitter), driven by a running vacuum component, can also be described in terms of a scalar field. This point will be investigated in detail next section.

5. Running Vacuum Cosmology: A Non-Canonical Scalar Field Description

Let us now consider a pure scalar field ϕ minimally coupled with gravity. In the framework of general relativity, the total action for a non-canonical field can be written as

$$S = S_g + S_\phi = \int d^4x\sqrt{-g}\left[-\frac{M_p^2}{2}R + \mathcal{L}(X, \phi)\right], \tag{39}$$

where $\mathcal{L}(X, \phi)$ is a functional of the standard kinetic term $X = \frac{1}{2}\partial^\mu\phi\partial_\mu\phi \equiv \dot{\phi}^2/2$.

In what follows, it will be assumed that the dominant energy contribution in the primeval universe comes from the non-canonical scalar field, which is described by the Lagrangian [49]

$$\mathcal{L}(X, \phi) \equiv X\left(\frac{X}{M^4}\right)^{\beta-1} - V(\phi), \tag{40}$$

where the scale M has dimension of mass, $V(\phi)$ is the potential, and β a real positive number. The energy–momentum tensor (EMT), $T^{\mu}_{\nu} = (\frac{\partial \mathcal{L}}{\partial X})\partial^{\mu}\phi\partial_{\nu}\phi - \delta^{\mu}_{\nu}\mathcal{L}$, is diagonal, with components

$$\rho_{\phi} \equiv \rho_k + V(\phi) = (2\beta - 1)X\left(\frac{X}{M^4}\right)^{\beta-1} + V(\phi), \tag{41}$$

$$p_{\phi} \equiv p_k - V(\phi) = X\left(\frac{X}{M^4}\right)^{\beta-1} - V(\phi), \tag{42}$$

where ρ_k is a non-canonical kinetic term, and a dot means a time derivative. It is also easy to show that the following equation of motion drives the evolution of the non-canonical scalar field:

$$\ddot{\phi} + \frac{3H\dot{\phi}}{2\beta - 1} + \left(\frac{V'}{\beta(2\beta - 1)}\right)\left(\frac{2M^4}{\dot{\phi}^2}\right)^{\beta-1} = 0, \tag{43}$$

which can also be obtained from the total energy conservation law ($u_{\mu}T^{\mu\nu}_{;\nu} = 0$).

Let us now prove that a canonical scalar field cannot describe both the smooth dynamic and thermodynamics of a running $\Lambda(H)$-term (from de Sitter to de Sitter), as discussed in the previous sections.

5.1. The Failure of the Canonical Case

The generic non-canonical Lagrangian Equation (40) can also be seen as non-linear extension of the ordinary scalar field Lagrangian, which is readily recovered by taking $\beta = 1$ in Equation (40). One finds

$$\mathcal{L}(X, \phi) \equiv X - V(\phi) = \frac{1}{2}\dot{\phi}^2 - V(\phi). \tag{44}$$

As should be expected, the energy density (41) and pressure (42) of the field now reduces to

$$\rho_{\phi} \equiv \rho_k + V(\phi) = \frac{\dot{\phi}^2}{2} + V(\phi), \tag{45}$$

$$p_{\phi} \equiv p_k - V(\phi) = \frac{\dot{\phi}^2}{2} - V(\phi), \tag{46}$$

where ρ_k and p_k are, respectively, the kinetic energy density and pressure of the ordinary canonical scalar field.

On the other hand, it is also widely known that such a scalar field can be interpreted as a mixture of *two interacting perfect fluids* with different equations of state, namely a material Zeldovich's stiff fluid [50], where ($\rho_k = p_k = \dot{\phi}^2/2$), plus a pure vacuum ($\Lambda$-term) obeying an EoS $\rho_v = V(\phi) = -p_v$ (see, e.g., ref. [51] for a more detailed discussion with examples). Therefore, as happens in the running vacuum model, by assuming that both components are gravitationally coupled, in principle, there will be no local energy–momentum conservation for each perfect fluid component separately. Only the total energy–momentum tensor of the system as a whole is conserved.

Nevertheless, under such an interpretation, whether the potential of the unstable vacuum dominates (maximum scale H_I) and decays spontaneously through a non-adiabatic decaying process conserving the specific entropy (as discussed in Section 3), the energy stored in the field will be transferred to a stiff component. In this way, the standard thermal bath formed by ultra-relativistic particles ($p_k = \rho_k/3$) is not generated. An inevitable consequence is that a new phase transition needs to be hypothesised (perhaps through a new coupling term), thereby forcing the formation of a thermal bath, with the universe finally entering in the standard radiation phase. Naturally, in this case, the two-fluid description without additional assumption is unable to faithfully describe the complete scenario driven by the running vacuum cosmology.

Such a scenario will happen naturally when the coherent field oscillation phase is absent, with the vacuum field (Λ term) continuously approaching its final, very low (but finite!) value, which is defined by the scale H_F. As we shall see next, such a picture allow us to represent an evolution driven by the same scalar field (from de Sitter to de Sitter) based on a non-canonical scalar field, and mimicking the running vacuum term, as discussed in previous sections.

5.2. The Non-Canonical Case

Let us now consider the generic non-standard case. To begin with, we remark from Equations (41) and (42) that the general EoS parameter for the kinetic term (the material fluid) can be defined as

$$\omega \equiv \frac{p_k}{\rho_k} = \frac{1}{2\beta - 1}. \tag{47}$$

Therefore, based on the two-fluid interpretation, the kinetic term for $\beta = 2$ exactly obeys the radiation EoS.[2] Thus, by assuming, as before, that the classical universe emerged from the quantum gravity regime as a de Sitter-like solution with scale H_I (see Figure 1), this means that $\rho_k(H_I) = p_k(H_I) = 0$, and the unstable potential $V(\phi)$ must decay directly in ultra-relativistic particles. Interestingly, the inflationary process is not adiabatic, since the primeval thermal bath is a consequence of the energy transfer from the potential $V(\phi(H))$ or, equivalently, the $\Lambda(H)$-term to the radiation-created component (see Section 3). In a certain sense, this inflationary scenario resembles some variants of the warm inflationary process, because the primeval accelerating expansion is not adiabatic. However, it is somewhat different, because the whole thermal bath here is a consequence of the running vacuum process (de Sitter instability), and not related to an initial singular state, as assumed in some warm inflationary scenarios [54–57].

Now, let us demonstrate that the general equation of motion for the running $\Lambda(H)$ cosmology Equation (16) is reproduced by the non-canonical scalar field based on the two-fluid interpretation, as proposed above. In this case, the **EFE** Equations (2) and (3) now take the form

$$\rho_k + V(\phi) = 3M_P^2 H^2\,, \tag{48}$$

$$p_k - V(\phi) = -M_P^2\left[2\dot{H} + 3H^2\right], \tag{49}$$

where ρ_k and p_k are the kinetic energy density and pressure of the fluid satisfying the EoS (47), while $V(\phi)$ must be determined in such a way that $\Lambda(H(\phi)) = M_P^{-2}V(\phi)$. Interestingly, the above equations imply that the cosmic dynamics is also driven by

$$\dot{H} + \frac{3(1+\omega)}{2}H^2[1 - \Omega_\Lambda(H)] = 0\,, \tag{50}$$

where the equation of state $p = \omega\rho$ ($\omega \geq 0$) was used, and $\Omega_\Lambda(H) = V(H(\phi))/3M_P^2H^2$ is the vacuum density parameter. The above equation should be compared with (5) in Section 2.

Once the two-fluid interpretation is assumed, the 'H-equation of motion' is independent of the particular expression of the potential. Nevertheless, an expression for $V(\phi)$ needs to be given, and the scalar field solution properly derived.

Inspired by the derived expression (13) for $\Lambda(H)$, let us discuss the potential contributions emulating the different cosmic phases, starting from the primeval de Sitter state.

5.3. From de Sitter to the Radiation-Vacuum Dominated Era

To begin with, let us consider $\beta = 2$. The energy density and pressure of the non-canonical scalar field becomes

$$\rho_\phi \equiv \rho_k + V(\phi) = \frac{3\dot{\phi}^4}{4M^4} + V(\phi), \tag{51}$$

$$p_\phi \equiv \frac{1}{3}\rho_k - V(\phi) = \frac{\dot{\phi}^4}{4M^4} - V(\phi), \tag{52}$$

with the kinetic component emulating the radiation EoS. Now, in order to describe the initial de Sitter configuration and its transition to a radiation dominated phase, let us also assume the following scalar field potential:

$$V(\phi) = V_I \frac{1+\nu(\alpha\phi)^4}{[1+(\alpha\phi)^4]^2}, \quad \phi \geq 0. \tag{53}$$

where $V_I = V(\phi = 0) = 3M_P^2 H_I^2$ and the constant α is a dimensional parameter, $[\alpha] = [\phi]^{-1}$. Its value is calculated in Appendix C

$$\alpha = \frac{\left[(1-2\nu)(1-\nu)^3\right]^{1/4}}{M}\sqrt{\frac{H_I}{2M_P}}. \tag{54}$$

In order to obtain the expression for the radiation energy density, let us assume the *ansatz* $\rho_k = (1-\nu)(\alpha\phi)^4 V(\phi)/[1+\nu(\alpha\phi)^4]$. Inserting this into Equation (51), we obtain

$$\rho_k(\phi) = V_I \frac{(1-\nu)(\alpha\phi)^4}{[1+(\alpha\phi)^4]^2}, \tag{55}$$

and

$$H(\phi) = \frac{1}{(1-\nu)}\frac{\dot{\phi}}{\phi} = \frac{H_I}{[1+(\alpha\phi)^4]^{1/2}}. \tag{56}$$

Now, by inserting the expression for $H(\phi)$ above into Equations (53) and (55), one can show that

$$V(H(\phi)) = 3\nu M_P^2 H^2 + 3(1-\nu)M_P^2 H^2 \left(\frac{H}{H_I}\right)^2, \tag{57}$$

$$\rho_k = 3(1-\nu)M_P^2 H^2 \left[1 - \left(\frac{H}{H_I}\right)^2\right], \tag{58}$$

thereby recovering the expressions (19) and (21) shown in Section 4. Hence, the proposed potential Equation (53) emulates completely the cosmological dynamics of the vacuum-radiation dominated era Equation (17). In particular, we see that there is no a thermal bath for $H = H_I$ since, in this case, $\rho_k(H_I) = 0$.

In Figure 2, we display the evolution of the basic dimensionless quantities, namely the potential scalar field and the kinetic term (ρ_k). The latter one describes the created component forming the thermal bath of effectively massless particles $(\omega = 1/3)$. As one may check, the transition from vacuum to the radiation-vacuum phase is also determined by the condition $\rho_k = V(\phi)$ [see discussion below Equation (24)].

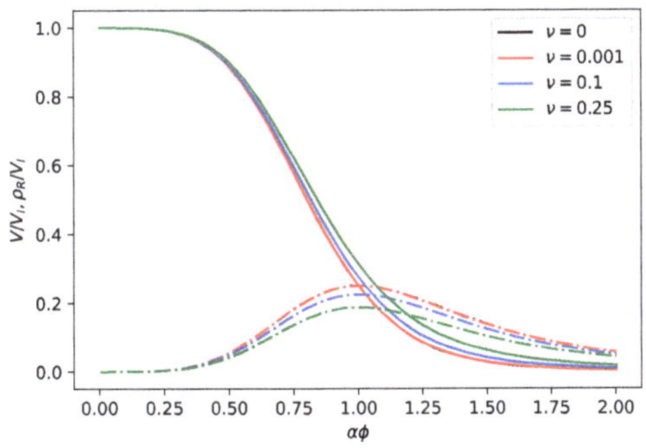

Figure 2. Dimensionless potential and kinetic energy (radiation) density. The solid lines describe the evolution of the potential with respect to $\alpha\phi$ and some selected values of the parameter ν, depicted in the inner rectangle. The dashed lines display the same behaviour for the radiation energy density. In the beginning of the classical evolution, we have only an emergent de Sitter universe for all values of ν ($V = V_I, \rho_R = 0$). Notice that all of the created particles are the leftover of the primeval de Sitter vacuum state ($V = V_I$) for $H = H_I$ and $\phi = 0$. For $\phi = \alpha^{-1}$, $\rho_R = V(\alpha\phi)$, $H = H_e$, and inflation ends [see also Equation (23) and Figure 1].

5.4. *From Matter-Vacuum to the Final de Sitter Era*

In this case, $\omega = 0$, which means that $\beta >> 1$ in Equation (47). Now, based on the expression (13) for $\Lambda(H)$, we propose the following potential $V(\phi(H))$ to the matter dominated era:

$$V(\phi) = V_F\left[1 + \nu(\sigma\phi)^{-3}\right], \tag{59}$$

where $V(\phi) = 3Mp^2H^2$, and σ is also a dimensional parameter as before, $[\sigma] \equiv [\phi]^{-1}$. As a function of the scales H_F and M, and the parameter ν, it is given by (see Appendix C)

$$\sigma = (1-\nu)\frac{H_F}{M^2}. \tag{60}$$

In order to obtain the expression for the matter energy density, we will again consider a similar *ansatz* of the transition from de Sitter (H_I) to the radiation-vacuum dominated phase, namely $\rho_k = (1-\nu)(\sigma\phi)^{-3}V(\phi)/[1+\nu(\sigma\phi)^{-3}]$. Now, keeping this in mind, it is easy to show that

$$\rho_k(\phi) = (1-\nu)V_F(\sigma\phi)^{-3}, \tag{61}$$

and

$$H(\phi) = \frac{1}{(1-\nu)}\frac{\dot{\phi}}{\phi} = H_F\left[1 + (\sigma\phi)^{-3}\right]^{1/2}. \tag{62}$$

By substituting the expression for $H(\phi)$ above in both Equations (59) and (61), one can show that

$$V(\phi(H)) = 3(1-\nu)M_P^2H_F^2 + 3\nu M_P^2H^2, \tag{63}$$

$$\rho_k = 3(1-\nu)M_P^2H^2\left[1 - \left(\frac{H_F}{H}\right)^2\right], \tag{64}$$

recovering the expressions (27) and (28) shown in Section 4. As it happens for the early universe, the potential (59) recovers the cosmological dynamics of the matter-vacuum era (25). Figure 3 shows the time behaviour of the basic dimensionless quantities as defined in Figure 2.

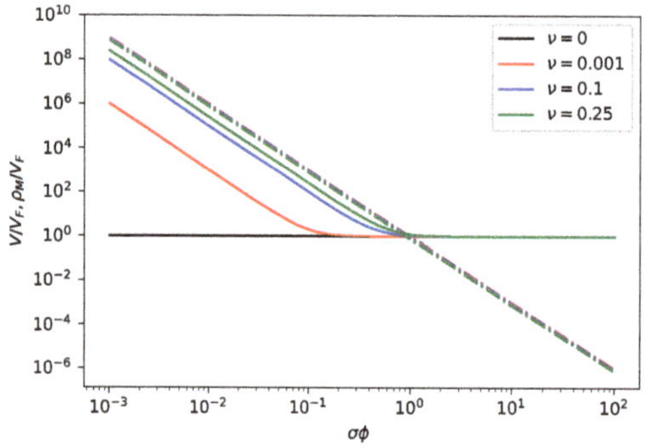

Figure 3. Dimensionless potential and matter energy density. As in Figure 2, different solid lines describe the evolution of the potential with respect to $\sigma\phi$ for some selected values of the parameter ν, depicted in the inner rectangle. The dashed lines display the same behaviour for the dimensionless matter energy density. Notice that, for all allowed values of ν, the model evolves in the long run to a de Sitter spacetime, described by the scale H_F [see also Equation (59) and Figure 1].

6. Final Remarks and Conclusions

Entropy is the quantity which allows us to describe the progress of non-equilibrium dissipative processes. A special ingredient of our scenario discussed here is that it takes into account the second law of thermodynamics from the very beginning. For instance, the non-equilibrium thermodynamics was explicitly used to establish an integral relation uniting $\Lambda(H)$ and the particle creation rate $\Gamma(H)$. The running vacuum emerges from an unknown quantum gravity regime as a primeval de Sitter vacuum. Due to a structural thermal instability, it transfers irreversibly energy, particles, and entropy to the spacetime, forming the primeval thermal bath at the expenses of its own energy density at scale, which is initially defined by H_I, which is smaller than the reduced Planck mass. In this connection, see also [45], where the dynamic evolution of this phase forming the thermal bath was investigated for $\nu = 0$.

Subsequently, the evolution departs slightly from the standard radiation- and matter-dominated phases sustained by a softer generation of entropy. Since the scale H_F is incredibly low, for all practical purposes, the final vacuum state will be eternal.

There are also two interesting aspects of the scenario proposed here. Firstly, the irreversible evolution of the running vacuum generating a complete cosmology can also be described by a single minimally coupled non-canonical scalar field when it is interpreted as a mixture of two interacting perfect fluids, as suggested long ago (see Section 5). Secondly, all of the energy scales appearing in the potential, and also in the non-canonical kinetic term, are sub-Planckian with no fine-tuning.

Finally, we stress that the mild deviation from the rigid (i.e., non-dynamical) vacuum model (ΛCDM) may also provide an explanation to the current H_0 and σ_8 tensions, perhaps simpler than other phenomenological dark energy interacting models. Studies along these lines are in progress and, naturally, still deserves a closer scrutiny.

Author Contributions: P.E.M.A. Investigation, calculations and Figures 2 and 3. R.C.S. Investigation, calculations and Figure 1. J.A.S.L. Conceptualization, methodology and writing. All authors have read and agreed to the published version of the manuscript.

Funding: This research received no external founding.

Data Availability Statement: No data were generated.

Acknowledgments: P.E.M.A. is grateful to a PhD fellowship from CAPES and J.A.S.L. is partially supported by CNPq under grant 310038/2019-7, CAPES (88881.068485/2014), and FAPESP (LLAMA Project No. 11/51676-9).

Conflicts of Interest: The authors declare no conflicts of interest.

Appendix A. "Adiabatic" Creation ($\dot{\sigma} = 0$) and Some Consequences

In this appendix, we demonstrate that the specific entropy (per particle) in our approach remains constant ($\dot{\sigma} = 0$), and also the validity of Equation (9).

The specific entropy is defined by

$$\sigma \equiv \frac{s}{n} \Rightarrow \dot{\sigma} = \sigma\left(\frac{\dot{s}}{s} - \frac{\dot{n}}{n}\right). \tag{A1}$$

From Equations (7) and (8), one may check that

$$\frac{\dot{n}}{n} = \frac{\dot{s}}{s} = \Gamma - 3H, \tag{A2}$$

and, therefore, $\dot{\sigma} = 0$.

Finally, by defining the expressions for the comoving number of particles, $N = na^3$, and entropy of the fluid, $S = sa^3$, it is straightforward to compute the time derivatives:

$$\frac{\dot{N}}{N} = \frac{1}{na^3}\frac{d(na^3)}{dt} = \frac{\dot{n}}{n} + 3H = \Gamma, \tag{A3}$$

$$\frac{\dot{S}}{S} = \frac{1}{sa^3}\frac{d(sa^3)}{dt} = \frac{\dot{s}}{s} + 3H = \Gamma, \tag{A4}$$

where the last equality in both equations comes from (A2). Both results were summarised in Equation (9). Note that, although depending on the symmetries of the FLRW geometry, Equation (9) is independent of Einstein's field equations.

Appendix B. Determination of γ Parameter and the Constant B

Let us now consider a quadratic correction in the creation rate $\Gamma(H) = 3\nu H$ (see Section 3):

$$\Gamma(H) = 3\nu H\left[1 + \frac{\gamma}{\nu}(H/H_I)^2\right], \tag{A5}$$

where the numerical factors are for mathematical convenience. In this case, a simple integration of (11) provides an expression for $\Lambda(H)$ in terms of the constant B, and also the set of free parameters (H_I, ν) appearing above. We find

$$\Lambda(H) = B + 3\nu H^2 + \frac{3\gamma H^2}{2}\left(\frac{H}{H_I}\right)^2. \tag{A6}$$

At low redshifts ($H << H_I$), the last term on the *r.h.s.* is negligible, and by using the late time boundary condition $\Lambda(H_F) = 3H_F^2$, a value $B = 3(1-\nu)H_F^2$ is obtained. In the opposite extreme, $H >> H_F$, the constant B is negligible, and using de Sitter condition at early times, $\Lambda(H_I) = 3H_I^2$, it follows that $\gamma = 2(1-\nu)$. Finally, by inserting such values, the result (13) is recovered.

Appendix C. Determination of α and σ Parameters

In this appendix, we determine the values of α and σ parameters appearing in the expressions of the scalar field potential to the vacuum-radiation and matter-vacuum phases. To begin with, we first consider the part of the derivation that is common for both quantities. From Einstein field Equations (48) and (49), the kinetic term of the non-canonical scalar field energy density can be written as

$$\rho_k = -\frac{2M_P^2\dot{H}}{(1+\omega)}, \tag{A7}$$

where the equality is also a consequence of the kinetic equation of state (47). Let us now consider each case separately.

(i) α parameter. By taking $\omega = 1/3$, and using the definition of ρ_k from (51), the above equation implies that $\dot{\phi}$ can be written as

$$\dot{\phi}^4 = -2M^4 M_P^2 \dot{H}, \tag{A8}$$

and replacing $\dot{H}$ from the cosmological 'H-equation of motion' (50), we obtain

$$\dot{\phi} = \sqrt{2}(1-\nu)^{1/4} M M_P^{1/2} H^{1/2} \left[1 - \frac{H^2}{H_I^2}\right]^{1/4}, \tag{A9}$$

where we have discarded $\dot{\phi} < 0$. Actually, $\phi \geq 0$ and its value is always increasing. Now, remembering that $H = \dot{\phi}/[(1-\nu)\phi]$ (56), the above equation can be rewritten as

$$\phi = \frac{\sqrt{2} M M_P^{1/2}}{(1-\nu)^{3/4} H^{1/2}} \left[1 - \frac{H^2}{H_I^2}\right]^{1/4}. \tag{A10}$$

Hence, recalling that $\phi_e = \alpha^{-1}$, and that at the end of inflation the Hubble parameter is $H(\phi_e) = H_I\sqrt{1 - 2\nu/2(1-\nu)}$, the above equation at $\phi = \phi_e$ becomes

$$\phi_e = \alpha^{-1} = \frac{\sqrt{2} M M_P^{1/2}}{(1-\nu)^{3/4} H_e^{1/2}} \left[1 - \frac{H_e^2}{H_I^2}\right]^{1/4} \tag{A11}$$

$$= \sqrt{2}\left[\frac{1}{(1-\nu)^3(1-2\nu)}\right]^{1/4} M \sqrt{\frac{M_P}{H_I}}. \tag{A12}$$

so that the expression for α is given by

$$\alpha = \frac{1}{M}\left[(1-\nu)^3(1-2\nu)\right]^{1/4} \sqrt{\frac{H_I}{2M_P}}. \tag{A13}$$

Note that beyond the reduced Planck mass, α depends on the relevant free scales of the two vacuum-radiation phase (H_I, M). This means that a large class of sub-Planckian values are accessible, even assuming $H_I = M_p$.

(ii) σ parameter: In this case, $\beta >> 1$ and $\omega = 0$ while the expression of $V(\phi)$ in the matter phase is given by (59). Now, following the same procedure of the previous case, we see from (A7) that

$$\rho_k = -2M_P^2 \dot{H} \;\Rightarrow\; \dot{\phi}^{2\beta} = -\frac{2^{(1+\beta)} M^{4(\beta-1)} M_P^2 \dot{H}}{2\beta - 1}. \tag{A14}$$

Let us rewrite $\dot{\phi}$ using the fact that $2\beta - 1 = 1/\omega$ (47):

$$\dot{\phi}^{\frac{1+\omega}{\omega}} = -\omega 2^{\frac{1+3\omega}{2\omega}} M_P^2 M^{4\left(\frac{1-\omega}{2\omega}\right)} \dot{H}. \tag{A15}$$

Substituting the cosmological 'H-equation of motion' for the matter dominated phase (25) in the expression above, one can write

$$\dot{\phi}^{\frac{1+\omega}{\omega}} = 3\omega(1-\nu) 2^{\frac{1+\omega}{2\omega}} M_P^2 M^{4\left(\frac{1-\omega}{2\omega}\right)} H^2 \left[1 - \left(\frac{H_F}{H}\right)^2\right]. \tag{A16}$$

In the same way we performed for the determination of α, we now utilise the relation between H and ϕ, and evaluate the resulting expression at $\phi = \sigma^{-1}$

$$\sigma^{-\frac{1+\omega}{\omega}} = \frac{3\omega . 2^{\frac{1+\omega}{2\omega}} M_P^2 M^{4\left(\frac{1-\omega}{2\omega}\right)}}{(1-\nu)^{\frac{1}{\omega}}} H_\sigma^{\frac{\omega-1}{\omega}} \left[1 - \left(\frac{H_F}{H_\sigma}\right)^2\right]. \tag{A17}$$

From Equation (62) one can see that $H(\phi = \sigma^{-1}) \equiv H_\sigma = \sqrt{2}H_F$. Thus, the expression for σ is given by

$$\sigma = \frac{1}{(3\omega M_P^2)^{\frac{\omega}{1+\omega}}} \left[\frac{1-\nu}{M^{2(1-\omega)} H_F^{\omega-1}}\right]^{\frac{1}{1+\omega}}. \tag{A18}$$

In the limit $\omega \to 0$, we obtain

$$\sigma = (1-\nu)\frac{H_F}{M^2}, \tag{A19}$$

which, differently from α, does not depend on the initial de Sitter scale H_I.

Notes

1 Such a result is not surprising because, from **EFE**, one finds $\ddot{a} = -(6M_P^2)^{-1}[(1+3\omega)\rho - 2\rho_\Lambda]a$. Thus, for $\omega = 1/3$, $\ddot{a} = 0$ implies $\rho \equiv \rho_R = \rho_\Lambda$ while for $\omega = 0$, one finds $\rho \equiv \rho_M = 2\rho_\Lambda$.

2 This kind of single-fluid EoS can also be obtained from a k-essence field with Lagrangian $L(g(\phi), X) = g(\phi)F(X)$ [52]. However, as remarked in [53], a two-fluid description with dissipation is possible only if another interacting k-essence is considered.

References

1. Gliner, E.B. Algebraic Properties of the Energy-momentum Tensor and Vacuum-like States of Matter. *Sov. J. Exp. Theor. Phys.* **1966**, *22*, 378.
2. Sakharov, A.D. The Initial Stage of an Expanding Universe and the Appearance of a Nonuniform Distribution of Matter. *Sov. J. Exp. Theor. Phys.* **1966**, *22*, 241.
3. Murphy, G.L. Big-Bang Model without Singularities. *Phys. Rev. D* **1973**, *8*, 4231–4233. [CrossRef]
4. Starobinsky, A.A. A new type of isotropic cosmological models without singularity. *Phys. Lett. B* **1980**, *91*, 99–102. [CrossRef]
5. Anderson, P. Effects of quantum fields on singularities and particle horizons in the early universe. *Phys. Rev. D* **1983**, *28*, 271–285. [CrossRef]
6. Anderson, P.R. Effects of quantum fields on singularities and particle horizons in the early universe. II. *Phys. Rev. D* **1984**, *29*, 615–627. [CrossRef]
7. Birrell, N.D.; Davies, P.C.W. *Quantum Fields in Curved Space*; Cambridge Monographs on Mathematical Physics; Cambridge University Press: Cambridge, UK, 1982. [CrossRef]
8. Vilenkin, A. Quantum origin of the universe. *Nucl. Phys. B* **1985**, *252*, 141–152. [CrossRef]
9. Mottola, E. Thermodynamic instability of de Sitter space. *Phys. Rev. D* **1986**, *33*, 1616–1621. [CrossRef]
10. Prigogine, I.; Geheniau, J.; Gunzig, E.; Nardone, P. Thermodynamics of cosmological matter creation. *Proc. Natl. Acad. Sci. USA* **1988**, *85*, 7428–7432. [CrossRef]
11. Prigogine, I.; Geheniau, J.; Gunzig, E.; Nardone, P. Thermodynamics and cosmology. *Gen. Relat. Gravit.* **1989**, *21*, 767–776. [CrossRef]
12. Calvão, M.O.; Lima, J.A.S.; Waga, I. On the thermodynamics of matter creation in cosmology. *Phys. Lett. A* **1992**, *162*, 223–226. [CrossRef]
13. Lima, J.A.S.; Calvao, M.O.; Waga, I. Cosmology, Thermodynamics and Matter Creation. *arXiv* **2007**, arXiv:0708.3397.
14. Lima, J.A.S.; Germano, A. On the equivalence of bulk viscosity and matter creation. *Phys. Lett. A* **1992**, *170*, 373–378. [CrossRef]
15. Weinberg, S. The cosmological constant problem. *Rev. Mod. Phys.* **1989**, *61*, 1–23. [CrossRef]
16. Velten, H.E.S.; vom Marttens, R.F.; Zimdahl, W. Aspects of the cosmological "coincidence problem". *Eur. Phys. J. C* **2014**, *74*, 3160. [CrossRef]
17. Riess, A.G.; Casertano, S.; Yuan, W.; Macri, L.M.; Scolnic, D. Large Magellanic Cloud Cepheid Standards Provide a 1% Foundation for the Determination of the Hubble Constant and Stronger Evidence for Physics beyond ΛCDM. *Astrophys. J.* **2019**, *876*, 85. [CrossRef]
18. Cunha, J.V.; Marassi, L.; Lima, J.A.S. Constraining H0 from the Sunyaev–Zel'dovich effect, galaxy cluster X-ray data and baryon oscillations. *Mon. Not. R. Astron. Soc. Lett.* **2007**, *379*, L1–L5. [CrossRef]
19. Lima, J.A.S.; Cunha, J.V. A 3% Determination of H_0 at Intermediate Redshifts. *Astrophys. J. Lett.* **2014**, *781*, L38. [CrossRef]

20. Birrer, S.; Treu, T.; Rusu, C.E.; Bonvin, V.; Fassnacht, C.D.; Chan, J.H.H.; Agnello, A.; Shajib, A.J.; Chen, G.C.; Auger, M.; et al. H0LiCOW—IX. Cosmographic analysis of the doubly imaged quasar SDSS 1206+4332 and a new measurement of the Hubble constant. *Mon. Not. R. Astron. Soc.* **2019**, *484*, 4726–4753. [CrossRef]
21. Aghanim, N.; Akrami, Y.; Ashdown, M.; Aumont, J.; Baccigalupi, C.; Ballardini, M.; Banday, A.J.; Barreiro, R.B.; Bartolo, N.; Basak, S.; et al. Planck 2018 results: VI. Cosmological parameters. *Astron. Astrophys.* **2020**, *641*, A6. [CrossRef]
22. Hildebrandt, H.; Köhlinger, F.; Van den Busch, J.L.; Joachimi, B.; Heymans, C.; Kannawadi, A.; Wright, A.H.; Asgari, M.; Blake, C.; Hoekstra, H.; et al. KiDS+VIKING-450: Cosmic shear tomography with optical and infrared data. *Astron. Astrophys.* **2020**, *633*, A69. [CrossRef]
23. Di Valentino, E.; Anchordoqui, L.A.; Akarsu, O.; Ali-Haimoud, Y.; Amendola, L.; Arendse, N.; Asgari, M.; Ballardini, M.; Basilakos, S.; Battistelli, E.; et al. Cosmology Intertwined III: $f\sigma_8$ and S_8. *Astropart. Phys.* **2021**, *131*, 102604. [CrossRef]
24. Lucca, M. Multi-interacting dark energy and its cosmological implications. *Phys. Rev. D* **2021**, *104*, 083510. [CrossRef]
25. Carneiro, S.; Tavakol, R. On vacuum density, the initial singularity and dark energy. *Gen. Relat. Gravit.* **2009**, *41*, 2287. [CrossRef]
26. Lima, J.A.S.; Basilakos, S.; Solà, J. Expansion history with decaying vacuum: A complete cosmological scenario. *Mon. Not. R. Astron. Soc.* **2013**, *431*, 923–929. [CrossRef]
27. Perico, E.L.D.; Lima, J.A.S.; Basilakos, S.; Solà, J. Complete cosmic history with a dynamical $\Lambda(H)$ term. *Phys. Rev. D* **2013**, *88*, 063531. [CrossRef]
28. Lima, J.A.S.; Basilakos, S.; Solà, J. Nonsingular decaying vacuum cosmology and entropy production. *Gen. Relat. Gravit.* **2015**, *47*, 40. [CrossRef]
29. Solà, J.; Gómez-Valent, A.; de Cruz Pérez, J. Hints of dynamical vacuum energy in the expanding universe. *Astrophys. J.* **2015**, *811*, L14. [CrossRef]
30. Gómez-Valent, A.; Solà, J. Relaxing the σ_8 tension through running vacuum in the Universe. *EPL (Europhys. Lett.)* **2017**, *120*, 39001. [CrossRef]
31. Solà, J.; Gómez-Valent, A.; de Cruz Pérez, J. The H_0 tension in light of vacuum dynamics in the universe. *Phys. Lett. B* **2017**, *774*, 317–324. [CrossRef]
32. Zilioti, G.J.M.; Santos, R.C.; Lima, J.A.S. From de Sitter to de Sitter: Decaying vacuum models as a possible solution to the main cosmological problems. *Adv. High Energy Phys.* **2018**, *2018*, 6980486. [CrossRef]
33. Özer, M.; Taha, M.O. A possible solution to the main cosmological problems. *Phys. Lett. B* **1986**, *171*, 363–365. [CrossRef]
34. Özer, M.; Taha, M.O. A model of the universe free of cosmological problems. *Nucl. Phys. B* **1987**, *287*, 776–796. [CrossRef]
35. Freese, K.; Adams, F.C.; Frieman, J.A.; Mottola, E. Cosmology with decaying vacuum energy. *Nucl. Phys. B* **1987**, *287*, 797–814. [CrossRef]
36. Chen, W.; Wu, Y.S. Implications of a cosmological constant varying as R^{-2}. *Phys. Rev. D* **1990**, *41*, 695–698. [CrossRef]
37. Abdel-Rahman, A.M.M. Singularity-free decaying-vacuum cosmologies. *Phys. Rev. D* **1992**, *45*, 3497–3511. [CrossRef] [PubMed]
38. Carvalho, J.C.; Lima, J.A.S.; Waga, I. Cosmological consequences of a time-dependent Λ term. *Phys. Rev. D* **1992**, *46*, 2404–2407. [CrossRef]
39. Waga, I. Decaying Vacuum Flat Cosmological Models: Expressions for Some Observable Quantities and Their Properties. *Astrophys. J.* **1993**, *414*, 436. [CrossRef]
40. Silveira, V.; Waga, I. Decaying Lambda cosmologies and power spectrum. *Phys. Rev. D* **1994**, *50*, 4890–4894. [CrossRef]
41. Lima, J.A.S. Thermodynamics of decaying vacuum cosmologies. *Phys. Rev. D* **1996**, *54*, 2571–2577. [CrossRef]
42. Lima, J.A.S.; Silva, A.I.; Viegas, S.M. Is the radiation temperature-redshift relation of the standard cosmology in accordance with the data? *Mon. Not. R. Astron. Soc.* **2000**, *312*, 747–752. [CrossRef]
43. Lima, J.A.S.; Trevisani, S.; Santos, R. Cosmic "adiabatic" photon creation: Temperature law and blackbody spectrum. *Phys. Lett. B* **2021**, *820*, 136575. [CrossRef]
44. Landau, L.D.; Lifshitz, E.M. *Fluid Mechanics*, 2nd ed.; Course of Theoretical Physics; Pergamon: Oxford, UK, 1987; Volume 6.
45. Lima, J.A.S.; Almeida, P.E.M. Noncanonical scalar fields and the primeval thermal bath. *Int. J. Mod. Phys. D* **2021**, *30*, 2142025. [CrossRef]
46. Shapiro, I.L.; Solà, J. The scaling evolution of the cosmological constant. *J. High Energy Phys.* **2002**, *2002*, 006. [CrossRef]
47. Basilakos, S.; Plionis, M.; Solà, J. Hubble expansion and structure formation in time varying vacuum models. *Phys. Rev. D* **2009**, *80*, 083511. [CrossRef]
48. Kamionkowski, M.; Kovetz, E.D. The Quest for B Modes from Inflationary Gravitational Waves. *Annu. Rev. Astron. Astrophys.* **2016**, *54*, 227–269. [CrossRef]
49. Unnikrishnan, S.; Sahni, V.; Toporensky, A. Refining inflation using non-canonical scalars. *J. Cosmol. Astropart. Phys.* **2012**, *2012*, 018. [CrossRef]
50. Zel'dovich, Y.B. The equation of state at ultrahigh densities and its relativistic limitations. *Zh. Eksp. Teor. Fiz.* **1961**, *41*, 1609–1615.
51. Chimento, L.P. Symmetry and inflation. *Phys. Rev. D* **2002**, *65*, 063517. [CrossRef]
52. Socorro, J.; Rosales, J.J. Quantum Fractionary Cosmology: K-Essence Theory. *Universe* **2023**, *9*, 185. [CrossRef]
53. Arroja, F.; Sasaki, M. Note on the equivalence of a barotropic perfect fluid with a k-essence scalar field. *Phys. Rev. D* **2010**, *81*, 107301. [CrossRef]
54. Berera, A.; Fang, L.Z. Thermally Induced Density Perturbations in the Inflation Era. *Phys. Rev. Lett.* **1995**, *74*, 1912–1915. [CrossRef] [PubMed]

55. Berera, A. Warm Inflation. *Phys. Rev. Lett.* **1995**, *75*, 3218–3221. [CrossRef] [PubMed]
56. Berera, A. Thermal properties of an inflationary universe. *Phys. Rev. D* **1996**, *54*, 2519–2534. [CrossRef]
57. Maia, J.M.F.; Lima, J.A.S. Extended warm inflation. *Phys. Rev. D* **1999**, *60*, 101301. [CrossRef]

universe

MDPI

Article

Constraint on the Cosmic Curvature in a Model with the Schwarzschild–de Sitter Metric from Supernovae and Gamma-Ray Burst Observational Data

Vladimir N. Yershov

Institute of Radio Engineering and Telecommunication Technologies, State University of Aerospace Instrumentation, 67 Bol'shaya Morskaya, 190000 St. Petersburg, Russia; vyershov@guap.ru

Abstract: In developing his cosmological model of 1917, de Sitter theoretically predicted the phenomenon of cosmological redshift (the de Sitter effect), which he did long before the discovery of this phenomenon in observations. The de Sitter effect is gravitational by its nature, as it is due to differences between the coordinate systems of the observer and the distant source. However, the relationship between the redshift and distance derived from the de Sitter metric is at odds with observations, since this relationship is nonlinear (quadratic) for small redshifts, while the observed relationship between the same quantities is strictly linear. This paper discusses the possibility that cosmological redshift is gravitational by its nature, as in de Sitter's 1917 model. At the same time, here, as in de Sitter's model, an elliptical space is used, the main characteristic of which is the identification of its antipodal points. But, unlike de Sitter's model, here, in order to ensure strict linear dependence of the redshift on distance, the origin of the reference system is transferred to the observer's antipodal point. The Schwarzschild–de Sitter metric used in this model allows you to estimate the curvature of space from observational data. To achieve this, a theoretical Hubble diagram is built within the framework of the model with the Schwarzschild–de Sitter metric, which is compared with observations from the *Pantheon+* catalogue of type Ia supernovae and the Amati catalogue of gamma-ray bursts in the redshift range of $0 < z < 8$. As a result of this comparison, we found that the lower estimate of the radius of curvature of space was quite large: 2.4×10^{15} Mpc. This means that the observational data indicate a negligible curvature of space.

Keywords: Schwarzschild–de Sitter metric; elliptical space; wormholes; gravitational redshift; type-Ia supernovae

1. Introduction

The first exact solutions of Einstein's field equations were spherically symmetric. The Schwarzschild metric [1] corresponds to the following interval in spherical coordinates r, θ, ϕ:

$$ds^2 = g_{tt}^{\rm Sch} c^2 dt^2 - \frac{dr^2}{g_{tt}^{\rm Sch}} - r^2(d\theta^2 + \sin^2\theta d\varphi^2)\,, \quad (1)$$

with its metric coefficient associated with time

$$g_{tt}^{\rm Sch} = 1 - \frac{r_g}{r}, \quad (2)$$

where $r_g = 2\,GM/c^2$ is the gravitational radius, M is the central mass, c is the speed of light, and G is the gravitational constant. This solution is symmetric with respect to the local centre of spherical mass distribution. Two other exact solutions to Einstein's field equations were found, one by Einstein [2]:

$$ds^2 = c^2 dt^2 - dr^2 - R^2 \sin^2\frac{r}{R}(d\theta^2 + \sin^2\theta d\varphi^2) \quad (3)$$

and the other by de Sitter [3]:

$$ds^2 = \cos^2 \frac{r}{R} c^2 dt^2 - dr^2 - R^2 \sin^2 \frac{r}{R} (d\theta^2 + \sin^2 \theta d\varphi^2)\,. \tag{4}$$

These two solutions are also spherically symmetrical, but their symmetry is not local because the solutions are not related to the preferred centre of curvature or the central mass distribution. The curvature R^{-2} is global by definition, so the symmetry of these models is spherical with respect to any arbitrary point in space.

In the Schwarzschild solution (1), the metric coefficient g_{tt} implies a transformation (dilation) of time along the radial coordinate r. It follows from the coefficient (2) that this transformation leads to a complete halt in the flow of time at the Schwarzschild radius $r = r_g$. A similar cessation of time is inherent in the de Sitter metric (4), since the time coordinate in this metric is also subject to dilatation along the radial coordinate r due to the metric coefficient

$$g_{tt}^{\mathrm{dS}} = \cos^2 \frac{r}{R} = \cos^2 \chi\,. \tag{5}$$

Here, the projective angle χ is such that

$$r = R\chi\,. \tag{6}$$

Thus, according to the expression (5), time stops at a large distance from the observer, corresponding to $\chi = \frac{\pi}{2}$, when $\cos\chi = 0$. At closer distances to the observer, time dilation leads to a decrease in the frequencies of the spectral lines of light sources. On the basis of this property, de Sitter theoretically predicted the possibility of cosmological redshift, which was called the de Sitter effect. This prediction was experimentally confirmed ten years later by Lemaître [4], and then by Hubble, who wrote that the observed "velocity-distance relation may represent the de Sitter effect" [5].

As can be seen from the comparison of metrics (1) and (4), the de Sitter effect is gravitational by its nature, since it is caused by the difference between the coordinate systems of the source and the observer in curved spacetime, similar to what happens in the Schwarzschild metric. But, unlike the gravitational redshift in the Schwarzschild metric, the de Sitter gravitational redshift is global and isotropic: its character remains the same, regardless of the choice of the observer's point in space and the direction of observation.

In Einstein's metric (3), the metric coefficient associated with time is identical to unity; therefore, in Einstein's cosmological model, there is no effect of time dilation, and time is universal for all space. Einstein's metric (3) corresponds to the Riemannian space $\mathbb{S}^3$ with a constant positive curvature $\lambda = R^{-2}$, due to the presence of a non-zero density of matter distribution over space, $\rho > 0$. In the case of the de Sitter metric (4), $\lambda = 3R^{-2}$, and $\rho_0 = 0$ (i.e., the space is empty). The distance d_{os} from the observer (o) to the source (s) is measured by the radial coordinate r:

$$\underset{|r=0}{o} \xrightarrow{r = R\chi \equiv d_{os}} s \xrightarrow{\qquad\qquad} a\,. \tag{7}$$

The notation 'a' is used here for the observer's antipodal point in space $\mathbb{S}^3$. This point corresponds to the projective angle $\chi = \pi$, which marks a very large distance from the observer. In principle, light from the source can reach the observer from this point as well, although in practice, this is unlikely due to the presence of horizons, which will be discussed below. In addition, there is coordinate ambiguity for $\chi > \pi/2$, as discussed by de Sitter in [3].

To avoid this ambiguity, de Sitter replaced the Riemannian sphere $\mathbb{S}^3$ with an elliptical (projective) space, the main property of which is the identification of its antipodal points ($o \equiv a$). The same elliptical space was used in subsequent cosmological models by Lemaître [4], Tolman [6] and Robertson [7–9], from where elliptical space transitioned into the standard cosmological model of cold dark matter ΛCDM with the cosmological term Λ.

In 1939, R.C. Tolman proposed a general method for finding exact solutions to Einstein's field equations [10]. One of the new solutions found by Tolman combines the Schwarzschild metric with the de Sitter metric (SdS):

$$ds^2 = g_{tt}c^2dt^2 - \frac{dr^2}{g_{tt}} - R^2 \sin^2 \frac{r}{R}(d\theta^2 + \sin^2 \theta d\varphi^2)\,. \tag{8}$$

In this metric, the metric coefficient associated with time

$$g_{tt} = 1 - \frac{r_g}{r} - \frac{r^2}{R^2} \tag{9}$$

includes the curvature parameter R^{-2}. Therefore, the SdS solution can be used to estimate the curvature of space. However, this cannot be carried out directly, because, as with the Schwarzschild metric, the SdS metric is local and anisotropic. This problem is solved by using the main property of elliptical space, i.e., the connectivity between its antipodal points. In this case, it is also necessary to transfer the origin of the coordinate system from the point where the observer is located to their antipodal point. We will discuss this in Section 2, and then, in Section 3, we will estimate the R parameter by calculating the theoretical distance moduli for the SdS metric and comparing them to the distance moduli derived from observations of distant sources such as supernovae and gamma-ray bursts.

2. Materials and Methods

2.1. Elliptical Space

The luminosity distances calculated with the use of the metric coefficient

$$g_{tt}^{\mathrm{dS}} = \cos^2 \frac{r}{R} \tag{10}$$

in the de Sitter metric are not consistent with the observations because here, the theoretical relationship between luminosity distances and redshifts is nonlinear (quadratic) near the origin. At the same time, the observational data show a strictly linear relationship between luminosity distances and redshifts at low redshifts.

Notwithstanding, one can avoid nonlinearity by moving the origin of the coordinate system from the observer point (o) to the antipodal point (a):

$$o \xleftarrow{\;d_{so}\;} s \xleftarrow{\;r=r_s\;} \underset{r=0|}{a \equiv o} \tag{11}$$

Compared with the diagram (7), in this coordinate system, distances are calculated as in the normal coordinate system associated with an observer. Here, r_s is the radial distance between the antipodal point (a) and the source location (s). In order to use the Schwarzschild–de Sitter metric in the calculations, let us bring to bear the main property of elliptical space, the connectivity between its antipodal points (see below). Our goal is to calculate the source-to-observer distance (d_{so}) in the SdS metric and compare it with the redshift z of the source. Using the above notations, we obtain the distance between the observer and their antipodal point in the form

$$r_o = r_s + d_{so}\,. \tag{12}$$

In this configuration, the centre of symmetry of the SdS metric is located at the observer's antipodal point (a). From the observer's perspective, this point looks like a distant sphere with a large (or infinite) radius. It encompasses the entire celestial sphere (4π steradian) around the observer. In this way, the SdS metric becomes spherically symmetric and isotropic for any arbitrary point in space, provided that the antipodal points are endowed with this metric.

Endowing an elliptical space with the SdS metric is possible if the mathematical identification of distant points is materialised through their physical connectivity by means of a structure with a metric that is an exact solution of Einstein's field equations. Such an exact solution to the field equations was found in 1935 by Einstein and Rosen [11]. This solution interconnects two different spaces or two different regions of the same space. Therefore, it was called the Einstein–Rosen bridge, but more often, it is called the "wormhole" [12]. In the simplest case, the exterior space around a wormhole is described by the Schwarzschild metric or by the Schwarzschild–de Sitter metric (if space has a curvature of R^{-2}). In both metrics, the redshift is gravitational by its nature, so it can be calculated using the metric coefficient (9). Such a calculation within the scheme (11) makes it possible to obtain theoretical cosmological redshifts of distant sources.

As shown by Morris, Thorne and Yurtsever [13], it is most likely that wormholes are microscopic objects with dimensions in the order of the Planck–Wheeler length, $\ell_P = \sqrt{G\hbar/c^3} = 1.62 \times 10^{-33}$ cm. This determines the scale of the difference between the idealised (mathematical) elliptical space and its physical counterpart, the space whose antipodal regions are connected through microscopic Einstein–Rosen wormholes.

There is a widespread belief that wormholes are unstable and that even the slightest disturbance associated with the passage of particles of matter or radiation through their throats destroys them. However, most publications on the stability or instability of wormholes are based on the concept of their traversability [14–16]. Other publications discuss this topic either in terms of exotic theories of gravity [17], or within the framework of the theory of quantum gravity [18], which does not yet exist. Therefore, such publications can be disregarded.

In classical general relativity, wormholes are the smallest possible objects. As such, they are not traversable and, therefore, they are stable by definition. Moreover, according to the ideas of Einstein, Rosen and Wheeler, the answer to the question of the relationship between the discrete (particles of matter) and the continuous (space) lies precisely in the possibility of the existence of microscopic structures within the framework of general relativity. Thus, the question of where matter particles and radiation can come from at all is clarified by the possibility of the existence of microscopic wormholes, proven by Einstein and Rosen.

A static wormhole is described by the Schwarzschild metric (1). But our goal here is to estimate the curvature of space R^{-2}, which is encoded in the Schwarzschild–de Sitter metric (8). Therefore, we use this metric, with its free parameter R to be evaluated using observations.

The observer is located near one of the throats of the wormhole (hereinafter referred to as the near throat), and the observed source is somewhere between the observer and the far throat of the observer's antipodal point. According to our choice of coordinate system, any distance is now measured from the origin of that coordinate system (a) to the source (r_s) and to the observer (r_o), with the origin being a large antipodal sphere surrounding the observer. Our goal is to find the relationship between the redshift of the source (z) and the distance from the source to the observer:

$$d_{so}(z) = r_o - r_s\,. \tag{13}$$

Since the origin of the coordinates for these calculations is located on the surface of a distant sphere with radius r_o, regardless of the direction to the source, this distance remains the same when observing any source (for example, a supernova). But the distance to the source r_s from the origin can vary, depending on the source redshift All distances are counted along the line of sight to the source (a supernova). The global (cosmological) redshift of the source is given by the Schwarzschild metric with a gravitational radius r_g, calculated from the origin a (any point on the antipodal sphere).

In this case, the effect of cosmological gravitational redshift is caused by the local distribution of masses in the vicinity of the observer due to the connectivity between the region of elliptical space adjacent to the observer and the region around the observer's antipodal point. In elliptical space, with its identified antipodal points, the centre of the local mass distribution (as well as all other points of this mass distribution, including the points of the Schwarzschild horizons of each mass) are geometrically transformed into a huge sphere around the observer. The radius of this sphere is so large that any local differences in the positions of the mass distributions correspond to negligible relative displacements of the distant sphere surfaces, including the surfaces of the Schwarzschild horizons.

In this way, a collective spherically symmetrical horizon of radius r_g is formed around each point in space, and all points in space are equivalent in this respect. So, the universe looks the same to any arbitrarily chosen observer. In such a scheme, all masses within some vicinity of the observer contribute to the global collective gravitational redshift through their antipodal points. From the point of view of an observer, the size of this neighbourhood can be very large. It includes matter in the form of local and distant galaxies, as well as possibly distant clusters of galaxies. However, compared to the gravitational radius r_g, the size of this region around the observer is negligible.

2.2. Redshift–Distance Relationship

In our scheme (11), there are two unknown distances, r_o and r_s. The latter quantity can be replaced by the source-to-observer distance $d_{so} = r_o - r_s$, which is the sought theoretical quantity to be compared with observational data. The gravitational radius r_g and the global radius of curvature R are the unknowns. Thus, in total, our model has three free parameters: r_o, r_g and R, which should be determined from observations.

Both the source and the observer are in the SdS metric (8) with its redshift-defining coefficient (9). In this case, the redshift z of the source with respect to the observer is calculated using the following expression:

$$z = \sqrt{\frac{g_{tt}^o}{g_{tt}^s}} - 1 \tag{14}$$

or, taking into account (9),

$$(z+1)^2 = \left[1 - \frac{r_g}{r_o} - \frac{r_o^2}{R^2}\right]\left[1 - \frac{r_g}{r_s} - \frac{r_s^2}{R^2}\right]^{-1}. \tag{15}$$

2.2.1. Approximate Solution

Expression (15) can be converted to the desired functional dependence of the distance $d_{so} = r_o - r_s$ on the source redshift z by regrouping the terms included in this expression and temporarily taking r_g as the distance unit: $r_g = 1$. Later, this parameter can be translated into some generally accepted units of distance measurement, for example, in light years or megaparsecs. By rearranging Equation (15), we obtain

$$(1+z)^2 = \left(1 - r_o^{-1} - r_o^2 R^{-2}\right)\left(1 - r_s^{-1} - r_s^2 R^{-2}\right)^{-1}. \tag{16}$$

Replacing r_s in (16) with $r_s = r_o - d_{so}$, we obtain an expression containing only the sought parameters r_o, d_{so} and R, as well as the redshift z, which is the input argument:

$$(r_o - d_{so})^{-1} = 1 - (1 - r_o^{-1} - r_o^2 R^{-2})(1+z)^{-2} - (r_o - d_{so})^2 R^{-2}. \tag{17}$$

Then, the distance d_{so} as a function of redshift z reads

$$d_{so} = r_o - \left[1 - (1 - r_o^{-1} - r_o^2 R^{-2})(1+z)^{-2} - (r_o - d_{so})^2 R^{-2}\right]^{-1}. \tag{18}$$

This simple solution is recursive; hence, it is approximate. Given that the value of R^2 in the denominator of the last term in (18) is known to be extremely large (from observations), the distances d_{so} can be calculated by successive approximations. In the zero-order approximation, the last term in expression (18) can be neglected:

$$d_{so}^{(0)} = r_o - \left[1 - (1 - r_o^{-1} - r_o^2 R^{-2})(1+z)^{-2}\right]^{-1}. \tag{19}$$

This depends only on z and can be substituted on the right-hand side of the expression (18) to calculate the required distances d_{so}.

2.2.2. Exact Solution

An exact solution to Equation (15) also exists (the author must acknowledge that it was proposed by Øyvind Grøn, the reviewer of this paper). The solution is achieved by converting Equation (15) into a reduced cubic equation,

$$r_s^3 + 3pr_s + 2q = 0 \tag{20}$$

using the coefficients $p = -\frac{1}{3}\left[1 - \frac{r_g}{r_o} - \frac{r_o^2}{R^2}\right](1+z)^{-2}R^2$ and $q = \frac{r_g R^2}{2}$. Since $|p^3| > |q^2|$, while $p < 0$, the discriminant of (20) is negative, $p^3 + q^2 < 0$, which means that (20) has three real solutions [19]:

$$r_s = 2\sqrt{-p}\cos\frac{\varphi}{3}, \qquad \text{and} \qquad r_s^{\pm} = -2\sqrt{-p}\cos\frac{\varphi \pm \pi}{3}, \tag{21}$$

with $\varphi = \arccos(-q/\sqrt{-p^3})$. Like in (18), the required source-to-observer distance is calculated as $d_{so} = r_o - r_s$ using one of the exact solutions (21):

$$d_{so} = r_o - 2\sqrt{-p}\cos\frac{\varphi}{3}, \qquad \text{or} \qquad d_{so}^{\pm} = r_o + 2\sqrt{-p}\cos\frac{\varphi \pm \pi}{3} \tag{22}$$

The choice between these three solutions can be made by comparing them with the approximate solution (18), since it is the only one. It turns out that the first of these solutions results in distances decreasing with the growth of z, while the third (d_{so}^{-}) leaves the distances approximately constant within the redshift range $0 < z < 0.1$. Both contradict observational facts. The only solution that corresponds to the observations is the second solution d_{so}^{+}. It also coincides with the approximate solution from Section 2.2.1.

2.2.3. Luminosity Distance

The luminosity distance comparable with that of the source, is obtained from the expressions (18) or (22) by multiplying these expressions by the scaling factor $(1+z)^2$:

$$d_L(z) = d_{so}(1+z)^2, \tag{23}$$

with one of the $(1+z)$-factors accounting for the loss of luminosity due to the cosmological redshift z, as well as for the lower rate at which the photons reach the observer because of the cosmological time dilatation due to the non-unit metric coefficient g_{tt}. The other $(1+z)$-factor takes into account the distortion of the photon's trajectory (coefficient g_{tt}^{-1} in the Schwarzschild–de Sitter metric). The expression (23) can be used to determine the free parameters of our model by comparing the theoretical distance moduli

$$\mu^{\text{theor}} = 5\log d_L + 25 \tag{24}$$

(in stellar magnitudes) with the distance moduli obtained from observations. The numerical coefficients in (24) correspond to the luminosity distances d_L, expressed in Mpc.

2.3. Observational Data

The Formulas (18)–(24) describe the theoretical relationship between distances to remote sources and redshifts. Distances are expressed as the photometric properties of sources. Thus, the assessment of theoretical parameters should be based on observational photometric data for a wide range of cosmological redshifts. One of the most accurate sets of observational data of this kind is the photometric catalogue *Pantheon+* [20,21], which contains data on 1701 type Ia supernovae in the redshift range of $0 < z_{SN} < 2.26$.

Yet, this range of redshifts is not wide enough to indicate differences between cosmological models, since these differences are manifested significantly only for redshifts $z > 3$. The range of redshifts can be expanded by using additional observational data from the gamma-ray burst (GRB) catalogue compiled by Amati et al. [22]. This catalogue contains 193 moduli of gamma-ray burst distances μ_{GRB}, calculated and calibrated using the Amati relation [23]. The range of redshifts covered by this catalogue, $0.03 < z_{GRB} < 8.1$, is significantly wider than in the case of type Ia supernovae.

The μ_{GRB} data are much noisier than μ_{SNe}, and they are also slightly systematically biased within the redshift range $0 < z < 0.7$, where all cosmological models and data must coincide. This bias is calculated to be -0.258 [mag] by minimising the Pearson's χ^2 for 27 GRBs within the mentioned redshift range (the GRB data are corrected for this bias).

3. Results

Calculations using the Formulas (18)–(24) give estimates of the parameters R, r_o and r_g within the framework of the SdS model. The parameters are estimated by minimising the χ^2 Pearson criterion [21]

$$\chi^2 = \Delta\mathbf{D}^T C^{-1} \Delta\mathbf{D}, \tag{25}$$

in which C is the covariance matrix and $\Delta\mathbf{D}$ is the vector of residuals between theoretical and observational distance moduli:

$$\Delta\mathbf{D}_i = \mu_i^{\text{theor}}(z_i) - \mu_i. \tag{26}$$

The length of this vector is $N = 1701$, which corresponds to the length of the *Pantheon+* catalogue, which provides the observational μ_i of type Ia supernovae. The theoretical distance moduli are calculated for the SdS model by using the Formulas (18), (22) and (24), as described in the previous section. For comparison purposes, $\mu_i^{\text{theor}}(z_i)$ is also calculated by using the standard formalism of the ΛCDM model, applying the cosmological parameters estimated by the authors of the *Pantheon+* catalogue [21].

Since here, we extend the range of available source redshifts by using an additional catalogue of calibrated GRBs, which provides only the uncertainties of the GRB distance moduli instead of the full covariance matrix, we can use a simplified χ-squared calculation

$$\chi^2 = (\operatorname{diag} C^T \Delta\mathbf{D})^2 = \sum_{i=1}^{N} \frac{\Delta\mathbf{D}_i^2}{\sigma_{\mu_i}^2}, \tag{27}$$

where $\sigma_{\mu_i}^2$ are the uncertainties of μ_i corresponding to the diagonal of the covariance matrix (see, e.g., [24], § IIIc for theoretical work or [25,26] for practical examples of using (27) for comparison of various cosmological models between each other).

The parameters of the standard ΛCDM model, as estimated in [21] by using the *Pantheon+* sample, are $H_0 = 73.6 \pm 1.1$ [km/s/Mpc], $\Omega_M = 0.334 \pm 0.018$ and $\Omega_k = 0$. The theoretical distance moduli curve corresponding to these parameters is indicated in Figure 1 by the thin solid curve. It matches fairly well with the observational type Ia distance moduli (pink circles in this plot), with Pearson's $\chi^2_{\Lambda\text{CDM}} = 906.1$.

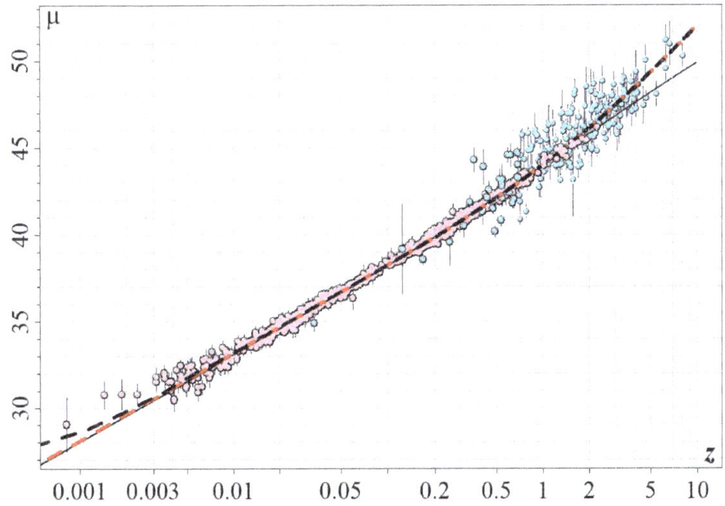

Figure 1. Comparison of the theoretical distance moduli μ (dashed and solid curves) with the observational distance moduli of the joint data sample, including 1701 type Ia supernovae from the Pantheon+ catalogue (pink circles) and 193 gamma-ray bursts from the Amati catalogue (blue circles). The thin solid curve shows the theoretical distance moduli calculated by using the parameters of the standard ΛCDM model. The bold dashed curves indicate the distance moduli based on the Schwarzschild–de Sitter metric for the approximate (red) and exact solutions (black). The abscissa corresponds to the redshifts z of type Ia supernovae and gamma-ray bursts.

The theoretical parameters r_g, r_o and R for the SdS model are obtained by minimising Pearson's χ^2, which is achieved by the global descent method with consecutive iterations. These parameters based on 1701 type Ia supernova data are given in the second column of Table 1. The one-sigma confidence intervals for these parameters are calculated by varying the parameters and taking those of their values that correspond to a Pearson's probability of 68.3%. These values are divided by the square root of the number of degrees of freedom, $\sqrt{N - n_p}$, where $n = 3$ is the number of free parameters, and $N = 1701$ is the number of sources in the sample.

The calculated theoretical distance moduli for the SdS model are shown as two dashed bold curves in Figure 1, with the red curve corresponding to the approximate solution of (15), while the black dashed curve represents the exact solution. The quality of the SdS-model fit into the observational data is characterised by the minimum value of the Pearson criterion, which turns out to be $\chi^2 = 887.6$ for the approximate solution and $\chi^2 = 871.3$ for the exact solution. These values indicate slightly better binding of the theory discussed here to the observational data, in comparison with the ΛCDM model.

The additional distance moduli extending the observational redshift range to $z \approx 8$ are 193 data points from the Amati catalogue. They are shown in Figure 1 as blue circles. With this joint data sample of SNe+GRB containing 1894 data points, the number of degrees of freedom is $N - n = 1891$. In the case of the approximate solution, the tolerance intervals of the parameters for the type Ia SNe and for the joint data sample of SNe+GRB remain approximately the same as for the SNe alone, because the photometric accuracy of SNe is much higher than that of GRBs. With the exact solution, the tolerances are smaller when the more accurate SN data are used for determining the model parameters.

Table 1. Fitting of the SdS theoretical model to the observational data.

Parameter	SNe	SNe+GRB	Units
	Approximate solution		
R	$(1.08^{+\infty}_{-0.83}) \times 10^5$	$(1.17^{+\infty}_{-0.94}) \times 10^5$	$[r_g]$ [1]
$r_o - 1$	$(9.91 \pm 0.15) \times 10^{-8}$	$(9.91^{+0.18}_{-0.06}) \times 10^{-8}$	$[r_g]$
r_g	$(2.13 \pm 0.03) \times 10^{10}$	$(2.13^{+0.03}_{-0.04}) \times 10^{10}$	[Mpc]
χ^2_{min}	887.6	2033.7	–
	Exact solution		
R	$(1.127^{+5.2}_{-0.85}) \times 10^5$	$(1.127^{+5.9}_{-0.89}) \times 10^5$	$[r_g]$
$r_o - 1$	$(9.914 \pm 0.009) \times 10^{-8}$	$(9.914 \pm 0.011) \times 10^{-8}$	$[r_g]$
r_g	$(2.12 \pm 0.03) \times 10^{10}$	$(2.12 \pm 0.04) \times 10^{10}$	[Mpc]
χ^2_{min}	871.3	2018.0	–

[1] Only the lower limit is determinable.

Nevertheless, the GRB data confirm the parameter values for r_o and r_g obtained from SNe (compare the second and third columns of Table 1). The first four rows of this Table provide the outcome of the approximate solution, and the last four rows correspond to the more accurate results of the exact solution.

The estimated value of the SdS model parameter $R = (1.127^{+5.2}_{-0.85}) \times 10^5$, shown in the second column of Table 1, is the lower estimate of the radius of curvature of the universe, since the upper tolerance interval of R is very wide (even being indefinite in the case of the approximate solution). The value of R (in units of r_g) combined with the scale factor r_g suggests that the spatial curvature of the universe is constrained from above by the value $1/R^2 \approx 1.75 \times 10^{-31}$ Mpc^{-2}.

4. Discussion

Observations of 1701 type Ia supernovae collected in the *Pantheon+* catalogue, combined with observations of 193 gamma-ray bursts from the Amati catalogue, make it possible to determine the lowest possible value of the radius of curvature of the universe in a cosmological model with the Schwarzschild–de Sitter metric. It turns out to be very large: 2.4×10^{15} Mpc. Accordingly, the curvature of space in this model is constrained from above by the value 1.75×10^{-31} Mpc^{-2}. That is, the observational data used in this work for these estimates are compatible with the possibility that space is flat.

In this paper, we discuss the origin of cosmological redshift in an alternative way: the redshift of remote galaxies is interpreted as a manifestation of the *static* Schwarzschild–de Sitter metric. Many studies will disagree with this interpretation because it is commonly believed that the universe cannot be static. The point is that the standard cosmological model, with its interpretation of redshift as the result of space expansion described by the *dynamic* Friedmann–Lemaître–Robertson–Walker (FLRW) metric, can explain a large variety of observational facts. Therefore, since the late 1990s, the standard model has been dubbed the concordance cosmological model, which puts all the observations together.

The current general view is that static cosmological models have long been a thing of the past and that it is not worth interpreting astrophysical observations in terms of static models, since they allegedly cannot explain such theoretically and observationally substantiated facts, such as Big Bang nucleosynthesis, the initial annihilation of matter and antimatter, the recombination of plasma in the process of the universe's expansion, the process of structure formation and the subsequent reionisation of the medium during the formation of the first stars and galaxies.

However, a static model is discussed here for good reasons: the observational facts obtained over the past two years by the James Webb Space Telescope (JWST) have disarranged the entire coherent picture of concordance cosmology and made us think about the nature of cosmological redshift.

The light-gathering power and high angular resolution of the JWST provide exquisitely detailed information about the physical processes that took place in the universe 13 billion years ago and earlier—a glimpse into the so-called "dark ages", when, according to the concept of the expanding universe, the process of recombination ended and the process of reionisation should have begun.

The generally accepted view was that the process of the formation of the first stars and the beginning of reionisation corresponded to redshifts $z \sim 20$. But the JWST observations have shown that at distances corresponding to these redshifts, there are a large number of massive galaxies (with masses $> 10^{10}\,M_{\odot}$) that formed less than 700 million years after the beginning of the universe. The observed large number density of massive galaxies with $z > 10$ [27] is at odds with the predictions of the theory of galaxy formation. The recently published UNCOVER catalogue of JWST galaxies [28,29] contains a very large number of galaxies with $z \approx 20$, while, according to the ΛCDM model, this is the redshift where the first stars and galaxies are expected to be formed. Therefore, at this redshift, corresponding to the universe's age of ~160–200 million years, there should be much fewer galaxies, which are expected to be shapeless and small in size.

These observational facts are currently explained by galaxy evolution. But there is a problem here, because the JWST observations show that galaxies with $z > 10$ are fully formed, with disks and bulges, containing large amounts of dust and heavy chemical elements. They are look-alikes of the late-universe galaxies, but with one exception: the physical sizes of remote galaxies are much smaller than those of the galaxies next to us. This is an illusion due to the geometry of the expanding scaling factor. By contrast, in the geometry of a static coordinate system, the galaxies have quite normal sizes, in accordance with their masses and luminosities.

Additional confirmation that the small angular sizes of high-redshift galaxies are not due to their evolution comes from a recently discovered contradiction between the reionisation rate of the early universe known from pre-JWST observations and the new reionisation rate based on the JWST observations. The observed large number density of high-redshift massive galaxies necessarily implies—within the framework of ΛCDM—an extremely high rate of star formation, since the ΛCDM-based time that has elapsed since the Big Bang is very short (a few hundred million years).

It follows that the density of ionising photons must then be very high. According to the calculations in [30], reionisation should not only have begun much earlier than the standard model suggests, but should have already been completed by the time the galaxies observed by the JWST telescope formed. This is at odds with other observational facts, like the large abundance of neutral hydrogen at redshifts $z \approx 13$–17. For example, the discovery of an anomalously strong neutral hydrogen absorption line in the redshift range $z \approx 13$–17 [31] indicates the presence of a large amount of neutral hydrogen, which is incompatible with the ΛCDM-based theory of reionisation. By contrast, in static or slowly-evolving coordinates, such a problem does not occur, since in these cases, the volume of space for $z \approx 13$–17 in which the observed galaxies are located turns out to be much larger than in the case of the ΛCDM model. Accordingly, the number density of galaxies per unit volume is lower in a static model, and there can be much more neutral hydrogen within this volume, which is consistent with observations and which removes all the contradictions.

Consequently, the reionisation process based on the JWST observations is now in very strong tension with the cosmic microwave background (CMB) and the Lyman-α forest [32]. This is a new kind of tension (CMB-based reionisation optical depth tension), which adds to the previous CMB-based issues discussed intensively in the literature, including Hubble constant tension, σ_8-tension, the CMB Cold Spot problem, and many other issues, which we are not going to discuss here. But, taken altogether, plus the lack of evolution time from the beginning of the universe for the high-redshift galaxies discovered by the JWST, this all indicates a profound crisis of the ΛCDM, which must somehow be resolved.

As for the other observational facts, in the past, static cosmological models were not only supported by the same observational phenomena as ΛCDM, but they predicted that such phenomena existed. In the literature, there are plenty of papers discussing how static-universe models solve challenging observational facts in alternative ways to ΛCDM. So, there is no need to review these topics in detail here.

And for those researchers who are convinced that without the expansion of the universe, it is impossible to explain all the observed facts, there is a palliative solution. The fact is that the static Schwarzschild–de Sitter metric is mathematically equivalent to the dynamic McVittie metric [33,34]. The latter is the Schwarzschild metric (like in the model discussed here) embedded in the FLRW metric of the expanding universe. Thus, all arguments regarding the Big Bang's nucleosynthesis, structure formation, etc., remain valid for the McVittie (≡SdS) metric, as for the metric of ΛCDM.

The only difference between McVittie's solutions and ΛCDM is that most of the cosmological redshift in the McVittie metric is due to the gravitational effect (as in the SdS metric discussed here), and only a small fraction is due to the effect associated with space expansion. Therefore, the Hubble constant H_0 in this case becomes very small, which is equivalent to a very large value of the parameter R obtained here in the SdS metric. Accordingly, the age of the universe is estimated to be several orders of magnitude larger than the 13.8 billion years suggested in the ΛCDM model. We will discuss this option elsewhere.

Nevertheless, we will mention explanations of some observational facts as they are seen within the framework of static cosmological models. For example, the cosmological redshift phenomenon was itself predicted theoretically by de Sitter for a static cosmological model well before the appearance of any dynamical cosmological models. Moreover, de Sitter warned in his 1917 paper that the lines of spectra systematically displaced towards the red might give rise to a spurious positive radial velocity interpretation [3]. In 1923, Eddington repeated this warning by writing the following: in de Sitter's theory, there is the general displacement of spectral lines to the red in distant objects due to the slowing down of atomic vibrations which would be erroneously interpreted as a motion of recession [35].

Then, in 1926, Eddington predicted thermalised background radiation to exist with $T = 3\,\mathrm{K}$ for a static-universe model [36]. Later, in 1937, a similar prediction with respect to the CMB temperature $T = 2.8\,\mathrm{K}$ within a static-universe framework was proposed by W. Nernst [37]. Only much later, in 1953, did G. Gamow make his prediction with respect to the CMB and its temperature $T = 7\,\mathrm{K}$ for the expanding-universe model [38]. The abundances of light elements in a static universe were explained by G.R. Burbidge and F. Hoyle [39,40], R. Salvaterra and A. Ferrara [41] and others, although there are some unresolved issues for both the static and dynamic universe models. For example, the expanding universe model predicts that the synthesis of deuterium from hydrogen occurred only during the short period of nucleosynthesis after the Big Bang. In the future, the synthesis of deuterium cannot occur—it can only be destroyed in the interior of stars [42]. For this reason, the observed abundance of deuterium must be gradually diminishing with time. The Big Bang nucleosynthesis theory predicts the same evolution for lithium. But, contrary to this theory, observations show that the abundance of lithium is growing with time. This fact is attributed to the phenomenon of cosmic-ray spallation [43].

As for the process of structure formation in the universe, it is believed that the initial inhomogeneities that subsequently give rise to structures such as galaxy clusters and filaments arose at the inflationary stage due to quantum fluctuations, and then due to baryonic acoustic oscillations at the initial stage of the expansion of the universe, when the density of matter was sufficiently high. It is believed that in a static universe, there can be no process of structure formation. However, according to the classical scenarios, structure formation in static universe models occurs due to the mechanism of gravitational instability. Initial fluctuations in the homogeneous gas of primordial hydrogen grow exponentially into large-scale structures [44,45]. In the early years of cosmology, when static cosmological models were considered to be on a par with expanding-universe models, these scenarios agreed with observations. But when Eddington published a paper in 1930 showing that Einstein's static model of the universe was unstable [46], static models fell out of fashion. Nevertheless, in 1970, N. Rosen, the collaborator of Einstein, proved that Eddington's judgment was not well founded and that static universe models are stable [47]. This reopened the possibility of exploring the structure-formation process in models with static metrics.

These studies continued throughout the 1980s to the beginning of the 2000s. It was found that in static models, matter aggregation is likely to be fractal [48,49], which was later confirmed by statistical studies of CMB maps [50] and of the matter distribution in the universe [51,52]. Opposite to this, in the standard model of an expanding universe, structures cannot be fractal. Therefore, the reasons for the existence of the largest structures in the universe (the filaments and wall-like super-clusters, with huge voids between them) still cannot be explained within the framework of ΛCDM.

Based on the newly recognised strong ΛCDM tension called "the reionisation optical depth tension", one can confidently conclude that the model-dependent cosmological parameters derived from the *"Planck"*-mission CMB measurements cannot be fully trusted. Thence, the H_0-tension is resolved and is supportive of the type Ia supernova value of $H_0 = 73.5\,\mathrm{km\,s^{-1}\,Mpc^{-1}}$. The corresponding age of the universe is then reduced to 12.4 Gyr (instead of the commonly accepted value of 13.8 Gyr deduced from the CMB *Plank* cosmology). This means that the galaxies with $z \approx 20$ observed by the JWST are as young as 160 million years old, which highlights to the extreme degree the impossibility of constructing a physically meaningful galaxy-formation model predicting a well-developed galaxy to be built in such a short time. This suggests that the universe's age, according to the ΛCDM model, is likely to be incorrect. The only alternative to solving this problem is to revise the standard cosmological model, which would make it possible to determine the real (larger) age of the universe.

Some people would be tempted to relegate the gravitational redshift hypothesis discussed here as an "amusing curiosity", not worthy of being seriously discussed. Then, they would also have to agree that the hypothesis of the late 1970s of the inflationary universe by Starobinsky, Linde, Guth and others, which, according to the authors themselves in those years, was considered an "amusing curiosity", also had no right to be discussed. However, that hypothesis (still a hypothesis that has not been confirmed by any observations) has now become the basis of the standard cosmological model.

The main problem with the two hundred or so existing inflation models is that they are quite successful in explaining current experimental data, but they do not allow for the prediction of the results of future measurements. In contrast to these hypotheses, the model discussed in this manuscript makes a clear testable prediction of the results of future measurements. The graph in Figure 1 shows that supernovae with redshifts $z = 3$ or more, when they are detected, will be dimmer, compared to what ΛCDM predicts for these redshifts. But in order to explain the additional dimming of supernovae from $z > 1$, in relation to the simple model of an expanding universe, the concept of an unobservable physical entity called "dark energy" was introduced into the standard model. Its energy density, as determined by the supernova distance moduli, turned out to be 10^{120} times less than the vacuum energy density determined experimentally in elementary particle physics.

The cosmological model presented here does not need to introduce unobservable physical entities. It predicts the photometric parameters of distant sources, depending on their redshifts, and the observed quantities follow the predictions exactly based on gravitational redshift formalism.

Funding: This research received no external funding.

Data Availability Statement: The data used in this manuscript are available at https://pantheonplussh0es.github.io (accessed on 10 July 2024) and in the table from the work [22].

Acknowledgments: This research made use of the following archives: The *Pantheon+* type Ia supernova distance moduli from [20,21] and the gamma-ray burst distance moduli prepared by Amati et al. [22]. I also acknowledge the use of the cosmoFns software package (version 1.1-1) created by Andrew Harris (University of Maryland, USA) and available at the following web page:

github.com/cran/cosmoFns (accessed on 10 July 2024). I would like to thank Paul Kuin, Leslie Morrison, Alice Breeveld, Victor Tostykh and Mat Page for engaging in useful discussions on the matters in this paper. I gratefully acknowledge the significant contribution to this work of one of the reviewers, Øyvind Grøn, who proposed an accurate solution to the problem by converting the main equation of this problem into a cubic one, which is noted in Section 2.2.2 of this text.

Conflicts of Interest: The author declares no conflicts of interest.

Abbreviations

The following abbreviations are used in this manuscript:

CMB	Cosmic Microwave Background (radiation)
FLRW	Friedmann–Lemaitre–Robertson–Walker (metric)
GRB	gamma-ray burst
JWST	James Webb Space Telescope
ΛCDM	lambda cold dark matter (cosmological model)
SN	supernova
dS	de Sitter (metric)
SdS	Schwarzschild–de Sitter (metric)

References

1. Schwarzschild, K. Über das Gravitationsfeld eines Massenpunktes nach der Einsteinschen Theorie. In *Sitzungsberichte der königlich preussischen Akademie der Wissenschaften*; Deutsche Akademie der Wissenschaften: Berlin, Germany, 1916; Volume 3, pp. 189–196.
2. Einstein, A. Kosmologische betrachtungen zur allgemeinen Relativitätstheorie. In *Sitzungsberichte der Königlich Preussischen Akademie der Wissenschaften*; Deutsche Akademie der Wissenschaften: Berlin, Germany, 1917; pp. 142–152.
3. de Sitter, W. On Einstein's theory of gravitation, and its astronomical consequences. Third paper. *Mon. Not. R. Astron. Soc.* **1917**, *78*, 3–28. [CrossRef]
4. Lemaître, G. Un univers homogène de masse constante et de rayon croissant rendant compte de la vitesse radiale des nébuleuses extra-galactiques. *Ann. Soc. Sci. Brux. A* **1927**, *47*, 49–59.
5. Hubble, E. A relation between distance and radial velocity among extragalactic nebulae. *Proc. Natl. Acad. Sci. USA* **1929**, *15*, 168–173. [CrossRef] [PubMed]
6. Tolman, R.C. On the estimation of distances in a curved universe with a non-static line element. *Proc. Natl. Acad. Sci. USA* **1930**, *16*, 511–520. [CrossRef] [PubMed]
7. Robertson, H.P. Kinematics and world structure. *Astrophys. J.* **1935**, *82*, 284–301. [CrossRef]
8. Robertson, H.P. Kinematics and world structure II. *Astrophys. J.* **1936**, *83*, 187–201. [CrossRef]
9. Robertson, H.P. Kinematics and world structure III. *Astrophys. J.* **1936**, *83*, 257–271. [CrossRef]
10. Tolman, R.C. Static solutions of Einstein's field equations for spheres of fluid. *Phys. Rev.* **1939**, *55*, 364–373. [CrossRef]
11. Einstein, A.; Rosen, N. The Particle Problem in the General Theory of Relativity. *Phys. Rev.* **1935**, *48*, 73–77. [CrossRef]
12. Morris, M.S.; Thorne, K.S. Wormholes in spacetime and their use for interstellar travel: A tool for teaching general relativity. *Am. J. Phys.* **1988**, *56*, 395–412. [CrossRef]
13. Morris, M.S.; Thorne, K.S.; Yurtsever, U. Wormholes, time machines, and the weak energy condition. *Phys. Rev. Lett.* **1988**, *61*, 1446–1449. [CrossRef] [PubMed]
14. Bronnikov, K.; Lipatova, L.; Novikov, I.; Shatskiy, A. Example of a stable wormhole in general relativity. *Grav. Cosmol.* **2013**, *19*, 269–274. [CrossRef]
15. Blázquez-Salcedo, J.-L.; Knoll, C.; Radu, E. Traversable wormholes in Einstein-Dirac-Maxwell theory. *Phys. Rev. Lett.* **2021**, *126*, 101102. [CrossRef] [PubMed]
16. Koiran, P. Infall time in the Eddington–Finkelstein metric, with application to Einstein–Rosen bridges. *Int. J. Mod. Phys.* **2021**, *30*, 2150106. [CrossRef]
17. Rosa, J.L. Double gravitational layer traversable wormholes in hybrid metric-Palatini gravity. *Phys. Rev. D* **2022**, *104*, 064002. [CrossRef]
18. Cox, P.H.; Harms, B.C.; Hou, S. Stability of Einstein-Maxwell-Kalb-Ramond wormholes. *Phys. Rev. D* **2016**, *93*, 044014. [CrossRef]
19. Korn, G.A.; Korn, T.M. *Mathematical Handbook for Scientists and Engineers*, 2nd ed.; Dover Civil and Mechanical Engineering; Dover Publications: Mineola, NY, USA, 2000; 1152p.
20. Scolnic, D.; Brout, D.; Carr, A.; Riess, A.G.; Davis, T.M.; Dwomoh, A.; Jones, D.O.; Ali, N.; Charvu, P.; Chen, R.; et al. The Pantheon+ Analysis: The Full Dataset and Light-Curve Release. *Astrophys. J.* **2022**, *938*, 113. [CrossRef]
21. Brout, D.; Scolnic, D.; Popovic, B.; Riess, A.G.; Zuntz, J.; Kessler, R.; Carr, A.; Davis, T.M.; Hinton, S.; Jones, D.; et al. The Pantheon+ Analysis: Cosmological Constraints. *Astrophys. J.* **2022**, *938*, 110. [CrossRef]
22. Amati, L.; D'Agostino, R.; Luongo, O.; Muccino, M.; Tantalo, M. Addressing the circularity problem in the $E_p - E_{ISO}$ correlation of gamma-ray bursts. *Mon. Not. R. Astron. Soc.* **2019**, *486*, L46–L51. [CrossRef]

23. Amati, L.; Guidorzi, C.; Frontera, F.; Della Valle, M.; Finelli, F.; Landi, R.; Montanari, E. Measuring the Cosmological Parameters with the $E_{p,i} - E_{iso}$ Correlation of Gamma-Ray Bursts. *Mon. Not. R. Astron. Soc.* **2008**, *391*, 577–584. [CrossRef]
24. Cash, W. Parameter estimation in astronomy through application of likelihood ratio. *Astrophys. J.* **1979**, *228*, 939–947. [CrossRef]
25. López-Corredoira, M.; Calvo-Torel, J.L. Fitting of supernovae without dark energy. *Int. J. Mod. Phys. D* **2022**, *31*, 2250104. [CrossRef]
26. Scolnic, S.M.; Jones, D.O.; Rest, A.; Pan, Y.C.; Chornock, R.; Foley, R.J.; Huber, M.E.; Kessler, R.; Narayan, G.; Riess, A.G.; et al. The complete light-curve sample of spectroscopically confirmed SNe Ia from Pan-STARRS1 and cosmological constraints from the combined Pantheon sample. *Astrophys. J.* **2018**, *859*, 101. [CrossRef]
27. Finkelstein, S.L.; Leung, G.C.; Bagley, M.B.; Dickinson, M.; Ferguson, H.C.; Papovich, C.; Akins, H.B.; Haro, P.A.; Davé, R.; Dekel, A.; et al. The Complete CEERS Early Universe Galaxy Sample: A Surprisingly Slow Evolution of the Space Density of Bright Galaxies at $z \sim 8.5$–14.5. *Astrophys. J. Lett.* **2024**, *969*, L2. [CrossRef]
28. Wang, B.; Leja, J.; Labbé, I.; Bezanson, R.; Whitaker, K.E.; Brammer, G.; Furtak, L.J.; Weaver, J.R.; Price, S.H.; Zitrin, A.; et al. The UNCOVER Survey: A First-look HST+JWST Catalog of Galaxy Redshifts and Stellar Population Properties Spanning $0.2 < z < 15$. *Astrophys. J.* **2024**, *270*, 12. [CrossRef]
29. Weaver, J.R.; Cutler, S.E.; Pan, R.; Whitaker, K.E.; Labbé, I.; Price, S.H.; Bezanson, R.; Brammer, G.; Marchesini, D.; Leja, J.; et al. The UNCOVER Survey: A First-look HST+JWST Catalog of 60,000 Galaxies near A2744 and beyond. *Astrophys. J. Suppl. Ser.* **2024**, *270*, 7. [CrossRef]
30. Muñoz, J.B.; Mirocha, J.; Chisholm, J.; Furlanetto, S.R.; Charlotte Mason, C. Reionization after JWST: A photon budget crisis? *arXiv* **2024**, arXiv:2404.07250.
31. Bowman, J.D.; Rogers, A.E.E.; Monsalve, R.A.; Thomas, J.; Mozdzen, T.J.; Nivedita Mahesh, N. An absorption profile centred at 78 megahertz in the sky-averaged spectrum. *Nature* **2018**, *555*, 67–70. [CrossRef]
32. Nakane, M.; Ouchi, M.; Nakajima, K.; Harikane, Y.; Ono, Y.; Umeda, H.; Isobe, Y.; Zhang, Y.; Xu, Y. Lyα Emission at $z = 7$–13: Clear Evolution of Lyα Equivalent Width Indicating a Late Cosmic Reionization History. *Astrophys. J.* **2024**, *967*, 28. [CrossRef]
33. McVittie, G.C. The mass-particle in an expanding Universe. *Mon. Not. R. Astron. Soc.* **1933**, *93*, 325–339. [CrossRef]
34. Lake, K.; Abdelqaderm, M. More on McVittie's legacy: A Schwarzschild-de Sitter black and white hole embedded in an asymptotically ΛCDM cosmology. *Phys. Rev. D* **2011**, *84*, 044045. [CrossRef]
35. Eddington, A.S. *The Mathematical Theory of Relativity*; Cambridge University Press: Cambridge, UK, 1923; p. 161.
36. Eddington, A.S. *Internal Constitution of the Stars*; Cambridge University Press: Cambridge, UK, 1926; p. 407.
37. Nernst, W. Weitere prüfung der annahme lines stationären zustandes im weltall. *Zeit. Phys.* **1937**, *106*, 633–661. [CrossRef]
38. Gamow, G. The expanding universe and the origin of galaxies. *Kgl. Dan. Vidensk. Selsk. Mat. Fys. Medd.* **1953**, *27*, 3–15.
39. Burbidge, G.R. Was there really a Big Bang? *Nature* **1971**, *233*, 36–40. [CrossRef] [PubMed]
40. Burbidge, G.R.; Hoyle, F. The origin of helium and the other light elements. *Astrophys. J.* **1998**, *509*, L1–L3. [CrossRef]
41. Salvaterra, R.; Ferrara, A. Is primordial ^{4}He truly from the Big Bang? *Mon. Not. R. Astron. Soc.* **2003**, *340*, L17–L20. [CrossRef]
42. Pagel, B.E.J. Abundances of elements of cosmological interest. *Phil. Trans. R. Soc. Lond. A* **1982**, *307*, 19–35. [CrossRef]
43. Spite, F.; Spite, M. Abundances of Lithium in unevolved halo stars and old disk stars: Interpretations and consequences. *Astron. Astrophys.* **1982**, *115*, 357–366.
44. Jeans, J. *Astronomy and Cosmogony*; Cambridge University Press: Cambridge, UK, 1928; 428p. [CrossRef]
45. Hoyle, F. On the fragmentation of gas clouds into galaxies and stars. *Astrophys. J.* **1953**, *118*, 513–528. [CrossRef]
46. Eddington, A.S. On the instability of Einstein's spherical world. *Mon. Not. R. Astron. Soc.* **1930**, *90*, 668–678. [CrossRef]
47. Rosen, N. Static universe and cosmic field. *Ann. Math. Pure Appl.* **1970**, *14*, 305–308. [CrossRef]
48. Baryshev, Y. The hierarchical structure of metagalaxy a review of problems. *Rep. Sp. Aph. Obs. Rus. Acad. Sci.* **1981**, *14*, 24–43.
49. Baryshev, Y.; Teerikorpi, P. *The Discovery of Cosmic Fractals*; World Scientific: Singapore, 2002; 408p.
50. Myllári, A.A.; Raikov, A.A.; Orlov, V.V.; Tarakanov, P.A.; Yershov, V.N.; Yezhkov, M.Y. Fractality of isotherms of the Cosmic Microwave Background based on data from the Planck Spacecraft. *Astrophysics* **2016**, *59*, 31–37. [CrossRef]
51. Coleman, P.H.; Pietronero, L. The fractal structure of the universe. *Phys. Rep.* **1992**, *213*, 311–389. [CrossRef]
52. Gaite, J. The fractal geometry of the cosmic web and its formation. *Adv. Astron.* **2019**, *2019*, 6587138. [CrossRef]

universe MDPI

Article

Coupled Quintessence Inspired by Warm Inflation

Paulo M. Sá [1,2]

[1] Departamento de Física, Faculdade de Ciências e Tecnologia, Universidade do Algarve, Campus de Gambelas, 8005-139 Faro, Portugal; pmsa@ualg.pt
[2] Instituto de Astrofísica e Ciências do Espaço, Faculdade de Ciências, Universidade de Lisboa, Campo Grande, 1749-016 Lisboa, Portugal

Abstract: We investigate a coupled quintessence cosmological model in which a dark-energy scalar field with an exponential potential interacts directly with a dark-matter fluid through a dissipative term inspired by warm inflation. The evolution equations of this model give rise to a three-dimensional dynamical system for which a thorough qualitative analysis is performed for all values of the relevant parameters. We find that the model is able to replicate the observed sequence of late-time cosmological eras, namely, a long enough matter-dominated era followed by a present era of accelerated expansion. In situations where there is a significant transfer of energy from dark energy to dark matter, temporary scaling-type solutions may arise, but, asymptotically, all solutions are dominated by dark energy.

Keywords: coupled quintessence; dark energy; dark matter; cosmology

1. Introduction

One of the most remarkable developments in modern cosmology was the discovery, in the late 1990s, that the Universe is undergoing a period of accelerated expansion [1,2] due to an unknown form of energy called dark energy. Today, this acceleration is firmly established by several other independent precise cosmological observations [3]. In particular, recent measurements of baryon acoustic oscillations from the first year of observations from the Dark Energy Spectroscopic Instrument (DESI) confirm, with unprecedented precision, the accelerated expansion of the Universe [4].

The simplest candidate for dark energy is a cosmological constant Λ, accounting for about 68% of the current total energy density of the Universe [5]. The other components of the standard cosmological model—dark matter, baryonic matter, and radiation—account for about 27%, 5%, and 0.005% of the total energy density, respectively [5,6].

A theoretically appealing alternative is to consider that the role of dark energy is played not by a cosmological constant but rather by a dynamic scalar field whose potential energy dominates the current phase of the evolution of the Universe [7], inducing the observed cosmic acceleration. Such a possibility seems to be preferred—albeit moderately—by DESI's first-year results when they are combined with data from other studies [4]. This quintessence scalar field can in principle be directly coupled to dark matter—whose precise nature is also unknown [8]—giving rise to coupled quintessence [9].

If the nature of dark energy and dark matter are currently unknown, much more so is the form of an eventual non-gravitational coupling between them. This circumstance has allowed for great freedom in the construction of coupled quintessence models (also known as interacting dark energy models) [9–38].

A popular choice in the literature identifies dark matter with a pressureless non-relativistic perfect fluid and considers the potential of the dark-energy scalar field ϕ to be of the exponential type and the interaction term between the dark components to be of the form $Q \propto \rho_{\rm DM}\dot{\phi}$, where $\rho_{\rm DM}$ is the energy density of the dark-matter fluid and an overdot denotes a derivative with respect to cosmic time t. Such an interaction term, which has been motivated by scalar–tensor theories, allows for the existence of late-time accelerated scaling solutions [10–12]. Because of its potential to address the cosmological coincidence problem [39] and the Hubble constant tension [40,41], this coupled quintessence model has attracted substantial attention over the years.

More recently, a generalized interaction term of the form $Q \propto \rho_{\rm DM}\dot{\phi}C(\phi)$, where $C(\phi) \propto \phi^n$ and n is a positive integer, was considered [35]. With such a coupling between dark energy and dark matter, the quintessence model no longer admits scaling attractor solutions, but, interestingly, for certain values of a relevant parameter, during the approach to the dark-energy-dominated final state, the solution mimics an accelerated scaling solution. Naturally, this result raises the question of how common accelerated scaling solutions are in coupled quintessence models. We deem this issue to deserve further investigation.

In this article, we investigate a coupled quintessence model with an interaction term Q, inspired by warm inflation.

According to the warm-inflation paradigm [42], energy is continuously transferred from the inflaton field ψ to a radiation bath, thus ensuring that the energy density of the latter, $\rho_{\rm R}$, is not diluted during the inflationary expansion and that a smooth transition to a radiation-dominated era can occur without the need for a separate post-inflationary reheating phase (for recent reviews on warm inflation, see [43,44]). As a result, during the inflationary period, the evolution equations for the inflaton field and radiation require an extra dissipative term, becoming

$$\ddot{\psi} + 3H\dot{\psi} + \frac{\partial V}{\partial \psi} = -\Gamma\dot{\psi}, \tag{1}$$

$$\dot{\rho}_{\rm R} + 4H\rho_{\rm R} = \Gamma\dot{\psi}^2, \tag{2}$$

where H is the Hubble parameter, $V = V(\psi)$ the potential of the inflaton field, and Γ the so-called dissipation coefficient, which, in general, is a function of the inflaton field and the temperature of the radiation bath, $\Gamma = \Gamma(\psi, T)$.

The above-described warm-inflation paradigm can be realized in realistic cosmological models, yielding results for the scalar spectral index n_s and the tensor-to-scalar ratio r, consistent with Planck observations (see [43,44] and references therein). Furthermore, it has been shown that warm inflation is favored by the de Sitter swampland conjectures [45–48].

Warm-inflation-type dissipation processes can be present at later stages of the Universe's evolution, giving rise, in particular, to a direct coupling between dark energy and dark matter. Such a possibility was recently implemented in the context of a steep-potential quintessence inflationary model, allowing for the unification of early and late stages of the evolution of the Universe through dissipative effects [49,50]. Here, we consider a coupled quintessence model in which, at late times, a dark-energy scalar field ϕ interacts directly with a dark-matter fluid through a dissipative term of the type $Q = \Gamma\dot{\phi}^2$. As a first approach, we choose the dissipation coefficient Γ to be constant, leaving more complex cases for future publications.

This article is organized as follows. In Section 2, we present our coupled quintessence cosmological model and write the corresponding evolution equations as a three-dimensional dynamical system. In Section 3, we investigate the stability properties of this dynamical system, identify the global attractors, and describe the trajectories that correspond to the relevant late-time cosmological solutions. Finally, in Section 4, we present our conclusions.

2. The Coupled Quintessence Cosmological Model

Assuming a flat Friedmann–Lemaître–Robertson–Walker metric,

$$ds^2 = -dt^2 + a^2(t)d\Sigma^2, \tag{3}$$

where $a(t)$ is the scale factor and $d\Sigma^2$ is the metric of the three-dimensional Euclidean space, the evolution equations of the coupled quintessence cosmological model are given by

$$H^2 = \frac{\kappa^2}{3}\left(\frac{\dot{\phi}^2}{2} + V + \rho_{\text{DM}}\right), \tag{4}$$

$$\dot{H} = -\frac{\kappa^2}{2}\left(\dot{\phi}^2 + \rho_{\text{DM}}\right), \tag{5}$$

$$\ddot{\phi} + 3H\dot{\phi} + \frac{\partial V}{\partial \phi} = \frac{Q}{\dot{\phi}}, \tag{6}$$

$$\dot{\rho}_{\text{DM}} + 3H\rho_{\text{DM}} = -Q. \tag{7}$$

In the above equations, ϕ is the quintessence dark-energy scalar field with potential $V(\phi)$, ρ_{DM} is the energy density of a pressureless dark-matter fluid, $H \equiv \dot{a}/a$ is the Hubble parameter, Q is the interaction term between dark energy and dark matter, overdots denote derivatives with respect to cosmic time t, and we use the notation $\kappa \equiv \sqrt{8\pi G} = \sqrt{8\pi}/m_{\text{P}}$, where G is the gravitational constant and m_{P} is the Planck mass.

Defining the energy density and pressure of the scalar field ϕ as

$$\rho_\phi = \frac{\dot{\phi}^2}{2} + V(\phi) \quad \text{and} \quad p_\phi = \frac{\dot{\phi}^2}{2} - V(\phi), \tag{8}$$

Equation (6) can be written as

$$\dot{\rho}_\phi + 3H(\rho_\phi + p_\phi) = Q. \tag{9}$$

In the evolution Equations (4)–(7), for the sake of simplicity, ordinary baryonic matter was neglected since it makes a small contribution to the total energy density of matter. Additionally, radiation was also excluded since we are only interested in late-time cosmological solutions.

At present, no fundamental underlying theory specifies the exact form of the interaction term Q between dark energy and dark matter. Hence, any approach to this problem is necessarily phenomenological, and the selection of the most suitable interaction model will be decided, ultimately, by precise cosmological observations.

In this spirit, one could consider the energy transfer between the dark components to be dictated by non-local quantities, for instance, by the Universe's expansion rate H, yielding an interaction term of the form $Q \propto Hf(\rho_\phi, \rho_{\text{DM}})$, where f is some function of ρ_ϕ and ρ_{DM}. Such an approach has been rather popular in the literature (for a review, see Refs. [51,52].

Alternatively, we could relate this energy transfer to local dissipative effects as in the warm inflationary scenario [43,44]. This is our option in this article. More specifically, we choose the coupling between dark energy and dark matter to be of the form

$$Q = \Gamma\dot{\phi}^2, \tag{10}$$

where Γ is a dissipation coefficient determined only by local properties of the dark-sector interactions. Here, we are assuming that the dissipative processes occurring in the early Universe are also present at later stages of evolution, namely, during the matter-dominated era and the current era of accelerated expansion driven by the potential energy of the quintessence field. This is a natural assumption. Indeed, if seemingly disparate phenomena, like inflation, dark matter, and dark energy, can be unified under the same theoretical framework (for such a triple unification, see, for instance, Ref. [30], one could also expect the interactions between the underlying particles/fields to somehow remain active during all stages of the Universe's evolution. These interactions would, then, give rise to the interaction term (10), with a dissipation coefficient Γ, depending both on the scalar field ϕ and the dark-matter energy density ρ_{DM} [49].

In this article, as a first approach, we choose Γ to be a constant (with dimension of mass), leaving more general dissipative coefficients for future work.

Note that, with an interaction term of the form (10), the evolution equation for the quintessence scalar field (6) can be written as

$$\ddot{\phi} + (3H - \Gamma)\dot{\phi} + \frac{\partial V}{\partial \phi} = 0, \tag{11}$$

revealing that the direct interaction between the dark components modifies the so-called "Hubble friction" term. In what follows, we will consider both possibilities that this term is enhanced or diminished due to the direct interaction with the dark-matter fluid.

In what concerns the potential of the dark-energy scalar field, we choose it to be of the exponential form,

$$V(\phi) = V_a e^{-\mu\kappa\phi}, \tag{12}$$

where V_a is a positive constant with dimension $(\text{mass})^4$ and μ is a dimensionless constant.

To study the cosmological solutions of the system of Equations (4)–(7), we resort to the powerful method of qualitative analysis of dynamical systems.

To this end, we introduce the dimensionless variables

$$x = \frac{\kappa\dot{\phi}}{\sqrt{6}H}, \quad y = \frac{\kappa\sqrt{V}}{\sqrt{3}H}, \quad \text{and} \quad z = \frac{H_*}{H\Omega_{\text{DM}} + H_*}, \tag{13}$$

as well as a new time variable τ, defined as

$$\frac{d\tau}{dt} = \frac{H}{1-z}, \tag{14}$$

where $\Omega_{\text{DM}} \equiv \kappa^2\rho_{\text{DM}}/(3H^2)$ is the dark-matter density parameter and H_* is a positive constant with dimension of mass (the dimensionless variable z should not be confused with the cosmological redshift, defined as $a_0/a(t) - 1$, where a_0 is the present-time value of the scale factor). Note that, from Equation (4), it immediately follows

$$\Omega_{\text{DM}} = 1 - x^2 - y^2. \tag{15}$$

Before proceeding, some comments are in order about the choice of the new variables z and τ (the choice of x and y is the usual one [53]). The interaction term between dark energy and dark matter, given by Equation (10), cannot be written as a function of the variables x and y only; therefore, an extra variable z is required to close the dynamical system. In choosing it, we must ensure that the surface $x^2 + y^2 = 1$ becomes an invariant manifold of the dynamical system, so that no trajectory can cross this surface and enter the region of (unphysical) negative values of Ω_{DM}. In addition, we also want to compactify the phase space in the z direction, say between $z = 0$, corresponding to $H = +\infty$, and $z = 1$, corresponding to $H = 0$. The simplest choice of z that satisfies these requirements is given in Equation (13). However, with such z, the interaction term appearing in the evolution equation for x becomes proportional to $(1-z)^{-1}$ and, consequently, diverges as $z \to 1$. We could study the properties of the dynamical system on the $z = 1$ plane by just considering the divergent term and neglecting the others. Instead, we opt to remove the singularity by choosing $d\tau$ proportional to $(1-z)^{-1}$, which amounts to multiplying the right-hand side of the evolution equations by $1 - z$. This procedure allows us to study the dynamical system on the $z = 1$ plane (without any divergent terms) and simultaneously preserves the stability properties in other regions of phase space.

In the variables x, y, z, and τ, the evolution Equations (4)–(7) of the coupled quintessence cosmological model reduce to a three-dimensional dynamical system, namely,

$$x_\tau = \left[-3x + \frac{\sqrt{6}}{2}\mu y^2 + \frac{3}{2}x(1 + x^2 - y^2) \right](1 - z) + \alpha x(1 - x^2 - y^2)z, \tag{16a}$$

$$y_\tau = \left[-\frac{\sqrt{6}}{2}\mu x + \frac{3}{2}(1 + x^2 - y^2) \right] y(1 - z), \tag{16b}$$

$$z_\tau = \left[\frac{3}{2}(1 - x^2 + y^2)(1 - z) + 2\alpha x^2 z \right] z(1 - z), \tag{16c}$$

where $\alpha = \Gamma / H_*$ is a dimensionless constant parameterizing the energy exchange between dark matter and dark energy.

Inspection of this dynamical system reveals that the surfaces $y = 0$, $z = 0$, and $z = 1$ are invariant manifolds. Furthermore, from the evolution equation for the dark-matter density parameter, obtained from Equations (15), (16a) and (16b),

$$\Omega_{\mathrm{DM},\tau} = \Omega_{\mathrm{DM}} \left[3(x^2 - y^2)(1 - z) - 2\alpha x^2 z \right], \tag{17}$$

we conclude that the surface $x^2 + y^2 = 1$ ($\Omega_{\mathrm{DM}} = 0$) is also an invariant manifold. Taking into account that Ω_{DM} is, by definition, non-negative, and restricting ourselves to expanding cosmologies, the phase space of the dynamical system (16) is then the half-cylinder $\{(x, y, z) \,|\, x^2 + y^2 \leq 1, y \geq 0, 0 \leq z \leq 1\}$.

In what concerns the parameter space, let us point out that the dynamical system (16) is invariant under the transformation $x \to -x$ and $\mu \to -\mu$, implying that, without any loss of generality, the parameter μ can be assumed to be positive. Furthermore, we assume that the parameter α can be either positive (energy is transferred from the dark-matter fluid to the dark-energy scalar field) or negative (energy is transferred in the opposite direction). Altogether, this means that the parameter space of our coupled quintessence model is $\{(\alpha, \mu) \,|\, \alpha \neq 0, \mu > 0\}$.

In the dimensionless variables x, y, and z, the dark-energy density parameter and the effective equation-of-state parameter are given by

$$\Omega_\phi \equiv \frac{\kappa^2}{3H^2}\rho_\phi = x^2 + y^2 \tag{18}$$

and

$$w_{\mathtt{eff}} \equiv \frac{p_\phi}{\rho_{\mathrm{DM}} + \rho_\phi} = x^2 - y^2, \tag{19}$$

respectively, where ρ_ϕ and p_ϕ are defined in Equation (8). The latter quantity may take values in the range $[-1, 1]$. When the evolution of the Universe is completely dominated by the potential $V(\phi)$, the effective equation-of-state parameter equals -1, and one recovers the cosmological constant case. When the dynamics is dominated by the dark-matter fluid or the scalar-field kinetic term, one obtains $w_{\mathtt{eff}} = 0$ or $w_{\mathtt{eff}} = 1$, respectively. Expansion is accelerated if $-1 \leq w_{\mathtt{eff}} < -1/3$.

The coupled phantom cosmological model inspired by warm inflation, in which $w_{\mathtt{eff}}$ may be less than -1, was studied in Ref. [54].

3. Cosmological Solutions

The dynamical system (16) has two critical lines and, depending on the values of the parameters α and μ, up to six critical points. Table 1 summarizes their properties, namely, the conditions for their existence and the physical behavior of the solutions in their vicinity, while Table 2 summarizes the stability properties of the critical points and lines.

Table 1. Properties of the critical points and critical lines of the dynamical system (16). Ω_ϕ, $\Omega_{\rm DM}$, and $w_{\tt eff}$ are, respectively, the dark-energy density parameter, the dark-matter density parameter, and the effective equation-of-state parameter. Accelerated expansion of the Universe occurs for $w_{\tt eff} < -1/3$.

Critical Point/Line	Existence	Ω_ϕ	$\Omega_{\rm DM}$	$w_{\rm eff}$	Acceleration
$A(1,0,0)$	$\mu > 0, \alpha \neq 0$	1	0	1	no
$B(-1,0,0)$	$\mu > 0, \alpha \neq 0$	1	0	1	no
$C(0,0,0)$	$\mu > 0, \alpha \neq 0$	0	1	0	no
$D\left(\frac{\mu}{\sqrt{6}}, \sqrt{1-\frac{\mu^2}{6}}, 0\right)$	$0 < \mu \leq \sqrt{6}, \alpha \neq 0$	1	0	$-1+\frac{\mu^2}{3}$	$0 < \mu < \sqrt{2}$
$E\left(\frac{\sqrt{6}}{2\mu}, \frac{\sqrt{6}}{2\mu}, 0\right)$	$\mu \geq \sqrt{3}, \alpha \neq 0$	$\frac{3}{\mu^2}$	$1-\frac{3}{\mu^2}$	0	no
$F\left(\frac{\mu}{\sqrt{6}}, \sqrt{1-\frac{\mu^2}{6}}, \frac{3(\mu^2-6)}{(2\alpha+3)\mu^2-18}\right)$	$0 < \mu < \sqrt{6}, \alpha < 0$ or $\mu = \sqrt{6}, \alpha \neq 0$	1	0	$-1+\frac{\mu^2}{3}$	$0 < \mu < \sqrt{2}$
$G(0,y,1)$	$\mu > 0, \alpha \neq 0, 0 \leq y \leq 1$	y^2	$1-y^2$	$-y^2$	$\frac{1}{\sqrt{3}} < y \leq 1$
$H\left(x, \sqrt{1-x^2}, 1\right)$	$\mu > 0, \alpha \neq 0, -1 \leq x \leq 1$	1	0	$-1+2x^2$	$\lvert x\rvert < \frac{1}{\sqrt{3}}$

Table 2. The stability of the critical points and critical lines of the dynamical system (16). None of the critical points A to F are stable. For positive α (transfer of energy from dark matter to dark energy), the critical line $H\left(x, \sqrt{1-x^2}, 1\right)$ is an attractor, while for negative α (transfer of energy from dark energy to dark matter), the attractor is the point $G(0,1,1) = H(0,1,1)$, lying at the intersection of the critical lines.

Critical Point/Line	Eigenvalues	Stability
$A(1,0,0)$	$\left\{3, 0, 3-\frac{\sqrt{6}}{2}\mu\right\}$	no
$B(-1,0,0)$	$\left\{3, 0, 3+\frac{\sqrt{6}}{2}\mu\right\}$	no
$C(0,0,0)$	$\left\{-\frac{3}{2}, \frac{3}{2}, \frac{3}{2}\right\}$	no
$D\left(\frac{\mu}{\sqrt{6}}, \sqrt{1-\frac{\mu^2}{6}}, 0\right)$	$\left\{3-\frac{\mu^2}{2}, -3+\frac{\mu^2}{2}, -3+\mu^2\right\}$	no
$E\left(\frac{\sqrt{6}}{2\mu}, \frac{\sqrt{6}}{2\mu}, 0\right)$	$\left\{\frac{3}{2}, -\frac{3}{4}\left(1-\sqrt{\frac{24}{\mu^2}-7}\right), -\frac{3}{4}\left(1+\sqrt{\frac{24}{\mu^2}-7}\right)\right\}$	no
$F\left(\frac{\mu}{\sqrt{6}}, \sqrt{1-\frac{\mu^2}{6}}, \frac{3(\mu^2-6)}{(2\alpha+3)\mu^2-18}\right)$	$\left\{\frac{\alpha\mu^4}{(2\alpha+3)\mu^2-18}, \frac{\alpha\mu^2(\mu^2-6)}{(2\alpha+3)\mu^2-18}, \frac{\alpha\mu^2(\mu^2-6)}{(2\alpha+3)\mu^2-18}\right\}$	no
$G(0,y,1)$	$\{0, 0, \alpha(1-y^2)\}$	$\mu > 0, \alpha < 0, y = 1$
$H\left(x, \sqrt{1-x^2}, 1\right)$	$\{0, -2\alpha x^2, -2\alpha x^2\}$	$\mu > 0, \alpha > 0, \lvert x\rvert \leq 1$ or $\mu > 0, \alpha < 0, x = 0$

The critical points A, B, C, D, and E, lying on the plane $z = 0$, are those already present in the uncoupled quintessence model. This follows from the fact that for $z = 0$, Equations (16a) and (16b) decouple from Equation (16c), yielding the two-dimensional dynamical system of Ref. [53], which describes a quintessence scalar field interacting only through gravity with a pressureless fluid. The critical point F and the critical lines G and H are new; they arise due to the introduction of a direct interaction term between dark energy and dark matter.

Inspection of the eigenvalues shown in Table 2 for all possible values of the parameters α and μ belonging to the parameter space reveals that none of the critical points A to F can be an attractor, because at least one of the corresponding eigenvalues is positive. The same applies to the critical lines G for $\alpha > 0$ $(y \neq 1)$ and H for $\alpha < 0$ $(x \neq 0)$. Thus, we conclude that the attractors can only be the critical lines G and H for $\alpha < 0$ and for $\alpha > 0$, respectively. Let us analyze these two cases in detail.

The critical points belonging to the critical line $H(x, \sqrt{1-x^2}, 1)$ have just one eigenvalue equal to zero (for $x \neq 0$) and, therefore, they are normally hyperbolic, meaning that the linear stability theory suffices to assess the behavior of the trajectories in their vicinity. Since for $\alpha > 0$, the other two eigenvalues are negative, we conclude that, for such values of the parameter α and $x \neq 0$, these critical points are stable, i.e., they attract the trajectories along the noncritical directions. By continuity arguments, the critical point $H(0,1,1)$ is also an attractor for $\alpha > 0$.

The stability of the critical points belonging to the critical line $G(0, y, 1)$ is not so straightforward to assess. Indeed, because at least two eigenvalues are equal to zero, the linear theory does not suffice, and one has to resort to alternative methods to investigate stability. In Appendix A, we show using the center manifold theory that, for $\alpha < 0$, the trajectories, when approaching the critical line G, drift in the y-direction, converging to the critical point $G(0,1,1) = H(0,1,1)$, which lies at the intersection of the critical lines G and H. Therefore, this point is a global attractor for $\alpha < 0$.

In summary, for $\alpha > 0$ the critical line $H(x, \sqrt{1-x^2}, 1)$ is the global attractor, while for $\alpha < 0$, all trajectories are attracted to the critical point $G(0,1,1) = H(0,1,1)$.

Taking into account that, at the critical lines of G and H, the dark-energy density parameter is $\Omega_\phi = y^2$ and $\Omega_\phi = 1$, respectively, (see Table 1), from the above results on the stability of the critical points and lines, it immediately follows that, asymptotically, all cosmological solutions describe a Universe completely dominated by dark energy, irrespective of the value of α, which parameterizes the energy exchange between the dark components of the Universe.

This result is somewhat unexpected since, at first sight, a significant transfer of energy from the dark-energy scalar field to the dark-matter fluid should favor the appearance of scaling solutions, i.e., solutions for which the ratio between the density parameters of dark matter and dark energy is nonzero. However, as shown above, there are no attractors corresponding to such scaling solutions, and the asymptotic dominance of dark energy is an unavoidable feature of the cosmological model under consideration. Still, there is a subtlety here; we will clarify it below after presenting a detailed description of the solutions of cosmological relevance.

Note that the critical point $G(0,1,1) = H(0,1,1)$ is deep inside the region of the phase space in which expansion is accelerated ($x^2 - y^2 < -1/3$), while the critical line $H(x, \sqrt{1-x^2}, 1)$ is only partially inside this region. This means that, for $\alpha < 0$, the attractor always corresponds to a Universe with accelerated expansion, while for $\alpha > 0$, the attractor corresponds to accelerated expansion only if $|x| < 1/\sqrt{3}$.

Agreement with observations requires the present era of accelerated expansion of the Universe to be preceded by a matter-dominated era, long enough to allow for structure formation. This sequence of cosmological eras corresponds to those trajectories of the dynamical system (16) that pass near the critical point C before proceeding to the global attractor at $G(0,1,1) = H(0,1,1)$ for $\alpha < 0$, or $H(x, \sqrt{1-x^2}, 1)$ for $\alpha > 0$ and $|x| < 1/\sqrt{3}$. In what follows, we will focus our attention on such trajectories since they correspond to solutions that agree, at least qualitatively, with cosmological observations.

Each set of values of the parameters α and μ corresponds to a specific trajectory in the phase space of the dynamical system (16). Let us recall that α parameterizes the direct transfer of energy between the two dark components of the model (from dark matter to dark energy if $\alpha > 0$, in the opposite direction if $\alpha < 0$) and μ parameterizes the steepness of the scalar-field potential.

For $\alpha > 0$, we can restrict our analysis to $0 < \mu < \sqrt{2}$ because only for these values of the parameter μ does the final state of evolution correspond to accelerated expansion. In this case, the trajectories we are interested in pass close to the critical point C (close enough to guarantee a matter-dominated era of appropriate duration), proceed to the vicinity of the critical point D, and then climb vertically in the direction of the $z = 1$ plane, converging asymptotically to the point with coordinates $x = \mu/\sqrt{6}$, $y = \sqrt{1 - \mu^2/6}$, and

$z = 1$, belonging to the critical line H. At this final state, the effective equation-of-state parameter is given by $w_{\mathtt{eff}} = -1 + 2x^2 \simeq -1 + \mu^2/3$ (see Table 1).

The evolution of the dark-energy density parameter Ω_ϕ, the dark-matter density parameter Ω_{DM}, and the effective equation-of-state parameter $w_{\mathtt{eff}}$ are illustrated in Figure 1 for the case $\alpha = 1$ and $\mu = 0.3$. Initial conditions $x_i = y_i = 9.21 \times 10^{-6}$ and $z_i = 10^{-7}$ were chosen to guarantee a long enough matter-dominated era, stretching from $\tau = -8$ to $\tau = -0.4$, which corresponds to a redshift between 3000 and 0.5, and also to guarantee that, at the present time $\tau = 0$, $\Omega_\phi(0) = 0.68$. The effective equation-of-state parameter tends, asymptotically, to -0.97. In Figure 2, the evolution of the effective equation-of-state parameter $w_{\mathtt{eff}}$ is shown for different values of μ. For comparison, we also show the curve corresponding to the ΛCDM concordance model, obtained by choosing $\alpha = 0$, $\mu = 0$, and $x_i = 0$.

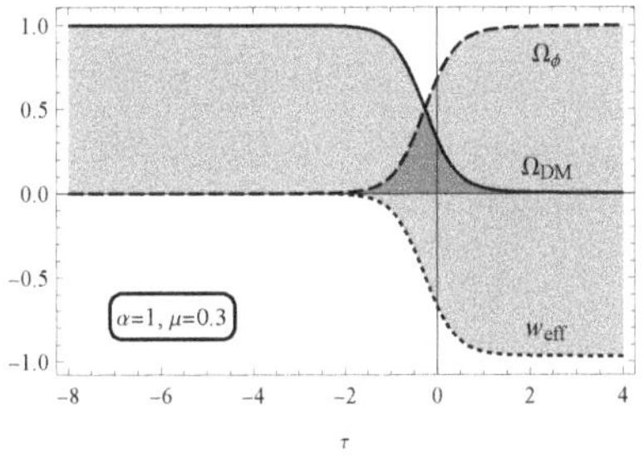

Figure 1. Evolution of Ω_ϕ, Ω_{DM}, and $w_{\mathtt{eff}}$ for the case $\alpha = 1$ and $\mu = 0.3$. A matter-dominated era, long enough to allow for structure formation, is followed by an era of dark-energy domination, during which the expansion of the Universe is accelerated.

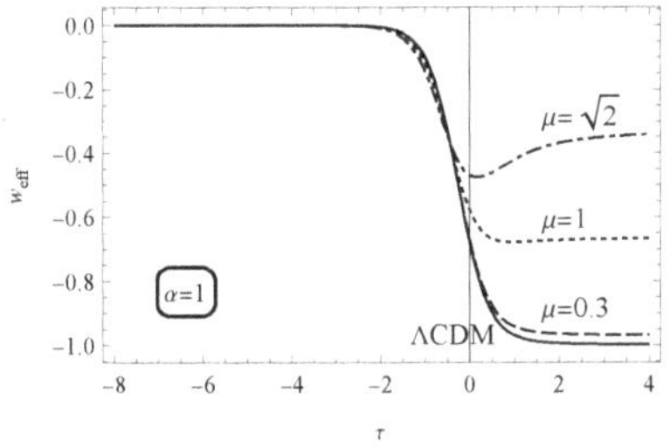

Figure 2. Evolution of the effective equation-of-state parameter $w_{\mathtt{eff}}$ for $\alpha = 1$ and different values of the parameter μ. For comparison, the curve corresponding to the ΛCDM concordance model is also shown. For small values of μ, the evolution of the coupled quintessence cosmological model is quite similar to ΛCDM.

Let us now turn to the case $\alpha < 0$. For $0 < \mu \leq \sqrt{3}$, the behavior of the trajectories is similar, initially, to the case $\alpha > 0$; they pass near C and D, but then they climb to the vicinity of the critical point F (lying above D), and, from there, they proceed to the final state at $G(0,1,1) = H(0,1,1)$. For $\sqrt{3} < \mu < \sqrt{6}$, the trajectories, after passing near C, proceed to E (instead of D), then approach F, before heading to the attractor at $G(0,1,1) = H(0,1,1)$. Finally, for $\mu \geq \sqrt{6}$, the trajectories pass near C, then approach E, from where they proceed directly to the attractor at $G(0,1,1) = H(0,1,1)$.

In all cosmological solutions with $\alpha < 0$, at the final state, the effective equation-of-state parameter is given by $w_{\mathtt{eff}} = -y^2 = -1$ (see Table 1). This corresponds to accelerated expansion irrespective of the value of the parameter μ. This result is illustrated in Figure 3, where $w_{\mathtt{eff}}$ is seen to asymptotically converge to -1 for different values of the parameter μ. Note, however, that this convergence proceeds at an exceedingly slow rate. Meanwhile, the solution mimics a scaling solution, for which the energy density of dark matter is a significant fraction of the total energy density, as shown in Figure 4. In both Figures 3 and 4, initial conditions were chosen to guarantee a long enough matter-dominated era and also $\Omega_\phi(0) = 0.68$, namely, $x_i = y_i = 9.21 \times 10^{-6}$ for $\mu = 0.3$, $x_i = y_i = 1.024 \times 10^{-5}$ for $\mu = 1$, $x_i = y_i = 1.455 \times 10^{-5}$ for $\mu = \sqrt{3}$, and $x_i = y_i = 2.04 \times 10^{-5}$ for $\mu = 2$ ($z_i = 10^{-7}$ in all cases).

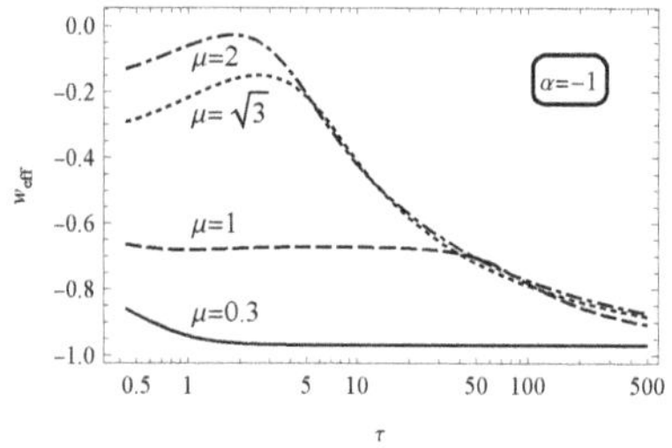

Figure 3. Evolution of the effective equation-of-state parameter $w_{\tt eff}$ for $\alpha = -1$ and different values of the parameter μ. For negative α, $w_{\tt eff}$ asymptotically converge to -1 irrespective of μ.

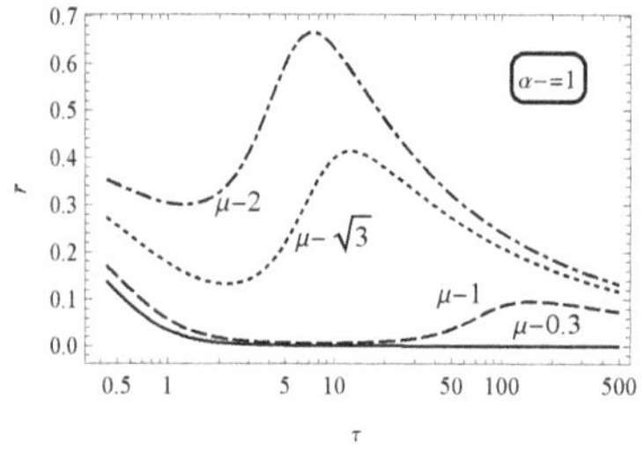

Figure 4. Evolution of the ratio $r \equiv \Omega_{\rm DM}/\Omega_\phi$ for $\alpha = -1$ and different values of the parameter μ. For negative α, the convergence of r to zero proceeds at an exceedingly slow rate, meaning that, in the near future, the solutions effectively behave as scaling solutions.

The asymptotic behavior of the above-described cosmological solutions can be better understood by analyzing the evolution of the interaction term Q, which, in terms of the variables x, y, and z, reads

$$Q = \frac{6\alpha H_*^3}{\kappa^2} \frac{(1-z)^2 x^2}{(1-x^2-y^2)^2 z^2}, \tag{20}$$

where H_* is the positive constant introduced in Equation (13).

As shown in Figure 5, for $\alpha > 0$, Q rapidly converges to a constant positive value, which guarantees a steady transfer of energy from the dark-matter fluid to the dark-energy scalar field and, consequently, reinforces the dominance of the latter in the evolution of the Universe. For $\alpha < 0$, there is initially a significant transfer of energy from dark energy to dark matter, keeping the energy density of the latter at an expressive level—a situation that mimics the behavior of a scaling solution—but, as time goes on, this transfer of energy approaches zero, the energy density of the dark-matter fluid becomes more and more negligible, and the Universe finally completes its transition to an era of total dark-energy domination. In Figure 5, initial conditions were chosen to guarantee a long enough matter-dominated era and also $\Omega_\phi(0) = 0.68$, namely, $x_i = y_i = 1.024 \times 10^{-5}$ and $z_i = 10^{-7}$.

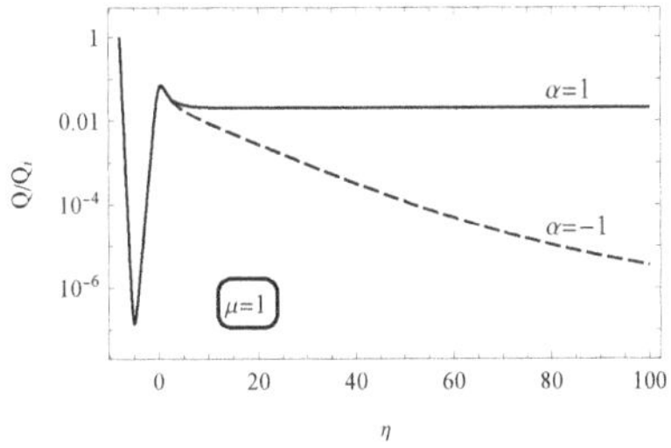

Figure 5. Evolution of the interaction term Q (divided by its value at the beginning of the numerical integration Q_i). For $\alpha > 0$, a steady transfer of energy from dark matter to dark energy reinforces the asymptotic dominance of the latter. For $\alpha < 0$, an initial significant transfer of energy from dark energy to dark matter helps the latter to be maintained at an expressive level, a situation that mimics the behavior of a scaling solution; however, as time passes, this energy transfer tends to zero and the Universe is allowed to complete its transition to an era of total dark-energy domination.

4. Conclusions

In this article, we have investigated a coupled quintessence cosmological model with an interaction term inspired by warm inflation.

With an appropriate choice of dimensionless variables, the evolution equations of the coupled quintessence model can be written as a three-dimensional dynamical system.

A stability analysis of the dynamical system's critical points and lines shows that, asymptotically, all cosmological solutions describe a Universe completely dominated by dark energy, irrespective of the values of the parameters α and μ.

However, a thorough study of the phase-space trajectories for different α and μ reveals that, in the case of negative α (corresponding to a transfer of energy from the dark-energy scalar field to the dark-matter fluid), for any value of μ, the approach to the final state of dark-energy dominance is so slow that, in the near future, the cosmological solutions effectively behave as scaling solutions.

Such behavior had already been observed previously in another coupled quintessence model [35] with an interaction term between dark energy and dark matter of the form $Q \propto \rho_{\rm DM}\dot{\phi}\phi^n$, where n is a positive integer. In that study, as in the present one, strictly speaking, there are no scaling attractor solutions; however, for certain values of a relevant parameter, there are solutions that, during the approach to the final state of cosmic evolution, behave, for all practical purposes, as accelerated scaling solutions. It seems thus that, in coupled quintessence cosmological models, scaling solutions are more ubiquitous than one might expect from a simple stability analysis of the critical points of the corresponding dynamical system.

The analysis of the dynamical system (16) also reveals the existence of a family of phase-space trajectories corresponding to an appropriate sequence of cosmic eras, namely, a matter-dominated era, long enough to allow for structure formation, followed by a present era of accelerated expansion. Such cosmologically relevant trajectories exist for both positive and negative values of the parameter α (for $\alpha > 0$, accelerated expansion at the current time requires $\mu < \sqrt{2}$). Thus, irrespective of the direction of energy transfer between dark matter and dark energy, the coupled quintessence cosmological model given by Equations (4)–(7) is able to replicate the observed late-time stages of the evolution of the Universe.

Within warm inflationary models, a variety of forms have been adopted for the dissipation coefficient, from the simplest, based on general phenomenological considerations, to the more elaborate ones, derived from quantum field theory. As a first approach to the study of coupled quintessence models with an interaction term inspired by warm inflation, we have considered Γ in Equation (10) to be constant. Dissipation coefficients depending on both the scalar field ϕ and the dark-matter energy density $\rho_{\rm DM}$ will be considered in future work.

Funding: This research was funded by Fundação para a Ciência e a Tecnologia (Portugal) through the research grants doi.org/10.54499/UIDB/04434/2020 and doi.org/10.54499/UIDP/04434/2020.

Data Availability Statement: No new data were created or analyzed in this study.

Acknowledgments: The author thanks A. J. Roberts for comments on the central manifold theory.

Conflicts of Interest: The author declares no conflicts of interest.

Abbreviations

The following abbreviations are used in this manuscript:

DESI	Dark Energy Spectroscopic Instrument
ΛCDM	Lambda Cold Dark Matter

Appendix A. Stability Analysis of the Critical Line $G(0, y, 1)$

Consider a specific point $G(0, y_c, 1)$ of the critical line $G(0, y, 1)$, where $y_c \neq 1$ is a constant.

The eigenvalues of the stability matrix of the dynamical system (16), evaluated at this point, are $\lambda_{1,2} = 0$ and $\lambda_3 = \alpha(1 - y_c^2)$. Because two eigenvalues are zero, the stability properties of the critical point cannot be determined within the linear theory; we must resort to alternative methods. Here, we will use the center manifold theory to assess the stability of the critical point $G(0, y_c, 1)$ (see Refs. [32,35] for examples of this theory's application in cosmological contexts similar to the present article).

Introducing new variables

$$u = x + \frac{\sqrt{6}\mu y_c^2}{2\alpha(y_c^2 - 1)}(z - 1), \quad v = y - y_c, \quad w = z - 1, \tag{A1}$$

the dynamical system (16) is brought to the form

$$u_\tau = \alpha(1 - y_c^2)u + f_1(u, v, w), \tag{A2a}$$

$$v_\tau = -\frac{3}{2}y_c(1 - y_c^2)w + f_2(u, v, w), \tag{A2b}$$

$$w_\tau = f_3(u, v, w), \tag{A2c}$$

where $f_i = \mathcal{O}(u^2, v^2, w^2, uv, uw, vw)$, $i = 1, 2, 3$.

The center manifold $u = h(v, w)$, which is a solution of the partial differential equation

$$\begin{aligned}\frac{\partial h}{\partial v}\left[-\frac{3}{2}y_c(1 - y_c^2)w + f_2\big(h(v, w), v, w\big)\right] + \frac{\partial h}{\partial w}f_3\big(h(v, w), v, w\big)\\ - \alpha(1 - y_c^2)h(v, w) - f_1\big(h(v, w), v, w\big) = 0,\end{aligned} \tag{A3}$$

is given by

$$h(v, w) = \frac{\sqrt{6}\mu}{2\alpha}v^2 w + \mathcal{O}(v^4, v^3 w, v^2 w^2, v w^3, w^4), \tag{A4}$$

for $y_c = 0$, and

$$h(v, w) = \frac{\sqrt{6}\mu y_c}{\alpha(1 - y_c^2)^2}vw - \frac{\sqrt{6}\mu y_c^2[3 + \alpha(1 - y_c^2)]}{2\alpha^2(1 - y_c^2)^2}w^2 + \mathcal{O}(v^3, v^2 w, v w^2, w^3), \tag{A5}$$

for $0 < y_c < 1$.

The flow on the center manifold is determined by the differential equations

$$v_\tau = -\frac{3}{2}vw \quad \text{and} \quad w_\tau = \frac{3}{2}w^2, \tag{A6}$$

for $y_c = 0$, and

$$v_\tau = -\frac{3}{2}y_c(1 - y_c^2)w \quad \text{and} \quad w_\tau = \frac{3}{2}(1 + y_c^2)w^2, \tag{A7}$$

for $0 < y_c < 1$. In the u-direction the flow is given by Equation (A2a).

Taking into account that in the neighborhood of the critical point $w < 0$ (also $v > 0$ for the case $y_c = 0$), we conclude that both v_τ and w_τ are positive, implying that the trajectories approach the critical point along the w-direction and move away from it along the v-direction. Along the u-direction, the trajectories approach the critical point if $\alpha < 0$ and move away from it if $\alpha > 0$.

In terms of the original variables x, y, and z, the above results mean that, for $\alpha < 0$, the trajectories, when approaching the critical line $G(0, y, 1)$, drift in the y-direction, asymptotically converging to the critical point $G(0, 1, 1) = H(0, 1, 1)$, which lies in the intersection of the critical lines G and H. Hence, this point is a global attractor for $\alpha < 0$.

References

1. Riess, A.G. et al. [Supernova Search Team] Observational evidence from supernovae for an accelerating Universe and a cosmological constant. *Astron. J.* **1998**, *116*, 1009–1038. [CrossRef]
2. Perlmutter, S. et al. [Supernova Cosmology Project] Measurements of Ω and Λ from 42 high-redshift supernovae. *Astrophys. J.* **1999**, *517*, 565–586. [CrossRef]
3. Astier, P.; Pain, R. Observational evidence of the accelerated expansion of the Universe. *C. R. Phys.* **2012**, *13*, 521–538. [CrossRef]
4. Adame, A.G. et al. [DESI Collaboration] DESI 2024 VI: Cosmological Constraints from the Measurements of Baryon Acoustic Oscillations. *arXiv* **2024**, arXiv:2404.03002
5. Aghanim, N. et al. [Planck Collaboration] Planck 2018 results VI. Cosmological parameters. *Astron. Astrophys.* **2020**, *641*, A6. [CrossRef]
6. Fixsen, D.J. The temperature of the cosmic microwave background. *Astrophys. J.* **2009**, *707*, 916–920. [CrossRef]
7. Caldwell, R.R.; Dave, R.; Steinhardt, P.J. Cosmological imprint of an energy component with general equation of state. *Phys. Rev. Lett.* **1998**, *80*, 1582–1585. [CrossRef]
8. Bertone, G.; Tait, T.M.P. A new era in the search for dark matter. *Nature* **2018**, *562*, 51–56. [CrossRef] [PubMed]
9. Amendola, L. Coupled quintessence. *Phys. Rev. D* **2000**, *62*, 043511. [CrossRef]
10. Amendola, L. Scaling solutions in general nonminimal coupling theories. *Phys. Rev. D* **1999**, *60*, 043501. [CrossRef]
11. Holden, D.J.; Wands, D. Self-similar cosmological solutions with nonminimally coupled scalar field. *Phys. Rev. D* **2000**, *61*, 043506. [CrossRef]
12. Billyard, A.P.; Coley, A.A. Interactions in scalar field cosmology. *Phys. Rev. D* **2000**, *61*, 083503. [CrossRef]
13. Zimdahl, W.; Pavón, D.; Chimento, L.P. Interacting quintessence. *Phys. Lett. B* **2001**, *521*, 133–138. [CrossRef]
14. Tocchini-Valentini, D.; Amendola, L. Stationary dark energy with a baryon-dominated era: Solving the coincidence problem with a linear coupling. *Phys. Rev. D* **2002**, *65*, 063508. [CrossRef]
15. Gumjudpai, B.; Naskar, T.; Sami, M.; Tsujikawa, S. Coupled dark energy: Towards a general description of the dynamics. *J. Cosmol. Astropart. Phys.* **2005**, *06*, 007. [CrossRef]
16. Zhang, X. Coupled Quintessence in a power-law case and the cosmic coincidence problem. *Mod. Phys. Lett. A* **2005**, *20*, 2575–2582. [CrossRef]
17. Capozziello, S.; Nojiri, S.; Odintsov, S.D. Unified phantom cosmology: Inflation, dark energy and dark matter under the same standard. *Phys. Lett. B* **2006**, *632*, 597–604. [CrossRef]
18. Böhmer, C.G.; Caldera-Cabral, G.; Lazkoz, R.; Maartens, R. Dynamics of dark energy with a coupling to dark matter. *Phys. Rev. D* **2008**, *78*, 023505. [CrossRef]
19. Henriques, A.B.; Potting, R.; Sá, P.M. Unification of inflation, dark energy, and dark matter within the Salam-Sezgin cosmological model. *Phys. Rev. D* **2009**, *79*, 103522. [CrossRef]
20. Lopez Honorez, L.; Mena, O.; Panotopoulos, G. Higher-order coupled quintessence. *Phys. Rev. D* **2010**, *82*, 123525. [CrossRef]
21. Bertolami, O.; Carrilho, P.; Páramos, J. Two-scalar-field model for the interaction of dark energy and dark matter. *Phys. Rev. D* **2012**, *86*, 103522. [CrossRef]
22. Amendola, L.; Barreiro, T.; Nunes, N.J. Multifield coupled quintessence. *Phys. Rev. D* **2014**, *90*, 083508. [CrossRef]
23. Tzanni, K.; Miritzis, J. Coupled quintessence with double exponential potentials. *Phys. Rev. D* **2014**, *89*, 103540. [CrossRef]
24. van de Bruck, C.; Mifsud, J.; Mimoso, J.P.; Nunes, N.J. Generalized dark energy interactions with multiple fluids. *J. Cosmol. Astropart. Phys.* **2016**, *11*, 031. [CrossRef]
25. Nunes, R.C.; Pan, S.; Saridakis, E.N. New constraints on interacting dark energy from cosmic chronometers. *Phys. Rev. D* **2016**, *94*, 023508. [CrossRef]
26. Barros, B.J. Kinetically coupled dark energy. *Phys. Rev. D* **2019**, *99*, 064051. [CrossRef]
27. Odintsov, S.D.; Oikonomou, V.K. Unification of inflation with dark energy in $f(R)$ gravity and axion dark matter. *Phys. Rev. D* **2019**, *99*, 104070. [CrossRef]

28. Pan, S.; Yang, W.; Di Valentino, E.; Saridakis, E.N.; Chakraborty, S. Interacting scenarios with dynamical dark energy: Observational constraints and alleviation of the H_0 tension. *Phys. Rev. D* **2019**, *100*, 103520. [CrossRef]
29. Sá, P.M. Unified description of dark energy and dark matter within the generalized hybrid metric-Palatini theory of gravity. *Universe* **2020**, *6*, 78. [CrossRef]
30. Sá, P.M. Triple unification of inflation, dark energy, and dark matter in two-scalar-field cosmology. *Phys. Rev. D* **2020**, *102*, 103519. [CrossRef]
31. Oikonomou, V.K. Unifying inflation with early and late dark energy epochs in axion $F(R)$ gravity. *Phys. Rev. D* **2021**, *103*, 044036. [CrossRef]
32. Sá, P.M. Late-time evolution of the Universe within a two-scalar-field cosmological model. *Phys. Rev. D* **2021**, *103*, 123517. [CrossRef]
33. Waeming, A.; Klangburam, T.; Pongkitivanichkul, C.; Samart, D. Dark matter and dark energy from a Kaluza–Klein inspired Brans–Dicke gravity with barotropic fluid. *Eur. Phys. J. C* **2022**, *82*, 409. [CrossRef]
34. da Fonseca, V.; Barreiro, T.; Nunes, N.J. A simple parametrisation for coupled dark energy. *Phys. Dark Universe* **2022**, *35*, 100940. [CrossRef]
35. Potting, R.; Sá, P.M. Coupled quintessence with a generalized interaction term. *Int. J. Mod. Phys. D* **2022**, *31*, 2250111. [CrossRef]
36. D'Agostino, R.; Luongo, O. Cosmological viability of a double field unified model from warm inflation. *Phys. Lett. B* **2022**, *829*, 137070. [CrossRef]
37. van de Bruck, C.; Poulot, G.; Teixeira, E.M. Scalar field dark matter and dark energy: A hybrid model for the dark sector. *J. Cosmol. Astropart. Phys.* **2023**, *07*, 019. [CrossRef]
38. Yang, W.; Pan, S.; Mena, O.; Di Valentino, E. On the dynamics of a dark sector coupling. *J. High Energy Astrophys.* **2023**, *40*, 19–40. [CrossRef]
39. Steinhardt, P.J. *Critical Problems in Physics*; Fitch, V.L., Marlow, D.R., Dementi, M.A.E., Eds.; Princeton University Press: Princeton, NJ, USA, 1997.
40. Di Valentino, E.; Mena, O.; Pan, S.; Visinelli, L.; Yang, W.; Melchiorri, A.; Mota, D.F.; Riess, A.G.; Silk, J. In the realm of the Hubble tension—A review of solutions. *Class. Quantum Grav.* **2021**, *38*, 153001. [CrossRef]
41. Riess, A.G.; Breuval, L. The Local Value of H_0. In Proceedings of the IAU Symposium 376, Budapest, Hungary, 17–21 April 2023; arXiv:2308.10954.
42. Berera, A. Warm inflation. *Phys. Rev. Lett.* **1995**, *75*, 3218–3221. [CrossRef]
43. Kamali, V.; Motaharfar, M.; Ramos, R.O. Recent Developments in Warm Inflation. *Universe* **2023**, *9*, 124. [CrossRef]
44. Berera, A. The Warm Inflation Story. *Universe* **2023**, *9*, 272. [CrossRef]
45. Motaharfar, M.; Kamali, V.; Ramos, R.O. Warm inflation as a way out of the swampland. *Phys. Rev. D* **2019**, *99*, 063513. [CrossRef]
46. Das, S. Warm inflation in the light of swampland criteria. *Phys. Rev. D* **2019**, *99*, 063514. [CrossRef]
47. Das, S. Note on single-field inflation and the swampland criteria. *Phys. Rev. D* **2019**, *99*, 083510. [CrossRef]
48. Bertolami, O.; Sá, P.M. Multi-field cold and warm inflation and the de Sitter swampland conjectures. *J. Cosmol. Astropart. Phys.* **2022**, *09*, 001. [CrossRef]
49. Lima, G.B.F.; Ramos, R.O. Unified early and late Universe cosmology through dissipative effects in steep quintessential inflation potential models. *Phys. Rev. D* **2019**, *100*, 123529. [CrossRef]
50. Das, S.; Hussain, S.; Nandi, D.; Ramos, R.O.; Silva, R. Stability analysis of warm quintessential dark energy model. *Phys. Rev. D* **2023**, *108*, 083517. [CrossRef]
51. Bolotin, Y.L.; Kostenko, A.; Lemets, O.A.; Yerokhin, D.A. Cosmological evolution with interaction between dark energy and dark matter. *Int. J. Mod. Phys. D* **2015**, *24*, 1530007. [CrossRef]
52. Wang, B.; Abdalla, E.; Atrio-Barandela, F.; Pavón, D. Dark matter and dark energy interactions: Theoretical challenges, cosmological implications and observational signatures. *Rep. Prog. Phys.* **2016**, *79*, 096901. [CrossRef]
53. Copeland, E.J.; Liddle, A.R.; Wands, D. Exponential potentials and cosmological scaling solutions. *Phys. Rev. D* **1998**, *57*, 4686–4690.
[CrossRef]
54. Halder, S.; Pan, S.; Sá, P.M.; Saha, T. Coupled phantom cosmological model motivated by the warm inflationary paradigm. *arXiv* **2024**, arXiv:2407.15804.

universe MDPI

Article

Stochastic Tunneling in de Sitter Spacetime

Taiga Miyachi [1,*], Jiro Soda [1,2] and Junsei Tokuda [3]

[1] Department of Physics, Kobe University, Kobe 657-8501, Japan; jiro@phys.sci.kobe-u.ac.jp
[2] International Center for Quantum-Field Measurement Systems for Studies of the Universe and Particles (QUP), KEK, Tsukuba 305-0801, Japan
[3] Particle Theory and Cosmology Group, Center for Theoretical Physics of the Universe, Institute for Basic Science (IBS), Daejeon 34126, Republic of Korea; jtokuda@ibs.re.kr
* Correspondence: tmiyachi@stu.kobe-u.ac.jp

Abstract: Tunneling processes in de Sitter spacetime are studied by using the stochastic approach. We evaluate the Martin–Siggia–Rose–Janssen–de Dominicis (MSRJD) functional integral by using the saddle-point approximation to obtain the tunneling rate. The applicability conditions of this method are clarified using the Schwinger–Keldysh formalism. In the case of a shallow potential barrier, we reproduce the Hawking–Moss (HM) tunneling rate. Remarkably, in contrast to the HM picture, the configuration derived from the MSRJD functional integral satisfies physically natural boundary conditions. We also discuss the case of a steep potential barrier and find an interesting Coleman–de Luccia (CDL) bubblelike configuration. Since the starting point of our analysis is the Schwinger–Keldysh path integral, which can be formulated in a more generic setup and incorporates quantum effects, our formalism sheds light on further studies of tunneling phenomena from a real-time perspective.

Keywords: tunneling; stochastic approach; de Sitter spacetime; MSRJD functional integral

1. Introduction

Tunneling is an important nonperturbative phenomenon in cosmology. Hence, it is crucial to understand tunneling processes in curved spacetime. In particular, tunneling in de Sitter spacetime is worth studying in detail. The reason is that the de Sitter spacetime is the simplest non-trivial curved spacetime. Moreover, the results obtained there can be applicable to tunneling phenomena during inflation. Since both Minkowski and de Sitter spacetime have the maximal symmetry, there exist a similarity. Indeed, the Euclidean instanton method for evaluating the tunneling rate in Minkowski spacetime [1,2] is applicable to tunneling processes in de Sitter spacetime [3]. However, the extension of the Euclidean method to the de Sitter spacetime has a difficulty in interpretation. In the case of the de Sitter spacetime, there are two saddle solutions: Coleman–De Luccia (CDL) [3] and Hawking–Moss (HM) instantons [4]. The latter instanton is a homogeneous solution. Therefore, it is not straightforward to interpret the tunneling rate as that for the tunneling process from a false vacuum to a true vacuum [5,6].

To circumvent the difficulty, a real-time formalism might be useful. In the case of the de Sitter spacetime in the flat chart, the stochastic formalism has been used for investigating time evolution of fluctuations during inflation [7,8]. In this formalism, Langevin equations describe the dynamics of long-wavelength modes. Stochastic noises stem from short-wavelength quantum modes. The stochastic formalism has also been used to analyze the tunneling processes in de Sitter spacetime [9–15]. However, most of the previous works treated only the HM transition by utilizing the Fokker–Plank (FP) equation for a homogeneous scalar field, which can be derived from the Langevin equation. Obviously, the HM transition cannot describe the bubble nucleation in contrast to CDL tunneling. To describe the bubble, we have to include the spatial gradient term, and the FP equation becomes a functional differential equation instead of a partial differential equation. Hence, at first glance, it seems difficult to treat the CDL bubble by the stochastic approach. However, we make an observation that there exists a path integral representation to solutions of the FP equation. In condensed matter physics, it is called the Martin–Siggia–Rose–Janssen–de Dominicis (MSRJD) functional integral [16–19].

In this paper, we use the MSRJD path integral formula to study the tunneling processes on the fixed de Sitter background. The applicability conditions of this method are clarified based on the Schwinger–Keldysh formalism. We consider a potential $V(\phi)$ which has a local minimum at $\phi = \phi_{\rm false}$, a true minimum at $\phi_{\rm true}$, and a local maximum at $\phi_{\rm top}$ as shown in Figure 1. We use the saddle-point method and evaluate the tunneling rate by computing the action for the tunneling configurations [20]. We investigate both HM and CDL tunneling processes. In the case of a shallow potential barrier, the conventional HM tunneling rate is reproduced. Remarkably, the tunneling configuration in the MSRJD functional integral method has physically natural boundary conditions in contrast to the Euclidean method. We then clarify that the HM tunneling rate is the tunneling rate of a (roughly Hubble sized) coarse-grained patch transitioning from $\phi_{\rm false}$ to $\phi_{\rm top}$. In the case of a steep potential barrier, we find an interesting CDL bubblelike configuration which is obtained by solving the solution of a scalar field equation in Euclidean anti-de Sitter space even though we do not work in the imaginary time. Our results show how the bubble nucleation process could be described in the stochastic approach.

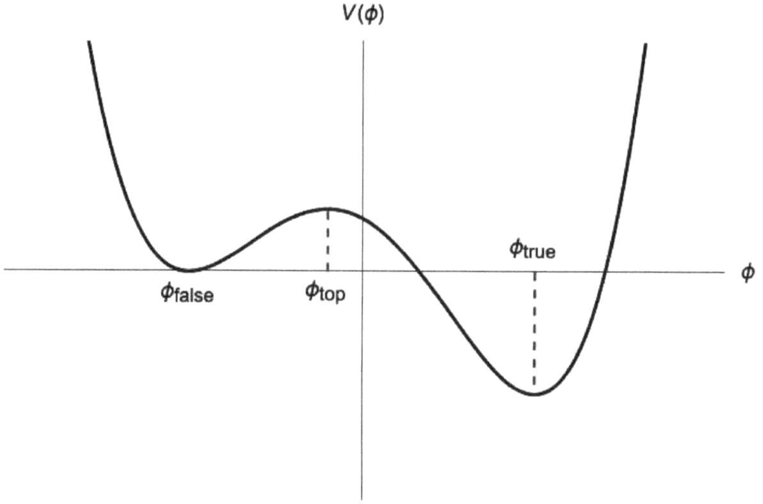

Figure 1. The schematic picture of the potential $V(\phi)$ which has a false vacuum at $\phi = \phi_{\rm false}$, true vacuum at $\phi = \phi_{\rm true}$ and the top of the potential at $\phi = \phi_{\rm top}$.

This paper is organized as follows. In Section 2, we present the setup and review the stochastic approach. In Section 3, the MSRJD functional integral is constructed from the Langevin equations. We also outline the derivation of the MSRJD functional integral starting from the Schwinger–Keldysh formalism. In Section 4, we apply the MSRJD functional integral to the case of the shallow potential barrier and study the HM tunneling process. In Section 5, we consider the case of the steep potential barrier and find the CDL bubblelike configuration. We derive the tunneling rate using the configuration and compare it with the one obtained in Section 4. We also compare our results with those in the Euclidean method. Section 6 is devoted to the conclusion. Some technical details are presented in appendices.

2. Stochastic Approach

In this section, we review the stochastic approach [7,8].

2.1. Setup

We consider a real scalar field ϕ in a (3 + 1)-dimensional de Sitter background. The background geometry is described by the metric

$$ds^2 = -dt^2 + a(t)^2 d\boldsymbol{x}^2 = a(\eta^2)(-d\eta^2 + d\boldsymbol{x}^2), \tag{1}$$

with the scale factor

$$a(t) = e^{Ht} = -\frac{1}{H\eta}, \tag{2}$$

where H is the Hubble constant. The action is given by

$$S = \int dt d^3x\, a^3 \left[\frac{1}{2}\dot{\phi}^2 - \frac{1}{2}a^{-2}(\nabla\phi)^2 - V(\phi)\right], \tag{3}$$

where a dot denotes a derivative with respect to t and $V(\phi)$ is a potential function. From this action, the canonical conjugate momentum Π is defined as $\Pi := a^3\dot{\phi}$ and the Hamilton's equations of motion are given by

$$\dot{\phi} = a^{-3}\Pi\,, \qquad \dot{\Pi} = a\nabla^2\phi - a^3 V'(\phi)\,. \tag{4}$$

2.2. Langevin Equation

First, we divide the quantum fields ϕ and Π into an infrared (IR) part and an ultraviolet (UV) part, respectively, as

$$\phi = \phi_{\rm IR} + \phi_{\rm UV}\,, \qquad \Pi = \Pi_{\rm IR} + \Pi_{\rm UV}\,. \tag{5}$$

The UV part is defined as

$$\begin{aligned}\phi_{\rm UV} &:= \int \frac{d^3k}{(2\pi)^3}\theta(k - k_c(t))\left[a_{\boldsymbol{k}}u_{\boldsymbol{k}}(t)e^{i\boldsymbol{k}\cdot\boldsymbol{x}} + a^\dagger_{\boldsymbol{k}}u^*_{\boldsymbol{k}}(t)e^{-i\boldsymbol{k}\cdot\boldsymbol{x}}\right],\\ \Pi_{\rm UV} &:= \int \frac{d^3k}{(2\pi)^3}\theta(k - k_c(t))a(t)^3\left[a_{\boldsymbol{k}}\dot{u}_{\boldsymbol{k}}(t)e^{i\boldsymbol{k}\cdot\boldsymbol{x}} + a^\dagger_{\boldsymbol{k}}\dot{u}^*_{\boldsymbol{k}}(t)e^{-i\boldsymbol{k}\cdot\boldsymbol{x}}\right],\end{aligned} \tag{6}$$

where $u_{\boldsymbol{k}}$ is a mode function that is specified later, and $a_{\boldsymbol{k}}$ and $a^\dagger_{\boldsymbol{k}}$ are annihilation and creation operators, respectively. Here, $\theta(k - k_c(t))$ is a step function, and $k_c(t)$ is the cut-off scale defined as

$$k_c(t) := \varepsilon a(t)H, \tag{7}$$

where ε is a constant. The value of ε is discussed later. Substituting Equation (6) into Hamilton's Equation (4), we obtain the following equations;

$$\dot{\phi}_{\rm IR} = a^{-3}\Pi_{\rm IR} + \xi^\phi\,, \qquad \dot{\Pi}_{\rm IR} = a\nabla^2\phi_{\rm IR} - a^3V'(\phi_{\rm IR}) + \xi^\Pi\,, \tag{8}$$

where ξ^ϕ and ξ^Π are defined as

$$\begin{aligned}\xi^\phi &:= \dot{k}_c(t)\int \frac{d^3k}{(2\pi)^3}\delta(k - k_c(t))\left[a_{\boldsymbol{k}}u_{\boldsymbol{k}}(t)e^{i\boldsymbol{k}\cdot\boldsymbol{x}} + a^\dagger_{\boldsymbol{k}}u^*_{\boldsymbol{k}}(t)e^{-i\boldsymbol{k}\cdot\boldsymbol{x}}\right],\\ \xi^\Pi &:= \dot{k}_c(t)\int \frac{d^3k}{(2\pi)^3}\delta(k - k_c(t))\left[a_{\boldsymbol{k}}a(t)^3\dot{u}_{\boldsymbol{k}}(t)e^{i\boldsymbol{k}\cdot\boldsymbol{x}} + a^\dagger_{\boldsymbol{k}}a(t)^3\dot{u}^*_{\boldsymbol{k}}(t)e^{-i\boldsymbol{k}\cdot\boldsymbol{x}}\right].\end{aligned} \tag{9}$$

To derive Equations (8) and (9), we assumed that the time evolution of UV modes could be well approximated by the free theory that is defined around the false vacuum. The mode function $u_{\boldsymbol{k}}(t)$ introduced in (9) then satisfies the following equation

$$\ddot{u}_k + 3H\dot{u}_k + a^{-2}k^2u_k + V''(\phi_{\text{false}})u_k = 0. \tag{10}$$

From this equation, we see the mode function only depends on $k := |\boldsymbol{k}|$. Hence, we simply denote the mode function as u_k. Note that this mode function is used for UV modes $k > k_c(t)$.

Let us consider statistical properties of ξ^ϕ and ξ^Π. The annihilation and creation operators $a_{\boldsymbol{k}}$ and $a^\dagger_{\boldsymbol{k}}$ satisfy the following commutation relations:

$$[a_{\boldsymbol{k}}, a^\dagger_{\boldsymbol{k}'}] = (2\pi)^3\delta^{(3)}(\boldsymbol{k} - \boldsymbol{k}'), \qquad [a_{\boldsymbol{k}}, a_{\boldsymbol{k}'}] = [a^\dagger_{\boldsymbol{k}}, a^\dagger_{\boldsymbol{k}'}] = 0. \tag{11}$$

Defining the vacuum state $|0\rangle$ as

$$a_{\boldsymbol{k}}\,|0\rangle = 0, \quad \forall \boldsymbol{k}, \tag{12}$$

we can calculate the one-point and two-point correlation functions of ξ^ϕ and ξ^Π as

$$\begin{aligned} &\langle 0|\xi^\alpha(t,\boldsymbol{x})|0\rangle = 0, \quad (\alpha,\beta = \phi,\Pi), \\ &\langle 0|\xi^\alpha(t,\boldsymbol{x})\xi^\beta(t',\boldsymbol{x}')|0\rangle = \frac{1}{2\pi^2}\dot{k}_c(t)k_c(t)^2\frac{\sin(k_c(t)r)}{k_c(t)r}g^{\alpha\beta}(t)\delta(t-t'), \end{aligned} \tag{13}$$

where $r := |\boldsymbol{x} - \boldsymbol{x}'|$, and

$$\begin{cases} g^{\phi\phi}(t) & := |u_{k_c}(t)|^2 \\ g^{\Pi\Pi}(t) & := a(t)^6|\dot{u}_{k_c}(t)|^2 \\ g^{\phi\Pi}(t) & = (g^{\Pi\phi})^* := a(t)^3u_{k_c}(t)\dot{u}^*_{k_c}(t). \end{cases} \tag{14}$$

Because of the Wick theorem, all higher correlation functions can be decomposed into two-point functions. In the stochastic approach, we replace the operator ξ^α by the real random field which has the following statistical properties

$$\begin{aligned} &\langle \xi^\alpha(t,\boldsymbol{x})\rangle_\xi = 0, \quad (\alpha,\beta = \phi,\Pi), \\ &\langle \xi^\alpha(t,\boldsymbol{x})\xi^\beta(t',\boldsymbol{x}')\rangle_\xi = \frac{1}{2\pi^2}\dot{k}_c(t)k_c(t)^2\frac{\sin(k_c(t)r)}{k_c(t)r}\mathrm{Re}[g^{\alpha\beta}(t)]\delta(t-t'), \end{aligned} \tag{15}$$

where $\langle\cdots\rangle_\xi$ represents an expectation value with the distribution function of ξ^α. With this prescription, Equation (8) can be interpreted as the Langevin equations with the noise ξ^α stemming from quantum fluctuations. Since ξ^α are now regarded as classical variables, we ignore the imaginary parts of $g^{\phi\Pi}$ and $g^{\Pi\phi}$ which come from non-commutativity. Accordingly, IR variables ϕ_{IR} and Π_{IR} are also regarded as classical stochastic variables in this prescription. The validity of this classical approximation is discussed in the next section.

2.3. Bunch–Davies Vacuum

Now we define the vacuum state $|0\rangle$ by specifying the mode function u_k. Equation (10) is the equation for the mode $k > k_c$. Let us assume that we can ignore the term V'' in solving (10). This can be justified when $|V''|$ is sufficiently small to satisfy $\max\left[(k_c/a)^2, H^2\right] \gg |V''|$. Under this approximation, (10) can be solved as

$$u_k = (-k\eta)^{\frac{3}{2}}\left[C_1H^{(1)}_{\frac{3}{2}}(-k\eta) + C_2H^{(2)}_{\frac{3}{2}}(-k\eta)\right], \tag{16}$$

where C_1, C_2 are constants of integration, and $H_\nu^{(1)}(z), H_\nu^{(2)}(z)$ are Hankel functions. We choose the Bunch–Davies vacuum, i.e., the vacuum that coincides with Minkowski's vacuum at past infinity ($\eta \to -\infty$) so that

$$u_k \to \frac{-H\eta}{\sqrt{2k}} e^{-ik\eta}, \quad (-k\eta \to \infty). \tag{17}$$

This determines C_1, C_2 as

$$C_1 = -\sqrt{\frac{\pi}{4k}} \frac{H}{k}, \qquad C_2 = 0\,. \tag{18}$$

Thus, we obtain the mode function

$$u_k = \frac{H\eta}{\sqrt{2k}} \left(\frac{1}{-ik\eta} - 1 \right) e^{-ik\eta}. \tag{19}$$

To check (17), we used the following formulae

$$H^{(1)}_{\frac{3}{2}}(z) = \sqrt{\frac{2}{\pi z}} \left(\frac{1}{iz} - 1 \right) e^{iz}, \qquad H^{(2)}_{\frac{3}{2}}(z) = \sqrt{\frac{2}{\pi z}} \left(\frac{1}{-iz} - 1 \right) e^{-iz}. \tag{20}$$

Substituting (19) into (14) and using (7), we obtain

$$\begin{cases} g^{\phi\phi} &= \frac{H^2\eta^2}{2k_c}\left(\frac{1}{k_c^2\eta^2}+1\right) = \frac{1}{2Ha^3}\left(\varepsilon^{-3}+\varepsilon^{-1}\right) \\ g^{\Pi\Pi} &= \frac{k_c}{2H^2\eta^2} = \frac{Ha^3}{2}\varepsilon \\ g^{\phi\Pi} &= g^{\Pi\phi *} = \frac{1}{2}\left(\frac{1}{k_c\eta}+i\right) = -\frac{1}{2}\varepsilon^{-1}+\frac{i}{2} \end{cases}. \tag{21}$$

From (8), we have to scale $g^{\alpha\beta}$ as $g^{\phi\phi} \to g^{\phi\phi}$, $g^{\Pi\Pi} \to g^{\Pi\Pi}/(H^2a^6)$ and $g^{\phi\Pi} \to g^{\phi\Pi}/(Ha^3)$ in order to compare these quantities. Then, $g^{\phi\phi}$ and $g^{\Pi\Pi}$ become dominant for the case $\varepsilon \ll 1$ and $\varepsilon \gg 1$, respectively.

2.4. Reduced Langevin Equation

Let us focus on physics at the super-horizon scales by taking $\varepsilon \ll 1$ and assume a shallow potential so that we can use the massless approximation to evaluate the noise correlations. In this setup, we can use (21), from which we find that $g^{\phi\phi}$ becomes dominant. We then ignore the noise for momentum ξ^{Π}. We also ignore the gradient term in (8) because we focus on the physics at super-horizon scales. Since the potential is assumed to be shallow, the time variation in $\phi_{\rm IR}(t)$ is then very small. Therefore, we may integrate the second equation of (8) as

$$\Pi_{\rm IR} \simeq -\frac{a^3}{3H} V'(\phi_{\rm IR}). \tag{22}$$

Substituting this relation into the first equation of (8), we obtain

$$\dot{\phi}_{\rm IR} = -\frac{V'(\phi_{\rm IR})}{3H} + \xi^{\phi}, \tag{23}$$

where the two-point correlation function of ξ^{ϕ} is given at the leading order by

$$\langle \xi^{\phi}(t,\boldsymbol{x})\xi^{\phi}(t',\boldsymbol{x}')\rangle_{\xi} = \frac{H^3}{4\pi^2} j_0(k_c r)\delta(t-t'), \quad j_0(k_c r) := \frac{\sin(k_c r)}{k_c r}. \tag{24}$$

Equation (23) is the Langevin equation with the noise correlation (24) derived in [8].

3. Path Integral Formalism for the Stochastic Approach

In this section, we introduce a path integral approach for the Langevin equation following [21]. From now on, we replace the IR field and its conjugate momentum as $(\phi_{\rm IR}, \Pi_{\rm IR}) \to (\phi_c, \Pi_c)$. As we see in Section 3.2, this notation corresponds to the Keldysh basis. In Section 3.2, we provide a first-principle derivation of the stochastic approach based on the Schwinger–Keldysh formalism following [22,23][1]. We briefly show the sketch of the derivation in the main text, and details of derivations are given in Appendix A.

3.1. MSRJD Functional Integral Representation

3.1.1. Zero-Plus-One-Dimensional Theories

Equation (23) can be regarded as the equation determining the position of a particle $\phi(t)$ with the white noise ξ^ϕ defined by Equation (24). For $k_c r \ll 1$, we have $\sin(k_c r)/(k_c r) \simeq 1$ as a good approximation. Let us discretize the field and noise as $\phi_c(t) \to \phi_i$, $\xi^\phi(t) \to \xi_i^\phi$, $(i \in \mathbf{Z})$. Then, Equation (23) reads

$$X_i := \phi_i - \phi_{i-1} + \Delta t\left(\frac{V'(\phi_i)}{3H} - \xi_i^\phi\right) = 0, \tag{25}$$

where Δt is the temporal discretization interval. Note that the discretization (25) is called Ito discretization[2]. Denoting a solution of (23) by $\phi_c[\xi]$, the expectation value of an observable $\langle \mathcal{O} \rangle_\xi$ can be formally expressed as

$$\langle \mathcal{O}[\phi_c(t)] \rangle_\xi = \int \mathcal{D}\phi_c \, \mathcal{O}[\phi_c] \, \langle \delta(\phi_c - \phi_c[\xi]) \rangle_\xi = \int \mathcal{D}\phi_c \, \mathcal{O}[\phi_c] \left\langle \left| \frac{\delta X}{\delta \phi_c} \right| \delta(X) \right\rangle_\xi . \tag{26}$$

where $\mathcal{D}\phi_c := \prod_i d\phi_i$ is the functional measure, $\delta(\phi_c - \phi_c[\xi]) := \prod_i \delta(\phi_i - \phi_i[\xi])$, and $|\delta X/\delta \phi_c|$ is the determinant of $\{\delta X_i/\delta \phi_j\}$. When we choose the above discretization (25), $\delta X/\delta \phi_c$ becomes a triangular matrix with a unit in the diagonal components, i.e., the functional determinant is unity. Thus, substituting (25) into (26), we obtain

$$\langle \mathcal{O}[\phi_c(t)] \rangle_\xi = \int \mathcal{D}\phi_c \, \mathcal{O}[\phi_c] \left\langle \prod_i \delta\left(\phi_i - \phi_{i-1} + \Delta t\left(\frac{V'(\phi_i)}{3H} - \xi_i^\phi\right)\right) \right\rangle_\xi . \tag{27}$$

Representing the delta functions in terms of a Fourier integral and taking the continuum limit, we obtain the following path integral representation

$$\langle \mathcal{O}[\phi_c(t)] \rangle_\xi = \int \mathcal{D}(\phi_c, \tilde{\phi}) \, \mathcal{O}[\phi_c] \left\langle \exp\left[i \int dt \tilde{\phi} \left(\dot{\phi}_c + \frac{V'(\phi_c)}{3H} - \xi^\phi \right) \right] \right\rangle_\xi . \tag{28}$$

After averaging over the noise assuming Gaussian statistics, we obtain the Martin–Siggia–Rose–Janssen–de Dominicis (MSRJD) functional integral

$$\langle \mathcal{O}[\phi_c(t)] \rangle_\xi = \int \mathcal{D}(\phi_c, \tilde{\phi}) \, \mathcal{O}[\phi_c] \exp\left[\int \mathrm{d}t \left(i\tilde{\phi} \left(\dot{\phi}_c + \frac{V'(\phi_c)}{3H} \right) - \frac{H^3}{8\pi^2} \tilde{\phi}^2 \right) \right] . \tag{29}$$

Putting $\mathcal{O}[\phi_c] = \delta(\phi_c - \phi_c(t))$ and taking the boundary conditions $\phi_c(t') = \phi_c'$ and $\phi_c(t) = \phi_c$, we obtain the transition probability $p(\phi_c, t|\phi_c', t')$

$$p(\phi_c, t|\phi_c', t') = \int_{\phi_c(t')=\phi_c'}^{\phi_c(t)=\phi_c} \mathcal{D}(\phi_c, \tilde{\phi}) \exp\left[\int \mathrm{d}t \left(i\tilde{\phi} \left(\dot{\phi}_c + \frac{V'(\phi_c)}{3H} \right) - \frac{H^3}{8\pi^2} \tilde{\phi}^2 \right) \right] . \tag{30}$$

Finally, introducing new variables $\Pi_\Delta := i\tilde{\phi}$, we obtain the "phase space" path integral

$$p(\phi_c, t|\phi_c', t') = \int_{\phi_c(t')=\phi_c'}^{\phi_c(t)=\phi_c} \mathcal{D}(\phi_c, \Pi_\Delta) \exp\left[\int \mathrm{d}t (\Pi_\Delta \dot{\phi}_c - H(\phi_c, \Pi_\Delta)) \right] , \tag{31}$$

where we defined the Hamiltonian as

$$H(\phi_c,\Pi_\Delta) := -\frac{V'(\phi_c)}{3H}\Pi_\Delta - \frac{H^3}{8\pi^2}\Pi_\Delta^2 \,. \tag{32}$$

We emphasize that this is just a change of integration variable, not the Wick rotation. A set of saddle-point equations is defined as the Hamilton equations for the Hamiltonian defined above.

3.1.2. Three-Plus-One-Dimensional Theories

Let us consider the case of the full Langevin Equations (8), (13) and (21). When the space is discretized, (8) can be seen as the dynamics of many body particles. Thus, the procedure in the previous subsection can be utilized. Taking the continuum limit, we obtain the following expression which is analogous to (30),

$$\begin{aligned} p(\phi_c(\boldsymbol{x}),t|\phi'_c(\boldsymbol{x}),t') = &\int_{\phi_c(t',\boldsymbol{x})=\phi'_c(\boldsymbol{x})}^{\phi_c(t,\boldsymbol{x})=\phi_c(\boldsymbol{x})} \mathcal{D}(\phi_c,\Pi_c,\phi_\Delta,\Pi_\Delta) \\ &\times \exp\left[i\int \mathrm{d}^4x\Big(\Pi_\Delta\dot{\phi}_c - \phi_\Delta\dot{\Pi}_c - H(\phi_c,\Pi_c,\phi_\Delta,\Pi_\Delta)\Big)\right]. \end{aligned} \tag{33}$$

We define the Hamiltonian $H(\phi_c,\Pi_c,\phi_\Delta,\Pi_\Delta)$ as

$$\begin{aligned} H(\phi_c,\Pi_c,\phi_\Delta,\Pi_\Delta) =& \frac{\Pi_c\Pi_\Delta}{a^3} - (a\nabla^2\phi_c - a^3V'(\phi_c))\phi_\Delta \\ &- \frac{i}{2}\sum_{\alpha,\beta}\int \mathrm{d}^4x' X_\alpha(x)G^{\alpha\beta}(x,x')X_\beta(x') \,, \end{aligned} \tag{34}$$

where the Greek indices $\alpha,\beta = (\phi,\Pi)$ label the Δ fields as $(X_\phi, X_\Pi) = (\Pi_\Delta, -\phi_\Delta)$, and $G^{\alpha\beta}$ denotes the correlations of noises:

$$G^{\alpha\beta}(x,x') := \langle\xi^\alpha(x)\xi^\beta(x')\rangle_\xi \,, \tag{35}$$

where the RHS is given in (15). We use the formula (33) in Sections 4.2 and 5. Note that we have not changed the integration variables as $(\phi_\Delta,\Pi_\Delta) \to (-i\phi_\Delta, -i\Pi_\Delta)$.

In Sections 4 and 5, we take the initial state and the final state to be the false vacuum and the field configurations after tunneling (true vacuum or bubble), respectively.

3.2. Transition Probability from the Schwinger–Keldysh Formalism

In the above prescription, we treat IR fields $(\phi_{\mathrm{IR}},\Pi_{\mathrm{IR}})$ as classical stochastic variables. To quantify the validity of this approximation, we derive the stochastic approach from the first principles based on the Schwinger–Keldysh formalism. Note that we use the label "IR" for IR fields in this subsection following the notation adopted in Appendix A.

Suppose that the system is in the false vacuum $|\Psi_{\mathrm{false}}\rangle$ at the past infinity $t = -\infty$, and it evolves into a certain quantum state $|\{\phi^{\mathrm{IR}}(T,\boldsymbol{x})\}_{\boldsymbol{x}\in\mathcal{D}}\rangle$ at the final time $t = T$, where $|\{\phi^{\mathrm{IR}}(T,\boldsymbol{x})\}_{\boldsymbol{x}\in\mathcal{D}}\rangle$ is an eigenstate of an IR field $\hat{\phi}^{\mathrm{IR}}(\boldsymbol{x})$ at spatial points $\boldsymbol{x}$ in the domain $\mathcal{D}$ with an eigenvalue $\phi^{\mathrm{IR}}(T,\boldsymbol{x})$. Here, $\hat{\phi}^{\mathrm{IR}}(\boldsymbol{x})$ is an IR field in the Schrödinger picture[3]. The transition probability p for this process is given by

$$p(\{\phi^{\mathrm{IR}}(T,\boldsymbol{x})\}_{\boldsymbol{x}\in\mathcal{D}}) = \left|\langle\{\phi^{\mathrm{IR}}(T,\boldsymbol{x})\}_{\boldsymbol{x}\in\mathcal{D}}|\hat{U}(t,-\infty)|\Psi_{\mathrm{false}}\rangle\right|^2 , \tag{36}$$

where $\hat{U}(t,t')$ describes the unitary time evolution from t' to t. The RHS is the Fourier component of the generating functional $Z[J^{\mathrm{IR}}]$ for IR fields, and we have

$$p(\{\phi^{\rm IR}(T,x)\}_{x\in\mathcal{D}}) = \prod_{x\in\mathcal{D}} \int {\rm d}J^{\rm IR}(T,x) Z[J^{\rm IR}(T)] e^{-iJ^{\rm IR}(T,x)\phi^{\rm IR}(T,x)} \Big|_{\{J^{\rm IR}(T,y)=0\}_{y\notin\mathcal{D}}} , \quad (37)$$

where the generating functional $Z[J^{\rm IR}(T)]$ is defined by

$$Z[J^{\rm IR}(T)] := {\rm Tr}\left[\hat{U}(T,-\infty)\,|\Psi_{\rm false}\rangle\,\langle\Psi_{\rm false}|\,\hat{U}^{\dagger}(T,-\infty)\prod_x e^{iJ^{\rm IR}(T,x)\phi^{\rm IR}(T,x)}\right]. \quad (38)$$

We can evaluate $Z[J^{\rm IR}(T)]$ nonperturbatively in the IR sector based on the method developed in [22,23]. Our strategy is that we integrate out UV sector $k > k_c(t)$ perturbatively in nonlinear couplings to obtain $Z[J^{\rm IR}(T)]$. This is analogous to tracing out environmental degrees of freedom to evaluate the reduced density matrix for the system under considerations. Each step we take can be summarized as follows:

1. First, we derive a path integral representation of $Z[J^{\rm IR}(T)]$. We then split the integration variables into UV fields and IR fields such that the integration contour of UV variables of ϕ is closed; see discussions around Equation (A4) for more details[4].
2. We perform the integration over UV variables and evaluate an IR effective action, the so-called Feynman–Vernon influence functional [25] perturbatively.
3. We substitute the obtained expression of $Z[J^{\rm IR}(T)]$ into (37), giving rise to the path integral expression for the transition probability p.

Technical details are shown in Appendix A. At step 2, we assume that the quantum state for nonzero modes is given by the Bunch–Davies vacuum state for a free field defined around the false vacuum, and that the exact zero mode provides a non-fluctuating classical background field configuration $\phi = \phi_{\rm false}$.

In this way, we can obtain the path integral expression for the probability p as

$$p(\{\phi^{\rm IR}(T,x)\}_{x\in\mathcal{D}}) = \int \mathcal{D}(\phi_c^{\rm IR},\phi_\Delta^{\rm IR},\Pi_c^{\rm IR},\Pi_\Delta^{\rm IR})\; e^{i\mathcal{S}}\,\delta[\mathcal{C}_{\rm fin}]\delta[\mathcal{C}_{\rm ini}]\,. \quad (39)$$

Here, $\delta[\mathcal{C}_{\rm fin}]$ and $\delta[\mathcal{C}_{\rm ini}]$ fix the boundary conditions of the path integral at the final time and the initial time, respectively:

$$\delta[\mathcal{C}_{\rm fin}] := \prod_{x\in\mathcal{D}} \delta(\phi_c^{\rm IR}(T,x) - \phi^{\rm IR}(T,x)) \prod_y \delta(\phi_\Delta^{\rm IR}(T,y))\,, \quad (40a)$$

$$\delta[\mathcal{C}_{\rm ini}] := \prod_x \delta\Big(\phi_c^{\rm IR}(-\infty,x) - \phi_{\rm false}\Big)\delta\Big(\Pi_c^{\rm IR}(-\infty,x)\Big)\delta[\mathcal{C}_{\rm detail}]\,, \quad (40b)$$

where $\delta[\mathcal{C}_{\rm detail}]$ is not relevant for the main points here, and we define it in Appendix A. A term $\mathcal{S}$ may be understood as an IR effective action,

$$\mathcal{S} := \int {\rm d}^4x \left[\Pi_\Delta^{\rm IR}\dot{\phi}_c^{\rm IR} - \phi_\Delta^{\rm IR}\dot{\Pi}_c^{\rm IR} - H(\phi_c^{\rm IR},\Pi_c^{\rm IR},\phi_\Delta^{\rm IR},\Pi_\Delta^{\rm IR}) - \delta H\right], \quad (41a)$$

$$\delta H := a^3(t)\left[V(\phi_c^{\rm IR} + (\phi_\Delta^{\rm IR}/2)) - V(\phi_c^{\rm IR} - (\phi_\Delta^{\rm IR}/2)) - V'(\phi_c^{\rm IR})\phi_\Delta^{\rm IR}\right] + \delta H_{\rm higher}\,, \quad (41b)$$

where H is given in (34). $\delta H_{\rm higher}$ denotes the higher-order terms in the coupling constant. Note that Equation (39) allows us to understand the transition probability p in terms of the stochastic dynamics: this point is discussed in Appendix A.3.

Validity of Equation (33)

Now, we find that $\mathcal{S}$ coincides with the term in the exponent of (33) up to the term δH. Hence, (39) shows that the result (33) in the previous section can be justified from the first principles when the term δH is negligible. Suppose that the potential $V(\phi)$ is sufficiently flat in the regime of interest such that the perturbation theory works when integrating out UV modes. In this case, we can systematically calculate δH_{higher} perturbatively, and we have $\delta H_{\text{higher}} = 0$ at the leading order. We set $\delta H_{\text{higher}} = 0$ in the present analysis. We should keep in mind however that the nonperturbative physics at UV scales $k > k_c(t) = \varepsilon a(t) H$ is lost by setting $\delta H_{\text{higher}} = 0$. From this viewpoint, we should choose ε as large as possible.

Even after setting $\delta H_{\text{higher}} = 0$, we have nonzero δH. In the classical approximation, we may regard the Δ-variables $(\phi_\Delta^{\text{IR}}, \Pi_\Delta^{\text{IR}})$ as tiny quantities and keep only the terms which are linear in the Δ-variables in $\mathcal{S}$, leading to $\delta H \simeq 0$. In the case $\varepsilon \ll 1$, this approximation would work thanks to the squeezing of quantum fluctuations at super-horizon scales. However, it is more subtle if we can set $\delta H = 0$ in the case $\varepsilon \gtrsim 1$. We revisit this issue in Section 5.

4. Hawking–Moss Tunneling

In this section, we discuss the Hawking–Moss (HM) tunneling from the perspective of the MSRJD functional integral. We calculate the tunneling rate defined by (31) and (33) in Sections 4.1 and 4.2, respectively.

Hereafter, for the numerical calculations, we consider the potential [26] in Figure 2;

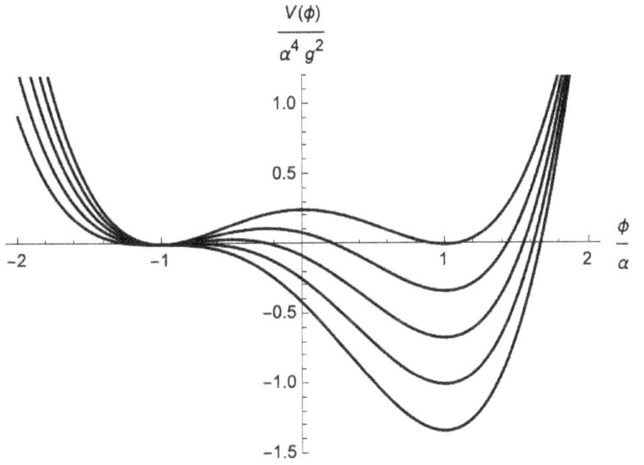

Figure 2. The potential of (42) for various β's. From top to bottom, the curves correspond to $\beta = 0, 0.25, 0.5, 0.75, 1$.

$$V(\phi) = \frac{g^2}{4}\left[(\phi^2 - \alpha^2)^2 + \frac{4\beta}{3}(\alpha\phi^3 - 3\alpha^3\phi - 2\alpha^4)\right], \tag{42}$$

for which the derivative with respect to ϕ is given by a simple form

$$V'(\phi) = g^2(\phi - \alpha)(\phi + \alpha)(\phi + \alpha\beta)\,. \tag{43}$$

For this potential, we have $\phi_{\text{false}} = -\alpha$, $\phi_{\text{true}} = \alpha$ and $\phi_{\text{top}} = -\beta\alpha$. The dimensionless parameter β must be $0 < \beta < 1$, otherwise $\phi = -\alpha$ is not a false vacuum. The potential at each point takes the following values:

$$V(-\alpha) = 0\,, \quad V(\alpha) = -\frac{4\beta g^2\alpha^4}{3}\,, \quad V(-\beta\alpha) = \frac{g^2\alpha^4}{12}(1-\beta)^3(3+\beta)\,. \tag{44}$$

The height of potential barrier ΔV between the false vacuum and the true vacuum is

$$\Delta V := V(-\beta\alpha) - V(-\alpha) = \frac{g^2\alpha^4}{12}(1-\beta)^3(3+\beta)\,. \tag{45}$$

The mass of fluctuations around a given point ϕ is

$$V''(\phi) = \alpha^2 g^2 \Big[3(\phi/\alpha)^2 + 2\beta(\phi/\alpha) - 1\Big] . \tag{46}$$

In our analysis, we work in the quantum field theory on the fixed de Sitter background. This assumption is valid when the condition $\max\big[|V(-\alpha)|, |V(-\beta\alpha)|, |V(\alpha)|\big] \ll 3M_{\rm pl}^2 H^2$ is satisfied. This is always satisfied when the following condition is imposed

$$H \ll \frac{3gM_{\rm pl}}{2}\left(\frac{H}{g\alpha}\right)^2 \approx 2.9 \times 10^{12}\,\mathrm{GeV}\left(\frac{g}{1.12 \times 10^{-4}}\right)\left(\frac{\sqrt{140}}{g\alpha/H}\right)^2 , \tag{47}$$

where we use $0 < \beta < 1$ and choose parameters $(g, g\alpha/H) = (1.12 \times 10^{-4}, \sqrt{140})$ as a benchmark point for later convenience. This shows that we can ignore the backreaction consistently with the weak coupling $g \ll 1$, provided that H is well below the Planck scale $M_{\rm pl} \approx 2.44 \times 10^{18}$ GeV. Note that under this condition, our discussion below is also applicable in the case of a quasi de Sitter background. However, it is non-trivial to generalize our analysis to the case when a non-negligible backreaction on H exists. This is because it is necessary to extend the first-principle derivation of the stochastic approach to account for the backreaction. This is in itself an interesting issue which we leave for future work.

4.1. The Case for the Reduced Langevin Equation

Let us start from the simpler formula (31). Hamilton's equations for the Hamiltonian $H(\phi_c, \Pi_\Delta)$ defined in (32) are given by

$$\begin{aligned} \dot{\phi}_c &= -\frac{V'(\phi_c)}{3H} - \frac{H^3}{4\pi^2}\Pi_\Delta \\ \dot{\Pi}_\Delta &= \frac{V''(\phi_c)}{3H}\Pi_\Delta . \end{aligned} \tag{48}$$

There exist many Hamiltonian flow lines in phase space (ϕ_c, Π_Δ) that represent the solution of Hamilton's equations as shown in Figure 3. Each flow line can be specified by the value of the Hamiltonian since it is conserved on the given flow line. In the Figure 3, there are two important flow lines corresponding to $H(\phi_c, \Pi_\Delta) = 0$. Since the Hamiltonian is the quadratic polynomial of Π_Δ, there are two Π_Δ satisfying $H(\phi_c, \Pi_\Delta) = 0$:

$$\Pi_\Delta = 0, \, -\frac{8\pi^2}{3H^4}V' . \tag{49}$$

In the case of $\Pi_\Delta = 0$, Equation (48) reduces to

$$\dot{\phi}_c = -\frac{V'(\phi_c)}{3H} . \tag{50}$$

Similarly, in the case of $\Pi_\Delta = -\frac{8\pi^2}{3H^4}V'(\phi_c)$, Equation (48) reduces to

$$\dot{\phi}_c = \frac{V'(\phi_c)}{3H} . \tag{51}$$

These two flow lines have three intersections which correspond to the false vacuum, the top of the potential hill and the true vacuum from left to right. On the flow line with $\Pi_\Delta = 0$, the false and true vacuums are stable and the top of the potential is unstable, while on the flow line with $\Pi_\Delta \neq 0$, the stability is reversed. The different signs of the potential forces account for the reverse of the stability at the intersections. We see that the tunneling configuration exists; starting from the false vacuum $(\phi_c, \Pi_\Delta) = (-\alpha, 0)$, going to the top of the potential $(\phi_c, \Pi_\Delta) = (-\beta\alpha, 0)$ through the flow line with $\Pi_\Delta \neq 0$ and finally going to the true vacuum through the flow line with $\Pi_\Delta = 0$.

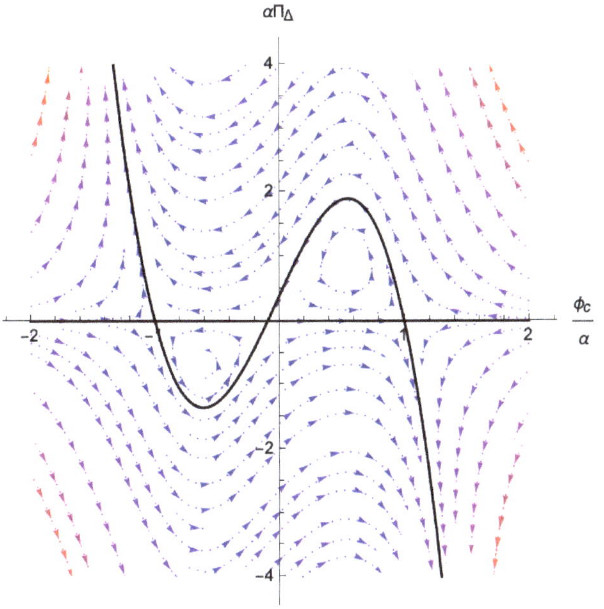

Figure 3. The Hamilton flow for (48) with $\alpha = H$, $g = 0.4$ and $\beta = 0.1$. The solid lines correspond to $H(\phi_c, \Pi_\Delta) = 0$. From left to right, the intersections of the solid lines correspond to the false vacuum, the top of the potential hill and the true vacuum.

We numerically solve the second equation with the initial conditions $\phi_c(t') = -\alpha$ and switch to the first equation when the field reaches the top of potential $\phi_c(t) = -\beta\alpha$. Thus, we obtain the tunneling configuration in Figure 4.

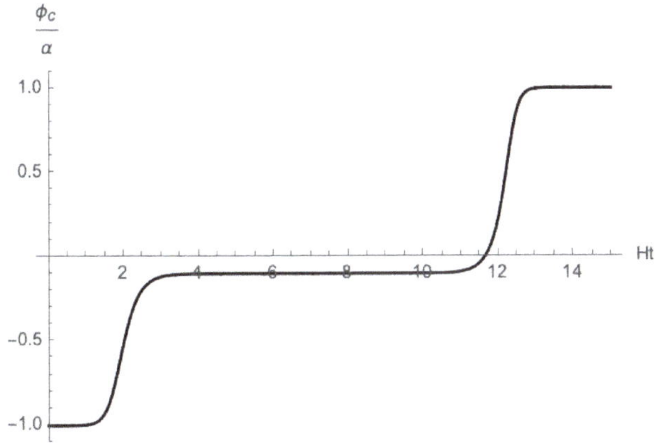

Figure 4. The tunneling configuration for the homogeneous fields is plotted for $\beta = 0.1$ and $\alpha g/H = \sqrt{10}$. We choose the initial condition as $\phi_c(10^{-15})/\alpha = -1 + 10^{-5}$ and solve the second equation of (50). When ϕ_c/α reach the top of potential $\phi_c/\alpha = -\beta$, we switch to the first equation of (50) with the initial condition $\phi_c/\alpha = -\beta + 10^{-5}$.

Now, we evaluate the transition probability (31) by substituting the above tunneling configurations. The action in (31) can be calculated without concrete solutions for ϕ_c;

$$I = \int_{t'}^{t} dt[\Pi_\Delta \dot{\phi}_c - H(\phi_c, \Pi_\Delta)] = -\frac{8\pi^2}{3H^4}\int_{t'}^{t_*} dt \dot{\phi}_c V'(\phi_c) = -\frac{8\pi^2}{3H^4}\Delta V, \quad (52)$$

where t_* is the time at which ϕ_c reaches the top of the potential, and ΔV is the difference of energy density between the top of the potential and the false vacuum. The trajectory from the top of the potential to the true vacuum does not contribute to the transition probability because $H(\phi_c, \Pi_\Delta) = 0$ and $\Pi_\Delta = 0$ on its trajectory. Thus, the transition probability of the tunneling can be evaluated as

$$p(\alpha, t| - \alpha, t') \sim \exp\left(-\frac{8\pi^2}{3H^4}\Delta V\right). \quad (53)$$

This shows a complete agreement with the result of the HM instanton on the fixed background [27].

Comments on this result are in order. It is pointed out in [27] that when the potential $V(\phi)$ has several degenerate local maxima, all of them yield the same tunneling action in the Euclidean method, despite differing distances from the false vacuum. In our formalism, however, a transition to a distant maximum is described by a saddle-point solution that passes through all the intermediate maxima, and the transition probability is given by the product of the factors on the RHS of (53) for each maximum. The aforementioned subtlety is absent in our formalism.

Since the analysis in Section 4.1 is based on (23) describing the stochastic dynamics of the IR field ϕ_c at a single spatial point, Equation (53) calculates the tunneling probability at a single spatial point, not on the whole universe[5]. Hence, physically speaking, (53) provides the tunneling probability of a coarse-grained patch with a physical radius $(\varepsilon H)^{-1}$. This point also becomes manifest in Section 4.2.

4.2. The Case for the Full Langevin Equation

Next, we evaluate the tunneling rate based on the formula (33). An advantage of (33) is that one can discuss various non-trivial dynamics in the whole universe covered by the flat chart in principle.

In the previous section, we focused on the dynamics at a single spatial point and found the HM tunneling process as a non-trivial solution of Hamilton's equations. In this section, we discuss the corresponding process in the global picture based on (33) by identifying a non-trivial configuration that satisfies Hamilton's equations in a good approximation. We then reproduce the result (52). We also estimate the characteristic time scale of the tunneling. In Section 4.2.1, we remark some technical complications that arise in the global picture.

We start with simplifying the expression (33). As discussed in Section 2, $G^{\phi\phi}$ becomes dominant, and the other terms can be neglected for $\varepsilon \ll 1$ in (34). Under this approximation, the exponent of the integrand of (33) is linear in ϕ_Δ. We can then perform the path integral $\int \mathcal{D}\phi_\Delta$ in (33), yielding the product of delta functions $\prod_x \delta(\dot{\Pi}_c - a\nabla^2\phi_c + a^3V'(\phi_c))$. These delta functions give an equation of motion for Π_c and eliminate the path integral $\int \mathcal{D}\Pi_c$. We solve this equation as $\Pi_c \simeq -a^3V'(\phi_c)/3H$, ignoring the spatial gradient and the time variation in ϕ_c. Furthermore, we change the integration variable as $\Pi_\Delta \to -i\Pi_\Delta$ in (33). Consequently, (33) reduces to

$$p(\phi_c(x), t|\phi'_c(x), t') \simeq \int_{\phi_c(t',x)=\phi'_c(x)}^{\phi_c(t,x)=\phi_c(x)} \mathcal{D}(\phi_c, \Pi_\Delta) \exp\left[\int d^4x(\Pi_\Delta\dot{\phi}_c - H_{\mathrm{HM}}(\phi_c, \Pi_\Delta))\right], \quad (54)$$

where the Hamiltonian H_{HM} is defined as

$$H_{\rm HM}(\phi_c, \Pi_\Delta) := -\frac{1}{3H}V'(\phi_c)\Pi_\Delta - \frac{H^3}{8\pi^2}\Pi_\Delta\overline{\Pi}_\Delta\,, \tag{55}$$

and $\overline{\Pi}_\Delta$ is given by

$$\overline{\Pi}_\Delta(t,\boldsymbol{x}) := \int {\rm d}^3x'\, j_0(k_c(t)|\boldsymbol{x}-\boldsymbol{x}'|)\Pi_\Delta(t,\boldsymbol{x}') = \int \frac{{\rm d}^3k}{(2\pi)^3}\frac{4\pi^2}{H^3}\frac{H^2}{2k^3}\delta(t-t_k)\Pi_\Delta(t,\boldsymbol{k})e^{i\boldsymbol{k}\cdot\boldsymbol{x}}\,. \tag{56}$$

Here, we define t_k by the condition $k_c(t_k) = k$. Hamilton's equations are

$$\dot{\phi}_c = -\frac{1}{3H}V'(\phi_c) - \frac{H^3}{4\pi^2}\overline{\Pi}_\Delta\,, \tag{57a}$$

$$\dot{\Pi}_\Delta = \frac{1}{3H}V''(\phi_c)\Pi_\Delta\,. \tag{57b}$$

Here, the term $\dot{\Pi}_\Delta(t,\boldsymbol{x})$ on the LHS of (57b) is defined as the Fourier transform of $\dot{\Pi}_\Delta(t,\boldsymbol{k})$ i.e., $\dot{\Pi}_\Delta(t,\boldsymbol{x}) := (2\pi)^{-3}\int {\rm d}^3k\,\dot{\Pi}_\Delta(t,\boldsymbol{k})e^{i\boldsymbol{k}\cdot\boldsymbol{x}}$. The same rule applies to other IR variables in the real space as in Equation (A24). Intuitively, this is because the stochastic formalism is formulated in the momentum space, and its Fourier transform provides the stochastic formalism in the real space; see Appendix A for more details.

Equation (57) admits a trivial configuration $\Pi_\Delta(t,\boldsymbol{k}) = 0$ under which we have $H_{\rm HM} = 0$, and (57a) becomes $\dot{\phi}_c = -V'(\phi_c)/(3H)$, describing the standard classical time evolution. Now, we are interested in the non-trivial configuration which satisfies (57). In the previous section, we focused on the non-trivial dynamics at the single spatial point $\boldsymbol{x} = \boldsymbol{x}_0$ and it was found that the tunneling configuration was obtained by $\Pi_\Delta(t,\boldsymbol{x}_0) = -\frac{8\pi^2}{3H^4}V'(\phi_c(t,\boldsymbol{x}_0))$. This configuration can be naturally extended to the one in the global picture by considering the following configuration

$$\Pi_\Delta(t,\boldsymbol{k}) = -\frac{8\pi^2}{3H^4}V'(\phi_c(t,\boldsymbol{x}_0))e^{-i\boldsymbol{k}\cdot\boldsymbol{x}_0}\,. \tag{58}$$

From now on, we show that this configuration satisfies a set of Hamilton Equations (57) in a good approximation. For this configuration, we have

$$\Pi_\Delta(t,\boldsymbol{x}) = -\frac{8\pi^2}{3H^4}V'(\phi_c(t,\boldsymbol{x}_0))f(r;t)\,, \tag{59a}$$

$$\dot{\Pi}_\Delta(t,\boldsymbol{x}) = -\frac{8\pi^2}{3H^4}V''(\phi_c(t,\boldsymbol{x}_0))\dot{\phi}_c(t,\boldsymbol{x}_0)f(r;t)\,, \tag{59b}$$

$$\overline{\Pi}_\Delta(t,\boldsymbol{x}) = -\frac{8\pi^2}{3H^4}V'(\phi_c(t,\boldsymbol{x}_0))j_0(k_c(t)r)\,, \tag{59c}$$

where $r := |\boldsymbol{x}-\boldsymbol{x}_0|$, and $f(r;t) := (2\pi)^{-3}\int {\rm d}^3k\,\theta(k_c(t)-k)\,e^{i\boldsymbol{k}\cdot(\boldsymbol{x}-\boldsymbol{x}_0)}$. The functions $j_0(k_c(t)r)$ and $f(r;t)$ rapidly oscillate for $r \gg k_c^{-1}(t)$. Hence, they exponentially decay to zero after they are smeared over the Hubble time: for instance, j_0 and f becomes the smooth function after the suitable coarse-graining as[6]

$$j_0(k_c(t)r) \xrightarrow{\text{time c.g.}} W(r;t) := \int {\rm d}t' w(t',t) j_0(k_c(t')r) \approx \begin{cases} 1 & (r \ll k_c^{-1}(t)) \\ 0 & (r \gg k_c^{-1}(t)) \end{cases},$$

$$f(r;t) \xrightarrow{\text{time c.g.}} W_f(r;t) := \int {\rm d}t'\, w(t',t) f(r;t') \approx \begin{cases} k_c^3(t)/(6\pi^2) & (r \ll k_c^{-1}(t)) \\ 0 & (r \gg k_c^{-1}(t)) \end{cases}. \tag{60}$$

Here, we perform the smearing by using a window function $w(t',t)$ which is approximately constant in the domain $|t-t'| \ll H^{-1}$ while it exponentially decays to zero at $|t-t'| \gg$

H^{-1}. We also impose the normalization condition $\int \mathrm{d}t' w(t',t) = 1$. We can choose $w(t',t) = H\pi^{-1/2}\exp[-H^2(t-t')^2]$ for example. Note that the property (60) is based on the behavior $j_0(k_c r) \simeq 1$ and $f(r;t) \simeq k_c^3/(6\pi^2)$ for $r \ll k_c^{-1}$ and the rapid oscillations of $j_0(k_c r)$ and $f_0(r;t)$ for $r \gg k_c^{-1}$. Hence, the property (60) is robust against the choice of $w(t',t)$.

Therefore, substituting Equations (59c) and (60) into Equation (57a) and performing the smearing in time, we obtain

$$\dot{\phi}_c(t,\boldsymbol{x}) = -\frac{1}{3H}\left[V'(\phi_c(t,\boldsymbol{x})) - 2V'(\phi_c(t,\boldsymbol{x}_0))j_0(k_c(t)r)\right] \tag{61a}$$

$$\xrightarrow{\text{time c.g.}} \begin{cases} -V'(\phi_c(t,\boldsymbol{x}))/(3H) & (r \gg k_c^{-1}(t)) \\ V'(\phi_c(t,\boldsymbol{x}_0))/(3H) & (r \ll k_c^{-1}(t)) \end{cases}. \tag{61b}$$

This exhibits the non-trivial dynamics only in the domain $r \lesssim k_c^{-1}$, i.e., a coarse-grained patch centered at a point $\boldsymbol{x}_0$ with a physical radius $(\varepsilon H)^{-1}$. Here, we used $\phi_c(t,\boldsymbol{x}) \simeq \phi_c(t,\boldsymbol{x}_0)$ for $r \ll k_c^{-1}(t)$. In the second line, it is assumed that $V'(\phi_c(t,\boldsymbol{x}_0))$ does not cancel the rapid oscillations of $j_0(k_c(t)r)$ so that their product $V'j_0$ decays exponentially at $r \gg k_c^{-1}$ after the suitable smearing. This assumption is valid thanks to the approximate constancy of super-horizon fluctuations $\phi_c(t,\boldsymbol{k})$ over the Hubble time: $\phi_c(t,\boldsymbol{k}) \simeq \phi_c(t',\boldsymbol{k})$ for $|t-t'| \lesssim H^{-1}$. Adopting the same assumption, from (59a) and (59b), we also obtain the smeared expressions of Π_Δ and $\dot{\Pi}_\Delta$ which turn out to satisfy (57b) under the condition (57a). Hence, we conclude that (59) describes a non-trivial saddle-point solution in a good approximation. Furthermore, after the smearing, we can find $H_{\mathrm{HM}} \approx 0$ for that configuration.

Now, we use the configuration (58) or (59) to obtain the tunneling configuration shown in Figure 4 again. We can calculate the action for this tunneling configuration similarly to (52). By substituting the exact expressions (59) and (61a), we obtain[7]

$$\begin{aligned} I &= -\frac{H^3}{8\pi^2}\int_{t'}^{t_*}\mathrm{d}t \int \mathrm{d}^3x\, \Pi_\Delta(t,\boldsymbol{x})\overline{\Pi}_\Delta(t,\boldsymbol{x}) \\ &= -\frac{8\pi^2}{3H^4}\int_{t'}^{t_*}\mathrm{d}t\, V'(\phi_c(t,\boldsymbol{x}_0))\frac{V'(\phi_c(t,\boldsymbol{x}_0))}{3H}\int \mathrm{d}^3x\, j_0(k_c(t)r)f(r;t) \\ &= -\frac{8\pi^2}{3H^4}\int_{t'}^{t_*}\mathrm{d}t\, \dot{\phi}_c(t,\boldsymbol{x}_0)V'(\phi_c(t,\boldsymbol{x}_0)) = -\frac{8\pi^2}{3H^4}\Delta V. \end{aligned} \tag{62}$$

In the first line, we used (57a) to eliminate $\dot{\phi}_c$ from the integrand. In the second line, we substituted the concrete configurations (59). In the third line, we used (61a) and performed the spatial integral as $\int \mathrm{d}^3x\, j_0(k_c(t)r)f(r;t) = 1$. Thus, we reproduced the previous result (52). Our analysis confirms that (62) gives the tunneling probability of a coarse-grained patch with a physical radius $(\varepsilon H)^{-1}$.

It is worth mentioning the time scale of the HM tunneling. We may naively calculate the typical time scale of the HM tunneling as $t_* - t' = \int_{\phi_{\mathrm{false}}}^{\phi_{\mathrm{top}}} \mathrm{d}\phi_c \frac{3H}{V'(\phi_c)}$. Here, the spacetime argument of the field is suppressed. However, this integral is in general divergent since $V' = 0$ at both ends of the integral. In the vicinity of the domain where $V' = 0$, we expect that the quantum fluctuations of the field would play an important role. Since the typical size of quantum fluctuations accumulated over the Hubble time is $H/(2\pi)$, we may define the regulated quantity t_{HM} as

$$t_{\mathrm{HM}} := \int_{\phi_{\mathrm{false}}+\frac{H}{2\pi}}^{\phi_{\mathrm{top}}-\frac{H}{2\pi}} \mathrm{d}\phi_c \frac{3H}{V'(\phi_c)}, \tag{63}$$

and we expect that t_{HM} would correctly characterize the typical time scale of the HM tunneling. In particular, for our potential (42), we have

$$t_{\rm HM} = \frac{3H}{2g^2\alpha^2(1+\beta)}\left[\left(\frac{3+\beta}{1-\beta}\right)\log\left|1-\frac{1-\beta}{H/2\pi\alpha}\right| - \log\left|\frac{1-\frac{2}{H/2\pi\alpha}}{1+\frac{1+\beta}{H/2\pi\alpha}}\right|\right] \sim \frac{1}{H}\cdot\mathcal{O}\Big((H/g\alpha)^2\Big). \quad (64)$$

Since we have $(H/g\alpha)^2 \sim (H^2/V'')_{\phi_{\rm false}\lesssim\phi_c\lesssim\phi_{\rm top}} \gg 1$ for the shallow potential, we conclude that the time scale of HM tunneling is much longer than the Hubble time.

4.2.1. Remark

So far it has been found that starting from (33), the configuration (58) approximately solves Hamilton's Equation (57) and reproduces the previous result (52) when it is substituted into the action. There is actually an important reason behind why we should choose the configuration (58) to evaluate the HM tunneling process.

To see this, it is important to realize an important difference between the current Hamilton Equation (57) and the previous one (48); the equation for ϕ_c now contains $\overline{\Pi}_\Delta$ rather than Π_Δ. Due to this difference, the structures of Hamilton's Equation (57) are understood as follows;

- The equation for ϕ_c (57a) is determined once $\overline{\Pi}_\Delta$ is specified;
- $\overline{\Pi}_\Delta(t,\boldsymbol{x})$ is understood as initial conditions for (57b) since $\overline{\Pi}_\Delta(t,\boldsymbol{x})$ contains only the boundary modes satisfying $k = k_c(t)$. Hence, $\Pi_\Delta(t,\boldsymbol{x})$ is obtained by solving (57b) for a given $\overline{\Pi}_\Delta$ under the condition (57a).

On top of that, substituting Hamilton's equation for ϕ_c into the action, we can rewrite the first line of (62) as

$$I = -\frac{1}{2}\int_{t'}^{t_*} \mathrm{d}t \int \frac{\mathrm{d}^3k}{(2\pi)^3}\,\Pi_\Delta(t,\boldsymbol{k})\Pi_\Delta(t,-\boldsymbol{k})\frac{H^2}{2k^3}\delta(t-t_k)\,. \quad (65)$$

The RHS depends only on the value of $\Pi_\Delta(t,\boldsymbol{k})|_{k=k_c(t)}$. Hence, we can evaluate the tunneling action once $\overline{\Pi}_\Delta$ is specified, provided that (65) or equivalently (33) is reliable.

We claim that (33) is valid when $\overline{\Pi}_\Delta$ is chosen so that $\Pi_\Delta(t>t_k,\boldsymbol{k})$ becomes a smooth function in time. Otherwise, it is not justified to ignore the higher-order corrections $\delta H_{\rm higher}$ which are neglected in (33). Intuitively, this states that the perturbation theory tends to be broken down around exotic configurations. When Π_Δ is a smooth function, we can relate $\overline{\Pi}_\Delta$ to Π_Δ after the suitable smearing in time:

$$\overline{\Pi}_\Delta(t,\boldsymbol{x}) \xrightarrow{\text{time c.g.}} \int \mathrm{d}^3x'\,\Pi_\Delta(t,\boldsymbol{x}')W(|\boldsymbol{x}-\boldsymbol{x}'|;t) \sim \mathcal{V}\Pi_\Delta(t,\boldsymbol{x})\,. \quad (66)$$

Here, we also used $\Pi_\Delta(t,\boldsymbol{x}) \sim \Pi_\Delta(t,\boldsymbol{x}')$ for $|\boldsymbol{x}-\boldsymbol{x}'| \lesssim k_c^{-1}$ and defined $\mathcal{V} := (4\pi/3)k_c^{-3}(t) \sim \int \mathrm{d}^3x'\, W(|\boldsymbol{x}-\boldsymbol{x}'|;t)$. We then require that Hamilton's equation for Π_Δ (57b) is satisfied in a good approximation to ensure the validity of (33). This constrains the choice of $\overline{\Pi}_\Delta$. Indeed, the configuration (58) approximately solves (57b) and satisfies the smoothness and henceforth (66) after the smearing. This would be the reason why our analysis with (58) based on (33) can reproduce the correct result.

4.3. *On the ε-Independence of the Hawking–Moss Tunneling*

In Sections 4.1 and 4.2, the tunneling probability of a coarse-grained patch with a physical radius $(\varepsilon H)^{-1}$ is calculated. Our results coincide with the HM tunneling and are independent of ε.

The ε-independence of the results would be a consequence of the scale independence of the dynamics of light scalar fields at super-horizon scales. To see this, it is useful to notice that a coarse-grained patch of physical radius $(\varepsilon_1 H)^{-1}$ at a time $t = T$ expands to a patch of larger physical radius $(\varepsilon_2 H)^{-1}$ at a later time $t = T + \delta t > T$ with $\varepsilon_2 = \varepsilon_1 \exp[-H\delta t] < \varepsilon_1 \ll 1$. The value of the IR field does not evolve from $t = T$ to $T + \delta t$

because the fluctuations of a light scalar field at super-horizon scales are approximately time-independent,

$$\int \frac{d^3k}{(2\pi)^3}\phi_c(t,\mathbf{k})e^{i\mathbf{k}.x}\theta(k_c(t)-k) \simeq \int \frac{d^3k}{(2\pi)^3}\phi_c(t+\delta t,\mathbf{k})e^{i\mathbf{k}.x}\theta(k_c(t)-k), \tag{67}$$

unless δt is much longer than H^{-1}. Hence, we can relate the coarse-grained dynamics of different $\varepsilon \ll 1$ by considering the time shift $t \to t+\delta t$ with the value of the IR field being kept fixed. This means that the tunneling probability from $\phi = \phi_{\text{false}}$ to ϕ_{true} of a patch with physical radius $(\varepsilon H)^{-1}$ should be independent of ε.

5. Coleman–de Luccia Tunneling

In the previous section, the result of the HM instanton was reproduced by using (33). We expect that the Formula (33) can describe not only the HM tunneling but also the CDL tunneling. In other words, we expect a flow line starting from the false vacuum to the bubble configurations as illustrated in Figure 5. In this section, we concretely show an interesting configuration by following the prescription in the previous section.

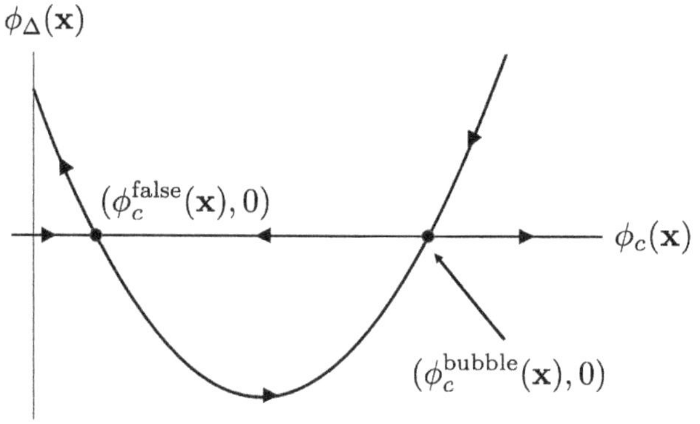

Figure 5. A schematic picture of Hamiltonian flows in the "field" phase space. We expect a flow line which starts from the false vacuum (left intersection) and reaches the bubble configuration (right intersection) through a nonzero $\phi_\Delta(x)$ path.

5.1. Coleman–de Luccia Bubble Solution

In the Euclidean method, the CDL instanton is dominant rather than the HM instanton when the potential barrier is steep [28–31]. By taking $\varepsilon \gg 1$, we can study nonperturbative physics at sub-horizon scales such as the formation of a bubble. In this case, $G^{\Pi\Pi}$ becomes dominant and the other noises can be ignored in (34). Under this approximation, the exponent of the integrand of (33) is linear in Π_Δ. We can then perform the path integral $\int \mathcal{D}\Pi_\Delta$ in (33), yielding the product of delta functions $\prod_x \delta(\dot{\phi}_c - a^{-3}\Pi_c)$. These delta functions eliminate the path integral $\int \mathcal{D}\Pi_c$[8]. Furthermore, we change the integration variable as $\phi_\Delta \to -i\phi_\Delta$ in (33), leading to

$$p \simeq \int \mathcal{D}(\phi_c,\phi_\Delta)\exp\left[\int d^4x\left(a^3\dot{\phi}_\Delta\dot{\phi}_c - H_{\text{CDL}}(\phi_c,\phi_\Delta)\right)\right] \tag{68}$$

with appropriate boundary conditions; we consider bubble configurations as boundary conditions later. Here, the Hamiltonian H_{CDL} is defined by

$$H_{\text{CDL}}(\phi_c,\phi_\Delta) := -\left(a\nabla^2\phi_c - a^3V'(\phi_c)\right)\phi_\Delta - \frac{a^3}{2}\phi_\Delta(x)\overline{\phi}_\Delta(x), \tag{69}$$

with

$$\overline{\phi}_\Delta(t,\boldsymbol{x}) := \frac{Hk_c^4(t)}{4\pi^2 a(t)} \int \mathrm{d}^3x'\, j_0\big(k_c(t)|\boldsymbol{x}-\boldsymbol{x}'|\big)\phi_\Delta(t,\boldsymbol{x}')\,. \tag{70}$$

We would like to find the appropriate configuration $\overline{\phi}_\Delta$ which describes the tunneling process. Following the choice made in Section 4, let us *suppose* that the appropriate choice is given by the configuration for which the Hamiltonian (69) vanishes:

$$\overline{\phi}_\Delta(x) = 0\,,\ -2(a^{-2}\nabla^2\phi_c - V'(\phi_c))\,. \tag{71}$$

Note that this condition is imposed up to the suitable smearing in time because $\overline{\phi}_\Delta(t,\boldsymbol{x})$ includes only the boundary Fourier mode $k = k_c(t)$. As discussed in Section 4.2.1, ϕ_Δ becomes sufficiently smooth for the appropriate $\overline{\phi}_\Delta$, and hence we also suppose

$$\overline{\phi}_\Delta \sim \frac{H^2\varepsilon}{3\pi}\phi_\Delta(x) \tag{72}$$

after the suitable averaging over the Hubble time, similarly to (66). We use the relation (72) to evaluate the action later. This would work at least for the purpose of estimating the action.

In principle, we do not need to use the estimate (72) in evaluating the action since it is determined once $\overline{\phi}_\Delta$ is specified based on the analogous logic discussed around (65). For this purpose, we need to specify the configurations $\{\phi_\Delta(t,\boldsymbol{k})|_{k=k_c(t)}\}$ in momentum space which result in the sufficiently smooth ϕ_Δ that solves the Hamilton equations. It is not easy to find such appropriate configurations precisely, however. Hence, we leave a more careful analysis on the choice of $\overline{\phi}_\Delta$ for future work. In this study, instead, we focus on how we can proceed with the analysis for the given configuration (71) while adopting the estimate (72) and how the bubble nucleation process could be described in the stochastic approach.

Using the relation (72), the Hamiltonian (69) and the configuration (71) become

$$H_{\mathrm{CDL}} \sim -\Big(a\nabla^2\phi_c - a^3V'(\phi_c)\Big)\phi_\Delta - \frac{H^2\varepsilon}{6\pi}a^3\phi_\Delta^2\,, \tag{73}$$

$$\phi_\Delta(x) = 0\,,\ -\frac{6\pi}{H^2\varepsilon}(a^{-2}\nabla^2\phi_c - V'(\phi_c))\,. \tag{74}$$

Hamilton's equations under the constraints (74) give the following equation of motions for ϕ_c:

$$\begin{aligned} &\ddot{\phi}_c + 3H\dot{\phi}_c = -(V'(\phi_c) - a^{-2}\nabla^2\phi_c), \quad (\phi_\Delta = 0), \\ &\ddot{\phi}_c + 3H\dot{\phi}_c = V'(\phi_c) - a^{-2}\nabla^2\phi_c, \qquad \Big(\phi_\Delta = -\frac{6\pi}{H^2\varepsilon}(a^{-2}\nabla^2\phi_c - V'(\phi_c))\Big). \end{aligned} \tag{75}$$

Similar to the previous case (50), the signs of the potential term and the gradient term are flipped in the equation of motion with the non-trivial $\phi_\Delta \neq 0$. Hence, we can expect the tunneling process to be realized. From (68), we find that only the solutions satisfying $\phi_\Delta \neq 0$ contribute to the action. Then, we focus on the second equation in (75).

Interestingly, this equation is the classical equation of motion in the Euclidean anti-de Sitter (AdS) space which is defined by the embedding equation

$$-X_0^2 + X_1^2 + X_2^2 + X_3^2 + X_4^2 = -H^{-2} \tag{76}$$

in five-dimensional Minkowski spacetime

$$\mathrm{d}s^2 = -\mathrm{d}X_0^2 + \mathrm{d}X_1^2 + \mathrm{d}X_2^2 + \mathrm{d}X_3^2 + \mathrm{d}X_4^2\,. \tag{77}$$

In fact, the d'Alembert operator $\partial_t^2 + 3H\partial_t + a^{-2}\nabla^2$ can be obtained from the following induced metrics;

$$\mathrm{d}s^2 = \mathrm{d}t^2 + e^{2Ht}\mathrm{d}\boldsymbol{x}^2 = \frac{1}{H^2\eta^2}(\mathrm{d}\eta^2 + \mathrm{d}\boldsymbol{x}^2) = H^{-2}(\mathrm{d}\rho^2 + \sinh^2\rho\,\mathrm{d}\Omega^2)\,, \tag{78}$$

where $\mathrm{d}\Omega^2 := \mathrm{d}\theta_1^2 + \sin^2\theta_1\mathrm{d}\theta_2^2 + \sin^2\theta_1\sin^2\theta_2\mathrm{d}\theta_3^2$ and $0 < \rho < \infty, 0 \le \theta_1 \le \pi, 0 \le \theta_2 \le \pi$ and $0 \le \theta_3 \le 2\pi$. The second metric is that in the Poincaré coordinates and the third one is that in the global coordinates. The explicit coordinate transformations are given by

$$\begin{aligned}
X_0 &= \frac{1}{2}\frac{\eta^2 + \boldsymbol{x}^2 + H^{-2}}{-\eta} = H^{-1}\cosh\rho, \\
X_1 &= \frac{x_1}{-H\eta} = H^{-1}\sinh\rho\sin\theta_1\cos\theta_2, \\
X_2 &= \frac{x_2}{-H\eta} = H^{-1}\sinh\rho\sin\theta_1\sin\theta_2\cos\theta_3, \\
X_3 &= \frac{x_3}{-H\eta} = H^{-1}\sinh\rho\sin\theta_1\sin\theta_2\sin\theta_3, \\
X_4 &= \frac{1}{2}\frac{\eta^2 + \boldsymbol{x}^2 - H^{-2}}{-\eta} = H^{-1}\sinh\rho\cos\theta_1.
\end{aligned} \tag{79}$$

For later convenience, we derive the relation between $(\eta, r := |\boldsymbol{x}|)$ and (ρ, σ_1) as

$$-H\eta = \frac{1}{\cosh\rho - \sinh\rho\cos\theta_1}, \qquad r = H^{-1}\frac{\sinh\rho\sin\theta_1}{\cosh\rho - \sinh\rho\cos\theta_1}. \tag{80}$$

From this, we can check that a point $(\eta, r) = (-H^{-1}, 0)$ is mapped to $\rho = 0$ for any $\theta_1 \in [0, \pi]$. Except for this, there is one-to-one correspondence between points in the (η, r)-plane with $\eta \le 0$ and $r \ge 0$ and points in the (ρ, θ_1)-plane with $\rho > 0$ and $0 \le \theta_1 \le \pi$. On a constant-η slice in the (η, r)-plane, the value of ρ for a given spatial point r is given by

$$\rho = \operatorname{arccosh}\left(\frac{1}{2}\frac{H^2(\eta^2 + r^2) + 1}{-H\eta}\right). \tag{81}$$

This function increases monotonically in r for a given η in the region $r > 0$.

From the global coordinate (78), we see the Euclidean AdS spacetime has $O(4)$ symmetry. Hence, we assume that the field ϕ_c depends only on ρ. Thus, the second equation of (75) reads

$$\frac{\mathrm{d}^2\phi_c}{\mathrm{d}\rho^2} + \frac{3}{\tanh\rho}\frac{\mathrm{d}\phi_c}{\mathrm{d}\rho} = \frac{V'(\phi_c)}{H^2}. \tag{82}$$

Note that this is the same equation utilized in the Euclidean method [3,5]. Imposing the boundary conditions [1,3]

$$\lim_{\rho\to\infty}\phi_c = -\alpha, \qquad \left.\frac{d\phi_c}{d\rho}\right|_{\rho=0} = 0, \tag{83}$$

we obtain the bubble solutions as shown in Figure 6. Note that the latter condition ensures the continuity of $\partial_\eta\phi_c(\eta, r)$ and $\partial_\eta^2\phi_c(\eta, r)$ at $(H\eta, Hr) = (-1, 0)$.

It is useful to fit the bubble configurations by the following fitting function (Figure 6);

$$\phi_c(\rho) = -\alpha\tanh\mu(\rho - \bar{\rho}), \tag{84}$$

where μ and $\bar{\rho}$ represent the thickness and the position of the bubble wall, respectively. Note that for $\beta \gtrsim 0.6$, the deviation from the true vacuum at $\rho = 0$ becomes significant [26], and the fitting (84) becomes bad.

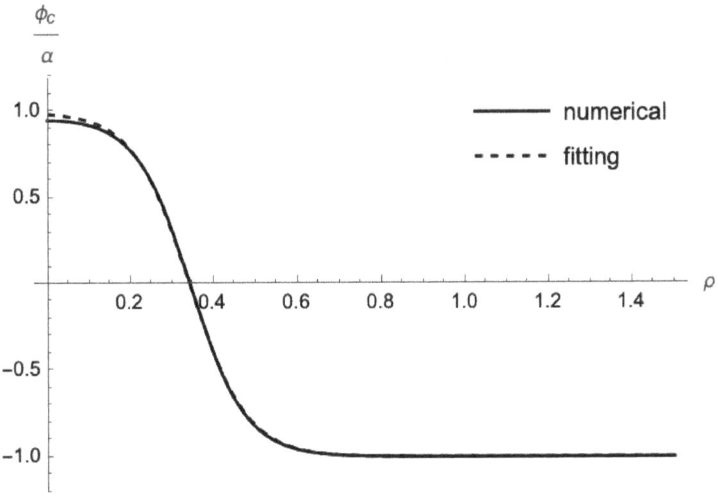

Figure 6. The tunneling configuration for the inhomogeneous field. The solid curve is the numerical solution for $\beta = 0.5$ and $\alpha g/H = \sqrt{140}$. We took the initial conditions as $\phi_c(10^{-15})/\alpha = 1 - 4.98284 \times 10^{-2}$ and $\phi_c'(10^{-15})/\alpha = 0$. The dashed curve is the fitting function (84) where $\mu = 7.32515$ and $\bar{\rho} = 0.34397$.

We also have to examine the boundary conditions in the Hamilton flow. As in the case of HM tunneling, we expect several intersections where $\phi_\Delta = 0$ (Figure 5). Since we have the concrete ϕ_Δ (74) and the bubble configurations (84), it is possible to obtain the curves in the (r, η)-plane on which $\phi_\Delta = 0$. Using (74) and (75), such curves are given by

$$F(\eta, r) := \partial_\eta^2 \phi_c - \frac{2}{\eta} \partial_\eta \phi_c = 0. \tag{85}$$

Note that this is equivalent to solving $a^{-2}\nabla^2\phi_c - V'(\phi_c) = 0$. Substituting (74) into (85) and using the fitting formula (84), we find four solutions which are expressed as four lines in the (η, r)-plane; two of them are placed at $\eta = -\infty$ with fixed r and $r = \infty$ with fixed η. These correspond to $\rho = \infty$ where $\phi = -\alpha$ (Figure 6). Also, we numerically find other two non-trivial hypersurfaces (Figure 7a). It can be seen that one of the curves is totally spacelike but another is partially timelike. Thus, it seems natural to take the $\eta = -\infty$ curve as an initial time slice on which ϕ is a false vacuum and a totally spacelike non-trivial curve as a final time slice on which the bubble is nucleated. We denote the region between the two spacelike curves as Σ.

As a matter of fact, the configurations on the final time slice has a bubble. It can be checked as follows; the value of ϕ_c at the origin exceeds the top of the potential hill. In the case of Figure 7a where we set $(\beta, \alpha g/H) = (0.5, \sqrt{140})$, we have $\phi_c/\alpha \sim -0.1 > -0.5 = -\beta$ at $Hr = 0, H\eta \sim -0.7$.

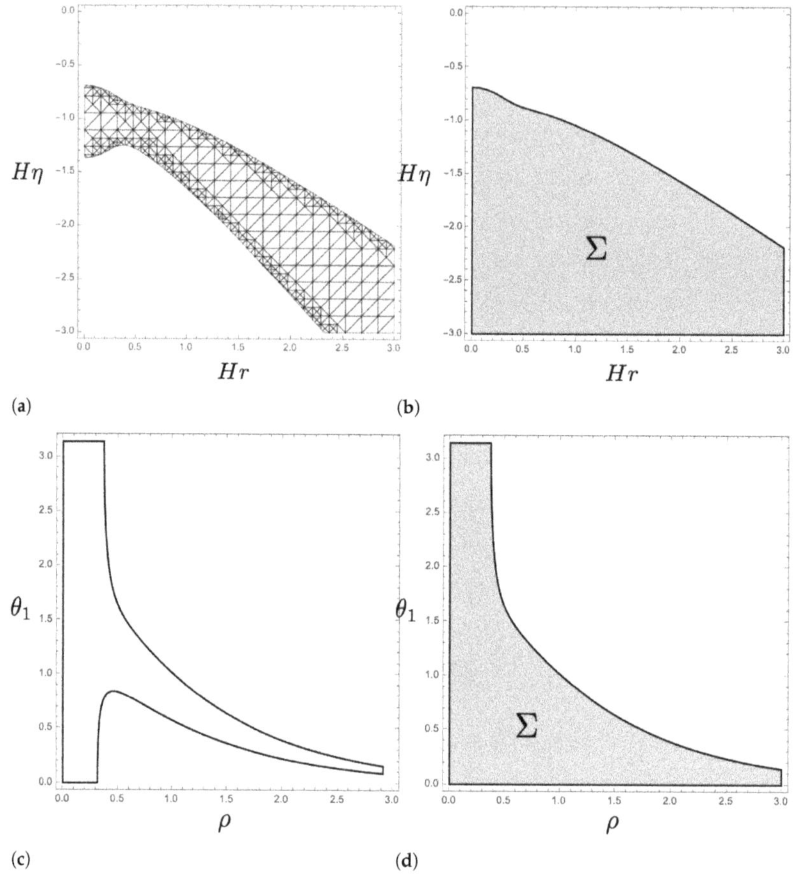

Figure 7. (**a**): The two non-trivial curves on which (85) is satisfied. We inserted the mesh to confirm whether the curves are spacelike or timelike. It can be seen that the upper line is spacelike but the lower one is partially timelike. (**b**): The integration region for (86). There is another spacelike curve at $\eta = -\infty$. (**c**): The two non-trivial curves in (ρ, θ_1) plane. (**d**): The integration region for (86) in the (ρ, θ_1) plane. All plots are with $\beta = 0.5$ and $\alpha g/H = \sqrt{140}$.

Now, we can evaluate the action for the bubble configurations for Σ (Figure 7b,d). We use the fitting function (84) for the numerical integration.

$$
\begin{aligned}
I \simeq & \int_\Sigma \mathrm{d}^4x \left[-a^3 \phi_\Delta \left(\ddot{\phi}_c + 3H\dot{\phi}_c\right) - H_{\mathrm{CDL}}\right] \\
= & -\frac{6\pi}{H^6 \varepsilon} \int_\Sigma \mathrm{d}\rho \mathrm{d}\theta_1 \mathrm{d}\theta_2 \mathrm{d}\theta_3 \frac{\sin^2\theta_1 \sin\theta_2 \sinh^3\rho}{(\cosh\rho - \sinh\rho\cos\theta_1)^4} \left(\partial_\eta^2 \phi_c - \frac{2}{\eta}\partial_\eta \phi_c\right)^2 \\
= & -\frac{24\pi^2}{H^2 \varepsilon} \int_\Sigma \mathrm{d}\rho \mathrm{d}\theta_1 \sin^2\theta_1 \sinh^3\rho \left[\frac{(\sinh\rho - \cosh\rho\cos\theta_1)^2}{(\cosh\rho - \sinh\rho\cos\theta_1)^2} \frac{\partial^2 \phi_c}{\partial \rho^2}\right. \\
& \left. + \left\{3\frac{\sinh\rho - \cosh\rho\cos\theta_1}{\cosh\rho - \sinh\rho\cos\theta_1} + \frac{\sin^2\theta_1}{\tanh\rho(\cosh\rho - \sinh\rho\cos\theta_1)^2}\right\} \frac{\partial \phi_c}{\partial \rho}\right]^2 \\
:= & -\frac{24\pi^2 \alpha^2}{H^2 \varepsilon} \tilde{I}(\mu, \bar{\rho}).
\end{aligned}
\quad (86)
$$

In the second line, we used $H_{\rm CDL} = 0$, (74) and (75). Also, we performed a coordinate transformation $(t, r) \to (\rho, \theta_1)$. In the third line, the following relations were used

$$-\frac{2}{\eta}\frac{\partial \phi_c(\rho)}{\partial \eta} = -\frac{2}{\eta}\frac{\partial \rho}{\partial \eta}\frac{\partial \phi_c(\rho)}{\partial \rho} = 2H^2(\cosh\rho - \sinh\rho\cos\theta_1)(\sinh\rho - \cosh\rho\cos\theta_1)\frac{\partial \phi_c(\rho)}{\partial \rho},$$

and

$$\begin{aligned}\frac{\partial^2 \phi_c(\rho)}{\partial \eta^2} &= H^2\bigg[(\sinh\rho - \cosh\rho\cos\theta_1)^2\frac{\partial^2}{\partial \rho^2} \\ &\quad + \left\{(\sinh\rho - \cosh\rho\cos\theta_1)(\cosh\rho - \sinh\rho\cos\theta_1) + \frac{\sin^2\theta_1}{\tanh\rho}\right\}\frac{\partial}{\partial \rho}\bigg]\phi_c(\rho).\end{aligned}$$

Here, the partial differentiation with respect to η was taken while keeping r constant.

5.2. Discussions

5.2.1. Appropriate Choice of ε

Since our results (86) depend on ε, we need to choose some specific value of ε to predict the tunneling rate. We emphasize that the ε dependence does not imply the pathology of our result. Rather our results should depend on ε because the nonperturbative effects from the UV modes are missed in our formalism, as discussed in Section 3.2. We then choose ε as large as possible to evaluate the tunneling rate.

Our analysis is valid only when the value of ε lies in a certain range. Below, we briefly discuss necessary conditions for ε to justify our analysis.

Lower Bound on ε

We need to choose a sufficiently large ε so that IR fields can describe the bubble configuration. This imposes that $(\varepsilon H)^{-1}$ should be smaller than the typical physical size of the bubble or the bubble wall. This condition is satisfied by taking $\varepsilon \gtrsim \mu$ for our parameter choices.

Upper Bound on ε

Our analysis is based on Equation (33), which is invalid when the quantum fluctuations of IR fields are too large. This imposes an upper bound on ε because the size of the fluctuations in ϕ_c at the scale $k \sim k_c(t)$ is proportional to ε:

$$\delta\phi_c|_{k\sim k_c} := \left[\int \frac{d^3k}{(2\pi)^3}\delta(\ln(k/k_c(t)))\langle|\hat{\phi}(t, \boldsymbol{k})|^2\rangle\right]^{1/2} \approx \frac{k_c/a}{2\pi} = \frac{\varepsilon H}{2\pi}, \tag{87}$$

where we assumed $(k_c/a)^2 \gg V''(\phi_c)$[9] and approximated ϕ as a massless scalar field.

As discussed in Section 3.2, (33) is justified when the term δH is negligible. Then, we define the ratio $\mathcal{Q}$ of the term δH to the final term on the RHS of (73) as[10]

$$\mathcal{Q} := \left|\frac{\delta H}{H^2\varepsilon\, a^3\phi_\Delta^2/(6\pi)}\right|_{\phi_\Delta = -\frac{6\pi}{H^2\varepsilon}(a^{-2}\nabla^2\phi_c - V'(\phi_c))} = \frac{3\pi^2}{2\varepsilon^2 H^4}\left|V'''(\phi_c)\left(a^{-2}\nabla^2\phi_c - V'(\phi_c)\right)\right|,$$

and we impose $\mathcal{Q} < 1$. In the second equality, we used $\delta H = (a^3/24)V'''(\phi_c)\phi_\Delta^3$. Similarly to (87), we can estimate the size of the quantum fluctuations of $a^{-2}\nabla^2\phi$. Then, we estimate $\mathcal{Q}$ at a large ε as $\mathcal{Q} \sim g^2\alpha\varepsilon/H$, where we also used $|V'''(\phi_c)| \sim g^2\alpha$ for $|\phi_c| \lesssim \alpha$. The condition $\varepsilon < H/(g^2\alpha)$ follows from $\mathcal{Q} < 1$.

Summary

In summary, it is necessary to choose ε satisfying the following condition to justify our analysis,

$$\mu \lesssim \varepsilon \lesssim \varepsilon_{\max} := H/\left(g^2\alpha\right) \approx 755 \times \left(\frac{1.12\times 10^{-4}}{g}\right)\left(\frac{\sqrt{140}}{g\alpha/H}\right). \tag{88}$$

This suggests that, at least when g is sufficiently tiny, we can reliably choose a large ε. Note that we cannot make the action I given in (86) arbitrarily small even if we take $\varepsilon_{\max} \to \infty$ in the free-theory limit $g \to 0$; for $\varepsilon = \varepsilon_{\max}$, we have $I|_{\varepsilon=\varepsilon_{\max}} \sim -g^{-1}(g\alpha/H)^3 \tilde{I}(\mu,\bar{\rho})$. Roughly speaking, the small $g\alpha/H$ corresponds to the bubble radius being larger than the Hubble radius. Since we are interested in the bubble being smaller than the Hubble radius, $g\alpha/H$ cannot be taken small arbitrarily.

A few more comments on the ε-dependence of our results are in order. Our result (86) shows that the value of I becomes larger for smaller choice of ε. This is because the size of the quantum fluctuations of modes $k \sim \varepsilon a(t)H$ is proportional to ε, while the nonperturbative physics caused by UV modes $k > \varepsilon a(t)H$ is truncated in our formalism. However, this interpretation implicitly assumes that the modes $k \sim \varepsilon_{\max}a(t)H$ are relevant for the tunneling process. We expect that this is the case because the kinetic energy of the fluctuations in modes $k \sim \varepsilon_{\max}a(t)H$ is estimated as $(\varepsilon_{\max}H)^4$, which is much smaller than the height of potential barrier $\Delta V \sim g^2\alpha^4$ for our parameter choices.

By contrast, very short scale physics which is insensitive to the detailed structure of $V(\phi)$ near its origin might be irrelevant for the tunneling process. Hence, we expect that the tunneling rate converges to some finite value in the limit $\varepsilon \to \infty$. It would be interesting to see whether the tunneling rate converges to a finite value as ε increases, and if so, from which value of ε this convergence begins.

5.2.2. Hawking–Moss vs. Coleman–de Luccia

From the Formula (33), we derived the two configurations which describe the tunneling from the false vacuum to the true vacuum. One is the HM configuration, and the other is the CDL-like configuration. Using these configurations, we evaluated the probability of the tunneling. Here, we compare these probabilities. Denoting the action (86) and (52) as I_{bubble} and I_{HM}, the ratio of the two actions is given by

$$\gamma := \frac{I_{\text{bubble}}}{I_{\text{HM}}} = \frac{9\alpha^2H^2}{\Delta V\varepsilon}\tilde{I}(\mu,\bar{\rho}) = \frac{1}{\varepsilon}\frac{108}{(1-\beta)^3(\beta+3)}\frac{H^2}{\alpha^2g^2}\tilde{I}(\mu,\bar{\rho}) \tag{89}$$

For example, if we choose $\beta = 0.5$ and $\alpha g/H = \sqrt{140}$ (see Figure 6), the fitting parameters become $\mu = 7.32515$ and $\bar{\rho} = 0.34397$, and the ratio becomes $\gamma = 12.9840/\varepsilon$. For these parameters, the upper bound of ϵ is given by (88), and then γ can be smaller than one for an appropriately small g. This means that the CDL bubble configuration is dominant rather than the HM configuration. Results for other parameters are also shown in Table 1.

Table 1. The values of γ for several parameter sets. The first and second ones are potential parameters. The third one is the initial condition for the equation of motion (82). The fourth and fifth ones are fitting parameters for (84). The sixth and seventh ones are the numerical results of (89) and (A48). The last one is the cut-off parameter ϵ for which (89) coincides with (A48).

β	$g^2\alpha^2/H^2$	$1-\phi_c(10^{-15})/\alpha$	μ	$\bar{\rho}$	$\varepsilon\gamma$	γ^E	ϵ_*
0.3	80	4.87210×10^{-5}	6.22563	1.05197	355.089	0.247004	1437.58
0.4	80	5.20273×10^{-3}	5.99645	0.65257	64.1185	0.0909796	704.758
0.5	80	3.89106×10^{-2}	5.63444	0.475247	33.9526	0.0599470	566.376

Table 1. *Cont.*

β	$g^2\alpha^2/H^2$	$1-\phi_c(10^{-15})/\alpha$	μ	$\bar{\rho}$	$\varepsilon\gamma$	γ^E	ϵ_*
0.3	140	3.50945×10^{-4}	8.13990	0.673305	52.2515	0.0486678	1073.64
0.4	140	8.96003×10^{-3}	7.82419	0.459659	19.2912	0.0239903	804.123
0.5	140	4.98284×10^{-2}	7.32515	0.34397	12.9840	0.0172063	754.606

5.2.3. Comparison with the Euclidean Method

We also computed the CDL action in the Euclidean method (Appendix B) and compared the ratios of the CDL action to the HM action (Table 1). For $\varepsilon > \varepsilon_*$, our tunneling process becomes more probable than the one predicted by the Euclidean method. As discussed in Section 5.2.1, we can consistently choose $\varepsilon > \varepsilon_*$ for certain choices of parameters (g, α, β). Perhaps our results may indicate the presence of a tunneling process which is more probable than the Euclidean method.

5.2.4. Bubble Nucleation Hypersurface and the Subsequent Evolution

We found the non-trivial spacelike hypersurface on which the bubble was nucleated (Figure 7a), where the hypersurface was given by the condition $F(\eta, r) = 0$ or equivalently $a^{-2}\nabla^2\phi_c - V'(\phi_c) = 0$. The subsequent evolution was then described by the one shown in the first line of (75), which is the standard classical equation of motion. The configuration on the hypersurface gave the initial data for the classical dynamics after the bubble nucleation.

The field value on the spacelike hypersurface does not reach the true vacuum but rather, it lies between the true vacuum and the top of the potential as mentioned in Section 5.1. This is possible because the location of the hypersurface is defined by the condition $a^{-2}\nabla^2\phi_c - V'(\phi_c) = 0$, where the gradient force is balanced with the potential force. Though it is non-trivial to solve the evolution starting from the general spacelike hypersurfaces, it is desirable to solve it to fully understand the formation of the true-vacuum bubble in our scenario.

We conclude this section by pointing out that our formalism naturally predicts that the condition $\overline{\phi}_\Delta = 0$ (or $\phi_\Delta = 0$ via (72)) defines the hypersurface on which the quantum dynamics is switched to the classical dynamics and the non-trivial field configuration is nucleated. This would be the generic prediction of our formalism that holds true once the appropriate configuration $\overline{\phi}_\Delta$ is specified.

6. Conclusions

We studied the tunneling processes on the de Sitter background by using the stochastic approach. A novel point is that we applied the MSRJD functional integral to the problem. In this formalism, the tunneling rate was obtained by evaluating the functional integral with the saddle-point approximation. Using this method, we investigated both the HM and the CDL tunnelings.

For the HM case, we first analyzed (31), which is the MSRJD functional integral of the stochastic Equation (23) on a single spatial point. We succeeded in deriving the tunneling solution in the "phase space" (Figure 3) which represented the tunneling process from the false vacuum, through the top of the potential hill, and finally to the true vacuum (Figure 4). This solution had natural tunneling boundary conditions which could not be obtained from the Euclidean method. The tunneling rate for this configuration coincided with that of the HM instanton at the leading order. In Section 4.2, we succeeded in re-deriving this result starting from (33), which described the stochastic dynamics in the global region covered by the flat chart. We also estimated the time scale of the HM tunneling as (63), which became much longer than the Hubble time scale. Our analysis clarified the physical picture of the HM instanton, i.e., the HM transition probability represents the transition probability of a coarse-grained patch with a physical radius $(\varepsilon H)^{-1}$. Some technical complications that arose in dealing with (33) and the way of handling them were remarked on in Section 4.2.1. Physically, these complications were due to the spatial correlations of stochastic noises.

For the CDL case, we found the configurations which described the bubble nucleation process. The tunneling rates for the bubble configurations depended on the cut-off scale dividing IR and UV fields. We argued that this dependence came from the truncation of the nonperturbative effects from UV modes. We also discussed the valid choice of the cut-off scale for which the MSRJD functional integral would be reliable. This consideration was based on the first-principle derivation of the stochastic approach from the Schwinger–Keldysh formalism. With an appropriate cut-off scale, it turned out that the CDL tunneling rate was larger than the HM one for the steep potential barrier. However, as mentioned above (73), it has not yet been investigated if our configuration can really be used for estimating the tunneling action. This aspect would be important for evaluating the tunneling action, being left for future work. Nonetheless, we believe that our study clarifies how we can proceed with the analysis and how the bubble nucleation process could be described in the stochastic approach; for instance, our formalism can naturally define the location of the hypersurface on which the quantum dynamics is switched to the classical dynamics and the non-trivial field configuration is nucleated.

It is interesting to understand the relation between the stochastic approach and the Euclidean method. We compared the CDL tunneling rate in our method with that in the Euclidean method and found the former became larger than the latter for the potential considered with a certain cut-off scale. It would be worth investigating the meaning of this result. In particular, we need to know a precise relation between the CDL instanton and our configuration. Intriguingly, our method is also related to that discussed in [32–36], where the bubble accidentally appears from the initial quantum fluctuations. In our method, however, we assumed that we could neglect the nonperturbative effects from UV modes, the quantum component of the potential of the form $V'''(\phi_c)\phi_\Delta^3$, and that the configuration (74) could be used to estimate the tunneling rate. The first two were summarized into the term δH. We also assumed a fixed-background spacetime. The relaxation of these assumptions would be important to seek the relations among the three methods. It would also be useful to apply our path integral method to the tunneling in flat space and make a comparison with the Euclidean method.

Once this relaxation is achieved, it would also be interesting to apply our method to tunneling phenomena where the backreaction to the background geometry is non-negligible and make comparisons with recently discussed methods. For instance, the formalism of the Wheeler–DeWitt equation [37–39] and the tunneling potential [40] must be related to ours because the HM exponent is reproduced. The tunneling in the black hole spacetime [41,42] is also interesting. In particular, [42] discussed the HM transition with a black hole from the viewpoint of the stochastic approach.

Our method using the saddle-point approximation effectively amounts to solving real-time quantum dynamics, which reduces to Starobinsky's stochastic approach [7,8] at leading order when the super-horizon dynamics in the de Sitter space is considered. However, since our starting point is the Schwinger–Keldysh path integral, which can be formulated in a more generic setup and incorporates all the quantum effects in principle, we believe that our formalism sheds light on further studies of tunneling phenomena from a real-time perspective.

One of the advantages of our method is that it is applicable to a dynamical setup. The application of our method to the inflation models such as the chain inflation [43] and the warm inflation [44] is also intriguing. Using our method, we can study the tunneling process in the inflationary background under dissipation and fluctuations coming from the circumference. We leave these issues for future work.

Author Contributions: Conceptualization, T.M., J.S. and J.T.; methodology, T.M., J.S. and J.T.; software, T.M.; investigation, T.M., J.S. and J.T.; writing—original draft, T.M., J.S. and J.T.; visualization, T.M. All authors have read and agreed to the published version of the manuscript.

Funding: T.M. was supported by JST SPRING, grant number JPMJSP2148 and JSPS KAKENHI grant number JP23KJ1543. J.S. was in part supported by JSPS KAKENHI grant numbers JP17H02894, JP17K18778, JP20H01902, JP22H01220. J.T. is supported by IBS under the project code IBS-R018-D1.

Data Availability Statement: Data are contained within the article.

Conflicts of Interest: The authors declare no conflict of interest.

Appendix A. Stochastic Approach from the First Principle

In this section, we evaluate the generating functional for IR fields $Z[J^{\rm IR}(T)]$ defined in Equation (38) to obtain (39). Note that each step we need to take for evaluating $Z[J^{\rm IR}(T)]$ is summarized in Section 3.2.

Appendix A.1. Path Integral Representation

A path integral representation of $Z[J^{\rm IR}(T)]$ takes the following standard form,

$$Z[J^{\rm IR}(T)] = \int \mathcal{D}(\phi_+,\phi_-,\Pi_+,\Pi_-)\, e^{iJ^{\rm IR}\cdot(\phi_+^{\rm IR}+\phi_-^{\rm IR})/2} \times e^{i(S_H^+-S_H^-)} \prod_x \delta(\phi_+(T,x)-\phi_-(T,x))\Psi_0[\phi_+]\Psi_0^*[\phi_-]\,, \tag{A1}$$

where $\Psi_0[\phi]$ is an initial wave functional. We specify it more concretely in Appendix A.2.1. $S_H^\pm = \int^{t=T} {\rm d}^4x(\Pi^\pm\dot\phi^\pm - H[\phi^\pm,\Pi^\pm])$ is a Hamiltonian action. After the rotation of the basis $(X_+,X_-)\to(X_c,X_\Delta) := (\frac{X_++X_-}{2}, X_+-X_-)$ with $X=\phi,\Pi$, (A1) is written as

$$Z[J^{\rm IR}(T)] = \int \mathcal{D}(\phi_c,\phi_\Delta,\Pi_c,\Pi_\Delta)\, e^{iJ^{\rm IR}\cdot\phi_c^{\rm IR}} e^{iS_H[\phi_c,\phi_\Delta,\Pi_c,\Pi_\Delta]} \prod_x \delta(\phi_\Delta(T,x))\rho_0[\phi_c,\phi_\Delta]\,. \tag{A2}$$

Here, we defined $S_H[\phi_c,\phi_\Delta,\Pi_c,\Pi_\Delta] := (S_H^+ - S_H^-)|_{(X_+,X_-)\to(X_c,X_\Delta)}$ with $X=\phi,\Pi$. We also defined $\rho_0[\phi_c,\phi_\Delta] := \Psi_0[\phi_c+(\phi_\Delta/2)]\Psi_0^*[\phi_c-(\phi_\Delta/2)]$.

Appendix A.2. Nonperturbative Generating Functional for IR Sector

Now, we explain how to calculate $Z[J^{\rm IR}(T)]$ nonperturbatively in the IR sector. This allows us to capture the nonperturbative physics of the sector $k\le k_c(t)$. Our strategy is to split the integration variables into UV modes $k>k_c(t)$ and IR modes $k\le k_c(t)$ for each time step t, and perform the integration over UV variables to obtain $Z[J^{\rm IR}(T)]$. One may simply split the integration variables $X_a(t,\boldsymbol{k})$[11], where $X=(\phi,\Pi)$ and $a=(c,\Delta)$, by the following replacement in (A2):

$$X_a(t,\boldsymbol{k}) \to \begin{cases} X_a^{\rm IR}(t,\boldsymbol{k}) & (t\ge t_k) \\ X_a^{\rm UV}(t,\boldsymbol{k}) & (t<t_k)\,, \end{cases} \tag{A3}$$

where t_k is defined by the condition $k_c(t_k)=k$. It is then tempting to perform the integration over UV variables $X_a^{\rm UV}$ by using the Schwinger–Keldysh (or *closed-time-path*) formalism. However, the expression obtained from (A2) after the replacement (A3) does not have the product of delta functions $\prod_{\boldsymbol{k}}\delta(\phi_\Delta^{\rm UV}(t_k-0^+,\boldsymbol{k}))$ at the final time $t=t_k-0^+$[12]. That is, the time contour for each UV ϕ-variable with modes $0<k\le k_c(T)$ is not closed at the final time. Hence, the usual Schwinger–Keldysh formalism does not apply for evaluating the integration over UV variables. To resolve this issue, Refs. [22,23] proposed a new way of splitting integration variables. The splitting is given by (A3) for $X=\Pi$, while for ϕ-variable, it is defined by the following rule rather than (A3):

$$\phi_c(t,\boldsymbol{k}) \to \begin{cases} \phi_c^{\rm IR}(t,\boldsymbol{k}) & (t>t_k) \\ \phi_c^{\rm UV}(t,\boldsymbol{k}) & (t\le t_k)\,, \end{cases} \qquad \phi_\Delta(t,\boldsymbol{k}) \to \begin{cases} \phi_\Delta^{\rm IR}(t,\boldsymbol{k}) & (t\ge t_k) \\ \phi_\Delta^{\rm UV}(t,\boldsymbol{k}) & (t<t_k)\,. \end{cases} \tag{A4}$$

In this replacement (A4), the variables $\phi_\Delta^{\rm UV}(t_k,\boldsymbol{k})$ are absent while the variables $\phi_c^{\rm UV}(t_k,\boldsymbol{k})$ are present. Therefore, in the expression obtained from (A2) after the new replacement, the time contour for every UV ϕ-variable including those for modes $\boldsymbol{k}$ with $0<k\leq k_c(T)$ is closed. This allows us to perform the integration over UV variables based on the Schwinger–Keldysh formalism as we see below. Note that there is no subtlety in the replacement of integration variables $X_a(t,\boldsymbol{k})$ for a zero mode $k=0$ and those for deep UV modes $k>k_c(T)$: in both replacement rules mentioned above, they are simply replaced by IR variables $X_a^{\rm IR}(t,\boldsymbol{k})$ and UV variables $X_a^{\rm UV}(t,\boldsymbol{k})$, respectively. The time contours for them are closed thanks to the product of delta functions in the original expression (A2).

After the new replacement, the term $S_H[\phi_c,\phi_\Delta,\Pi_c,\Pi_\Delta]$ in the exponent of (A2) is decomposed into three pieces: purely IR terms $S_H^{\rm IR}:=S_H[\phi_c^{\rm IR},\phi_\Delta^{\rm IR},\Pi_c^{\rm IR},\Pi_\Delta^{\rm IR}]$, purely UV terms $S_H^{\rm UV}:=S_H[\phi_c^{\rm UV},\phi_\Delta^{\rm UV},\Pi_c^{\rm UV},\Pi_\Delta^{\rm UV}]$ and the remaining IR–UV mixing terms $S_{\rm mixed}$:

$$S_H[\phi_c,\phi_\Delta,\Pi_c,\Pi_\Delta]\to S_H^{\rm IR}+S_H^{\rm UV}+S_{\rm mixed}\,. \tag{A5}$$

In terms of these quantities, $Z[J^{\rm IR}(T)]$ can be formally written as [22,23]

$$Z[J^{\rm IR}(T)]=\int\mathcal{D}(\phi_c^{\rm IR},\phi_\Delta^{\rm IR},\Pi_c^{\rm IR},\Pi_\Delta^{\rm IR})\,e^{iJ^{\rm IR}\cdot\phi_c^{\rm IR}}e^{iS_H^{\rm IR}}e^{i\Gamma}\times\prod_{k\leq k_c(T)}\delta(\phi_\Delta^{\rm IR}(T,\boldsymbol{k}))\prod_{0<k\leq k_c(T)}\delta(\phi_c^{\rm IR}(t_k,\boldsymbol{k})), \tag{A6}$$

where $i\Gamma$ is the effective action for IR fields,

$$e^{i\Gamma}:=\int\mathcal{D}(\phi_c^{\rm UV},\phi_\Delta^{\rm UV},\Pi_c^{\rm UV},\Pi_\Delta^{\rm UV})\,e^{i(S_H^{\rm UV}+S_{\rm mixed})}\prod_{k>0}\delta(\phi_\Delta^{\rm IR}(t_f(k),\boldsymbol{k}))\rho_0[\phi_c^{\rm UV},\phi_\Delta^{\rm UV};\phi_c^{\rm IR},\phi_\Delta^{\rm IR}], \tag{A7}$$

with $t_f(k):=\min[T,t_k]$. Here, $\rho_0[\phi_c,\phi_\Delta]\to\rho_0[\phi_c^{\rm UV},\phi_\Delta^{\rm UV};\phi_c^{\rm IR},\phi_\Delta^{\rm IR}]$ due to the replacement of variables. The IR variables $(\phi_c^{\rm IR},\phi_\Delta^{\rm IR})$ in the argument of ρ_0 contain only a zero mode. Since the time path for each UV ϕ-variable is closed in (A7), we can calculate $i\Gamma$ perturbatively as usual by writing the connected diagrams with n external IR fields that are connected by UV propagators. The diagrammatic rules such as vertex factors and symmetry factors follow the standard rules of the Scheinger–Keldysh formalism for given vertexes in $S_H^{\rm UV}+S_{\rm mixed}$.

Interestingly, due to the non-trivial replacement rule (A4), the $\Pi\dot{\phi}$ terms in S_H also contribute to $S_{\rm mixed}$ [11,22–24]. Referring to such contributions and the remaining terms in $S_{\rm mixed}$ as $S_{\rm tr}$ and $S_{\text{mix-int}}$, respectively, we have $S_{\rm mixed}=S_{\rm tr}+S_{\text{mix-int}}$ with [13]

$$S_{\rm tr}:=\int_{-\infty}^{T}{\rm d}t\int\frac{{\rm d}^3k}{(2\pi)^3}\delta(t-t_k)\Big[\phi_\Delta^{\rm IR}(t,-\boldsymbol{k})\Pi_c^{\rm UV}(t,\boldsymbol{k})-\Pi_\Delta^{\rm IR}(-\boldsymbol{k},t)\phi_c^{\rm UV}(t,\boldsymbol{k})\Big]\,. \tag{A8}$$

The bi-linear vertexes exist only at the UV→IR transition time t_k for given modes with $0<k\leq k_c(T)$. Physically, these vertexes set the initial fluctuations for such modes in (A6).

Appendix A.2.1. Integrate out Short-Wavelength Modes

Now, we calculate $i\Gamma$ perturbatively by specifying the setup more concretely. We organize the perturbation theory around the false vacuum state ρ_0 for which the expectation value of the zero mode is $\phi_{\rm false}$. We assume that the fluctuations of zero modes are tiny enough so that the couplings between zero-mode fluctuations and those of nonzero modes $k>0$ are switched off in the past infinity $t=-\infty$. We then take the initial state of nonzero modes to be the Bunch–Davies vacuum state for a free field, where the time evolution of a free field is defined by the quadratic action expanded around the homogeneous background $\phi_{\rm false}$. These assumptions can be implemented by writing the initial quantum state as

$$\rho_0[\phi_c^{\rm UV},\phi_\Delta^{\rm UV};\phi_c^{\rm IR},\phi_\Delta^{\rm IR}]=\rho_{\rm BD}[\phi_c^{\rm UV},\phi_\Delta^{\rm UV};\phi_{\rm false}]\,\rho_0[\phi_c^{\rm IR}-\phi_{\rm false},\phi_\Delta^{\rm IR}] \tag{A9}$$

which is properly normalized as

$$1 = \prod_{k>0} \int \mathrm{d}\phi_c^{\mathrm{UV}}(-\infty, \boldsymbol{k}) \int \mathrm{d}\phi_\Delta^{\mathrm{UV}}(-\infty, \boldsymbol{k})\, \rho_{\mathrm{BD}}[\phi_c^{\mathrm{UV}}, \phi_\Delta^{\mathrm{UV}}; \phi_{\mathrm{false}}]$$
$$= \int \mathrm{d}\phi_c^{\mathrm{IR}}(-\infty, \boldsymbol{0}) \int \mathrm{d}\phi_\Delta^{\mathrm{IR}}(-\infty, \boldsymbol{0})\, \rho_0[\phi_c^{\mathrm{IR}} - \phi_{\mathrm{false}}, \phi_\Delta^{\mathrm{IR}}]\,. \tag{A10}$$

$\rho_{\mathrm{BD}}[\phi_c^{\mathrm{UV}}, \phi_\Delta^{\mathrm{UV}}; \phi_{\mathrm{false}}]$ is the density matrix for nonzero modes. An initial state for the zero mode is given by $\rho_0[\phi_c^{\mathrm{IR}} - \phi_{\mathrm{false}}, \phi_\Delta^{\mathrm{IR}}]$, for which we have

$$\langle \Psi(-\infty)|\, \hat{\phi}(\boldsymbol{k}=\boldsymbol{0})\, |\Psi(-\infty)\rangle = \int \mathrm{d}\phi_c^{\mathrm{IR}} \rho_0[\phi_c^{\mathrm{IR}} - \phi_{\mathrm{false}}, \phi_\Delta^{\mathrm{IR}}]\phi_c^{\mathrm{IR}} = \phi_{\mathrm{false}}(2\pi)^3\delta(\boldsymbol{0})\,. \tag{A11}$$

Combining this with $\langle \Psi(-\infty)|\, \hat{\phi}(\boldsymbol{k})\, |\Psi(-\infty)\rangle = 0$ for nonzero modes $k > 0$, we have

$$\langle \Psi(-\infty)|\, \hat{\phi}(\boldsymbol{k})\, |\Psi(-\infty)\rangle = \phi_{\mathrm{false}}(2\pi)^3\delta(\boldsymbol{k})\,. \tag{A12}$$

Hence, the VEV of the IR field $\hat{\phi}^{\mathrm{IR}}(t, \boldsymbol{x})$ in the past infinity $t = -\infty$ is given by ϕ_{false}.

The interaction vertexes for calculating $i\Gamma$ are the bi-linear vertexes (A8) and nonlinear vertexes coming from interacting potential $V(\phi)$. We assume that the potential is sufficiently flat under the region we are interested in. It is then natural to separate $i\Gamma$ into the leading-order pieces and higher-order terms in the coupling constant as follows,

$$e^{i\Gamma} = \rho_0[\phi_c^{\mathrm{IR}} - \phi_{\mathrm{false}}, \phi_\Delta^{\mathrm{IR}}]\, e^{i\Gamma_{\mathrm{LO}}^{k>0}} e^{-i\int \mathrm{d}^4x\, \delta H_{\mathrm{higher}}}\,,$$
$$e^{i\Gamma_{\mathrm{LO}}^{k>0}} := \int \mathcal{D}(\phi_c^{\mathrm{UV}}, \phi_\Delta^{\mathrm{UV}}, \Pi_c^{\mathrm{UV}}, \Pi_\Delta^{\mathrm{UV}})\, e^{iS_{H,\mathrm{free}}^{\mathrm{UV}} + S_{\mathrm{tr}}} \prod_{k>0} \delta(\phi_\Delta^{\mathrm{IR}}(t_f(k), \boldsymbol{k}))\rho_{\mathrm{BD}}[\phi_c^{\mathrm{UV}}, \phi_\Delta^{\mathrm{UV}}; \phi_{\mathrm{false}}]\,. \tag{A13}$$

Here, $S_{H,\mathrm{free}}^{\mathrm{UV}}[\phi_c^{\mathrm{UV}}, \phi_\Delta^{\mathrm{UV}}, \Pi_c^{\mathrm{UV}}, \Pi_\Delta^{\mathrm{UV}}; \phi_{\mathrm{false}}]$ is the free part of the Hamiltonian action defined around the false vacuum as explained above. Higher-order corrections to Γ in the coupling constant are represented by the term $\delta H_{\mathrm{higher}}$, which can be evaluated perturbatively. We have $\delta H_{\mathrm{higher}} = 0$ at the leading order.

To calculate $i\Gamma_{\mathrm{LO}}^{k>0}$ by writing down the connected diagrams with external IR fields, we only need to consider the bi-linear vertexes S_{tr} given in (A8). The number of diagrams are only four, and $i\Gamma_{\mathrm{LO}}^{k>0}$ can be calculated as [22,23]

$$i\Gamma_{\mathrm{LO}}^{k>0} = \frac{-1}{2}\int \mathrm{d}t \int \mathrm{d}t' \int \frac{\mathrm{d}^3k}{(2\pi)^3}\, X_\alpha(t, \boldsymbol{k}) g^{\alpha\beta}(t, t', k) X_\beta(t', -\boldsymbol{k})$$
$$= \frac{-1}{2}\int \mathrm{d}^4x \int \mathrm{d}^4x'\, X_\alpha(x) G^{\alpha\beta}(x, x') X_\beta(x')\,. \tag{A14}$$

Here, the Greek indices $\alpha, \beta = (\phi, \Pi)$ label the IR-Δ fields: $(X_\phi, X_\Pi) = (\Pi_\Delta^{\mathrm{IR}}, -\phi_\Delta^{\mathrm{IR}})$. Substituting Equation (A13) into (A6), we find

$$Z[J^{\mathrm{IR}}(T)] = \int \mathcal{D}(\phi_c^{\mathrm{IR}}, \phi_\Delta^{\mathrm{IR}}, \Pi_c^{\mathrm{IR}}, \Pi_\Delta^{\mathrm{IR}})\, e^{iJ^{\mathrm{IR}}\cdot\phi_c^{\mathrm{IR}}} \rho_0[\phi_c^{\mathrm{IR}} - \phi_{\mathrm{false}}, \phi_\Delta^{\mathrm{IR}}]\, e^{i(S_H^{\mathrm{IR}} + \Gamma_{\mathrm{LO}}^{k>0})} e^{-i\int \mathrm{d}^4x\, \delta H_{\mathrm{higher}}}$$
$$\times \prod_{k \le k_c(T)} \delta(\phi_\Delta^{\mathrm{IR}}(T, \boldsymbol{k})) \prod_{0<k\le k_c(T)} \delta(\phi_c^{\mathrm{IR}}(t_k, \boldsymbol{k}))\,. \tag{A15}$$

Now, we introduce $i\Gamma_{k=0}$ as

$$e^{i\Gamma_{k=0}[\phi_\Delta^{\mathrm{IR}}, \Pi_\Delta^{\mathrm{IR}}]} := \int \mathrm{d}\phi_c^{\mathrm{IR}} \rho_0[\phi_c^{\mathrm{IR}} - \phi_{\mathrm{false}}, \phi_\Delta^{\mathrm{IR}}] e^{-i\Pi_\Delta^{\mathrm{IR}}(\phi_c^{\mathrm{IR}} - \phi_{\mathrm{false}}(2\pi)^3\delta(\boldsymbol{0}))}\,. \tag{A16}$$

By definition, we have

$$\int \mathrm{d}\phi_c^{\mathrm{IR}} \rho_0[\phi_c^{\mathrm{IR}} - \phi_{\mathrm{false}}, \phi_\Delta^{\mathrm{IR}}]\, e^{-i\Pi_\Delta^{\mathrm{IR}}(\phi_c^{\mathrm{IR}} - \phi_{\mathrm{false}}(2\pi)^3\delta(\mathbf{0}))}$$
$$= \int \mathrm{d}\phi_c^{\mathrm{IR}}\, e^{-i\Pi_\Delta^{\mathrm{IR}}\phi_c^{\mathrm{IR}}} e^{i\Gamma_{k=0}[\phi_\Delta^{\mathrm{IR}}, \Pi_\Delta^{\mathrm{IR}}]} \delta\Big(\phi_c^{\mathrm{IR}} - \phi_{\mathrm{false}}(2\pi)^3\delta(\mathbf{0})\Big)\,. \tag{A17}$$

Substituting (A17) into (A15), we find

$$Z[J^{\mathrm{IR}}(T)] = \int \mathcal{D}(\phi_c^{\mathrm{IR}}, \phi_\Delta^{\mathrm{IR}}, \Pi_c^{\mathrm{IR}}, \Pi_\Delta^{\mathrm{IR}})\, e^{iJ^{\mathrm{IR}}\cdot\phi_c^{\mathrm{IR}}} e^{i(S_H^{\mathrm{IR}} + \Gamma_{\mathrm{LO}}^{k>0} + \Gamma_{k=0})} e^{-i\int \mathrm{d}^4x\, \delta H_{\mathrm{higher}}}$$
$$\times \prod_{k\leq k_c(T)} \delta(\phi_\Delta^{\mathrm{IR}}(T, \boldsymbol{k}))\delta(\phi_c^{\mathrm{IR}}(t_k, \boldsymbol{k}) - \phi_{\mathrm{false}}(2\pi)^3\delta(\boldsymbol{k}))\,, \tag{A18}$$

by defining $t_k|_{k=0} = -\infty$.

Appendix A.2.2. Long-Wavelength Sector

For later convenience, let us focus on the term $\int \mathrm{d}t\, \Pi_c^{\mathrm{IR}}(t, \boldsymbol{k})\dot{\phi}_\Delta^{\mathrm{IR}}(t, -\boldsymbol{k})$ in S_H^{IR}. We use the following trick to perform the integration by parts:

$$\delta(\phi_\Delta^{\mathrm{IR}}(T, \boldsymbol{k}))\, e^{i\int_{t_k}^T \mathrm{d}t\, \Pi_c^{\mathrm{IR}}(t,\boldsymbol{k})\dot{\phi}_\Delta^{\mathrm{IR}}(t,-\boldsymbol{k})}$$
$$= \delta(\phi_\Delta^{\mathrm{IR}}(T, \boldsymbol{k})) \int \mathrm{d}\Pi_c^{\mathrm{IR}}(t_k - 0^+, \boldsymbol{k})\, e^{i\int_{t_k-0^+}^T \mathrm{d}t\, \Pi_c^{\mathrm{IR}}(t,\boldsymbol{k})\dot{\phi}_\Delta^{\mathrm{IR}}(t,-\boldsymbol{k})} \delta(\Pi_c^{\mathrm{IR}}(t_k - 0^+, \boldsymbol{k}))$$
$$= \delta(\phi_\Delta^{\mathrm{IR}}(T, \boldsymbol{k})) \int \mathrm{d}\Pi_c^{\mathrm{IR}}(t_k - 0^+, \boldsymbol{k})\, e^{-i\int_{t_k-0^+}^T \mathrm{d}t\, \dot{\Pi}_c^{\mathrm{IR}}(t,\boldsymbol{k})\phi_\Delta^{\mathrm{IR}}(t,-\boldsymbol{k})} \delta(\Pi_c^{\mathrm{IR}}(t_k - 0^+, \boldsymbol{k}))\,. \tag{A19}$$

Thanks to this trick, (A18) can be written as

$$Z[J^{\mathrm{IR}}(T)] = \int \mathcal{D}(\phi_c^{\mathrm{IR}}, \phi_\Delta^{\mathrm{IR}}, \Pi_c^{\mathrm{IR}}, \Pi_\Delta^{\mathrm{IR}}) \int \widetilde{\mathcal{D}}\Pi_c^{\mathrm{IR}}\, e^{iJ^{\mathrm{IR}}\cdot\phi_c^{\mathrm{IR}}} e^{i(S_{H,(d)}^{\mathrm{IR}} + S_{H,(s)}^{\mathrm{IR}} + \Gamma_{\mathrm{LO}}^{k>0} + \Gamma_{k=0})} e^{-i\int \mathrm{d}^4x\, \delta H_{\mathrm{higher}}}$$
$$\times \prod_{k\leq k_c(T)} \delta(\phi_\Delta^{\mathrm{IR}}(T, \boldsymbol{k}))\delta(\phi_c^{\mathrm{IR}}(t_k, \boldsymbol{k}) - \phi_{\mathrm{false}}(2\pi)^3\delta(\boldsymbol{k}))\delta(\Pi_c^{\mathrm{IR}}(t_k - 0^+, \boldsymbol{k}))\,, \tag{A20}$$

where $\widetilde{\mathcal{D}}\Pi_c^{\mathrm{IR}} := \prod_{k\leq k_c(T)} \mathrm{d}\Pi_c^{\mathrm{IR}}(t_k - 0^+, \boldsymbol{k})$. Below, we simply omit this integration measure to simplify the notation. We also decompose S_H^{IR} as $S_H^{\mathrm{IR}} = S_{H,(d)}^{\mathrm{IR}} + S_{H,(s)}^{\mathrm{IR}}$ up to the boundary terms, which vanish thanks to the trick (A19) as

$$S_{H,(d)}^{\mathrm{IR}} = \int \mathrm{d}^4x \left[\Pi_\Delta^{\mathrm{IR}}\left(\dot{\phi}_c^{\mathrm{IR}} - a^{-3}\Pi_c^{\mathrm{IR}}\right) - \phi_\Delta^{\mathrm{IR}}\left(\dot{\Pi}_c^{\mathrm{IR}} - a(t)\nabla^2\phi_c^{\mathrm{IR}} + a^3(t)\, V'(\phi_c^{\mathrm{IR}})\right)\right], \tag{A21}$$
$$S_{H,(s)}^{\mathrm{IR}} = -\int \mathrm{d}^4x\, a^3(t)\left[V(\phi_c^{\mathrm{IR}} + (\phi_\Delta^{\mathrm{IR}}/2)) - V(\phi_c^{\mathrm{IR}} - (\phi_\Delta^{\mathrm{IR}}/2)) - V'(\phi_c^{\mathrm{IR}})\phi_\Delta^{\mathrm{IR}}\right]. \tag{A22}$$

Here, the spacetime argument of IR variables are omitted. We introduce the IR variables and their time derivatives as

$$X^{\mathrm{IR}}(t, \boldsymbol{x}) := \int \frac{\mathrm{d}^3k}{(2\pi)^3}\, \theta(k_c(t) - k)\, X^{\mathrm{IR}}(t, \boldsymbol{k})\, e^{i\boldsymbol{k}.\boldsymbol{x}}\,, \tag{A23}$$
$$\dot{X}^{\mathrm{IR}}(t, \boldsymbol{x}) := \int \frac{\mathrm{d}^3k}{(2\pi)^3}\, \theta(k_c(t) - k)\, \dot{X}^{\mathrm{IR}}(t, \boldsymbol{k})\, e^{i\boldsymbol{k}.\boldsymbol{x}}\,, \tag{A24}$$

where X^{IR} represents the IR variables: $X^{\mathrm{IR}} = (\phi_c^{\mathrm{IR}}, \phi_\Delta^{\mathrm{IR}}, \Pi_c^{\mathrm{IR}}, \Pi_\Delta^{\mathrm{IR}})$. Precisely speaking, for $X^{\mathrm{IR}} = \Pi_c^{\mathrm{IR}}$, we should replace $k_c(t)$ by $k_c(t + 0^+)$ so that we have $k_c(t) = k$ at $t = t_k - 0^+$. One may be concerned about the subtlety of the time derivatives $\dot{\phi}_c^{\mathrm{IR}}(t, \boldsymbol{x})$ and $\dot{\Pi}_c^{\mathrm{IR}}(t, \boldsymbol{x})$ in (A21): in general we have $\dot{X}^{\mathrm{IR}}(t, \boldsymbol{x}) \neq \frac{\partial}{\partial t}(X^{\mathrm{IR}}(t, \boldsymbol{x}))$, where the LHS is defined by (A24), due to the presence of a time-dependent step function. However, there is no subtlety thanks to the initial conditions for all IR modes $k \le k_c(T)$ of the form

$$\phi_c^{\rm IR}(t_k, \boldsymbol{k}) = \phi_{\rm false}(2\pi)^3\delta(\boldsymbol{k})\,, \qquad \Pi_c^{\rm IR}(t_k - 0^+, \boldsymbol{k}) = 0\,, \tag{A25}$$

which are imposed by the delta functions in the final line in (A20). The corresponding conditions in the real space for all $\boldsymbol{x}$ are

$$\dot{X}^{\rm IR}(t,\boldsymbol{x}) = \partial_t(X^{\rm IR}(t,\boldsymbol{x})) \quad \text{for} \quad X^{\rm IR} = (\phi_c^{\rm IR},\, \Pi_c^{\rm IR})\,, \tag{A26a}$$

$$\phi_c^{\rm IR}(-\infty,\boldsymbol{x}) = \phi_{\rm false}\,, \quad \Pi_c^{\rm IR}(-\infty,\boldsymbol{x}) = 0\,. \tag{A26b}$$

To implement these conditions in the path integral, we define $\delta[\mathcal{C}_{\rm ini}]$ by (40b) with

$$\delta[\mathcal{C}_{\rm detail}] := \prod_{-\infty\le t\le T}\ \prod_{X^{\rm IR}=(\phi_c^{\rm IR},\Pi_c^{\rm IR})}\ \prod_{\boldsymbol{x}} \delta\Big(\dot{X}^{\rm IR}(t,\boldsymbol{x}) - \partial_t(X^{\rm IR}(t,\boldsymbol{x}))\Big)\,. \tag{A27}$$

We find from Equations (37) and (A20) that the transition probability can be written in terms of the path integral as

$$p(\{\phi^{\rm IR}(T,\boldsymbol{x})\}_{T,\boldsymbol{x}\in\mathcal{D}}) = \int \mathcal{D}(\phi_c^{\rm IR},\phi_\Delta^{\rm IR},\Pi_c^{\rm IR},\Pi_\Delta^{\rm IR})\ e^{i\mathcal{S}}e^{i\Gamma_{k=0}}\delta[\mathcal{C}_{\rm fin}]\delta[\mathcal{C}_{\rm ini}]\,, \tag{A28}$$

where $\delta[\mathcal{C}_{\rm fin}]$ is defined in (40a), and the term $\mathcal{S}$ in the exponent is given by

$$\mathcal{S} = (S^{\rm IR}_{H,(d)} + S^{\rm IR}_{H,(s)} + \Gamma^{k>0}_{\rm LO}) - \int {\rm d}^4x\, \delta H_{\rm higher} \tag{A29}$$

which agrees with the definition of $\mathcal{S}$ given in (41). When the zero-mode fluctuations are negligible, we can set $\Gamma_{k=0} = 0$. This is the setup adopted in the main text. Then, (A28) coincides with (39).

Appendix A.3. Stochastic Interpretation of the Tunneling

It is useful to see how the stochastic equations emerge as a consequence of performing the path integral over IR variables nonperturbatively. This leads to the stochastic interpretation of the tunneling probability. A final result of this section is (A39).

For simplicity, let us ignore higher-order corrections $\delta H_{\rm higher}$ for a while. We start by introducing the auxiliary fields (ξ_ϕ, ξ_Π) as

$$e^{i\left(\Gamma^{k>0}_{\rm LO}+\Gamma_{k=0}+S^{\rm IR}_{H,(s)}\right)} = \int \mathcal{D}(\xi_\phi,\xi_\Pi)\, P[\xi_\phi,\xi_\Pi] e^{i\int {\rm d}^4x\left(\xi_\Pi(x)\phi_\Delta^{\rm IR}(x) - \xi_\phi(x)\Pi_\Delta^{\rm IR}(x)\right)}\,. \tag{A30}$$

We substitute (A30) into (A20). Then, the exponent becomes linear in $\phi_\Delta^{\rm IR}(t,\boldsymbol{k})$ or $\Pi_\Delta^{\rm IR}(t,\boldsymbol{k})$ with $t_0 \le t < T$:

$$\begin{aligned} Z[J^{\rm IR}(T)] \simeq &\int \mathcal{D}(\phi_c^{\rm IR},\phi_\Delta^{\rm IR},\Pi_c^{\rm IR},\Pi_\Delta^{\rm IR}) \int \widetilde{\mathcal{D}}\Pi_c^{\rm IR}\, e^{iJ^{\rm IR}\cdot\phi_c^{\rm IR}} \int \mathcal{D}(\xi_\phi,\xi_\Pi)\, P[\xi_\phi,\xi_\Pi] \\ &\times \exp\left\{ i\int_{-\infty}^T {\rm d}t \int \frac{{\rm d}^3k}{(2\pi)^3}\left[\Pi_\Delta^{\rm IR}(t,-\boldsymbol{k})\mathcal{L}_\phi(t,\boldsymbol{k}) - \phi_\Delta^{\rm IR}(t,-\boldsymbol{k})\mathcal{L}_\Pi(t,\boldsymbol{k})\right]\Theta(k_c(t)-k)\right\} \\ &\times \prod_{k\le k_c(T)} \delta(\phi_\Delta^{\rm IR}(T,\boldsymbol{k}))\delta(\phi_c^{\rm IR}(t_k,\boldsymbol{k}) - \phi_{\rm false}(2\pi)^3\delta(\boldsymbol{k}))\delta(\Pi_c^{\rm IR}(t_k - 0^+,\boldsymbol{k}))\,, \end{aligned} \tag{A31}$$

where

$$\mathcal{L}_\phi(t,\boldsymbol{k}) := \dot{\phi}_c^{\rm IR}(t,\boldsymbol{k}) - a^{-3}(t)\Pi_c^{\rm IR}(t,\boldsymbol{k}) - \xi_\phi(t,\boldsymbol{k})\,, \tag{A32a}$$

$$\mathcal{L}_\pi(t,\boldsymbol{k}) := \dot{\Pi}_c^{\rm IR}(t,\boldsymbol{k}) + a(t)k^2\phi_c^{\rm IR}(t,\boldsymbol{k}) + a^3(t)\,V'(\phi_c^{\rm IR})(t,\boldsymbol{k}) - \xi_\Pi(t,\boldsymbol{k})\,. \tag{A32b}$$

Consequently, we can perform the path integral $\int \mathcal{D}(\phi_\Delta^{\rm IR}, \Pi_\Delta^{\rm IR})$, leading to

$$Z[J^{\rm IR}(T)] \simeq \prod_{k \leq k_c(T)} \int {\rm d}\phi_c^{\rm IR}(T, \boldsymbol{k})\, e^{iJ^{\rm IR}\cdot\phi_c^{\rm IR}} \int \mathcal{D}(\xi_\phi, \xi_\Pi)\, P[\xi_\phi, \xi_\Pi]\bigg|_{\mathcal{L}_\phi = \mathcal{L}_\Pi = 0\ \&\ ({\rm A25})} . \quad ({\rm A33})$$

Now, we regard auxiliary fields (ξ_ϕ, ξ_Π) as noise variables whose correlations are given by

$$\begin{aligned} \langle \xi_\alpha(t, \boldsymbol{k}) \xi_\beta(t', \boldsymbol{k}') \rangle &:= \int \mathcal{D}(\xi_\phi, \xi_\Pi)\, P[\xi_\phi, \xi_\Pi]\, \xi_\alpha(t, \boldsymbol{k}) \xi_\beta(t', \boldsymbol{k}') \\ &= \left[\frac{i\delta}{\delta X_\alpha(t, \boldsymbol{k})} \frac{i\delta}{\delta X_\beta(t', \boldsymbol{k}')} \right] e^{i\left(\Gamma_{\rm LO}^{k>0} + \Gamma_{k=0} + S_{H,(s)}^{\rm IR}\right)} \bigg|_{\phi_\Delta^{\rm IR} = \Pi_\Delta^{\rm IR} = 0} , \end{aligned} \quad ({\rm A34})$$

where $\alpha, \beta = (\phi, \Pi)$ and $(X_\phi, X_\Pi) = (\Pi_\Delta^{\rm IR}, -\phi_\Delta^{\rm IR})$ following the notation in (A14). We also have the corresponding expression in the momentum space. If we ignore $S_{H,(s)}$ and $i\Gamma_{k=0}$, we have

$$\langle \xi_\alpha(t, \boldsymbol{k}) \xi_\beta(t', \boldsymbol{k}') \rangle = (2\pi)^3 \delta(\boldsymbol{k} + \boldsymbol{k}') {\rm Re}\left[g^{\alpha\beta}(t) \right] \delta(t - t') \delta(t - t_k) . \quad ({\rm A35})$$

Equation (A33) shows that the effective dynamics of IR fields are described by the set of Langevin equations $\mathcal{L}_\phi = \mathcal{L}_\Pi = 0$ with the initial condition (A25) and the noise correlations (A35).

In the real space, the dynamics is described by the set of Langevin equations,

$$\dot{\phi}_c^{\rm IR}(x) = a^{-3}(t) \Pi_c^{\rm IR}(x) + \xi_\phi(x) , \quad ({\rm A36a})$$

$$\dot{\Pi}_c^{\rm IR}(x) = a(t) \nabla^2 \phi_c^{\rm IR}(x) - a^3 V'(\phi_c^{\rm IR}(x)) + \xi_\Pi(x) , \quad ({\rm A36b})$$

with the initial conditions (A26) and the noise correlations

$$\langle \xi_\alpha(x) \xi_\beta(y) \rangle = G^{\alpha\beta}(x, y) . \quad ({\rm A37})$$

Equation (A33) takes the following form in the real space:

$$Z[J^{\rm IR}(T)] \simeq \prod_{\boldsymbol{x}} \int {\rm d}\phi_c^{\rm IR}(T, \boldsymbol{x})\, e^{iJ^{\rm IR}\cdot\phi_c^{\rm IR}} \int \mathcal{D}(\xi_\phi, \xi_\Pi)\, P[\xi_\phi, \xi_\Pi]\bigg|_{({\rm A26})\ \&\ ({\rm A36})} . \quad ({\rm A38})$$

Substituting (A38) into (37), we have

$$\begin{aligned} p(\{\phi^{\rm IR}(T, \boldsymbol{x})\}_{T, \boldsymbol{x} \in \mathcal{D}}) \simeq & \prod_{\boldsymbol{y}} \int {\rm d}\phi_c^{\rm IR}(T, \boldsymbol{y}) \prod_{\boldsymbol{x} \in \mathcal{D}} \delta(\phi_c^{\rm IR}(T, \boldsymbol{x}) - \phi^{\rm IR}(T, \boldsymbol{x})) \\ & \times \int \mathcal{D}(\xi_\phi, \xi_\Pi)\, P[\xi_\phi, \xi_\Pi]\bigg|_{({\rm A26})\ \&\ ({\rm A36})} . \end{aligned} \quad ({\rm A39})$$

This expression provides a stochastic interpretation of the transition probability; the dynamics of $\phi_c^{\rm IR}$ are described by the set of Langevin equations, and the transition probability is simply understood as the probability to realize the field configuration $\phi_c^{\rm IR}(T, \boldsymbol{x}) = \phi^{\rm IR}(T, \boldsymbol{x})$ in the domain $\boldsymbol{x} \in \mathcal{D}$ at the final time T starting from the initial configuration $\phi_c^{\rm IR}(-\infty, \boldsymbol{x}) = \phi_{\rm false}$. In particular, the set of Langevin equations is given by (8) with the noise correlations (15) when we ignore the terms $\delta H_{\rm higher}$, $S_{H,(s)}^{\rm IR}$, and $\Gamma_{k=0}$. Note that such terms simply correct the Langevin equations[14] and do not invalidate the above stochastic interpretation.

Appendix B. Coleman–de Luccia Tunneling in Euclidean Method

In this appendix, we provide the CDL action in the de Sitter background with the Euclidean method.

Generally, the ways of Euclideanization for the de Sitter spacetime depend on the coordinates. In this paper, we discuss the field theory on the metric (1), and then we choose the following coordinate again[15].

$$ds^2 = \frac{1}{H^2\eta^2}(-d\eta^2 + d\boldsymbol{x}^2). \tag{A40}$$

By taking the wick rotation as $\eta \to -i\tau$, the coordinate becomes

$$-ds^2 = \frac{1}{H^2\tau^2}(d\tau^2 + d\boldsymbol{x}^2). \tag{A41}$$

This is the Euclidean AdS space discussed in Section 5, and we can take the global coordinate

$$-ds^2 = H^{-2}(d\rho^2 + \sinh^2\rho d\Omega^2). \tag{A42}$$

We also impose that the field ϕ only depends on ρ. Therefore, the Euclidean action I^E and the equation of motion becomes

$$I^E = 2\pi^2 \int_0^\infty d\rho \sinh^3\rho \left[\frac{1}{2H^2}(\partial_\rho\phi)^2 + \frac{V(\phi)}{H^4}\right], \tag{A43}$$

$$\frac{d^2\phi}{d\rho^2} + \frac{3}{\tanh\rho}\frac{d\phi}{d\rho} = \frac{V'(\phi)}{H^2}. \tag{A44}$$

The second equation is the same as (82), and we can obtain the same bounce solutions discussed in Section 5. For numerical calculations, we consider the potential (42) and fitting function (84). Also, we take the following non-dimensionalizations;

$$\phi = \alpha\tilde{\phi}, \tag{A45}$$

$$V(\phi) = g^2\alpha^4\tilde{V}(\tilde{\phi}). \tag{A46}$$

Therefore, the Euclidean action (A43) becomes

$$\begin{aligned} I^E &= 2\pi^2\frac{\alpha^2}{H^2}\int_0^\infty d\rho \sinh^3\rho \left[\frac{1}{2}(\partial_\rho\tilde{\phi})^2 + \frac{g^2\alpha^2}{H^2}\tilde{V}(\tilde{\phi})\right] \\ &:= 2\pi^2\frac{\alpha^2}{H^2}\tilde{I}^E. \end{aligned} \tag{A47}$$

The ratio of the Coleman–de Luccia Euclidean action I^E_{bounce} to the Hawking–Moss action I_{HM} is given by

$$\gamma_E = \frac{I^E_{\text{bounce}}}{I_{\text{HM}}} = \frac{H^2}{g^2\alpha^2}\frac{9\tilde{I}^E_{\text{bounce}}}{(1-\beta)^3(\beta+3)}. \tag{A48}$$

Notes

1. For earlier discussions of a stochastic approach based on the Schwinger–Keldysh formalism, see, e.g., [11,24].
2. Generally, we can choose other discretization like Stratonovich's discretization. This ambiguity does not affect the result since the amplitudes of the noise do not depend on fields within the range of our approximation.
3. Precisely, $\hat{\phi}^{\text{IR}}(\boldsymbol{x})$ is defined in terms of the the Schrödinger picture field $\hat{\phi}(\boldsymbol{k})$ in momentum space as $\hat{\phi}^{\text{IR}}(\boldsymbol{x}) := \int \frac{d^3k}{(2\pi)^3}\hat{\phi}(\boldsymbol{k})\, e^{i\boldsymbol{k}.\boldsymbol{x}}\theta(k_c(t)-k)$. We have suppressed the trivial time dependence stemming from the step function.
4. There is a subtlety that modes satisfying $k \geq k_c(T)$ are initially regarded as the "UV" degrees of freedom (DoFs) while they become "IR" DoFs due to the accelerating expansion of the spacetime. However, by adopting this splitting procedure, we can use the Schwinger–Keldysh (or *closed-time-path*) formalism to evaluate the integration over UV variables first as usual.
5. This is consistent with the observation made in [6] that an HM solution corresponds to the transition over a region of a Hubble horizon volume.

6 Note that this behavior may motivate one to replace the term $j_0(k_c(t)r)$ in (55) by the step function

$$j_0(k_c(t)r) \to \theta(1 - k_c(t)r)\,. \tag{A49}$$

This is the approximation adopted in [8].

7 Note that we can also estimate the action I by performing the coarse-graining in time suitably and substituting the smeared expression of Π_Δ into the action:

$$I \simeq -\frac{8\pi^2}{3H^4}\int_{t'}^{t_*} \mathrm{d}t\, V'(\phi_c(t,x_0)) \int \mathrm{d}^3x\, \dot{\phi}_c(t,x) W_f(r;t) \simeq -\frac{8\pi^2}{3H^4}\int_{t'}^{t_*} \mathrm{d}t\, \dot{\phi}_c(t,x_0) V'(\phi_c(t,x_0)) = -\frac{8\pi^2}{3H^4}\Delta V\,,$$

reproducing the previous result (52). In the first approximate equality, we used $H_{\mathrm{HM}} = 0$, which is well satisfied after the smearing. In the second one, we used $\phi_c(t,x) \simeq \phi_c(t,x_0)$ for $r \ll k_c^{-1}(t)$ and the condition $\int \mathrm{d}^3x\, W_f(r;t) = 1$ which follows from the definition of W_f. This estimate may be more analogous to the calculation of the action in Section 4.1 since the approximate locality of the dynamics is restored after the smearing.

8 This is the reason why we illustrate flow lines in the (ϕ_c, ϕ_Δ)-plane in Figure 5.

9 This is compatible with the condition $\varepsilon < \alpha/H$ when the weak coupling $g \ll 1$ is considered.

10 We can also compare δH with the second and the third term on the RHS of (73). Such considerations do not change our estimate of $\varepsilon_{\mathrm{max}}$ in (88).

11 In the main text, we use the notation $X_{\boldsymbol{k}}(t)$ to write variables in momentum space. However, we adopt the notation $X(t,\boldsymbol{k})$ in this appendix since we have many subscripts such as c and Δ.

12 Here, 0^+ is the infinitesimal time step which is introduced to obtain the path integral representation of the unitary time evolution as usual.

13 Precisely speaking, we need to perform the UV–IR splitting in the path integral with an infinitesimal discrete time step for deriving S_{tr} correctly. After taking the continuum limit, we obtain (A8); see also [23]. Furthermore, the form of $S_{\text{mix-int}}$ depends on the model. In our case, we have

$$\begin{aligned} S_{\text{mix-int}} = \int \mathrm{d}^4x \big[& V(\phi_+^{\mathrm{UV}} + \phi_+^{\mathrm{IR}}) - V(\phi_+^{\mathrm{UV}}) - V(\phi_+^{\mathrm{IR}}) \\ & - V(\phi_-^{\mathrm{UV}} + \phi_-^{\mathrm{IR}}) + V(\phi_-^{\mathrm{UV}}) + V(\phi_-^{\mathrm{IR}}) \big]_{\phi_\pm^{\mathrm{UV}} = \phi_c^{\mathrm{UV}} \pm (\phi_\Delta^{\mathrm{UV}}/2),\, \phi_\pm^{\mathrm{IR}} = \phi_c^{\mathrm{IR}} \pm (\phi_\Delta^{\mathrm{IR}}/2)}\,. \end{aligned} \tag{A50}$$

14 For instance, leading-order corrections to the stochastic dynamics at the super-horizon scales are calculated in [22].

15 The analytic continuation of this coordinate is examined in [5].

References

1. Coleman, S.R. The Fate of the False Vacuum. 1. Semiclassical Theory. *Phys. Rev. D* **1977**, *15*, 2929–2936; Erratum in *Phys. Rev. D* **1977**, *16*, 1248. [CrossRef]
2. Callan, C.G., Jr.; Coleman, S.R. The Fate of the False Vacuum. 2. First Quantum Corrections. *Phys. Rev. D* **1977**, *16*, 1762–1768. [CrossRef]
3. Coleman, S.R.; De Luccia, F. Gravitational Effects on and of Vacuum Decay. *Phys. Rev. D* **1980**, *21*, 3305. [CrossRef]
4. Hawking, S.W.; Moss, I.G. Supercooled Phase Transitions in the Very Early Universe. *Phys. Lett. B* **1982**, *110*, 35–38. [CrossRef]
5. Rubakov, V.A.; Sibiryakov, S.M. False vacuum decay in de Sitter space-time. *Theor. Math. Phys.* **1999**, *120*, 1194–1212. [CrossRef]
6. Brown, A.R.; Weinberg, E.J. Thermal derivation of the Coleman-De Luccia tunneling prescription. *Phys. Rev. D* **2007**, *76*, 064003. [CrossRef]
7. Starobinsky, A.A. Stochastic de Sitter (Inflationary) Stage in the Early Universe. *Lect. Notes Phys.* **1986**, *246*, 107–126. [CrossRef]
8. Starobinsky, A.A.; Yokoyama, J. Equilibrium state of a selfinteracting scalar field in the De Sitter background. *Phys. Rev. D* **1994**, *50*, 6357–6368. [CrossRef] [PubMed]
9. Goncharov, A.S.; Linde, A.D.; Mukhanov, V.F. The Global Structure of the Inflationary Universe. *Int. J. Mod. Phys. A* **1987**, *2*, 561–591. [CrossRef]
10. Linde, A.D. Hard art of the universe creation (stochastic approach to tunneling and baby universe formation). *Nucl. Phys. B* **1992**, *372*, 421–442. [CrossRef]
11. Tolley, A.J.; Wyman, M. Stochastic tunneling in DBI inflation. *J. Cosmol. Astropart. Phys.* **2009**, *10*, 006. [CrossRef]
12. Noorbala, M.; Vennin, V.; Assadullahi, H.; Firouzjahi, H.; Wands, D. Tunneling in Stochastic Inflation. *J. Cosmol. Astropart. Phys.* **2018**, *09*, 032. [CrossRef]
13. Hashiba, S.; Yamada, Y.; Yokoyama, J. Particle production induced by vacuum decay in real time dynamics. *Phys. Rev. D* **2021**, *103*, 045006. [CrossRef]
14. Camargo-Molina, J.E.; Rajantie, A. Phase transitions in de Sitter: The stochastic formalism. *arXiv* **2022**, arXiv:2204.02875.

15. Camargo-Molina, J.E.; González, M.C.; Rajantie, A. Phase Transitions in de Sitter: Quantum Corrections *arXiv* **2022**, arXiv:2204.03480.
16. Martin, P.C.; Siggia, E.D.; Rose, H.A. Statistical Dynamics of Classical Systems. *Phys. Rev. A* **1973**, *8*, 423–437. [CrossRef]
17. De Dominicis, C.; Brezin, E.; Zinn-Justin, J. Field Theoretic Techniques and Critical Dynamics. 1. Ginzburg-Landau Stochastic Models Without Energy Conservation. *Phys. Rev. B* **1975**, *12*, 4945. [CrossRef]
18. Janssen, H.K. On a Lagrangean for classical field dynamics and renormalization group calculations of dynamical critical properties. *Z. Phys. B Condens. Matter* **1976**, *23*, 377–380. [CrossRef]
19. De Dominicis, C.; Peliti, L. Field Theory Renormalization and Critical Dynamics Above t(c): Helium, Antiferromagnets and Liquid Gas Systems. *Phys. Rev. B* **1978**, *18*, 353–376. [CrossRef]
20. Elgart, V.; Kamenev, A. Rare event statistics in reaction-diffusion systems. *Phys. Rev. E* **2004**, *70*, 041106. [CrossRef]
21. Altland, A.; Simons, B.D. *Condensed Matter Field Theory*, 2nd ed.; Cambridge University Press: Cambridge, UK, 2010. [CrossRef]
22. Tokuda, J.; Tanaka, T. Statistical nature of infrared dynamics on de Sitter background. *J. Cosmol. Astropart. Phys.* **2018**, *02*, 014. [CrossRef]
23. Tokuda, J.; Tanaka, T. Can all the infrared secular growth really be understood as increase of classical statistical variance? *J. Cosmol. Astropart. Phys.* **2018**, *11*, 022. [CrossRef]
24. Morikawa, M. Dissipation and Fluctuation of Quantum Fields in Expanding Universes. *Phys. Rev. D* **1990**, *42*, 1027–1034. [CrossRef] [PubMed]
25. Feynman, R.P.; Vernon, F.L., Jr. The Theory of a general quantum system interacting with a linear dissipative system. *Ann. Phys.* **1963**, *24*, 118–173. [CrossRef]
26. Weinberg, E.J. *Classical Solutions in Quantum Field Theory: Solitons and Instantons in High Energy Physics*; Cambridge Monographs on Mathematical Physics, Cambridge University Press: Cambridge, UK, 2012. [CrossRef]
27. Weinberg, E.J. Hawking-Moss bounces and vacuum decay rates. *Phys. Rev. Lett.* **2007**, *98*, 251303. [CrossRef] [PubMed]
28. Jensen, L.G.; Steinhardt, P.J. Bubble Nucleation and the Coleman-Weinberg Model. *Nucl. Phys. B* **1984**, *237*, 176–188. [CrossRef]
29. Jensen, L.G.; Steinhardt, P.J. Bubble Nucleation for Flat Potential Barriers. *Nucl. Phys. B* **1989**, *317*, 693–705. [CrossRef]
30. Hackworth, J.C.; Weinberg, E.J. Oscillating bounce solutions and vacuum tunneling in de Sitter spacetime. *Phys. Rev. D* **2005**, *71*, 044014. [CrossRef]
31. Batra, P.; Kleban, M. Transitions Between de Sitter Minima. *Phys. Rev. D* **2007**, *76*, 103510. [CrossRef]
32. Braden, J.; Johnson, M.C.; Peiris, H.V.; Pontzen, A.; Weinfurtner, S. New Semiclassical Picture of Vacuum Decay. *Phys. Rev. Lett.* **2019**, *123*, 031601; Erratum in *Phys. Rev. Lett.* **2022**, *129*, 059901. [CrossRef]
33. Blanco-Pillado, J.J.; Deng, H.; Vilenkin, A. Flyover vacuum decay. *J. Cosmol. Astropart. Phys.* **2019**, *12*, 001. [CrossRef]
34. Hertzberg, M.P.; Rompineve, F.; Shah, N. Quantitative Analysis of the Stochastic Approach to Quantum Tunneling. *Phys. Rev. D* **2020**, *102*, 076003. [CrossRef]
35. Tranberg, A.; Ungersbäck, G. Bubble nucleation and quantum initial conditions in classical statistical simulations. *J. High Energy Phys.* **2022**, *09*, 206. [CrossRef]
36. Hertzberg, M.P.; Yamada, M. Vacuum Decay in Real Time and Imaginary Time Formalisms. *Phys. Rev. D* **2019**, *100*, 016011. [CrossRef]
37. Kristiano, J.; Lambaga, R.D.; Ramadhan, H.S. Coleman-de Luccia Tunneling Wave Function. *Phys. Lett. B* **2019**, *796*, 225–229. [CrossRef]
38. Cespedes, S.; de Alwis, S.P.; Muia, F.; Quevedo, F. Lorentzian vacuum transitions: Open or closed universes? *Phys. Rev. D* **2021**, *104*, 026013. [CrossRef]
39. Maniccia, G.; De Angelis, M.; Montani, G. WKB Approaches to Restore Time in Quantum Cosmology: Predictions and Shortcomings. *Universe* **2022**, *8*, 556. [CrossRef]
40. Espinosa, J.R. Fresh look at the calculation of tunneling actions including gravitational effects. *Phys. Rev. D* **2019**, *100*, 104007. [CrossRef]
41. Gregory, R.; Moss, I.G.; Oshita, N. Black Holes, Oscillating Instantons, and the Hawking-Moss transition. *J. High Energy Phys.* **2020**, *07*, 024. [CrossRef]
42. Gregory, R.; Moss, I.G.; Oshita, N.; Patrick, S. Hawking-Moss transition with a black hole seed. *J. High Energy Phys.* **2020**, *09*, 135. [CrossRef]
43. Freese, K.; Spolyar, D. Chain inflation: 'Bubble bubble toil and trouble'. *J. Cosmol. Astropart. Phys.* **2005**, *07*, 007. [CrossRef]
44. Berera, A. Warm inflation. *Phys. Rev. Lett.* **1995**, *75*, 3218–3221. [CrossRef] [PubMed]

universe

MDPI

Communication

Radial Oscillations of Dark Matter Stars Admixed with Dark Energy

Camila Sepúlveda and Grigoris Panotopoulos *

Departamento de Ciencias Físicas, Universidad de la Frontera, Casilla 54-D, Temuco 4811186, Chile; c.sepulveda41@ufromail.cl
* Correspondence: grigorios.panotopoulos@ufrontera.cl

Abstract: We imagine spherically symmetric configurations made of both dark matter and dark energy in the halo of spiral galaxies. Adopting a polytropic equation of state for dark matter and the Extended Chaplygin gas equation of state for dark energy, we model the same object with three different dark matter–dark energy compositions. We compute the frequencies and the corresponding eigenfunctions of the ten lowest modes, integrating the equations for the radial perturbations by imposing the appropriate boundary conditions at the center and the surface of the object. Also, a comparison between the different models is made.

Keywords: dark matter; dark energy; relativistic stars; radial oscillations; asteroseismology; equations of state

1. Introduction

The concordance cosmological model, the ΛCDM model, describes the formation and structure of the Universe at large scales, and is based on cold dark matter and Λ, the cosmological constant. The existence of dark matter, still of unknown nature and origin, originates from old studies that showed that the Universe is mostly composed of an invisible material that affects galaxies and galaxy clusters only gravitationally. The pioneer works belong to F. Zwicky [1] in the 30s, and several decades later to V. Rubin, who studied the Andromeda's rotation curve and discovered that at any distance from the bulge of the galaxy, the rotation speed was constant [2]. These studies showed that approximately **85%** of the mass of the Universe corresponds to dark matter [3].

In addition to dark matter, there is another dark component in the Universe, called dark energy. This is also of unknown nature as of today, but has a repulsive rather than attractive character, since the current accelerating expansion of the Universe is produced by it [4–6]. Current modern data seem to show that approximately 70% of the Universe is made of dark energy, 25% of dark matter, and 5% of ordinary matter [7].

Compact stars [8], the final products of very massive stars, are extremely dense objects with a mass of the order of one solar mass, and a radius of a few kilometers, such as neutron stars [9] and strange quark stars [10–15], although the latter class of compact objects remain hypothetical for the time being. In the same way, there are studies that contemplate the possibility that relativistic exotic stars may exist, composed of dark matter [16–18], or dark energy even [19–22]. In a recent study, we considered configurations composed of both dark matter and dark energy adopting the two-fluid formalism [23]. In that study we gave a relativistic description based on the field equations of General Relativity.

To determine the radius and mass of those stars, we require the EoS of each fluid component, i.e., one for dark matter and another for dark energy, under the condition that there is no direct interaction between the fluids other than the gravitational force. Dark matter may be viewed as a Bose–Einstein condensate [24], and so we adopt the polytropic equation of state for dark matter [25,26]. Regarding dark energy, we consider the Extended

Chaplygin gas equation of state [27,28], which was introduced with the aim of unifying dark energy and non-relativistic matter working with a single fluid instead of two.

Regarding plausible scenarios and astrophysical mechanisms that may produce dark stars in the Universe, it has been proposed that the first stars to exist in the Cosmos, contrary to ordinary main sequence stars, were powered by dark matter heating rather than fusion [29,30]. Dark matter particles might have accumulated in the interior of the first stars and been annihilated, producing a heat source that can power such stars. Moreover, if Bose–Einstein condensation took place in the very early Universe, at very high redshifts, as suggested by the analysis performed in [16], at the moment of star formation most of the DM was already in a condensate form. Hence, one may naturally expect the formation of stellar type objects made of pure condensate dark matter stars [16].

Asteroseismology is a widely used technique, since the frequencies of the oscillation modes of pulsating stars are very sensitive to the inner structure and composition of the object. It has been proven to be a powerful tool at our disposal, since it allows us to probe the interior of the Sun, solar-like stars, and other classes of stars [31,32], such as nuclear reactions, their compositions, equations of state, differential rotation rates, and meridional circulation. Currently, asteroseismology has been extended to the study of compact stars [33–36], like white dwarfs [37] and neutron stars [38]. In addition, there are studies of exotic stars such as strange quark stars [39], dark matter stars [39–41], and dark energy stars [21]. For more details, the interested reader may consult for instance [42–50].

Regarding excitation and detectability, there are numerous astrophysical mechanisms that may excite oscillation modes of stars, such as tidal effects in binaries, starquakes caused by cracks during supernova explosions, magnetic reconfiguration, or any other form of instability [38,51–55]. The Kepler and CoRoT missions have already measured the oscillation spectra of solar-like, white dwarf, and red giant stars [56–61]. Actual wave detectors, like LIGO [38], do not detect the radial oscillations studied here, due to their low sensitivity at the kHz frequency range. However, the third-generation of ground-based gravitational wave detectors are expected to have a much higher sensitivity (by an order of magnitude), such as the Cosmic Explorer [62] and the Einstein Telescope [63]. These detections could give us information related to neutron star masses, frequencies, tidal Love numbers, amplitudes of the modes, damping times, and moments of inertia.

In the present work, we compute the frequencies and corresponding eigenfunctions of the 10 lowest radial oscillation modes of spherical configurations from the dark sector, namely both dark matter and dark energy. To that end, we model an object of a given mass and radius with three different models, which differ in their internal structure and composition.

The present article is structured as follows: after this introductory section, we present the structure equations governing the hydrostatic equilibrium and the equations of state employed here. Radial oscillations are discussed in Section 3, and we present and discuss our main results in Section 4. Finally, we conclude our work in Section 5.

Throughout the manuscript, we adopt the mostly positive metric signature $(-,+,+,+)$ and we work in natural geometrical units, setting $G = c = 1$ for the Newton's constant and speed of light in vacuum, respectively. All units used here are measured in GeV: $1\ \text{m} = 5.068 \times 10^{15}/\ \text{GeV}$, $1\ \text{kg} = 5.610 \times 10^{26}\ \text{GeV}$, $1\ \text{s} = 1.519 \times 10^{24}/\ \text{GeV}$, and the Planck mass $M_{pl} = 1.22 \times 10^{19}$ GeV.

2. Relativistic Spheres in General Relativity

We assume that the objects are non-rotating and are electrically neutral. Also, we work with relativistic configurations in four dimensions, with a vanishing cosmological constant.

2.1. Structure Equations

The metric tensor in Schwarzschild coordinates (t, r, θ, ϕ) for static and spherically symmetric space-times is given by

$$ds^2 = -e^{\nu}dt^2 + e^{\lambda}dr^2 + r^2(d\theta^2 + \sin^2\theta\, d\phi^2). \tag{1}$$

To obtain interior solutions describing hydrostatic equilibrium of stars composed of dark matter and dark energy, we need to integrate the Tolman–Oppenheimer–Volkoff (TOV) equations [64,65] for two-fluid formalism [66,67], which are given by

$$m'(r) = 4\pi r^2 \rho(r), \tag{2}$$

$$\nu'(r) = 2\,\frac{m(r) + 4\pi r^3 p(r)}{r^2(1 - 2m(r)/r)}, \tag{3}$$

$$p'_M(r) = -[\rho_M(r) + p_M(r)]\,\frac{\nu'(r)}{2}, \tag{4}$$

$$p'_E(r) = -[\rho_E(r) + p_E(r)]\,\frac{\nu'(r)}{2}, \tag{5}$$

which we consider a spherical configuration composed of dark matter and dark energy, whose pressure and energy density are p_M and ρ_M, for dark matter, and p_E and ρ_E for dark energy, respectively. $m(r)$ is the mass function, and a prime denotes differentiation with respect to the radial coordinate r.

The total pressure p and total density ρ are the sum of the pressures and densities of both fluids, i.e.,

$$p = p_M + p_E, \tag{6}$$

$$\rho = \rho_M + \rho_E. \tag{7}$$

The TOV equations are to be integrated imposing the following initial conditions at the centre of the star

$$m(0) = 0, \tag{8}$$

$$p_M(0) = p_{cM}, \tag{9}$$

$$p_E(0) = p_{cE}, \tag{10}$$

where p_{cM} and p_{cE} are the central pressure for dark matter and dark energy, respectively. Moreover, the the surface conditions of the star that must be satisfied are

$$p(R) = 0, \tag{11}$$

$$m(R) = M, \tag{12}$$

with R being the radius of the star and M its mass.

2.2. Equations of State

To solve the TOV equations, we will require the equations of state. Since each fluid is independent, dark matter and dark energy have their own equation of state.

Dark matter may be interpreted as a Bose–Einstein condensate [24], since it solves the core–cusp problem [68]. Hence, a polytropic EoS of the following form is obtained [16,25,26]

$$p_M(r) = k\,\rho_M^2(r), \tag{13}$$

where k is the constant of proportionality, determined by [16,26,39]

$$k = \frac{2\pi l}{m^3}, \tag{14}$$

with m being the mass of the dark matter particle, and l being the scattering length that determines the elastic cross-section at low energies, $\sigma = 4\pi l^2$ [16,26,39]. Each model is characterized by its own numerical value of k, which can be seen in Table 1. Nowadays, the DM self-interacting cross section must satisfy the constraint [69–71]

$$1.75 \times 10^{-4}\ \text{cm}^2/\text{g} < \frac{\sigma}{m} < (1-2)\ \text{cm}^2/\text{g}. \tag{15}$$

The numerical values of k displayed in Table 1 may be achieved if, e.g., $l \sim 1\ fm$ and $m \sim 1$ GeV, and the bounds on σ are satisfied. We comment in passing that regarding the fuzzy DM model based on a free ultra-light scalar field, recent works have put an upper bound on its mass, $m < 10^{-23}$ eV [72,73], violating the 10^{-22} eV from the linear matter power spectrum.

Table 1. Numerical values of the parameters A, B, k and f for each model. For all three models considered in this work we assume a sphere of mass M = 1.216 $M_\odot$ and radius R = 8.145 km.

	A	B	k	f
Model 1	$\sqrt{0.40}$	$2.299 \times 10^{-4}/\text{km}^2$	45.140 GeV^{-4}	0.60
Model 2	$\sqrt{0.55}$	$2.595 \times 10^{-4}/\text{km}^2$	55.567 GeV^{-4}	0.75
Model 3	$\sqrt{0.70}$	$2.848 \times 10^{-4}/\text{km}^2$	66.942 GeV^{-4}	0.90

As far as dark energy is concerned, following [20], we consider the Extended Chaplygin equation of state [27,28].

$$p_E(r) = -\frac{B^2}{\rho_E(r)} + A^2 \rho_E(r), \tag{16}$$

where A is dimensionless and B has dimensions of energy density and pressure, both positive constants. Note that the equation of state of Chaplygin gas is given by $p_{Ch} = -B^2/\rho_{Ch}$ [74], while the additional barotropic term, $A^2\rho_E$, provides us with another viable dark energy model [75]; as it predicts a transition from deceleration to acceleration, it fits supernovae data, while at the same time the equation-of-state parameter w interpolates between 0 and -1, unifying thus non-relativistic matter and DE. What is more, at low red shift, the energy density becomes a constant, mimicking thus a positive cosmological constant, see Equation (11) of [75]. Table 1 shows the numerical values for A and B that we have considered.

It must be taken into account that given the form of the EoS, the pressure and density cannot become zero at the same time. Therefore, a vanishing pressure at the surface of the star implies a non-vanishing value for the energy density, which is computed to be $\rho_s = B/A$.

Additionally, it is convenient to introduce a new dimensionless factor related to the central conditions of the stars defined as

$$f = \frac{p_{cM}}{p_{cM} + p_{cE}}. \tag{17}$$

The numerical values for A, B, k, and f that we have chosen for three different models are displayed in Table 1 in order to compare stars of the same mass and radius but with different inner composition and obtain realistic solutions. The values are M = 1.216 $M_\odot$ for the mass, and R = 8.145 km for the radius. The chosen values are motivated by the fact that there are a few observed objects with known masses and radii around those values, see for instance Table 1 of [76].

In a recent work, we studied some properties of dark stars made of dark matter and dark energy [23], and here we propose to study radial oscillations and frequencies for this class of objects.

3. Radial Oscillations

The equations for radial perturbations of the stars, given below, are satisfied by the radial displacement, Δr, and the pressure perturbation, Δp, defined by

$$\xi \equiv \frac{\Delta r}{r}, \tag{18}$$

$$\eta \equiv \frac{\Delta p}{p}, \tag{19}$$

which satisfy the following first order differential equations [77,78]:

$$\xi'(r) = -\left(\frac{3}{r} + \frac{p'}{\zeta}\right)\xi - \frac{1}{r\Gamma}\eta, \tag{20}$$

$$\eta'(r) = \omega^2\left[r\left(1 + \frac{\rho}{p}\right)e^{\lambda-\nu}\right]\xi - \left[\frac{4p'}{p} + 8\pi\zeta r e^{\lambda} - \frac{r(p')^2}{p\zeta}\right]\xi - \left[\frac{\rho p'}{p\zeta} + 4\pi\zeta r e^{\lambda}\right]\eta, \tag{21}$$

where ζ is defined to be

$$\zeta \equiv p + \rho, \tag{22}$$

e^{λ} and e^{ν} are the metric potentials, which are computed as follows

$$e^{\lambda}(r) = \frac{1}{1 - \frac{2m(r)}{r}}, \tag{23}$$

$$\nu(r) = \ln\left(1 - \frac{2M}{R}\right) + 2\int_R^r \frac{m(x) + 4\pi x^3 p(x)}{x^2(1 - 2m(x)/x)}\,dx, \tag{24}$$

with M the mass of the star and R its radius, while $\omega = s\omega_0$ is the frequency oscillation mode, where s is a dimensionless number, while the constant ω_0 is computed by

$$\omega_0 = \sqrt{\frac{M}{R^3}}. \tag{25}$$

Finally

$$\Gamma = c_s^2\left(1 + \frac{\rho}{p}\right), \tag{26}$$

$$c_s^2 = \frac{dp}{d\rho}, \tag{27}$$

with Γ being the relativistic adiabatic index, and c_s is the the speed of sound.

The unknown frequencies are determined solving the Sturm–Liouville boundary value problem imposing the following boundary conditions [38], both at the center $r = 0$ and at the surface of the star $r = R$

$$\left.\frac{\eta}{\xi}\right|_{r=0} = -3\Gamma(0), \tag{28}$$

$$\left.\frac{\eta}{\xi}\right|_{r=R} = \left[-4 + \left(1 - \frac{2M}{R}\right)^{-1}\left(-\frac{M}{R} - \frac{\omega^2 R^3}{M}\right)\right]. \tag{29}$$

since $\xi'(r)$ must be finite as $r \to 0$, and $\eta'(r)$ must be finite at the surface as ρ, so $p \to 0$. Furthermore, since the system of equations is linear, the scaling factor is free and therefore the oscillating functions may be normalized, setting $\xi(0) = 1$ [78].

Before we proceed, a comment is in order here. When dealing with radial oscillations of objects made of two fluid components, in general there are two approaches [79]. In the first, a consistent general relativistic formalism to deal with the couplings between oscillation amplitudes and Lagrangian perturbations for each fluid was developed in [36,80]. Unfortunately, dealing with a system of highly coupled and non-linear differential equations requires very time-consuming numerical calculations. In the second approach, which is in fact simpler, only the radial oscillation of the whole admixed star, namely treated as a single fluid sphere, is studied without explicitly considering the gravitational coupling between the two fluid components, see, e.g., [39,81]. This is the approach we are taking in the discussion to follow.

4. Results and Discussion

We have studied radial oscillations and frequencies of dark stars made of dark matter and dark energy for their first 10 lowest modes shown in Table 2. For this, we have selected three different models, which all have the same radius R = 8.145 km and mass M = 1.216 $M_\odot$ but different dark energy mass fractions. The precise numerical values of the parameters A, B , k, and f, are shown in Table 1. The three models considered in this work are as follows

$$\text{Model 1} \rightarrow 49\% \text{ DE mass fraction} \quad (30)$$
$$\text{Model 2} \rightarrow 44\% \text{ DE mass fraction} \quad (31)$$
$$\text{Model 3} \rightarrow 41\% \text{ DE mass fraction} \quad (32)$$

Table 2. Frequencies (in kHz) of radial oscillation modes for the three models discussed here, see text for more details.

Mode Order n	Model 1	Model 2	Model 3
0	4.56	5.50	6.25
1	12.03	13.81	15.28
2	18.74	21.39	23.61
3	25.30	28.83	31.79
4	31.81	36.21	39.91
5	38.29	43.57	48.01
6	44.76	50.92	56.09
7	51.21	58.25	64.16
8	57.66	65.58	72.23
9	64.11	72.90	80.29

In Figure 1, we show the eigenfunctions $\eta(r)$ and $\xi(r)$ versus normalized (dimensionless) radial coordinate $x = r/R$ for low, intermediate, and highly excited modes for all three models considered here. We recall that in a Sturm–Liouville boundary value problem, the number of zeros of the eigenfunctions corresponds to the overtone number n. In the graphs, it can be observed that this is indeed the case. For instance, the fundamental mode $n = 0$ has no zeros, while the first excited mode $n = 1$ and the second excited mode have one and two zeros, respectively. And so on and so forth for higher modes.

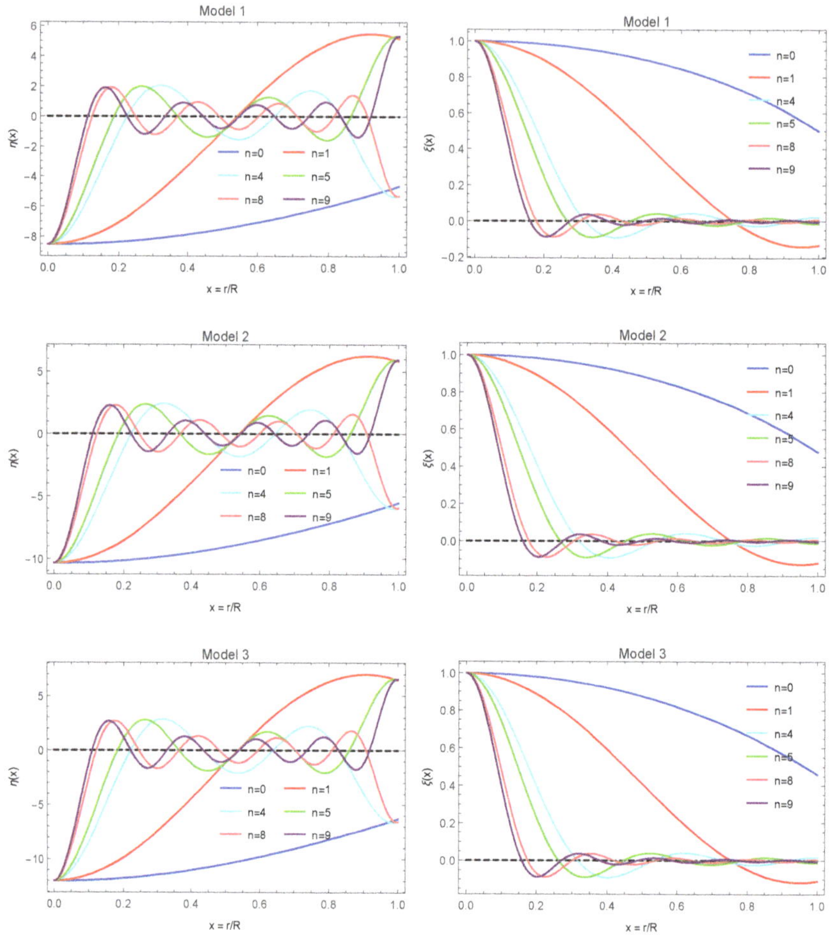

Figure 1. Perturbations $\eta(r)$ (left panels) and $\xi(r)$ (right panels) vs. $x = r/R$ for low order modes $n = 0$ (blue) and $n = 1$ (red), intermediate modes $n = 4$ (cyan) and $n = 5$ (green), and highly excited modes $n = 8$ (pink) and $n = 9$ (purple). Shown are Model 1 (top panels), Model 2 (center panels), and Model 3 (bottom panels).

In Table 2, where the numerical values of the frequencies are shown, we see that the frequencies are higher for Model 3 and lower for Model 1. In Figure 2, the large frequency separations defined as [38]

$$\Delta \nu_n \equiv \nu_{n+1} - \nu_n, \tag{33}$$

are displayed for all three models, where $n = 0, 1, 2, \ldots$ is the overtone number. As usual, for highly excited modes, the large frequency separations tend to a constant which is different from one model to another. They are also larger for Model 3 and smaller for Model 1. It is worth mentioning that Model 3 is characterized by the largest sound speed, see Figure 5 below, while regarding the fundamental mode, the ξ function of Model 3 decreases faster than the corresponding function of the other two models. Those features are in agreement with the results of [38], where radial oscillations of neutron stars were studied, and a similar behavior was observed.

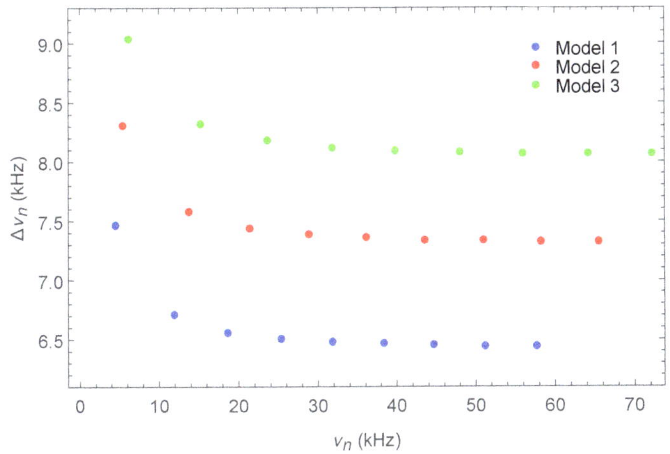

Figure 2. Large frequency separation $\Delta\nu_n$ vs. frequencies ν_n, both in kHz, for the lowest 10 modes. Blue is for Model 1, red for Model 2 and green for Model 3.

Figures 3 and 4 show the perturbations η and ξ, respectively, for the three models and for the fundamental mode and the first excited mode. Depending on the model, the values of $\eta(0)$ start with different values, whereas for ξ they all start with $\xi(0) = 1$. In the three models, the values of $\xi(r)$ are very similar for each mode. However, in $\eta(r)$ they differ, except for the zeros, which for each mode the three models are the same, as is ξ.

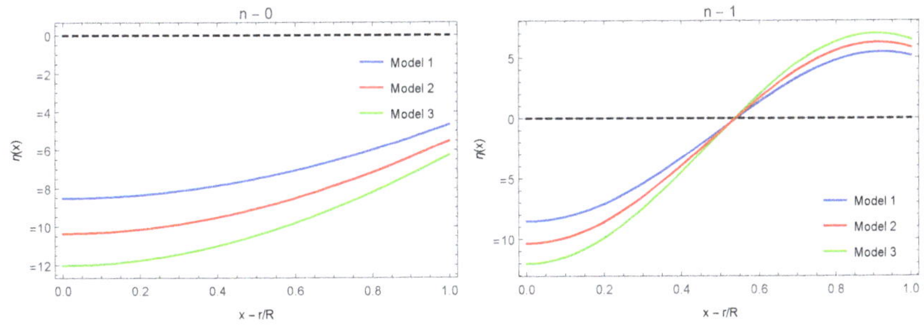

Figure 3. $\eta(r)$ perturbations for the fundamental mode $n = 0$ (**left**) and the first excited mode $n = 1$ (**right**). Blue represents Model 1, red represents Model 2, and green represents Model 3.

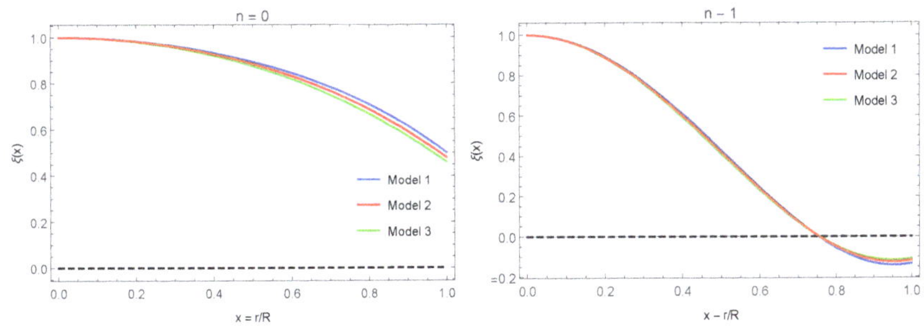

Figure 4. $\xi(r)$ perturbations for the fundamental mode $n = 0$ (**left**) and the first excited mode $n = 1$ (**right**). Blue denotes Model 1, red Model 2, and green Model 3.

We obtained that the dark energy mass fraction differs in all models, obtaining 49%, 44%, and 41% in Models 1, 2, and 3, respectively. If radial oscillations are discovered in dark stars, it will be possible to use their frequencies to infer the properties of their internal structure, and check if there is evidence for dark matter and dark energy inside the star, and their percentages or mass fraction.

Furthermore, we have checked that the results comply with the criteria for realistic solutions. The Buchdahl limit $C \equiv M/R < 4/9$ [82] is satisfied since $C = 0.222$. Stability, $\Gamma > 4/3$ [83], energy conditions $\rho \geq 0$ and $\rho \geq |p|$ [21,84–87], and causality, $0 < c_s^2 < 1$ [21], are fulfilled as well, as can be seen in Figure 5. The values for the speed of sound of Model 3 are the highest and those of Model 1 are the lowest.

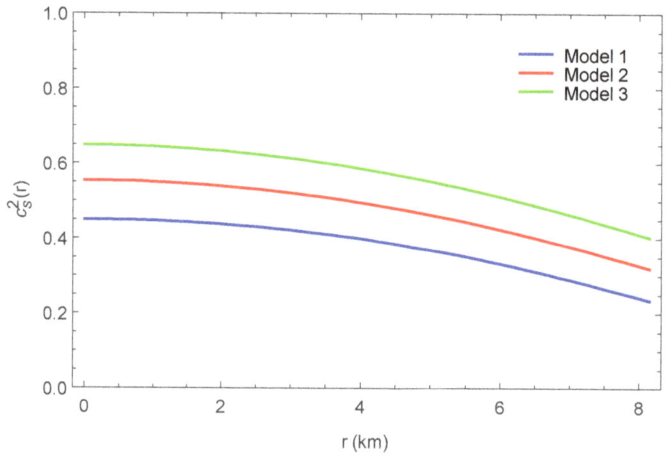

Figure 5. Speed of sound c_s^2 vs. radial coordinate r in km. Blue represents Model 1, red Model 2, and green Model 3.

5. Conclusions

To summarize our work, in the present article, we studied radial oscillations of spherical, non-rotating configurations made of both dark matter and dark energy. For the latter, we adopted the Extended Chaplygin gas EoS, while for the former, viewed as Bose–Einstein condensate, we assumed a polytropic EoS with index $n = 1$. We modeled the same object with mass $M = 1.216\ M_\odot$ and radius $R = 8.145$ km in three different dark matter–dark energy compositions. We integrated the equations for the perturbations, imposing the appropriate boundary conditions, both at the center and the surface of the objects, and we computed the frequencies as well as the corresponding eigenfunctions for the 10 lowest modes, namely fundamental (f mode) and nine excited modes. The large frequency separations are displayed as well for all three models considered here. A direct comparison between the three models is made showing (i) the two eigenfunctions of the fundamental and first excited mode as well as (ii) the speed of sound of the interior solution. We find that as the objects contain less and less dark energy, the sound speed and the frequencies increase.

Author Contributions: Conceptualization: G.P., Methodology: G.P., Formal analysis: G.P., Investigation: C.S. and G.P., Writing original draft preparation: C.S., Software: C.S., Writing-review and editing: C.S. and G.P., Visualization: C.S. and G.P. All authors have read and agreed to the published version of the manuscript.

Funding: The author C.S. acknowledges financial support from Universidad de la Frontera.

Data Availability Statement: No new data were created or analyzed in this study. Data sharing is not applicable to this article.

Conflicts of Interest: The authors declare no conflict of interest.

References

1. Zwicky, F. The redshift of extragalactic nebulae. *Helv. Phys. Acta* **1933**, *6*, 110.
2. Rubin, V.C.; Ford, W.K., Jr. Rotation of the Andromeda Nebula from a Spectroscopic Survey of Emission Regions. *Astrophys. J.* **1970**, *159*, 379–403. [CrossRef]
3. Munoz, C. Dark matter detection in the light of recent experimental results. *Int. J. Mod. Phys. A* **2004**, *19*, 3093–3170. [CrossRef]
4. Riess, A.G. et al. [Supernova Search Team]. Observational evidence from supernovae for an accelerating universe and a cosmological constant. *Astron. J.* **1998**, *116*, 1009–1038. [CrossRef]
5. Perlmutter, S. et al. [Supernova Cosmology Project]. Measurements of Ω and Λ from 42 high redshift supernovae. *Astrophys. J.* **1999**, *517*, 565–586. [CrossRef]
6. Freedman, W.L.; Turner, M.S. Measuring and understanding the universe. *Rev. Mod. Phys.* **2003**, *75*, 1433–1447. [CrossRef]
7. Aghanim, N. et al. [Planck]. Planck 2018 results. VI. Cosmological parameters. *Astron. Astrophys.* **2020**, *641*, A6; Erratum in *Astron. Astrophys.* **2021**, *652*, C4. [CrossRef]
8. Shapiro, S.L.; Teukolsky, S.A. *Black Holes, White Dwarfs, and Neutron Stars: The Physics of Compact Objects*; Wiley: New York, NY, USA, 1983; p. 645.
9. Lattimer, J.M.; Prakash, M. Neutron Star Observations: Prognosis for Equation of State Constraints. *Phys. Rep.* **2007**, *442*, 109–165. [CrossRef]
10. Alcock, C.; Farhi, E.; Olinto, A. Strange stars. *Astrophys. J.* **1986**, *310*, 261–272. [CrossRef]
11. Alcock, C.; Olinto, A. Exotic Phases of Hadronic Matter and their Astrophysical Application. *Ann. Rev. Nucl. Part. Sci.* **1988**, *38*, 161–184. [CrossRef]
12. Madsen, J. Physics and astrophysics of strange quark matter. *Lect. Notes Phys.* **1999**, *516*, 162–203. [CrossRef]
13. Weber, F. Strange quark matter and compact stars. *Prog. Part. Nucl. Phys.* **2005**, *54*, 193–288. [CrossRef]
14. Yue, Y.L.; Cui, X.H.; Xu, R.X. Is psr b0943+10 a low-mass quark star? *Astrophys. J. Lett.* **2006**, *649*, L95–L98. [CrossRef]
15. Leahy, D.; Ouyed, R. Supernova SN2006gy as a first ever Quark Nova? *Mon. Not. R. Astron. Soc.* **2008**, *387*, 1193. [CrossRef]
16. Li, X.Y.; Harko, T.; Cheng, K.S. Condensate dark matter stars. *J. Cosmol. Astropart. Phys.* **2012**, *6*, 1. [CrossRef]
17. Maselli, A.; Pnigouras, P.; Nielsen, N.G.; Kouvaris, C.; Kokkotas, K.D. Dark stars: Gravitational and electromagnetic observables. *Phys. Rev. D* **2017**, *96*, 023005. [CrossRef]
18. Panotopoulos, G.; Lopes, I. Dark stars in Starobinsky's model. *Phys. Rev. D* **2018**, *97*, 024025. [CrossRef]
19. Singh, K.N.; Ali, A.; Rahaman, F.; Nasri, S. Compact stars with exotic matter. *Phys. Dark Univ.* **2020**, *29*, 100575. [CrossRef]
20. Tello-Ortiz, F.; Malaver, M.; Rincón, Á.; Gomez-Leyton, Y. Relativistic anisotropic fluid spheres satisfying a non-linear equation of state. *Eur. Phys. J. C* **2020**, *80*, 371. [CrossRef]
21. Panotopoulos, G.; Rincón, Á.; Lopes, I. Radial oscillations and tidal Love numbers of dark energy stars. *Eur. Phys. J. Plus* **2020**, *135*, 856. [CrossRef]
22. Panotopoulos, G.; Rincón, Á.; Lopes, I. Slowly rotating dark energy stars. *Phys. Dark Univ.* **2021**, *34*, 100885. [CrossRef]
23. Sepúlveda, C.; Panotopoulos, G. On exotic objects made of dark energy and dark matter: Mass-to-radius profiles and tidal Love numbers. *Galaxies* **2023**, *11*, 101. [CrossRef]
24. Boehmer, C.G.; Harko, T. Can dark matter be a Bose-Einstein condensate? *J. Cosmol. Astropart. Phys.* **2007**, *6*, 025. [CrossRef]
25. Fan, J. Ultralight Repulsive Dark Matter and BEC. *Phys. Dark Univ.* **2016**, *14*, 84–94. [CrossRef]
26. Chavanis, P.H. Phase transitions between dilute and dense axion stars. *Phys. Rev. D* **2018**, *98*, 023009. [CrossRef]
27. Pourhassan, B.; Kahya, E.O. Extended Chaplygin gas model. *Results Phys.* **2014**, *4*, 101–102. [CrossRef]
28. Pourhassan, B.; Kahya, E.O. FRW cosmology with the extended Chaplygin gas. *Adv. High Energy Phys.* **2014**, *2014*, 231452. [CrossRef]
29. Freese, K.; Spolyar, D.; Bodenheimer, P.; Gondolo, P. Dark Stars: A New Study of the FIrst Stars in the Universe. *New J. Phys.* **2009**, *11*, 105014. [CrossRef]
30. Spolyar, D.; Bodenheimer, P.; Freese, K.; Gondolo, P. Dark Stars: A new look at the First Stars in the Universe. *Astrophys. J.* **2009**, *705*, 1031–1042. [CrossRef]
31. Turck-Chieze, S.; Couvidat, S. Solar neutrinos, helioseismology and the solar internal dynamics. *Rep. Prog. Phys.* **2011**, *74*, 086901. [CrossRef]
32. Chaplin, W.J.; Miglio, A. Asteroseismology of Solar-Type and Red-Giant Stars. *Ann. Rev. Astron. Astrophys.* **2013**, *51*, 353. [CrossRef]
33. Kokkotas, K.D.; Schmidt, B.G. Quasinormal modes of stars and black holes. *Living Rev. Rel.* **1999**, *2*, 2. [CrossRef] [PubMed]
34. Paschalidis, V.; Stergioulas, N. Rotating Stars in Relativity. *Living Rev. Rel.* **2017**, *20*, 7. [CrossRef] [PubMed]
35. Clemente, F.D.; Mannarelli, M.; Tonelli, F. Reliable description of the radial oscillations of compact stars. *Phys. Rev. D* **2020**, *101*, 103003. [CrossRef]
36. Kain, B. Radial oscillations and stability of multiple-fluid compact stars. *Phys. Rev. D* **2020**, *102*, 023001. [CrossRef]
37. Córsico, A.H.; Althaus, L.G.; Bertolami, M.M.M.; Kepler, S.O. Pulsating white dwarfs: New insights. *Astron. Astrophys. Rev.* **2019**, *27*, 7. [CrossRef]

38. Sagun, V.; Panotopoulos, G.; Lopes, I. Asteroseismology: Radial oscillations of neutron stars with realistic equation of state. *Phys. Rev. D* **2020**, *101*, 063025. [CrossRef]
39. Panotopoulos, G.; Lopes, I. Radial oscillations of strange quark stars admixed with condensed dark matter. *Phys. Rev. D* **2017**, *96*, 083013. [CrossRef]
40. Leung, S.C.; Chu, M.C.; Lin, L.M. Dark-matter admixed neutron stars. *Phys. Rev. D* **2011**, *84*, 107301. [CrossRef]
41. Leung, S.C.; Chu, M.C.; Lin, L.M. Equilibrium Structure and Radial Oscillations of Dark Matter Admixed Neutron Stars. *Phys. Rev. D* **2012**, *85*, 103528. [CrossRef]
42. Lopes, I.P. A New Look at the Eckart-Scuflaire-Osaki Classification Scheme of Stellar Oscillations. *Astrophys. J.* **2000**, *542*, 1071–1074. [CrossRef]
43. Lopes, I.P. Nonradial adiabatic oscillations of stars. *Astron. Astrophys.* **2001**, *373*, 916–931. [CrossRef]
44. Kokkotas, K.D.; Ruoff, J. Radial oscillations of relativistic stars. *Astron. Astrophys.* **2001**, *366*, 565. [CrossRef]
45. Miniutti, G.; Pons, J.A.; Berti, E.; Gualtieri, L.; Ferrari, V. Non-radial oscillation modes as a probe of density discontinuities in neutron stars. *Mon. Not. R. Astron. Soc.* **2003**, *338*, 389. [CrossRef]
46. Passamonti, A.; Bruni, M.; Gualtieri, L.; Sopuerta, C.F. Coupling of radial and non-radial oscillations of relativistic stars: Gauge-invariant formalism. *Phys. Rev. D* **2005**, *71*, 024022. [CrossRef]
47. Passamonti, A.; Bruni, M.; Gualtieri, L.; Nagar, A.; Sopuerta, C.F. Coupling of radial and axial non-radial oscillations of compact stars: Gravitational waves from first-order differential rotation. *Phys. Rev. D* **2006**, *73*, 084010. [CrossRef]
48. Savonije, G.J. Non-radial oscillations of the rapidly rotating Be star HD 163868. *Astron. Astrophys.* **2007**, *469*, 1057. [CrossRef]
49. Flores, C.V.; Lugones, G. Radial oscillations of color superconducting self-bound quark stars. *Phys. Rev. D* **2010**, *82*, 063006. [CrossRef]
50. Brillante, A.; Mishustin, I.N. Radial oscillations of neutral and charged hybrid stars. *Europhys. Lett.* **2014**, *105*, 39001. [CrossRef]
51. Franco, L.M.; Link, B.; Epstein, R.I. Quaking neutron stars. *Astrophys. J.* **2000**, *543*, 987. [CrossRef]
52. Andersson, N.; Jones, D.I.; Kokkotas, K.D.; Stergioulas, N. R mode runaway and rapidly rotating neutron stars. *Astrophys. J. Lett.* **2000**, *534*, L75. [CrossRef]
53. Tsang, D.; Read, J.S.; Hinderer, T.; Piro, A.L.; Bondarescu, R. Resonant Shattering of Neutron Star Crusts. *Phys. Rev. Lett.* **2012**, *108*, 011102. [CrossRef]
54. Hinderer, T.; Taracchini, A.; Foucart, F.; Buonanno, A.; Steinhoff, J.; Duez, M.; Kidder, L.E.; Pfeiffer, H.P.; Scheel, M.A.; Szilagyi, B.; et al. Effects of neutron-star dynamic tides on gravitational waveforms within the effective-one-body approach. *Phys. Rev. Lett.* **2016**, *116*, 181101. [CrossRef] [PubMed]
55. Chirenti, C.; Gold, R.; Miller, M.C. Gravitational waves from f-modes excited by the inspiral of highly eccentric neutron star binaries. *Astrophys. J.* **2017**, *837*, 67. [CrossRef]
56. Michel, E.; Baglin, A.; Auvergne, M.; Catala, C.; Samadi, R. CoRoT measures solar-like oscillations and granulation in stars hotter than the Sun. *Science* **2008**, *322*, 558–560. [CrossRef] [PubMed]
57. Mosser, B.; Belkacem, K.; Goupil, M.J.; Miglio, A.; Morel, T.; Barban, C.; Baudin, F.; Hekker, S.; Samadi, R.; Ridder, J.D.; et al. Red-giant seismic properties analyzed with CoRoT. *Astron. Astrophys.* **2010**, *517*, A22. [CrossRef]
58. Mosser, B.; Belkacem, K.; Goupil, M.J.; Michel, E.; Elsworth, Y.; Barban, C.; Kallinger, T.; Hekker, S.; DeRidder, J.; Samadi, R.; et al. The universal red-giant oscillation pattern: An automated determination with CoRoT data. *Astron. Astrophys.* **2011**, *525*, L9. [CrossRef]
59. Bischoff-Kim, A.; Østensen, R.H. Asteroseismology of the Kepler field DBV White Dwarf—It's a hot one! *Astrophys. J. Lett.* **2011**, *742*, L16. [CrossRef]
60. Corsico, A.H.; Althaus, L.G.; Bertolami, M.M.M.; Bischoff-Kim, A. Asteroseismology of the Kepler V777 Her variable white dwarf with fully evolutionary models. *Astron. Astrophys.* **2012**, *541*, A42. [CrossRef]
61. Hekker, S. CoRoT and Kepler results: Solar-like oscillators. *Adv. Space Res.* **2013**, *52*, 1581–1592. [CrossRef]
62. Evans, M.; Adhikari, R.X.; Afle, C.; Ballmer, S.W.; Biscoveanu, S.; Borhanian, S.; Brown, D.A.; Chen, Y.; Eisenstein, R.; Gruson, A.; et al. A Horizon Study for Cosmic Explorer: Science, Observatories, and Community. *arXiv* **2021**, arXiv:2109.09882.
63. Punturo, M.; Abernathy, M.; Acernese, F.; Allen, B.; Andersson, N.; Arun, K.; Barone, F.; Barr, B.; Barsuglia, M.; Beker, M.; et al. The Einstein Telescope: A third-generation gravitational wave observatory. *Class. Quant. Grav.* **2010**, *27*, 194002. [CrossRef]
64. Oppenheimer, J.R.; Volkoff, G.M. On massive neutron cores. *Phys. Rev.* **1939**, *55*, 374–381. [CrossRef]
65. Tolman, R.C. Static solutions of Einstein's field equations for spheres of fluid. *Phys. Rev.* **1939**, *55*, 364–373. [CrossRef]
66. Sandin, F.; Ciarcelluti, P. Effects of mirror dark matter on neutron stars. *Astropart. Phys.* **2009**, *32*, 278–284. [CrossRef]
67. Ciarcelluti, P.; Sandin, F. Have neutron stars a dark matter core? *Phys. Lett. B* **2011**, *695*, 19–21. [CrossRef]
68. Harko, T. Bose-Einstein condensation of dark matter solves the core/cusp problem. *J. Cosmol. Astropart. Phys.* **2011**, *5*, 022. [CrossRef]
69. Markevitch, M.; Gonzalez, A.H.; Clowe, D.; Vikhlinin, A.; David, L.; Forman, W.; Jones, C.; Murray, S.; Tucker, W. Direct constraints on the dark matter self-interaction cross-section from the merging galaxy cluster 1E0657-56. *Astrophys. J.* **2004**, *606*, 819–824. [CrossRef]
70. Robertson, A.; Massey, R.; Eke, V. What does the Bullet Cluster tell us about self-interacting dark matter? *Mon. Not. Roy. Astron. Soc.* **2017**, *465*, 569–587. [CrossRef]

71. Young, B.L. A survey of dark matter and related topics in cosmology. *Front. Phys.* **2017**, *12*, 121201. Erratum in *Front. Phys.* **2017**, *12*, 121202. [CrossRef]
72. Meinert, J.; Hofmann, R. Axial Anomaly in Galaxies and the Dark Universe. *Universe* **2021**, *7*, 198. [CrossRef]
73. Bañares-Hernández, A.; Castillo, A.; Camalich, J.M.; Iorio, G. Confronting fuzzy dark matter with the rotation curves of nearby dwarf irregular galaxies. *Astron. Astrophys.* **2023**, *676*, A63. [CrossRef]
74. Kamenshchik, A.Y.; Moschella, U.; Pasquier, V. An Alternative to quintessence. *Phys. Lett. B* **2001**, *511*, 265–268. [CrossRef]
75. Panotopoulos, G.; Lopes, I.; Rincón, Á. Lagrangian formulation for an extended cosmological equation-of-state. *Phys. Dark Univ.* **2021**, *31*, 100751. [CrossRef]
76. Aziz, A.; Ray, S.; Rahaman, F.; Khlopov, M.; Guha, B.K. Constraining values of bag constant for strange star candidates. *Int. J. Mod. Phys. D* **2019**, *28*, 1941006. [CrossRef]
77. Chanmugan, G. Radial oscillations of zero-temperature white dwarfs and neutron stars below nuclear densities. *Astrophys. J.* **1977**, *217*, 799. [CrossRef]
78. Vath, H.M.; Chanmugan, G. Radial oscillations of neutron stars and strange stars. *Astron. Astrophys.* **1992**, *260*, 250–254.
79. Jiménez, J.C.; Fraga, E.S. Radial Oscillations of Quark Stars Admixed with Dark Matter. *Universe* **2022**, *8*, 34. [CrossRef]
80. Kain, B. Dark matter admixed neutron stars. *Phys. Rev. D* **2021**, *103*, 043009. [CrossRef]
81. Panotopoulos, G.; Lopes, I. Radial oscillations of strange quark stars admixed with fermionic dark matter. *Phys. Rev. D* **2018**, *98*, 083001. [CrossRef]
82. Buchdahl, H.A. General Relativistic Fluid Spheres. *Phys. Rev.* **1959**, *116*, 1027. [CrossRef]
83. Moustakidis, C.C. The stability of relativistic stars and the role of the adiabatic index. *Gen. Rel. Grav.* **2017**, *49*, 68. [CrossRef]
84. Mak, M.K.; Harko, T. An Exact Anisotropic Quark Star Model. *Chin. J. J. Astron. Astrophys.* **2002**, *2002*, 248. [CrossRef]
85. Deb, D.; Chowdhury, S.R.; Ray, S.; Rahaman, F.; Guha, B.K. Relativistic model for anisotropic strange stars. *Ann. Phys.* **2017**, *387*, 239. [CrossRef]
86. Deb, D.; Chowdhury, S.R.; Ray, S.; Rahaman, F. A New Model for Strange Stars. *Gen. Rel. Grav.* **2018**, *50*, 112. [CrossRef]
87. Balart, L.; Panotopoulos, G.; Rincón, Á. Regular charged black holes, energy conditions and quasinormal modes. *arXiv* **2023**, arXiv:2309.01910.

universe

MDPI

Review

From Barthel–Randers–Kropina Geometries to the Accelerating Universe: A Brief Review of Recent Advances in Finslerian Cosmology

Amine Bouali [1,2], Himanshu Chaudhary [3,4], Lehel Csillag [3,5], Rattanasak Hama [6], Tiberiu Harko [3,7,*], Sorin V. Sabau [8,9], Shahab Shahidi [10]

1 Laboratory of Physics of Matter and Radiation, Mohammed I University, Oujda BP 717, Morocco; a1.bouali@ump.ac.ma
2 Higher School of Education and Training, Mohammed I University, Oujda BP 717, Morocco
3 Department of Physics, Babeș-Bolyai University, Kogălniceanu Street 1, 400084 Cluj-Napoca, Romania; himanshu.chaudhary@ubbcluj.ro; (H.C.); lehel.csillag@ubbcluj.ro (L.C.)
4 Research Center of Astrophysics and Cosmology, Khazar University, 41 Mehseti Street, Baku AZ1096, Azerbaijan
5 Department of Mathematics and Computer Science, Transilvania University, Iuliu Maniu Street 50, 500091 Brașov, Romania
6 Faculty of Science and Industrial Technology, Prince of Songkla University, Surat Thani Campus, Surat Thani 84000, Thailand; rattanasak.h@psu.ac.th
7 Astronomical Observatory, 19 Ciresilor Street, 400487 Cluj-Napoca, Romania
8 School of Biological Sciences, Department of Biology, Tokai University, Sapporo 005-8600, Japan; sorin@tokai.ac.jp
9 Graduate School of Science and Technology, Physical and Mathematical Sciences, Tokai University, Sapporo 005-8600, Japan
10 School of Physics, Damghan University, Damghan 41167-36716, Iran; s.shahidi@du.ac.ir
* Correspondence: tiberiu.harko@aira.astro.ro

Abstract: We present a review of recent developments in cosmological models based on Finsler geometry, as well as geometric extensions of general relativity formulated within this framework. Finsler geometry generalizes Riemannian geometry by allowing the metric tensor to depend not only on position but also on an additional internal degree of freedom, typically represented by a vector field at each point of the spacetime manifold. We examine in detail the possibility that Finsler-type geometries can describe the physical properties of the gravitational interaction, as well as the cosmological dynamics. In particular, we present and review the implications of a particular implementation of Finsler geometry, based on the Barthel connection, and of the (α, β) geometries, where α is a Riemannian metric, and β is a one-form. For a specific construction of the deviation part β, in these classes of geometries, the Barthel connection coincides with the Levi–Civita connection of the associated Riemann metric. We review the properties of the gravitational field, and of the cosmological evolution in three types of geometries: the Barthel–Randers geometry, in which the Finsler metric function F is given by $F = \alpha + \beta$, in the Barthel–Kropina geometry, with $F = \alpha^2/\beta$, and in the conformally transformed Barthel–Kropina geometry, respectively. After a brief presentation of the mathematical foundations of the Finslerian-type modified gravity theories, the generalized Friedmann equations in these geometries are written down by considering that the background Riemannian metric in the Randers and Kropina line elements is of Friedmann–Lemaitre–Robertson–Walker type. The matter energy balance equations are also presented, and they are interpreted from the point of view of the thermodynamics of irreversible processes in the presence of particle creation. We investigate the cosmological properties of the Barthel–Randers and Barthel–Kropina cosmological models in detail. In these scenarios, the additional geometric terms arising from the Finslerian structure can be interpreted as an effective geometric dark energy component, capable of generating an effective cosmological constant. Several cosmological solutions—both analytical and numerical—are obtained and compared against observational datasets, including Cosmic Chronometers, Type Ia Supernovae, and Baryon Acoustic Oscillations, using a Markov Chain Monte Carlo (MCMC) analysis. A direct comparison with the standard ΛCDM model is also carried out. The results indicate that Finslerian cosmological models provide a satisfactory fit to the observational data, suggesting they represent a viable alternative to the standard cosmological model based on general relativity.

Keywords: Barthel connection; (α, β) metrics; Finslerian cosmology; posterior inference; MCMC statistical analysis

1. Introduction—Early History and Modern Developments

In 1918, the same year in which Herman Weyl proposed his important extension of Riemannian geometry [1,2], by introducing the concept of non-metricity to mathematics and physics, another equally interesting generalization of the Riemannian framework was proposed by Paul Finsler [3]. The work of Weyl was clearly influenced by the progresses in the physics of his time, and especially by the development of general relativity by Einstein [4,5] and Hilbert [6], and it has been extensively used in physics, especially recently [7]. On the other hand, Finsler geometry, which represents, at least from a physicist's perspective, a much more drastic extension of Riemannian geometry, was not considered from the point of view of applications in fundamental physics for at least 40 years.

Actually, Finslerian geometry was already anticipated by Riemann [8], who already assumed the existence of metric structures in a general space based on the distance element $ds = F(x^1, x^2, \ldots, x^n; dx^1, dx^2, \ldots, dx^n) = F(x^1, x^2, \ldots, x^n;, y^1, y^2, \ldots, y^n)$, or simply $F(x, y)$, where the condition that for $y = 0$, F is a positive definite function defined on the tangent bundle TM must also be imposed. In order to assure the independence of the length of curves on their parameterization, it is further required that F is a homogeneous function of degree one in y. Riemann geometry is just a special case of the general distance element ds, obtained under the assumption $F^2 = g_{IJ}(x)dx^I dx^J$. Thus, Finsler geometry is not a proper generalization of Riemann's geometry, but it "...is just Riemannian geometry without the quadratic restriction" [9]. However, it is an established practice especially in the physics literature to describe Finsler geometry as a generalization of Riemann's geometry, and in the present work, we will also adopt this view, mostly for the simplicity of the presentation.

In a simple description, Finsler geometry can be considered as a geometry in which the metric tensor g_{ij} is a function of both the coordinates x, defined on the base manifold M, and of the tangent vectors y, so that $g_{IJ} = g_{IJ}(x, y)$. In a physically intuitive sense, Finsler geometry can be considered as a geometry in which the metric tensor, determining the distance between two neighboring points, is an arbitrary function of both coordinates and velocities. For presentations of Finsler geometry from a mathematical perspective, see [10–13].

The remarkable success of general relativity in the description of gravitational phenomena and its many observational confirmations have significantly strengthened the position of Riemannian geometry as the "true" geometry of nature. General relativity, in its Riemannian formulation, successfully passes all the Solar System tests, inside which it can explain the gravitational phenomenology with high precision. Important confirmations of the validity of general relativity include such diverse phenomena as the perihelion precession of Mercury, light bending by the Sun, frame dragging, the Nordtvedt effect in lunar motion, and the Shapiro time delay [14]. The experimental detection of gravitational waves represents a brilliant confirmation of the predictions of general relativity [15], which opened a new window in the Universe, and led to new perspectives on the black hole properties, their dynamical evolution, and the mass distribution of neutron stars [16]. Inspired by the immense success of general relativity, the Riemannian geometric framework was also extended, leading to important developments in mathematics like the introduction of the concept of torsion [17,18], or of the concept of absolute parallelism [19]. These new geometries have found important applications in physics, and they have been used to model the gravitational interaction from new perspectives.

From a mathematical point of view, Ehlers Pirani and Schild aimed to develop a constructive axiomatization of gravity. In their seminal 1972 paper, they built up the kinematical structure of general relativity, purely based on axioms, which have a somewhat empirical content [20]. Although the axiomatization is quite dense and technical, recent pedagogical reviews are present in the literature [21–23]. We briefly outline the four steps of the axiomatization in the following [24]:

1. In the first step, the basic physical/empirical entities are postulated, such as the timelike worldlines for freely falling massive particles, the lightlike worldlines of light rays, and radar echoes between massive particle worldlines. These empirical elements provide enough structure to define a system of coordinates and allow for the construction of a differentiable structure on the spacetime set M, turning it into a smooth manifold.
2. The conformal structure is established by requiring that, at each point in spacetime, the set of all possible directions (i.e., tangent vectors) splits into two components when the directions corresponding to massless (lightlike) trajectories are removed. This splitting reflects the causal distinction between future and past. Additionally, in a sufficiently small neighborhood V around the worldline of a massive particle, for any point $p \in V$ not lying on the particle's path, the function that maps p to the product of the radar emission time t_e and reception time t_r, i.e., $p \mapsto t_e t_r$, must be at least twice differentiable.
3. Imposing that through each point in spacetime, and for each timelike direction, there exists one unique timelike (massive) trajectory passing through that point, which results in a projective structure. Each of these trajectories must admit a parametrization such that, in local coordinates near the point, the motion satisfies $\ddot{x} = 0$. This expresses the fact that particles move along straight lines in free fall.
4. In the final step, compatibility between the conformal and projective structures is required. In particular, light rays must be special cases of particle geodesics in the limit of zero mass. This determines the metric up to a conformal factor, which leads to a Weyl structure. Through some technical steps, eliminating the second clock effect leads to a Lorentzian structure, i.e., a pseudo-Riemannian manifold.

It turns out that if one weakens the twice differentiability assumption on the map $p \mapsto t_e t_r$, there are Finsler functions which satisfy all the other EPS axioms. A well-known example is given by

$$F(x,y) = e^{2\sigma(x,y)} \sqrt{|a_{IJ}(x) y^I y^J|}, \tag{1}$$

where $\sigma(x,y)$ is a non-singular 0-homogeneous function in y satisfying

$$\partial_K \sigma - y^M \Gamma^N_{MK}(x) \frac{\partial \sigma(x,y)}{\partial y^N} = 0, \tag{2}$$

with Γ^N_{MK} being the Christoffel symbols associated with the Riemannian metric $a_{IJ}(x)$. This geometry is of Berwald type, i.e., it is the closest to a Riemannian geometry, but it is not trivial.

Despite the systematic, rigorous and physically appealing nature of Finsler geometry, the applications in physics of this geometry appeared relatively late. A first step in this direction was taken in the work of Randers [25], who tried to formulate a unified theory of electromagnetism and gravity. Initially, the theory was formulated in a higher-dimensional Riemannian geometry. The Finslerian nature of the Randers geometry was recognized by Ingarden [26]. Essentially, Randers geometry is a specific example of a Finsler geometry, with the fundamental function $F(x,y) = \left(a_{IJ}(x) dx^I dx^J\right)^{1/2} + b_K(x) y^K$, where $b_K(x)$ is an arbitrary vector field. Presently, Randers spaces are considered as Finsler spaces $F^n = (M, \alpha + \beta)$, equipped with the Cartan nonlinear connection N, i.e., they are denoted with $RF^n = (M, \alpha + \beta, N)$ [27]. A comprehensive overview of several Finsler metrics and of some of their physical applications is given in Table 1.

Table 1. Examples of Finsler metrics with their defining functions and characteristic properties.

Name	Finsler Function $F(x,y)$	Properties
Semi-Riemannian	$F(x,y) = \sqrt{\lvert a_{IJ}(x)y^I y^J \rvert}$	Quadratic in y, reversible, i.e., $F(x,y) = F(x,-y)$.
Randers	$F(x,y) = \sqrt{a_{IJ}(x)y^I y^J} + b_I(x)y^I$	Non-reversible, i.e., $F(x,y) \neq F(x,-y)$; appears in EM analogs and Lorentz-violating physics.
Kropina	$F(x,y) = \frac{a_{IJ}y^I y^J}{b_I y^I}, \quad b_I y^I \neq 0$	Singular on hypersurfaces; non-reversible.
Matsumoto	$F(x,y) = \frac{\sqrt{a_{IJ}y^I y^J}}{\sqrt{a_{IJ}y^I y^J - b_K y^K}}$	Non-reversible; used in irreversible mechanics; singular on hypersurfaces.
Bogoslovsky	$F(x,y) = (-\eta_{IJ}y^I y^J)^{\frac{1-b}{2}} (n_K y^K)^b$	Breaks full Lorentz invariance; $0 < b < 1$; used in very special relativity.
Funk	$F(x,y) = \frac{\sqrt{\delta_{IJ}y^I y^J - \delta_{IJ}x^I y^J}}{1-\delta_{IJ}x^I x^J}$	Defined on unit ball; forward complete; non-reversible.
Locally Minkowskian	$F(x,y) = F(y)$	General flat case; depends only on y, independent of x.
α-β metrics	$F = \phi(\alpha,\beta), \quad \alpha = \sqrt{a_{IJ}y^I y^J}, \ \beta = b_K y^K$	Unifies and generalizes Randers, Kropina, and Matsumoto metrics via the scalar function ϕ.
General Lagrangian	$F(x,y) = \sqrt{L(x,y)}$, L 2-homogeneous in y	More general case; has some applications in mechanics.
Optical/media	$F(x,y) = n(x,y)\lVert y \rVert$	Includes anisotropic effects in optics through the direction-dependent refractive index $n(x,y)$.
Non-reversible Finsler	General F such that $F(x,y) \neq F(x,-y)$	Includes Funk, Randers, and non-reversible geometries.

Finsler geometry has found applications in several areas of physics, including the geometric formulation of quantum mechanics [28–31], kinetic theory of gases [32,33], dispersion relations [34] and optics [35]. A more exotic application is related to an extension of the standard model of particle physics, where a Finsler structure can be given for the nonminimal fermion sector [36]. It is also interesting to note that Lorentz-violating scalar fields can also be described by Riemann–Finsler geometry [37]. In the latter cases, the Finsler geometry considered is of Berwald type: the closest one to the Riemannan geometry, which is still not trivial and compatible with the weakened Ehlers–Pirani–Schild axioms.

The first attempts at formulating a Finslerian theory of gravitation, but still in the framework of unified field theories, belonged to Horváth [38], and Horváth and Moór [39]. Early Finslerian-type gravitational theories were also formulated in [40,41], where a set of Finslerian-type gravitational field equations were proposed, representing a straightforward extension of the Einstein field equations, and given by

$$R_{\mu\nu} - \frac{1}{2}g_{\mu\nu}R + \lambda g_{\mu\nu} = \chi T_{\mu\nu}, K_{\mu\nu} - \frac{1}{2}g_{\mu\nu}K + \lambda g_{\mu\nu} = \chi T_{\mu\nu}, \tag{3}$$

and

$$S_{\mu\nu} - \frac{1}{2}g_{\mu\nu}S - \lambda^{(i)}g_{\mu\nu} = -\chi^{(i)}T^i_{\mu\nu}, \tag{4}$$

respectively. In Equations (3) and (4), χ denotes the gravitational constant, λ is the cosmological constant, and $\lambda^{(i)}$ and $\chi^{(i)}$ represent the internal cosmological and gravitational constants. The quantities $R_{\mu\nu}$, R, $K_{\mu\nu}$, K and $S_{\mu\nu}$ and S denote the hh-, hv- and vv-Ricci curvature tensors, and the hh-, hv- and vv-scalar curvatures of the Finsler space (M,F). Moreover, $T_{\mu\nu}$ is the energy–momentum tensor of the baryonic matter, and $T^i_{\mu\nu}$ denotes the internal energy–momentum tensor. In [42] another Finslerian-type approach to gravity was introduced and developed. The main focus of this work was the Finslerian description and interpretation of the particle motion in a gravitational field. For the Finslerian extensions of general relativity along the line of [42], see [43]. Finslerian-type generalized Schwarzschild metrics have been investigated in [44,45]. For further studies of black holes and wormholes in Finsler geometry, see [46–55].

Miron [56] has proposed the vector bundle point of view to formulate a system of Einstein-type gravitational field equations. The starting point of this approach is to interpret the field y as a fiber at the point x of the base manifold M. The total space of this vector bundle is obtained as a unification of the fields x and y [57]. A nonlinear connection naturally induces an adapted basis in the fiber $T_{(x,y)}TM$ of the double tangent bundle TTM and a corresponding dual basis in the fiber $T^*_{(x,y)}TM$ of the cotangent bundle T^*TM. These are given by

$$X_A = \left(\frac{\delta}{\delta x^\lambda} = \frac{\partial}{\partial x^\lambda} - N^i_\lambda \frac{\partial}{\partial y^i}, \frac{\partial}{\partial y^i}\right), X^A = \left(dx^\kappa, \delta y^i = dy^i + N^i_\lambda dx^\lambda\right), \tag{5}$$

respectively.

In Equation (5), the indices A, B run over $(A, B) \in \{0, 1, 2, 3, ..., 7\}$, while the Greek indices λ, κ take values in $\{0, 1, 2, 3\}$. The functions N^i_λ define the nonlinear connection. By construction, the basis (5) is adapted to a Finsler-type metric on TM, given by

$$G = g_{\lambda\kappa}(x, y)dx^\kappa dx^\lambda + g_{ij}(x, y)\delta y^i \delta y^j, \tag{6}$$

where $g_{\lambda k}(x, y)$ and $g_{ij}(x, y)$ are the horizontal and vertical metric components, respectively.

In the total space of the vector bundle, the Einstein field equations are assumed to have their standard general relativistic form, $\mathcal{R}_{AB} - (1/2)\mathcal{R}g_{AB} = \tau_{AB}$, where τ_{AB} is the matter energy–momentum tensor. The field equations can be decomposed, and can be written as [56]

$$R_{\lambda\nu} - \frac{1}{2}(R + S)g_{\lambda\nu} = \tau_{\lambda\nu}, \overset{1}{P}_{i\lambda} = \tau_{i\lambda}, \overset{2}{P}_{\lambda i} = -\tau_{\lambda i}, \tag{7}$$

$$S_{ij} - \frac{1}{2}(R + S)g_{ij} = \tau_{ij}. \tag{8}$$

An interesting Finslerian-type gravitational theory was proposed in [58]. The basic idea of this approach is the assumption that the Einstein vacuum field equations can be obtained from the condition $H = H^i_i = 0$, where H^i_k, the Finslerian deviation tensor, is obtained from the first and second derivatives of the quantity $G^l = \gamma^l_{jk}\dot{x}^j\dot{x}^k/2$ as

$$H^i{}_k = 2\frac{\partial G^i}{\partial x^k} - \frac{\partial^2 G^i}{\partial x^j \partial \dot{x}^k}\dot{x}^j + 2\frac{\partial^2 G^i}{\partial \dot{x}^j \partial \dot{x}^k}G^j - \frac{\partial G^i}{\partial \dot{x}^j}\frac{\partial G^j}{\partial \dot{x}^k}. \tag{9}$$

If the metric is Riemannian, we obtain the limiting case of the general relativistic gravitational field equations. But Finslerian solutions of the gravitational field equations proposed in [58] can also be obtained.

An important class of Finsler geometries is represented by Berwald–Finsler spaces. A system of gravitational field equations defined in a Berwald geometry was proposed and investigated in [59]. The derivation of the field equations in the Berwald–Finsler geometry is essentially based on the Bianchi identities satisfied by the Chern curvature tensor. In general, the geometric part of the gravitational field equation is not symmetric, and this indicates that the principle of the local Lorentz invariance is not satisfied in this theory. The field equations as proposed in [59] are given by

$$\left[Ric_{\mu\nu} - \frac{1}{2}g_{\mu\nu}S\right] + \left\{\frac{1}{2}B_\alpha{}^\alpha{}_{\mu\nu} + B_\mu{}^\alpha{}_{\nu\alpha}\right\} = 8\pi G T_{\mu\nu}, \tag{10}$$

where

$$A_{\lambda\mu\nu} \equiv \frac{F}{4}\frac{\partial}{\partial y^\lambda}\frac{\partial}{\partial y^\mu}\frac{\partial}{\partial y^\nu}(F^2), \tag{11}$$

is the Cartan tensor, and $B_{\mu\nu\alpha\beta} = -A_{\mu\nu\lambda} R_{\theta\ \alpha\beta}^{\ \lambda} y^{\theta}/F$. The Cartan tensor measures the deviation of the Finsler geometry from the Riemannian one on a given manifold.

With the use of a variational principle, Finsler-type gravitational field equations have been derived from a Finsler–Lagrange function L in [60]. The action considered in [60] is given by

$$S[L] = \int_{\Sigma \subset TM} \mathrm{vol}(\Sigma) R_{|\Sigma}, \tag{12}$$

where $\Sigma = \{(x,\dot{x}) \in TM | F(x,\dot{x}) = 1\}$ denotes the unit tangent bundle, and vol(Σ), the volume form on Σ, is defined by using the Finsler metric. The field equations obtained from the action (12) are given by

$$2R - \frac{L}{3} g^{Lij} R_{\cdot i \cdot j} + \frac{2L}{3} g^{Lij} \left[(\nabla P_i)_{\cdot j} + P_{i|j} - P_i P_j \right] = 0, \tag{13}$$

where P is the Landsberg tensor, and

$$g^L_{ij} = \frac{1}{2} \frac{\partial^2 L}{\partial \dot{x}^i \partial \dot{x}^j} = \frac{1}{2} L_{\cdot i \cdot j}. \tag{14}$$

Moreover, $R_{.i.j}$ is the geodesic deviation operator, and R denotes its trace.

When extended to gravitational systems much bigger than the Solar System, which involve the presence of galactic and cosmological scales, Einstein's theory of general relativity faces a number of very serious challenges, whose possible solutions could be obtained only if we introduce a fundamental modification in our understanding of the gravitational force.

One of the most intriguing discoveries of the recent times is the observational proof that our Universe is in fact in a phase of accelerating expansion [61–65]. The transition from deceleration to acceleration occurred at a small redshift z, given by $z \approx 0.5$. Explaining this observation requires a profound modification of the theoretical foundations of Einstein's general relativity.

The simplest explanation of the recent exponential de Sitter-type expansion can be obtained by reintroducing the cosmological constant Λ in the Einstein field equations, introduced by Einstein in 1917 [66], and later rejected by him as the biggest blunder of his life. Einstein's main goal was to use Λ to construct a static, general relativistic cosmological model of the Universe.

The cosmological constant has a complicated history [67], but presently, it is adopted as one of the basic physical (geometrical?) parameters to build up the standard cosmological paradigm, the ΛCDM model, which is currently used for the interpretation of the observational data. The ΛCDM model also includes in its theoretical structure another basic, equally mysterious component, dark matter [68]. Intensive searches for the dark matter particle have yielded no results, and thus, the only evidence for the existence of dark matter is gravitational. However, despite these theoretical shortcomings, the ΛCDM model fits the observational data very well [69–72].

But on the theoretical level, the ΛCDM model is confronted with the uncertainty of its foundations: no (convincing) physical theory exists that could provide it with a solid basis. The first major problem is related to the geometrical or physical interpretation of the cosmological constant [67,73,74]. If we interpret the cosmological constant as the vacuum energy density ρ_{vac} at the Planck scale, we are led to the "worst prediction in physics" [75]. The vacuum energy density can be computed as

$$\rho_{vac} \approx \frac{\hbar}{c} \int_{k_{dS}}^{k_{Pl}} \sqrt{k^2 + \left(\frac{mc}{\hbar}\right)^2} d^3k \approx \rho_{Pl} = \frac{c^5}{\hbar G^2} = 10^{93} \frac{\mathrm{g}}{\mathrm{cm}^3}, \tag{15}$$

and this result differs by a factor of 10^{-120} from the observed value of the energy density associated with Λ, $\rho_\Lambda = \Lambda c^2/8\pi G \approx 10^{-30}$ g/cm^3 [71].

Presently, the ΛCDM standard paradigm is faced with several important problems. An important challenge to the ΛCDM model is the "Hubble tension", which has its origins in the differences in the estimations of the values of the Hubble constant H_0 (representing the present-day value of the Hubble function H), obtained on one hand from the CMB measurements [72], and on the other hand from the local observations of the Type Ia Supernovae [76–78]. The value of H_0 obtained by the SHOES collaboration for H_0 is $H_0 = 74.03 \pm 1.42$ km/s/Mpc [76]. The early Universe determinations, using the Planck satellite data, give for H_0 the value $H_0 = 67.4 \pm 0.5$ km/s/Mpc [72], which differs by $\sim 5\sigma$ from the SH0ES estimations.

Several other theoretical problems whose solution cannot be found in the framework of the ΛCDM paradigm still exist. Some of these problems are represented by the question of the smallness of Λ and its fine tuning. The question of why the transition from the decelerating to the accelerating phases took place recently is still waiting for a response. And perhaps the most important question, if the cosmological constant is really required to construct successful cosmological models, does not have an answer yet.

Hence, considering alternative approaches for the description of the gravitational interaction may allow us to solve the observational problems of cosmology, and reconsider the cosmological constant from a new perspective. There are three major possibilities that have been proposed for the extension of general relativity: the dark component approach, the dark gravity approach, and the dark coupling approach [79]. In the dark gravity approach, it is assumed that the geometry of the Universe, as well as the description of the gravitational interaction, requires a significant departure from the formalism of the Riemann geometry. Geometries that go beyond the Riemannian one, for example, Weyl geometry, geometries with torsion, or teleparallel geometries, have been investigated in [80–90]. For reviews of dark gravity-type theories, and their applications, see [91–95].

As a dark gravity candidate, Finsler geometry has an important scientific potential yet to be explored, despite the fact that it has already been considered in various contexts as an important alternative to Riemann geometry, and the standard ΛCDM paradigm.

Randers geometry was extensively applied in the study of the gravitational phenomena in [96–116]. In [100], generalized Friedmann equations in a Randers–Finsler geometry of the form

$$\dot{H} + H^2 + \frac{3}{4}HZ_t = -\frac{4\pi G}{3}(\rho + 3p), \tag{16}$$

$$\dot{H} + 3H^2 + \frac{11}{4}HZ_t = 4\pi G(\rho - p), \tag{17}$$

were obtained, where H, ρ, and p are the Hubble functions, and the baryonic matter energy density and pressure, respectively. The quantity $Z_t = \dot{u}_0$, with u_0 denoting the time component of the four-velocity u_μ. The generalized Friedmann equations above give the relation $3H^2 + 3HZ_t = 8\pi G\rho$. The term HZ_t, induced by the Finsler–Randers geometry, leads to the existence of new phases in the cosmic evolution of the Universe. The system of Friedmann Equations (16) and (17) was used in [101] for the investigation of particle creation processes in Finsler-type geometries.

A scalar–tensor theory that arises effectively from the Lorentz fiber bundle of a Finsler-type geometry was proposed in [117], where its cosmological implications were also investigated. The action in the presence of matter considered in this work is

$$\mathcal{S} = \frac{1}{16\pi G}\int \sqrt{|\det \mathbf{G}|}\,\mathcal{L}_G dx^{(N)} + \int \sqrt{|\det \mathbf{G}|}\,\mathcal{L}_M dx^{(N)}, \tag{18}$$

where $dx^{(N)} = d^4x \wedge \phi^{(1)} \wedge \phi^{(2)}$. Several Lagrangian densities were adopted, given by

$$\mathcal{L}_G = \tilde{\mathcal{R}} - \frac{1}{\phi}V(\phi), \tag{19}$$

where $V(\phi)$ is the potential for the scalar ϕ, and

$$\tilde{\mathcal{R}} = R - \frac{2}{\phi}\Box\phi + \frac{1}{2\phi^2}\partial_\mu\phi\partial^\mu\phi, \tag{20}$$

where $\Box$ is the d'Alembert operator, and $\tilde{\mathcal{R}}$ denotes the curvature for the particular situation of a holonomic basis $[X_M, X_N] = 0$. In the case of a non-holonomic basis, the adopted Lagrangian density is given by $\overline{\mathcal{L}}_G = \tilde{R}$. The generalized Friedmann equations are given by

$$3H^2 = 8\pi G\rho_m - \frac{\dot{\phi}^2}{4\phi^2} + \frac{1}{\phi}\left(\frac{V(\phi)}{2} - 3H\dot{\phi}\right), \tag{21}$$

$$\dot{H} = -4\pi G(\rho_m + P_m) + \frac{\dot{\phi}^2}{4\phi^2} + \frac{1}{2\phi}(H\dot{\phi} - \ddot{\phi}), \tag{22}$$

$$\ddot{\phi} + 3H\dot{\phi} = -16\pi G\phi\mathbb{C}_m - 6\phi\left(\dot{H} + 2H^2\right) + \phi V'(\phi) + \frac{\dot{\phi}^2}{2\phi}, \tag{23}$$

and

$$3H^2 = 8\pi G\rho_m - (1 + A)\frac{\dot{\phi}^2}{4\phi^2} - 3H\frac{\dot{\phi}}{\phi}, \tag{24}$$

$$\dot{H} = -4\pi G(\rho_m + P_m) + (1+A)\frac{\dot{\phi}^2}{4\phi^2} + \frac{1}{2\phi}(H\dot{\phi} - \ddot{\phi}), \tag{25}$$

$$(1 + A)(\ddot{\phi} + 3H\dot{\phi}) = -16\pi G\phi\mathbb{C}_m - 6\phi\left(\dot{H} + 2H^2\right) + \frac{\dot{\phi}^2}{2\phi}(1 + A + \phi A') - \dot{\phi}\dot{A}, \tag{26}$$

respectively. Here, $A(\phi)$ denotes a real function of ϕ. From the above equations, one can reconstruct the thermal history of the Universe, and obtain a sequence of matter and dark-energy-dominated phases. The effective dark energy equation of state has a parameter that can be either phantom or quintessence type. A phantom-divide crossing during the cosmological evolution also appears.

Berwald–Finsler geometries have been investigated as potential candidates for the description of the gravitational fields [118,119]. Spatially homogeneous and isotropic Berwald spacetimes, obtained from a Finsler Lagrangian constructed from a zero-homogeneous function defined on the tangent bundle, and which include the velocity dependence of the Finsler Lagrangian, were discussed in [119]. Cosmological Berwald geometries can also be used for the description of the dynamics of the Universe.

In a series of recent papers [120–123], a systematic investigation of the applications of Finsler geometry in cosmology was considered. The basic geometric ingredients used for the construction of gravitational theories were the (α, β) metrics, and the osculating Barthel connection. One of the basic properties of the Finsler geometry is that an arbitrary point vector field y is associated with each point of the spacetime manifold, with the metric g becoming a function of both x, the coordinates defined on the base manifold, and y, $g = g(x, y)$. In the osculating approach, one assumes the existence of a vector field $Y = Y(x)$, which allows the construction of a Riemannian metric $g(x) = g(x, Y(x))$. In the case of the (α, β) metrics, the connection associated with these metrics, the Barthel connection, is nothing but the Levi–Civita connection associated with $g(x)$.

The introduction of the osculating Barthel connection in the (α, β) spaces of the Finsler geometry leads to a significant simplification of the mathematical formalism, and opens the possibility of constructing unique and well-defined physical models that could be successfully used for the description of the gravitational interaction. In the present review, we briefly introduce, from the perspective of the cosmological applications, the Barthel–Randers, Barthel–Kropina, and conformal Barthel–Kropina models, and we also present their mathematical and theoretical foundations.

From the generalized Friedmann equations of the osculating Barthel–Randers–Kropina cosmological models, obtained by assuming that the background Riemannian metric is of the Friedmann–Lemaitre–Robertson–Walker (FLRW) type, an effective geometric dark energy component can be always generated, which results from the presence of extra terms in the cosmological evolution equations.

A central problem for the physical acceptability of the Finslerian-type models is how successfully they could describe the observational data. The cosmological tests, and comparisons with observational data of these dark energy models, considered in the above-mentioned works are investigated in detail. In the current review, we perform a detailed analysis of the three cosmological models of Barthel–Randers type, introduced in [120], constraining the model parameters. In our investigation, we use 15 Hubble data points (Cosmic Chronometers), the Pantheon Supernovae Type Ia data sample, and the most recent Baryon Acoustic Oscillation (BAO) measurements from the Dark Energy Spectroscopic Instrument (DESI) Data Release 2. The statistical analysis is performed by using Markov Chain Monte Carlo simulations. The results of the statistical analysis of the Barthel–Randers cosmological models are compared with a similar analysis of the Barthel–Kropina cosmological models performed in [122], and with the standard ΛCDM model. The Akaike information criterion (AIC) and the Bayesian information criterion (BIC) are used as the model selection tools. The statefinder diagnostics, consisting of the study of the jerk and snap parameters, and the $Om(z)$ diagnostics are also considered for the comparative study of the Barthel–Randers, Barthel–Kropina and ΛCDM cosmologies. Our statistical results indicate that the osculating Barthel-type (α, β) Finslerian dark energy models give a good description of the observational data, and thus, they can be considered a viable alternative to the ΛCDM model, even if not all of them are favorable compared to ΛCDM.

The present paper is organized as follows. We review some basic concepts and definitions of Finsler geometry in Section 2. In Section 3, we introduce the mathematical foundations of the Barthel–Randers–Kropina gravitational theories, by reviewing the basic concepts of (α, β) metrics, osculating spaces, and Barthel connections. The basic principles of constructing (α, β) cosmological models are also presented. The generalized Friedmann equations of the three basic models considered in the present work, Barthel–Randers, Barthel–Kropina, and conformal Barthel–Kropina, are also written, together with the corresponding energy balance equations. The thermodynamic interpretation of the cosmological models with a non-vanishing matter energy–momentum tensor is also presented and briefly discussed. The cosmological implications of the Barthel–Randers and Barthel–Kropina models are discussed in Section 4, where constraints on the free parameters of these models are obtained by using a combination of observational datasets, including Type Ia Supernovae, Baryon Acoustic Oscillations, and Hubble parameter measurements. A detailed comparison of the Barthel–Randers and Barthel–Kropina cosmological models is performed in Section 5. A discussion of the main results and the relevance of the statistical analysis is presented in Section 6. Finally, we discuss and conclude our results in Section 7.

2. Fundamentals of Finsler Geometry

The assumption that spacetime can be described mathematically as a four-dimensional differentiable manifold M, endowed with a pseudo-Riemannian tensor g_{IJ}, where $I, J, ... = 0, 1, 2, 3$, is one of the fundamental assumptions of modern theoretical physics. The next fundamental concept, the interval between two events located at the points x^I and $x^I + dx^I$ on the worldline of a standard clock, is defined by the chronological hypothesis as $ds = \left(g_{IJ}dx^I dx^J\right)^{1/2}$ [124,125]. A very important metrical generalization of the Riemannian geometry is the geometry anticipated by Riemann [8], but which was later on systematically developed by Finsler [3].

In a simple interpretation, Finsler spaces are metric spaces with the interval ds between two neighboring points $x = (x^I)$ and $x + dx = (x^I + dx^I)$ given by

$$ds = F(x, dx), \tag{27}$$

where F, the Finsler metric function, must be positively homogeneous of degree one in dx, and thus satisfy the condition

$$F(x, \lambda dx) = \lambda F(x, dx) \;\; \text{for} \;\; \lambda > 0. \tag{28}$$

The Finsler metric function F can be written in terms of the canonical coordinates of the tangent bundle $(x, y) = (x^I, y^I)$, where $y = y^I(\partial/\partial x^I)$ is any tangent vector y at x. Then we can introduce the Finsler metric tensor g_{IJ} defined according to

$$g_{IJ}(x, y) = \frac{1}{2}\frac{\partial^2 F^2(x, y)}{\partial y^I \partial y^J}. \tag{29}$$

Hence Equation (27) can be written as

$$ds^2 = g_{IJ}(x, y) y^I y^J. \tag{30}$$

Riemann spaces are some particular cases of the Finsler spaces, and they correspond to $g_{IJ}(x, y) = g_{IJ}(x)$ and $y^I = dx^I$, respectively.

Given a Finsler function F, one can obtain the geodesic equations in the form [124,125]

$$\frac{d^2 x^I}{d\lambda^2} + 2G^I(x, y) = 0, \tag{31}$$

where the functions $G^I(x, y)$ denote the spray coefficients

$$G^I(x, y) = \frac{1}{4} g^{IJ}\left(\frac{\partial^2 F^2}{\partial x^K \partial y^J} y^K - \frac{\partial F^2}{\partial x^J}\right). \tag{32}$$

Equivalently, more reminiscent of Riemannian geometry, they can be rewritten as

$$\frac{d^2 x^I}{d\lambda^2} + \Gamma^I_{JK}(x, y)\frac{dx^J}{d\lambda}\frac{dx^K}{d\lambda} = 0, \tag{33}$$

where

$$\Gamma^I_{JK}(x, y) y^J y^K = 2G^I(x, y) = \frac{1}{2} g^{IJ}\left(\frac{\partial^2 F^2}{\partial x^K \partial y^J} y^K - \frac{\partial F^2}{\partial x^J}\right). \tag{34}$$

Here, the quantities $\Gamma^I_{JK}(x, y)$ are the local coefficients of the Finsler connection used. We point out that the theory of Finsler connections is more complex than in the Riemannian case since the only metrical, torsion-free connection is the Levi–Civita one; hence, the existence of such a connection implies that the Finsler metric is in fact Riemannian. The obvious conditions imposed for the existence of Finsler connections are then non-metricity and torsion-free (this is the case of the Chern connection) or metrical connection with surviving torsion (this is the case of the Cartan connection). The precise form of the Finsler connections depends on the concrete form of the non-metricity and surviving torsion, but these details are beyond the aim of the present paper (one can consult [12,13,126] or other textbooks on Finsler geometry). The local coefficients of a Finsler connection D are defined as

$$D_{\frac{\delta}{\delta x^K}} \frac{\delta}{\delta x^J} = \Gamma^I(x, y)_{JK}\frac{\delta}{\delta x^I}, \quad D_{\frac{\partial}{\partial x^K}} \frac{\partial}{\partial x^J} = V^I(x, y)_{JK}\frac{\partial}{\partial x^I}, \tag{35}$$

where $\frac{\delta}{\delta x^I}$ is the adapted basis induced by the Cartan nonlinear connection. We prefer to keep the notation $C^I_{JK} = \frac{1}{2}\frac{\partial^3 F^2}{\partial y^I \partial y^J \partial y^K}$ for the Cartan tensor of (M, F). For instance, the Cartan connection of a Finsler space (M, F) is the only Finsler connection on TTM that is metrical with $h(hh)$ and $v(vv)$-torsion vanishing (the rest of the torsion survives). The local coefficients of the Cartan connection are given by

$$\Gamma^I_{JK}(x,y) = \frac{1}{2} g^{IL}(x,y)\left[\frac{\delta g_{LJ}(x,y)}{\delta x^K} + \frac{\delta g_{LK}(x,y)}{\delta x^J} - \frac{\delta g_{JK}(x,y)}{\delta x^L}\right],$$
$$V^I_{JK}(x,y) = \frac{1}{2} g^{IL}(x,y)\left[\frac{\partial g_{LJ}(x,y)}{\partial y^K} + \frac{\partial g_{LK}(x,y)}{\partial y^J} - \frac{\partial g_{JK}(x,y)}{\partial x^L}\right]. \tag{36}$$

Likewise, the Chern connection is defined to be the only Finsler connection on TTM which is almost compatible and torsion-free. The local coefficients of the Chern connection are given by $(\Gamma^I_{JK}(x,y), 0)$. However, despite the rather complicated form of the connection coefficients $\Gamma^I_{JK}(x,y)$, due to the homogeneity of the geometrical objects involved, Equation (33) takes the simple form

$$\frac{d^2 x^I}{d\lambda^2} + \gamma^I_{JK}(x,y)\frac{dx^J}{d\lambda}\frac{dx^K}{d\lambda} = 0, \tag{37}$$

where

$$\gamma^I_{JK}(x,y) = \frac{1}{2} g^{IL}(x,y)\left[\frac{\partial g_{LJ}(x,y)}{\partial x^K} + \frac{\partial g_{LK}(x,y)}{\partial x^J} - \frac{\partial g_{JK}(x,y)}{\partial x^L}\right], \tag{38}$$

are the formal Christoffel coefficients of the Finsler space (M, F).

Berwald spaces are a special class of Finsler geometry, and they can be obtained by assuming that the Berwald connection coefficients are independent of the fiber coordinate y [10].

With the help of the spray coefficients G^I, one can define the vector field S on $TM \setminus 0$ according to

$$S = y^I \frac{\partial}{\partial x^I} - 2G^I \frac{\partial}{\partial y^I}.$$

S is the spray induced by F. A curve γ defined on M is a geodesic of F if and only if its canonical lift $\hat{\gamma}(t) = (\gamma(t), \dot{\gamma}(t))$ to TM is an integral curve of S.

The notion of Finsler metrics that we have recalled here belong to the class of *classic Finsler metrics*; in other words, at each point $x \in M$, the function $F_x : T_x M \to \mathbb{R}$ is a function defined on the tangent space $T_x M$ of a differentiable manifold M satisfying the following conditions:

(i) F_x is C^∞ on $\widetilde{T_x M} = T_x M \setminus \{0\}$.
(ii) F_x is positively 1-homogeneous: $F_x(\lambda y) = \lambda F_x(y)$, for all $\lambda > 0$ and $y \in T_x M$.
(iii) For each $x \in M$, the Hessian matrix (29) is positive definite on $\widetilde{T_x M}$.

At each point $x \in M$, the indicatrix $\{y \in T_x M : F_x(y) = 1\}$ is a closed, strictly convex, smooth hypersurface around the origin of $T_x M$.

A more general notion is the notion of conic Finsler metrics, that is, Finsler norms defined only on a conic domain of $T_x M$. Let us recall that $A_x \subset T_x M$ is called a conic domain of $T_x M$ if A_x is an open, non-empty subset of $T_x M$ such that if $v \in A_x$, then $\lambda v \in A_x$, for all $\lambda > 0$. We remark that the origin of $T_x M$ does not belong to A_x except for the case $A_x = T_x M$.

We can now define a Finsler norm defined only on a conic domain $A_x \subset T_x M$ with the properties (i)–(iii) given above for all $y \in A_x$. At each point $x \in M$, the indicatrix $S_x := \{y \in A_x \subset T_x M : F_x(y) = 1\}$ is a hypersurface embedded in A_x as a closed subset.

Let $A \subset TM$ be an open subset of the tangent bundle $\pi : TM \to M$ such that $\pi(A) = M$, and A is conic in TM; that is, for each $x \in M$, the set $A_x := A \cap T_xM$ is a conic domain in T_xM. A function $F : A \to \mathbb{R}$ is a conic Finsler metric if its restriction $F_x : A_x \to \mathbb{R}$ satisfies the conditions (i)–(iii) above, for each $x \in M$. The local and global geometry of conic Finsler spaces can be now developed in a similar way with the case of classical Finsler metrics (see [127,128] and references therein).

3. Osculating (α, β)-Type Cosmological Models

In the present section, we review the mathematical and theoretical foundations of the Barthel–Randers–Kropina-type cosmological models, which we will discuss in the framework of the general (α, β) geometries. We also write down the generalized Friedmann equations obtained for the case of the Friedmann–Lemaitre–Robertson–Walker Riemannian metric. The thermodynamic interpretation of the models is also discussed.

3.1. Mathematical Foundations of the (α, β) Finslerian Cosmologies

We begin our discussion of the cosmology of the Finslerian-type geometries with a brief discussion of the mathematical properties of the (α, β) metrics, the Barthel connection, and the osculating geometries.

3.1.1. Kropina and (α, β) Geometries

As we have already mentioned, a special type of Finsler space is the Randers space [25], with the metric function given by

$$F = \left[a_{IJ}(x)y^I y^J\right]^{1/2} + b_I(x)y^I, \tag{39}$$

where a_{IJ} is the metric tensor of a Riemannian space, and $b_I(x)y^I$ is a linear 1-form defined on the tangent bundle TM, and we use the parametrization $dx^I = y^I$.

Another remarkable class of Finsler spaces are the Kropina spaces [129,130], which are Finsler spaces with metrics of the form

$$F(x,y) = \frac{a_{IJ}(x)y^I y^J}{b_I(x)y^I}. \tag{40}$$

By generalizing these results, Matsumoto [131,132] defined the notion of the (α, β) metrics in the following way: given a Finsler metric function $F(x,y)$, it is called an (α, β) metric if F is a positively homogeneous function $F(\alpha, \beta)$ of first degree in two variables $\alpha(x,y) = \left[a_{IJ}(x)y^I y^J\right]^{1/2}$ and $\beta(x,y) = b_I(x)y^I$, respectively.

In our analysis, we suppose that α is a pseudo-Riemannian or Riemannian metric, with the properties that it is non-degenerate (regular), or positive definite. We point out that the Kropina-type metrics are conic Finsler metrics defined only on some conic domain of the tangent bundle. By extension, we can regard any (α, β) as conic Finsler metrics, allowing the Riemannian part to be a pseudo-Riemannian metric.

In the special cases of the Randers and Kropina metrics, the Finsler metric functions are given by $F = \alpha + \beta$, and $F = \dfrac{\alpha^2}{\beta}$, respectively (see also Table 1). Therefore the Randers and Kropina metrics belong to the general class of the (α, β) metrics.

We can define the general (α, β) metrics as metrics having the metric function given by

$$F(\alpha, \beta) = \alpha\phi(\beta/\alpha) = \alpha\phi(s), \tag{41}$$

where $s = \beta/\alpha$, and $\phi = \phi(s)$ is a C^∞ positive function on an open interval $(-b_o, b_o)$.

Many examples of (α, β) metrics have been considered in the literature, the most studied one being the Randers metric $F = \alpha + \beta$ (see [133] and references within). The Kropina metrics are also classic examples of (α, β) metrics (see [134] for physical motivations of introducing this metric). More recently, the local and global aspects of Kropina metrics have been extensively studied in [128,135–138].

3.1.2. The Barthel Connection

Let (M^n, F) be a Finsler space, defined on a base manifold M^n. We can also define on M^n a vector field $Y(x) \neq 0$. We introduce now a specific mathematical object $(M^n, F(x,y), Y(x))$, which represents a Finsler space $(M^n, F(x,y))$ with a tangent vector field $Y(x)$ also defined. If the vector Y does not vanish at any point on M, then from the Finslerian metric $\hat{g}(x,y)$, one obtains the Y-Riemann metric $\hat{g}_Y(x) = \hat{g}(x,Y)$.

Obviously, one can evaluate any geometrical object in a Finsler space (M,F) in the specific direction given by a fixed vector field Y. This remark leads to the natural idea of evaluating the Chern connection coefficients $\Gamma^I_{JK}(x,y)$ given in (36) at $y = Y(x)$. This is the so-called Barthel connection introduced by Barthel himself in [139,140] and developed later by Ingarden and Matsumoto [134].

The Barthel connection is an affine connection on $(M, \hat{g}_Y)$, metrical with torsion

$$T^I_{JK}(x) = V^I_{LJ}(x, Y(x))Y^L_K - V^I_{LK}(x, Y(x))Y^L_J, \tag{42}$$

where $Y^L_J(x) := \frac{\partial Y^L}{\partial x^J} - N^L_J(x, Y(x))$ and $N^L_J(x,y)$ is the Cartan nonlinear connection of the Finsler space (M,F).

The local coefficients of the Barthel connection can be written as

$$b^I_{JK}(x) = F^I_{JK}(x, Y(x)) + V^I_{JL}(x, Y(x))Y^L_J(x). \tag{43}$$

The absolute differential of the vector Y is defined according to [141]

$$DY^I = dY^I + Y^K b^I_{KH}(x,Y)dx^H, \tag{44}$$

where by $b^I_{KH}(x,Y)$ we have denoted the coefficients of the Barthel connection. By using the homogeneity of the geometrical objects involved, we obtain

$$DY^I = dY^I + Y^K\left(\hat{\gamma}^I_{KH} - \hat{\gamma}^R_{KS}Y^S\hat{C}^I_{RH}\right)dx^H. \tag{45}$$

The Barthel connection has a number of interesting properties. It depends on the vector field on which it acts, a property that does not exist in Riemann geometry. Therefore, the Barthel connection has very different properties, and so it significantly differs from the connections in Riemann geometry. Another important property is related to the direction and magnitude dependence of the Barthel connection. For anisotropic metrics, like most of the Finsler metrics, all geometric properties depend on the direction. For the Barthel connection, there is no dependence on the magnitude of the vector field, but only on its direction. The Barthel connection, keeping the metric function unchanged by the parallel transport, is the simplest connection with this property. For Finsler vector fields, depending on both x and y, the Barthel connection allows the transition to the Cartan geometry of the Finsler spaces. Moreover, the Barthel connection can be considered as the connection of a point in Finsler space.

Finally, we would like to mention that unlike the Levi–Civita connection of the Riemannian geometry, the Finsler connections (Cartan or Chern connections) are not defined on the base manifold M, but they live on the total space of the tangent bundle [12,13]. On the other hand, even though they are completely different, the Barthel and Levi–Civita connections are affine connections on the base manifold M. This is a very important characteristic of Barthel-type geometries that leads to significant differences between the geometrical theories of gravity as formulated on Riemann and Finsler manifolds.

3.1.3. The Y-Osculating Riemann Geometry

In 1936, Nazim [142] introduced and developed the concept of osculating Riemann spaces of Finsler geometries. This concept was later studied by Varga [143]. In the osculating approach, one associates to a complex geometric object, like a Finsler geometry and connection, a simpler mathematical structure, for example, an affine or a linear connection, or a Riemann metric. Hence, in this approach, it is assumed that the osculating geometry approximates the more complicated one at a certain level. Thus, with the use of the osculating formalism, one can derive mathematical results that significantly simplify the description of the properties of the mathematically complex geometries.

Let a nowhere-vanishing local section Y of $\pi_M : TM \to M$ be given. Geometric objects existing on TM can be pulled back to M. Since $g_{IJ} \circ Y$ is a function defined on U, a new metric, defined according to

$$\widehat{g}_{IJ}(x) := g_{IJ}(x,y)\big|_{y=Y(x)}, \quad x \in U, \tag{46}$$

can be introduced. The pair $(U, \widehat{g}_{IJ})$ represents a Riemannian manifold, while $\widehat{g}_{IJ}(x)$ is the Y- osculating Riemannian metric, defined on (M,F).

For the osculating Riemannian metric given by Equation (46), we define the Christoffel symbols of the first kind as

$$\widehat{\gamma}_{IJK}(x) := \frac{1}{2}\left\{\frac{\partial}{\partial x^J}[g_{IK}(x,Y(x))] + \frac{\partial}{\partial x^K}[g_{IJ}(x,Y(x))] - \frac{\partial}{\partial x^I}[g_{JK}(x,Y(X))]\right\}. \tag{47}$$

With the use of the rule of the derivative of the composed functions, we explicitly find

$$\widehat{\gamma}_{IJK}(x) = \gamma_{IJK}(x,y)\big|_{y=Y(x)} + 2\left(C_{IJL}\frac{\partial Y^L}{\partial x^K} + C_{IKL}\frac{\partial Y^L}{\partial x^J} - C_{JKL}\frac{\partial Y^L}{\partial x^I}\right)\bigg|_{y=Y(x)}, \tag{48}$$

where C_{IJL} denotes the Cartan tensor. Hence, if a global section Y of TM exists, so that $Y(x) \neq 0, \forall x \in M$, one can always define the osculating Riemannian manifold $(M, \widehat{g}_{ij})$.

The case of the (α,β) metrics.

Let us focus now on the case of the (α,β) metrics. The Hessian matrix of an (α,β)-metric $F = F(\alpha,\beta)$ is given by

$$g_{IJ}(x,y) = \frac{L_\alpha}{\alpha}h_{IJ} + \frac{\partial L_{\alpha\alpha}}{\alpha^2}y_I y_J + \frac{\partial L_{\alpha\beta}}{\alpha}(y_I b_J + y_J b_i) + L_{\beta\beta}b_I b_J, \tag{49}$$

and the Cartan tensor by

$$2C_{IJK} = \frac{L_{\alpha\beta}}{\alpha}(h_{IJ}p_K + h_{JK}p_I + h_{KI}p_J) + L_{\beta\beta\beta}p_I p_J p_K, \tag{50}$$

where the notation $L := \frac{F^2}{2}$ is customary. Here, $p_I = b_I - \frac{\beta}{\alpha}y_I$, and $h_{IJ} = g_{IJ} - \frac{y_I y_J}{F^2}$ is the angular tensor of (M,F).

We choose the vector field as $Y = b$, with $b^I = a^{IJ}b_J$. Since the vector field Y is globally non-vanishing on M, we obtain the result that β has no zero points on M. Hence, we can define for the (α,β) metrics the b- osculating Riemannian manifold $(M, \widehat{g}_{IJ})$, with the Riemannian metric given by $\widehat{g}_{IJ}(x) := g_{IJ}(x,b)$, where $b^I = a^{IJ}b_J$.

With respect to α, the length $\widetilde{b}$ of b is obtained as $\widetilde{b}^2 = b_I b^I = \alpha^2(x,b)$. Moreover, we have $Y_I(x,b) = b_I$, respectively.

The b-osculating Riemannian metric can be explicitly written as

$$\widehat{g}_{IJ}(x) = \frac{L_\alpha}{\widetilde{b}}\bigg|_{y=b(x)} g_{IJ} + \left(\frac{L_{\alpha\alpha}}{\widetilde{b}^2} + 2\frac{L_{\alpha\beta}}{\widetilde{b}} + L_{\beta\beta} - \frac{L_\alpha}{\widetilde{b}^3}\right)\bigg|_{y=b(x)} b_I b_J. \tag{51}$$

Moreover, we obtain $\beta(x,b) = \tilde{b}^2$, and $p_I(x,b) = 0$. Hence, by using the expression of the Cartan tensor for an (α,β) metric, we obtain the important result that $C_{IJK}(x,b) = 0$. For $Y = b$, we find

$$\widehat{\gamma}_{IJK}(x) = \gamma_{IJK}(x,y)\big|_{y=b(x)}. \tag{52}$$

Thus, we have obtained the fundamental result that in the case of an (α,β)-metric, the Barthel connection, representing the linear b-connection, where $b^I = (a^{IJ}b_J)$, is the Levi–Civita connection of the b-Riemannian metric. Hence, after the evaluation of the fundamental Finsler tensor $g_{ij}(x,y)$ of (M,F) at the point $(x,Y(x))$, we obtain a Riemannian metric $\widehat{g}_Y$ on M, with its Levi–Civita connection.

The curvature tensor.

The Barthel connection with local coefficients $(b^A_{BC}(x))$ is an affine connection. We define the curvature tensor of an affine connection, having local coefficients $(\Gamma^A_{BC}(x))$, as

$$R^A_{BCD} = \frac{\partial\Gamma^A_{BD}}{\partial x^C} - \frac{\partial\Gamma^A_{BC}}{\partial x^D} + \Gamma^E_{BD}\Gamma^A_{EC} - \Gamma^E_{BC}\Gamma^A_{ED}. \tag{53}$$

Thus, it follows that the curvature of the Barthel connection can be obtained from Equation (53) by taking $(\Gamma^A_{BC}(x)) = (b^A_{BC}(x))$. For the Kropina metric with $F = \alpha^2/\beta$, the Barthel connection is equal to the Levi–Civita connection of the osculating metric $\widehat{g}_{AB}(x) = g_{AB}(x,b(x))$, with $b_I(x)$ denoting the components of the one-form β. Moreover, g_{AB} is the fundamental tensor of F.

Since $b^A_{BC} = \widehat{\gamma}^A_{BC}$, where $\widehat{\gamma}^A_{BC}$ are the coefficients of the Levi–Civita connection, for the curvature tensors of the Kropina metric, we find the expressions

$$\widehat{R}^A_{BCD} = \frac{\partial\widehat{\gamma}^A_{BD}}{\partial x^C} - \frac{\partial\widehat{\gamma}^A_{BC}}{\partial x^D} + \widehat{\gamma}^E_{BD}\widehat{\gamma}^A_{EC} - \widehat{\gamma}^E_{BC}\widehat{\gamma}^A_{ED}, \tag{54}$$

and

$$\widehat{R}_{BD} = \sum_A\left[\frac{\partial\widehat{\gamma}^A_{BD}}{\partial x^A} - \frac{\partial\widehat{\gamma}^A_{BA}}{\partial x^D} + \sum_E\left(\widehat{\gamma}^E_{BD}\widehat{\gamma}^A_{EA} - \widehat{\gamma}^E_{BA}\widehat{\gamma}^A_{ED}\right)\right], \tag{55}$$

respectively, where the indices A,B,C,D,E take the values $\{0,1,2,3\}$, and we have defined $\widehat{R}_{BD} = \widehat{R}^A_{BAD}$, and $\widehat{R}^B_D = \widehat{g}^{BC}\widehat{R}_{CD}$, respectively [120–123]. The generalized Ricci scalar is defined according to $\widehat{R} = \widehat{R}^B_B$.

3.2. Building Cosmological Models in (α,β) Geometries

We proceed now to the investigation of the possibility of using the Finslerian geometric framework, and the general (α,β) metrics, to build specific cosmological models. Our basic quantities in this respect are the Riemannian metric a_{IJ}, which determines $\alpha = a_{IJ}dx^I dx^J = a_{IJ}y^I y^J$, and the coefficient b_I of the one-form $\beta = b_I dx^I = b_I y^I$. To construct a cosmological model in the (α,β)-type Finsler geometries, we introduce several assumptions, detailed below.

The Universe is homogeneous and isotropic.

As a first approximation of the structure and matter distribution in the Universe, we assume the validity of the cosmological principle. The cosmological principle implies the homogeneity of the Universe. The homogeneity of the Universe imposes the fundamental constraint that on large cosmological scales, all the physical and geometrical properties of the Universe depend globally on the cosmological time only.

The Riemannian metric a_{IJ} is the FLRW metric.

In the following, we will restrict our study to the case in which the metric a in the definition of $\alpha = a_{IJ}dx^I dx^J$ is the flat, homogeneous and isotropic Friedmann–Lemaitre–Robertson–

Walker (FLRW) metric. In a system of coordinates $(x^0 = ct, x^1 = x, x^2 = y, x^3 = z)$ defined on the base manifold M, the Riemannian FLRW metric is given by

$$ds_R^2 = g_{IJ}dx^I dx^J = c^2dt^2 - a^2\left(x^0\right)\left(dx^2 + dy^2 + dz^2\right), \tag{56}$$

where t is the universal cosmological time, c is the speed of light, and $a(x^0)$ is the scale factor, describing the expansionary properties of the Universe. From an observational point of view, the Hubble function, defined as $H = (1/a(x^0))(da(x^0)/dx^0)$, plays an important role, since it allows the in depth comparison of the astrophysical data with the theoretical predictions.

The Finsler metric depends on x^0 only.

The cosmological principle, together with the homogeneity postulates, requires that together with $a_{IJ} = a_{IJ}(x^0)$, the components of the 1-form β are also functions of the cosmological time only, $b_I = b_I(x^0)$.

The 1-form b has vanishing space-like components.

The cosmological principle, requiring the homogeneity of the Universe, as well as the isotropy condition that follows from the choice of the Riemann metric as the FLRW metric, with only diagonal components, leads to a strong mathematical constraint on the components of the coefficients of the 1-form b. The isotropy of the metric imposes the condition that all the space-like components of A identically vanish, so that $b_1 = b_2 = b_3 = 0$. If this restriction is not satisfied, then after performing a spatial rotation in the (x, y, z) three-dimensional space, we can construct a preferred direction, oriented, for example, in the direction of the x coordinate. But the possibility of such a transformation and the existence of a preferred direction would contradict the isotropic condition implemented via the FLRW metric, as well as the observed large-scale spatial isotropy of the Universe. Hence, in the present approach to Finslerian cosmology, we assume that the vector b has only one independent component, so that $b_0(x^0)$. Therefore, in a homogeneous and isotropic cosmology, the 1-form field b takes the simple form

$$(b_I) = \left(a\left(x^0\right)\eta\left(x^0\right), 0, 0, 0\right) = \left(b^I\right), \tag{57}$$

where we have introduced the auxiliary function $\eta(x^0)$ to obtain a representation of b_0 in terms of the scale factor of the Universe.

Matter moves along the Hubble flow.

We assume that similarly to the standard general relativistic cosmology, defined in the Riemannian geometric setting, in Finslerian cosmology, we can also introduce a comoving frame in which the cosmological observers, as well as ordinary matter, move along with the Hubble flow, defined by the metric $a_{IJ}(x)$. If we introduce the Riemannian four-velocities u^I of the matter particles, defined as $u^I = dx^I/ds_R$, then the existence of a moving frame implies that the space-like components of the matter four-velocity vanish identically, and the four-velocity has only a non-zero temporal component $u^I = (1, 0, 0, 0)$, which can be normalized to one.

The matter content of the Universe is a perfect fluid.

We postulate that cosmological matter in the Universe consists of a perfect fluid, whose thermodynamic properties can be described by two basic thermodynamic quantities only, given by the energy density ρc^2, and the thermodynamic pressure p, respectively. We also assume that the thermodynamic quantities can be defined in the usual way, by using the standard definitions of statistical physics and thermodynamics. From assumptions c and d

it follows that the matter energy–momentum tensor has only two non-zero components, $\widehat{T}_0^0 = \rho c^2$, and $\widehat{T}_I^I = -p$, $I = 1,2,3$, and thus, it can be represented in the form

$$\widehat{T}_I^J = \begin{pmatrix} \rho c^2 & 0 & 0 & 0 \\ 0 & -p & 0 & 0 \\ 0 & 0 & -p & 0 \\ 0 & 0 & 0 & -p \end{pmatrix}. \tag{58}$$

Geometric quantities.

Once the above conditions and assumption have been adopted, we obtain the expressions of the Finsler metric, and of α and β in the FLRW cosmological background of the osculating Barthel–(α, β) geometry as follows:

(iii) $(a_{IJ}) = \begin{pmatrix} 1 & 0 & 0 & 0 \\ 0 & -a^2(x^0) & 0 & 0 \\ 0 & 0 & -a^2(x^0) & 0 \\ 0 & 0 & 0 & -a^2(x^0) \end{pmatrix}$;

(iv) $\alpha|_{y=b(x)} = a(x^0)\eta(x^0)$;

(v) $\beta|_{y=b(x)} = \left[a(x^0)\eta(x^0)\right]^2$;

(vi) $\left(h_{IJ}|_{y=b(x)}\right) = \begin{pmatrix} 0 & 0 & 0 & 0 \\ 0 & -a^2(x^0) & 0 & 0 \\ 0 & 0 & -a^2(x^0) & 0 \\ 0 & 0 & 0 & -a^2(x^0) \end{pmatrix}$.

where $I, J \in \{0,1,2,3\}$, and $h_{IJ}(x,y) := a_{IJ} - \frac{y_I}{\alpha}\frac{y_J}{\alpha}$ is the angular metric of (M, α). From the above expressions, it turns out that the Finsler metric g is also diagonal. This result follows in a natural way from the definition of the (α, β) metric. Thus, in this Finslerian modification of general relativity, we maintain one of the essential properties of the FLRW metric (56), namely homogeneity and isotropy of the Riemannian geometry.

Gravitational field equations.

We postulate that the Einstein gravitational field equations, which geometrically describe the properties of the gravitational interaction, can be formulated in a general (α, β) Finslerian geometry as

$$\widehat{G}_{IJ} = \widehat{R}_{IJ} - \frac{1}{2}\widehat{g}_{IJ}\widehat{R} = \frac{8\pi G}{c^4}\widehat{T}_{IJ}, \tag{59}$$

where G is the gravitational constant. These equations are the natural extension of the Riemannian Einstein equations in the Finslerian framework. They reduce to the standard general relativistic Einstein equations in the limiting case of the Riemann geometry.

Flowchart of the algorithmic construction of osculating Barthel-type Finslerian gravitational theories.

The steps necessary to construct a specific Barthel–Kropina-type cosmological model, and the underlying gravitational theory, are presented algorithmically in the form of a flowchart in Figure 1.

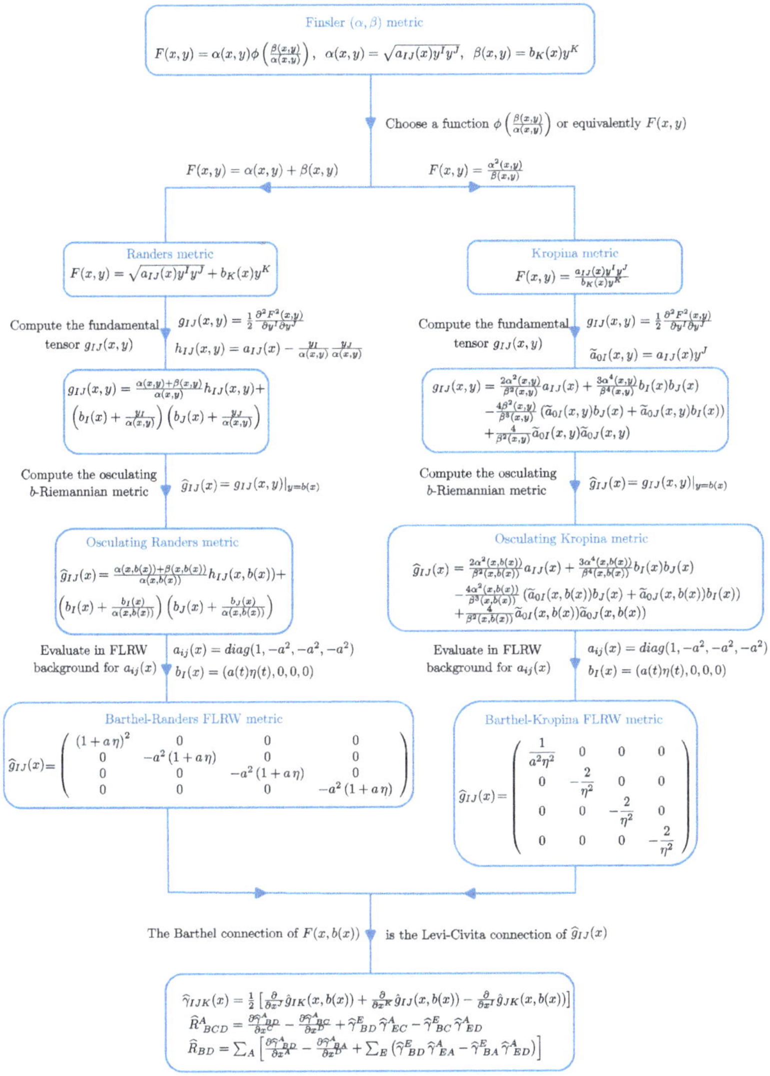

Figure 1. Flowchart of the algorithmic approach for the construction of the (α,β) gravitational models with the Barthel connection.

Although specific examples are restricted to the cases of Barthel–Randers and Barthel–Kropina geometries, the formalism can be easily extended to any other choices of the Finslerian function $F=\alpha\phi(s)$, and the implementation of the geometrical model into a gravitational theoretical framework can be performed easily. In all these cases, some extensions of standard general relativity can be obtained, leading to gravitational field equations that contain extra Finslerian terms, which in a cosmological context can be interpreted as describing an effective geometric dark energy, generated by the nonlocal geometric structure of the spacetime, with the metric tensor nonlocalized due to the presence of the vector y, representing an internal degree of freedom.

There are a large number of (α, β) metrics, whose physical and gravitational properties are worth investigating, and which could offer new and important insights into the gravitational and cosmological phenomena. The flowchart of the algorithmic approach presented in Figure 1 may simplify and clarify the necessary steps taken for the investigation of these classes of theories.

3.3. Barthel–Randers Cosmology

For the case of the Randers geometry, we have $F = \alpha + \beta$. By using the above assumptions, it turns out that the generalized Friedmann equations in this geometry take the form [120]

$$3H^2 = 8\pi G\phi^2\rho - \frac{3}{4}\frac{\dot{\phi}^2}{\phi^2} - 3H\frac{\dot{\phi}}{\phi}, \tag{60}$$

and

$$2\dot{H} + 3H^2 = -\frac{8\pi G}{c^2}\phi^2 p - \frac{\ddot{\phi}}{\phi} + \frac{5}{4}\frac{\dot{\phi}^2}{\phi^2} - H\frac{\dot{\phi}}{\phi}, \tag{61}$$

where we have denoted

$$\phi\left(x^0\right) = 1 + a\left(x^0\right)\eta\left(x^0\right). \tag{62}$$

Equations (60) and (61) give the dynamical evolution of H as

$$\dot{H} = -4\pi G\phi^2\left(\rho + \frac{p}{c^2}\right) - \frac{\ddot{\phi}}{2\phi} + \frac{\dot{\phi}^2}{\phi^2} + H\frac{\dot{\phi}}{\phi}. \tag{63}$$

Equation (60) can be rewritten as

$$3\left(H + \frac{\dot{\phi}}{2\phi}\right)^2 = 8\pi G\phi^2\rho. \tag{64}$$

By introducing the new Hubble function defined as

$$\widetilde{H} = H + \frac{\dot{\phi}}{2\phi}, \tag{65}$$

we obtain the final form of the generalized Friedmann equations in the Barthel–Randers cosmology as

$$3\widetilde{H}^2 = 8\pi G\phi^2\rho, \tag{66}$$

and

$$2\dot{\widetilde{H}} + 3\widetilde{H}^2 = -\frac{8\pi G}{c^2}\phi^2 p + \frac{\dot{\phi}^2}{\phi^2} + 2H\frac{\dot{\phi}}{\phi}, \tag{67}$$

respectively.

The energy conservation equation.

The energy conservation equation of the matter in the present model of the Barthel–Randers cosmology can be obtained by assuming that similarly to the standard Riemannian general relativistic case, the covariant divergence of the energy–momentum tensor vanishes, $\widehat{\nabla}_\mu T^{\mu\nu} = 0$, with the covariant derivative calculated with the help of the Barthel–Randers connection $\widehat{\gamma}^\mu{}_{\nu\alpha}$. Hence, the conservation equation in Barthel–Randers cosmology can be written as [120]

$$\dot{\rho} + 3\left(H + \frac{\dot{\phi}}{2\phi}\right)\left(\rho + \frac{p}{c^2}\right) = 0, \tag{68}$$

or, in an alternative form, as

$$\frac{d}{dt}\left(\rho a^3\right) + \frac{p}{c^2}\frac{d}{dt}a^3 + \frac{3}{2}a^3\left(\rho + \frac{p}{c^2}\right) = 0. \tag{69}$$

The conservation equation is not independent, and can be also derived directly with the use of the Friedman equations. By taking the time derivative of Equation (60), and after substituting $\dot{H}$ from Equation (61), we find

$$8\pi G\phi^2\left(\dot{\rho}+\frac{3Hp}{c^2}\right)-\frac{3}{8\phi^3}(\dot{\phi}^3-2H\phi\dot{\phi}^2-20H^2\phi^2\dot{\phi}^2)$$
$$+4\pi G\phi\dot{\phi}(3p+4\rho)+9H^3=0. \tag{70}$$

By substituting the expression of H^2 from the Friedmann Equation (60), we recover Equation (70).

3.4. Barthel–Kropina Cosmology

In the Kropina geometry, the Finslerian metric function is given by $F=\alpha^2/\beta$. The generalized Friedmann equations can be obtained directly from the Einstein equations, and are given by [121]

$$\frac{3(\eta')^2}{\eta^2}=\frac{8\pi G}{c^2}\frac{1}{a^2\eta^2}\rho, \tag{71}$$

and

$$2\frac{\eta''}{\eta}+2\mathcal{H}\frac{\eta'}{\eta}-3\frac{(\eta')^2}{\eta^2}=\frac{8\pi G}{c^4}\frac{p}{a^2\eta^2}, \tag{72}$$

respectively, where $\mathcal{H}$ denotes the generalized Hubble function of the Barthel–Kropina cosmology, defined according to $\mathcal{H}=(1/a(x^0))(da(x^0)/dx^0)$.

In the above equations, and in the following, a prime denotes the derivative with respect to x^0, a dot denotes the derivative with respect to the cosmological time t, and the standard Hubble function is given by $H=c\mathcal{H}$. After eliminating the term $-3(\eta')^2$ with the help of Equations (71) and (72), it becomes [121]

$$a\eta\frac{d}{dx^0}(\eta' a)=\frac{4\pi G}{c^4}\left(\rho c^2+p\right). \tag{73}$$

In the Barthel–Kropina geometry, the full system of the generalized Friedmann equations is represented by two ordinary differential equations with four unknowns (a,η,ρ,p). By considering an equation of state for the baryonic matter, $p=p(\rho)$, the number of unknowns in the system of generalized Friedmann equations becomes three, and the system is still underdetermined. Therefore, to obtain solvable cosmological models, and to close the system, we must impose a supplementary independent relation on two of the model parameters.

Energy balance equation.

One of the basic consequences of the standard Friedmann cosmology is the conservation of the matter energy–momentum tensor. But as one can easily observe from the Friedmann Equations (71) and (72), this property does not hold anymore in the Barthel–Kropina cosmology. The matter non-conservation equation as well as the energy density balance equation can be obtained after the multiplication of Equation (71) with a^3, and applying the time derivation operator d/dx^0 on the result. By using the second generalized Friedmann equation in the obtained relation, the energy non-conservation equation in the Barthel–Kropina cosmology is obtained as [121]

$$\frac{8\pi G}{c^4}\left[\frac{d}{dx^0}\left(\rho c^2a^3\right)+p\frac{d}{dx^0}a^3\right]=6a^5\left[\mathcal{H}(\eta')^2+(\eta'+\mathcal{H}\eta)\eta''+\mathcal{H}^2\eta\eta'\right]. \tag{74}$$

Equation (74) can be written in a form similar to the standard general relativistic conservation equation as

$$\frac{4\pi G}{c^4}\left[\frac{d}{dx^0}\left(\rho c^2a^3\right)+p\frac{d}{dx^0}a^3\right]=3a^5\left[\frac{8\pi G}{2c^4}\left(\frac{5}{3}\rho c^2+p\right)\frac{\mathcal{H}}{a^2}+\eta'\eta''\right]. \tag{75}$$

The general relativistic limit.

An interesting and important property of the generalized Friedmann equations of the Barthel–Kropina cosmological model, given by Equations (71) and (72), is that they admit a general relativistic limit, in which they take the form of the standard Friedmann equations of general relativity. The general relativistic limit is given by [121]

$$\eta \to \pm\frac{1}{a}, \beta \to (1,0,0,0). \tag{76}$$

Then from Equations (71) and (72), we immediately reobtain the Friedmann equations of standard general relativity

$$\frac{3(a')^2}{a^2} = \frac{8\pi G}{c^2}\rho, \; 2\frac{a''}{a} + \frac{(a')^2}{a^2} = -\frac{8\pi G}{c^4}p. \tag{77}$$

From Equation (77), it follows that the energy density ρ is conserved, with the conservation equation given by $\dot{\rho} + 3H(\rho + p/c^2) = 0$.

3.5. Conformal Barthel–Kropina Cosmology

Let us consider that an (α, β)-metric with $F = F(\alpha, \beta)$ is given. The conformal transformation of the metric is defined as [123]

$$\widetilde{F}(x,y) := e^{\sigma(x)}F(x,y) = \widetilde{F}\left(\widetilde{\alpha}, \widetilde{\beta}\right). \tag{78}$$

The metric (78) is again an $\left(\widetilde{\alpha}, \widetilde{\beta}\right)$ metric, with

$$\widetilde{\alpha} = e^{\sigma(x)}\alpha, \; \widetilde{\beta} = e^{\sigma(x)}\beta. \tag{79}$$

The fundamental tensor of $\widetilde{F}$ is calculated with the help of the Hessian [123]

$$\widetilde{g}_{IJ} := \frac{1}{2}\frac{\partial^2\widetilde{F}^2}{\partial y^I \partial y^J}. \tag{80}$$

The conformal transformation of the Kropina metric is obtained as

$$\widetilde{F} := e^{\sigma(x)}\frac{\alpha^2}{\beta} = \frac{\widetilde{\alpha}^2}{\widetilde{\beta}},$$

where $\widetilde{\alpha} = e^{\sigma(x)}\alpha$, $\widetilde{\beta} = e^{\sigma(x)}\beta$. The osculating Riemannian metric is obtained as

$$\widetilde{\widehat{g}}_{IJ}(x) = e^{2\sigma(x)}\widehat{g}_{IJ}(x), \tag{81}$$

where $\widehat{g}_{IJ}(x)$ is given by

$$\widehat{g}_{IJ}(x, b(x)) \;=\; \frac{2\alpha^2}{\beta^2}a_{IJ}(x) + \frac{3\alpha^4}{\beta^4}b_I b_J - \frac{4\alpha^2}{\beta^3}(b_I b_J + b_J b_I) + \frac{4}{\beta^2}b_I b_J, \tag{82}$$

where $b_I := a_{IJ}b^J = \widetilde{a}_{0I}(x, b(x))$ (see Figure 1). For cosmological applications we choose the conformal factor as $\sigma(x) = \phi(x^0)$. Hence in the study of the dynamical evolution of the Universe, we will restrict our investigations to conformal transformations of the Kropina metric that depend only on time.

The generalized Friedmann equations.

In the conformal Barthel–Kropina geometry, the generalized Friedmann equations take the form [123]

$$\frac{3(\eta')^2}{\eta^2} = \frac{8\pi G}{c^2}\frac{1}{a^2\eta^2}\tilde{\rho} + 3(\phi')^2 - 6\frac{\eta'}{\eta}\phi', \tag{83}$$

and

$$\frac{2}{\eta^2}[-3(\eta')^2 + 2\eta\eta'\mathcal{H} + 2\eta\eta''] = \frac{16\pi G}{c^4}\frac{1}{a^2\eta^2}\tilde{p}$$
$$-4\left[\phi'' + \frac{1}{2}(\phi')^2\right] + \left(\frac{\eta'}{\eta} - \mathcal{H}\right)\phi', \tag{84}$$

respectively. We eliminate now the term $-3(\eta')^2/\eta^2$ between Equations (83) and (84), and thus we find the relation

$$2\frac{1}{a\eta}\frac{d}{dx^0}(a\eta') = \frac{4\pi G}{c^4}\frac{1}{a^2\eta^2}\left(\tilde{\rho}c^2 + \tilde{p}\right) - \left(\phi'' - (\phi')^2\right) - \frac{11}{4}\frac{\eta'}{\eta}\phi' - \frac{1}{4}\phi'\mathcal{H}. \tag{85}$$

The general relativistic limit of the system (83) and (84) is obtained by taking $\eta \to 1/a$, and $(b_I(x)) = (1,0,0,0)$, respectively. Consequently, $\beta = y^0$. Thus, in this limit, the generalized Friedmann cosmological evolution equations of the conformal Barthel–Kropina model become [123]

$$3\mathcal{H}^2 = \frac{8\pi G}{c^4}\tilde{\rho}c^2 + 3(\phi')^2 + 6\mathcal{H}\phi', \tag{86}$$

and

$$2\mathcal{H}' + 3\mathcal{H}^2 = -\frac{8\pi G}{c^4}\tilde{p} + 2\left[\phi'' + \frac{1}{2}(\phi')^2\right] + \mathcal{H}\phi', \tag{87}$$

respectively. For $\phi = 0$ we fully recover the standard Friedmann equations of general relativity.

3.6. Thermodynamic Interpretation of the (α, β) Cosmologies

It is a general property of several Finslerian-type cosmological models that the matter energy–momentum is not conserved. For example, Equation (68) illustrates this situation for the case of the Barthel–Randers cosmological models. Hence, contrary to the general relativistic case, in the Barthel–Randers-type cosmological model, as well as in the Barthel–Kropina and conformally transformed Barthel–Kropina cosmologies, the baryonic matter content of the Universe is not conserved anymore. This intriguing property of the models raises the problem of the physical interpretation of the non-conservation of the matter energy–momentum tensor, and the problem of the cosmological significance of this effect.

A possible physical understanding of the energy–momentum non-conservation can be obtained by interpreting this effect by using the thermodynamics of irreversible processes, and assuming that it describes particle creation and annihilation in a cosmological environment. In the following, we briefly introduce first the foundations of the thermodynamics of irreversible processes, and then we illustrate the general formalism by considering the specific case of the Barthel–Randers-type cosmological model. The non-conservation of the energy–momentum tensor is a specific feature of several modified gravity theories, and specifically in approaches to gravity involving the presence of geometry–matter coupling [144]. Examples of such theories are the $f(R, L_m)$ [145] and the $f(R, T)$ [146] theories.

The non-conservation of the matter energy–momentum tensor, as shown, for example, for the Barthel–Randers case by Equation (68), can thus be interpreted as showing that due to the existence of the Finslerian geometric effects, during cosmological evolution, matter creation processes may take place. This indicates the possibility of creating matter from geometry. Quantum field theories in curved spacetime also predict the same particle creation effect, as initially proposed and investigated in [147–152]. In quantum field theory, particle creation is due to the time variation of the gravitational field. In an anisotropic Bianchi Type I metric, quantum particle creation was considered in [149], and for a quantum scalar field with a non-zero mass, the renormalized expression of the energy–momentum tensor was determined. Hence, the osculating Finsler–Barthel-type gravity theories, in which the creation of matter is also allowed by the general formalism, could also be interpreted as providing an effective, semiclassical description of the quantum effects in the gravitational field. It is important to note, however, that the nature of the particles is not necessarily known, unless quantum theoretical effects are taken into account.

3.6.1. Irreversible Thermodynamics and Matter Creation

Equation (68), obtained within the framework of the Barthel–Randers cosmology, shows that the covariant divergence of the energy–momentum tensor, which is a function of the equilibrium quantities of the thermodynamic system represented by the baryonic matter, is different from zero. Similar effects appear for the case of other thermodynamical quantities, like, for example, the particle and entropy fluxes. Hence, in the presence of matter creation, all the balance equilibrium equations must be modified to account for this effect [153–155]. We will present first the general formalism of irreversible thermodynamics in the presence of matter creation, by adopting a cosmological perspective. We will consider all the results in the Riemann space, with metric $a_{AB}(x)$. Moreover, we interpret the Finslerian effects as generating a set of specific physical events in the background Riemann geometry. Hence, in the following, all the physical and geometrical quantities will be expressed with the help of the FLRW metric (56). Consequently, all considered physical and geometrical quantities are functions of the cosmological time t only.

Particle balance equations.

To describe particle dynamics, we introduce the particle number density n, and the four-velocity u^I of matter. From these quantities, we construct the particle flux $N^I \equiv nu^I$. All these physical parameters are defined in Riemannian geometry. The particle balance equation in the presence of matter creation is given by

$$\nabla_I N^I = \dot{n} + 3Hn = n\Gamma, \tag{88}$$

where ∇_I is the covariant derivative defined in the Riemann space with the help of the Levi–Civita connection associated with the FLRW metric (56), while Γ denotes the matter creation rate. For $\Gamma \ll 3H$, the source term in the particle balance equation is negligible, and we reobtain the standard particle conservation law of general relativity.

The entropy flux.

We now introduce the entropy density $\tilde{s}$, and the entropy per particle $\sigma = \tilde{s}/n$. From these quantities we construct the entropy flux vector defined as $S^I \equiv \tilde{s}u^I = n\sigma u^I$. The divergence of the entropy flux gives the relation

$$\nabla_I S^I = n\dot{\sigma} + n\sigma\Gamma \geq 0, \tag{89}$$

where the positivity condition is a direct consequence of the second law of thermodynamics. The case $\sigma =$ constant gives the relation

$$\nabla_I S^I = n\sigma\Gamma = \tilde{s}\Gamma \geq 0. \tag{90}$$

Equation (90) shows that if the entropy per particle can be taken as a constant, then the variation of the entropy is only due to the matter generated via the transfer of the energy of the gravitational field to matter. Since $\tilde{s} > 0$, the particle production rate Γ must satisfy the basic thermodynamic condition $\Gamma \geq 0$. From a physical point of view, this condition can be interpreted as permitting the creation of matter from the gravitational field, but suppressing the opposite process.

The creation pressure.

If particle creation takes place, the matter energy–momentum tensor must also be modified to take into account the presence of the irreversible processes and the second law of thermodynamics. Generally, in the thermodynamical description of open systems, the energy–momentum tensor can be represented as [156]

$$T^{IJ} = T^{IJ}_{\rm eq} + \Delta T^{IJ}, \tag{91}$$

where $T^{IJ}_{\rm eq}$ is the equilibrium thermodynamic component [156], and ΔT^{IJ} describes the supplementary terms induced by particle creation. For a homogeneous and isotropic spacetime geometry, ΔT^{IJ}, giving the particle creation contribution to T^{IJ}, can be generally represented in the form

$$\Delta T^0_0 = 0, \quad \Delta T^J_{\ I} = -P_c \delta^J_{\ I}, \tag{92}$$

where P_c denotes the creation pressure, an effective thermodynamic quantity, which in a macroscopic physical system describes phenomenologically particle creation. Moreover, in a fully covariant representation, the tensor ΔT^{IJ} is given by [156]

$$\Delta T^{IJ} = -P_c h^{IJ} = -P_c\left(a^{IJ} - u^I u^J\right), \tag{93}$$

where $h^{IJ} = a^{IJ} - u^I u^J$ is the projection operator defined in the FLRW geometry. Thus, we straightforwardly obtain the relation $u_I \nabla_J \Delta T^{IJ} = 3HP_c$.

In the presence of particle production by the gravitational field, from the scalar component of the energy balance equation $u_I \nabla_J T^{IJ} = 0$, obtained from Equation (91), we obtain the time variation of the energy density ρ of the cosmological fluid as

$$\dot{\rho} + 3H\left[\rho + \frac{1}{c^2}(p + P_c)\right] = 0. \tag{94}$$

The thermodynamic quantities describing baryonic matter must also satisfy the Gibbs law [154]

$$nTd\left(\frac{\tilde{s}}{n}\right) = nTd\sigma = d\rho - \frac{\rho + p/c^2}{n}dn, \tag{95}$$

where T is the temperature of the cosmological matter.

3.6.2. Application: Particle Creation in Barthel–Randers Cosmology

As an example of the application of the irreversible thermodynamic of open systems, we consider now the interpretation of the Barthel–Randers cosmology as a cosmological theory also describing particle creation during the evolution of the Universe.

Creation pressure in Barthel–Randers cosmology.

After some simple algebraic transformations, the particle energy balance Equation (68) of the Barthel–Randers cosmology can be rewritten as

$$\dot{\rho} + 3H\left[\left(\rho + \frac{p}{c^2}\right)\left(1 + \frac{\dot{\phi}}{2H\phi}\right)\right] = 0. \tag{96}$$

By comparing Equations (94) and (96), we find the expression of the Barthel–Randers creation pressure as

$$P_c = \frac{\dot{\phi}}{2H\phi}\left(\rho c^2 + p\right) = \frac{c^2\rho\dot{\phi}}{2H\phi}(1 + \omega), \tag{97}$$

where $\omega = p/\rho c^2$. The energy density balance Equation (68) can now be derived equivalently from the divergence of the total energy momentum tensor T^{IJ}, given by

$$T^{IJ} = \left(\rho c^2 + p + P_c\right)u^I u^J - (p + P_c)a^{IJ}, \tag{98}$$

where all the mathematical operations must be performed in the Riemann space with the FLRW metric (56).

The particle creation rate.

By assuming adiabatic particle production, which requires that $\dot{\sigma} = 0$, the Gibbs relation (95) gives

$$\dot{\rho} = \left(\rho + \frac{p}{c^2}\right)\frac{\dot{n}}{n} = \left(\rho + \frac{p}{c^2}\right)(\Gamma - 3H). \tag{99}$$

By combining this relation with the energy balance Equation (96), we obtain the particle creation rate as a function of the creation pressure, the Hubble function and the equilibrium thermodynamic quantities as

$$\Gamma = \frac{-3HP_c}{\rho c^2 + p}. \tag{100}$$

By using Equation (97), we find for the Barthel–Randers cosmological particle creation rate the expression

$$\Gamma = -\frac{3}{2}\frac{\dot{\phi}}{\phi} = -\frac{3}{2}\frac{d}{dt}\ln\phi = -\frac{3}{2}\frac{\dot{a}\eta + a\dot{\eta}}{1 + a\eta}. \tag{101}$$

If $1 + a\eta > 0$, the condition $\Gamma \geq 0$, which is equivalent with the existence of particle creation, leads to the constraint

$$\dot{\phi} = \dot{a}\eta + a\dot{\eta} = a(\dot{\eta} + H\eta) \leq 0, \tag{102}$$

on the scale factor a, and the temporal component of the one-form η. The condition does not depend explicitly on the equation of state of the baryonic matter, but an indirect dependence via the Hubble function does exist. The condition $\dot{\phi} < 0$ implies the existence of a negative creation pressure, as it follows from Equation (97). Therefore particle production is thermodynamically allowed only if the creation pressure is negative, $P_c < 0$.

The Barthel–Randers particle balance equation can thus be reformulated to take the form

$$\dot{n} + 3Hn = -\frac{3}{2}n\frac{\dot{\phi}}{\phi} = \Gamma n, \tag{103}$$

and it can be integrated to give the expression for the particle number density

$$n = \frac{n_0}{\phi^{3/2}a^3}, \tag{104}$$

where n_0 is a constant of integration.

For a Universe consisting of pressureless dust with $p = 0$, the creation pressure is given by

$$P_c = \frac{\dot{\phi}}{2H\phi}\rho c^2 = \frac{\rho c^2}{2H}\frac{d}{dt}\ln\phi, \tag{105}$$

and it depends linearly on the baryonic matter density.

The divergence of the entropy flux vector S^I takes the form

$$\nabla_I S^I = \frac{-3n\sigma H P_c}{\rho c^2 + p} = -\frac{3}{2}n\sigma\frac{\dot{\phi}}{\phi} = n\sigma\Gamma. \tag{106}$$

The matter temperature.

The temperature T of the baryonic matter represents an important characteristic of physical systems. To obtain the temperature evolution in the presence of particle creation, we consider the general thermodynamic case in which the density and the pressure are functions of the particle number density, and of the temperature. Hence ρ and p can be generally represented in parametric form $\rho = \rho(n,T)$ and $p = p(n,T)$, respectively. Then, we obtain

$$\dot{\rho} = \left(\frac{\partial \rho}{\partial n}\right)_T \dot{n} + \left(\frac{\partial \rho}{\partial T}\right)_n \dot{T}. \tag{107}$$

By using the energy and particle balance equations, we find first the relation

$$-3H\left(\rho c^2 + p + P_c\right) = \left(\frac{\partial \rho}{\partial n}\right)_T n(\Gamma - 3H) + \left(\frac{\partial \rho}{\partial T}\right)_n \dot{T}. \tag{108}$$

As a second step we use the thermodynamic relation [156]

$$T\left(\frac{\partial p}{\partial T}\right)_n = \rho c^2 + p - n\left(\frac{\partial \rho}{\partial n}\right)_T, \tag{109}$$

which together with Equation (108) gives the temperature evolution of the matter in the Barthel–Randers cosmology in the presence of particle production as

$$\frac{\dot{T}}{T} = \frac{1}{c^2}\left(\frac{\partial p}{\partial \rho}\right)_n \frac{\dot{n}}{n} = \omega \frac{\dot{n}}{n}. \tag{110}$$

From the particle balance equation, we obtain the ratio $\dot{n}/n$ as

$$\frac{\dot{n}}{n} = -3\left(\frac{\dot{\phi}}{\phi} + \frac{\dot{a}}{a}\right). \tag{111}$$

By assuming that $\dot{\phi} < 0$, we obtain for the temperature evolution of the particles in the Barthel–Randers Universe the relation

$$T = T_0 \frac{\phi^{3\omega/2}}{a^3}. \tag{112}$$

Generally, ω is an arbitrary scale factor-dependent parameter $\omega = \omega(a)$ [157]. In a thermodynamically consistent cosmological model, all geometrical and physical quantities must be well-defined and regular for all $\omega(a)$.

3.6.3. Creation of Exotic Matter

In our discussion of the thermodynamics of open systems, as presented in the previous section, we have assumed that matter is created in the form of baryonic matter, satisfying the energy condition $\rho c^2 + p \geq 0$, and the restriction on the parameter of the matter equation of state $\omega \geq 0$. But we cannot a priori exclude the possibility case in which an exotic fluid with $\omega < 0$ is created during the cosmological evolution of the Barthel–Randers Universe. The open system thermodynamic approach to particle creation considered previously is also applicable if $\omega < 0$. For this case the creation pressure becomes negative for $\dot{\phi} > 0$, with the particle creation rate Γ becoming positive. Therefore, matter creation processes generating exotic particles can also be included in the Barthel–Randers cosmological model.

A particular and important case is represented by an exotic fluid satisfying the equation of state $\rho_{(ex)}c^2 + p_{(ex)} = 0$, with $\omega = -1$, which is equivalent to the presence of a cosmological constant. Then from Equation (94), we obtain

$$\dot{\rho}_{(ex)} = -3HP_c. \tag{113}$$

If the exotic particle creation process is adiabatic, with $\dot{\sigma} = 0$, from the Gibbs relation, we find

$$\dot{\rho}_{(ex)} = \left(\rho_{(ex)}c^2 + p_{(ex)}\right)\frac{\dot{n}_{(ex)}}{n_{(ex)}} = 0. \tag{114}$$

The above two equations independently give

$$\dot{\rho}_{(ex)} = P_c = 0, \tag{115}$$

which describe a Universe with constant exotic matter density and vanishing creation pressure. On the other hand, if $\omega = -2$, and $\rho_{(ex)} > 0$, then $P_c = -(c^2\dot{\phi}/2H\phi)\rho_{(ex)}$, $\Gamma = (3/2)\dot{\phi}/\phi$, and matter production can take place if the conditions $\dot{\phi} > 0$ and $\phi > 0$ are satisfied.

Thus, in a Barthel–Randers Universe, the creation of exotic matter or possibly of scalar fields is allowed in a way consistent with the laws of thermodynamics. The creation processes take place in the Riemannian geometry characterized by the FLRW metric.

4. Cosmological Implications of Barthel–Randers and Barthel–Kropina Models

In this section, we investigate the cosmological implications of the Barthel–Randers and Barthel–Kropina models by exploring three distinct variants of the Barthel–Randers framework alongside the Barthel–Kropina cosmological model. To constrain the free parameters of these models, we employ a combination of observational datasets, including Type Ia Supernovae, Baryon Acoustic Oscillations, and Hubble parameter measurements. By extracting the posterior distributions of the model parameters through Bayesian inference, we are able to assess the observational viability of the Barthel–Randers and Barthel–Kropina models. This comparison with the standard ΛCDM cosmology enables us to explore possible deviations from the conventional expansion history and to evaluate whether these geometrically extended models offer a competitive or improved description of the Universe's evolution.

4.1. Specific Cosmological Models

In this subsection, we present the normalized Hubble functions associated with both the Barthel–Randers [120] and Barthel–Kropina [121] cosmological models. Our starting point is the family of cosmological scenarios proposed in [120], where three distinct variants of the Barthel–Randers model were introduced. Each of these variants is characterized by a specific choice of the function $\varphi(z)$. The normalized Hubble function for all three models can be expressed in a unified form as

$$h(z) = \frac{2\varphi(z)\sqrt{(1+z)^3\sqrt{\varphi(z)}\,\Omega_{m0} + \varphi(z)^2\Omega_{\Lambda 0} + (1+z)^4\Omega_{r0}}}{(1+z)\frac{d\varphi(z)}{dz} - 2\varphi(z)}, \tag{116}$$

where z is the redshift, and Ω_{m0}, $\Omega_{\Lambda 0}$, and Ω_{r0} are the present-day matter, dark energy, and radiation density parameters, respectively. The dark energy density parameter $\Omega_{\Lambda 0}$ is obtained through the constraint

$$\Omega_{\Lambda 0} = \left(1 - \frac{1}{2}\varphi'(0)\right)^2 - \Omega_{m0} - \Omega_{r0}, \tag{117}$$

where $\varphi'(0)$ is the derivative of the function $\varphi(z)$ evaluated at $z = 0$. The form of $\varphi(z)$ determines the behavior of each specific model and encodes the influence of the underlying Finslerian geometry. In the following, we present the explicit forms of $\varphi(z)$ corresponding to each of the three models proposed in [120]:

1. **Linear model:** $\varphi(z) = 1 + \beta z$.
2. **Logarithmic model:** $\varphi(z) = 1 + \ln(1 + \beta z)$.
3. **Exponential model:** $\varphi(z) = e^{2\beta z}$.

We can obtain the corresponding normalized Hubble function by plugging the corresponding form of $\varphi(z)$ into Equation (116).

In the case of the Barthel–Kropina geometry, we adopt the model proposed by [121]. The corresponding normalized Hubble function is expressed as a system of differential equations in the redshift representation as follows:

$$\frac{df(z)}{dz} = -\frac{u(z)}{(1+z)h(z)}, \tag{118}$$

$$\begin{aligned}\frac{du(z)}{dz} = \frac{1+f(z)}{2(1+z)h(z)}\Big[&2h(z)\left(\frac{2}{1+f(z)} - 3\omega(1+f(z))\right)u(z) \\ &+3\left(\omega - \frac{1}{(1+f(z))^2}\right)u(z)^2 + 3\omega f(z)(2+f(z))h(z)^2\Big],\end{aligned} \tag{119}$$

$$\frac{dh(z)}{dz} = \frac{1}{2(1+z)h(z)}\left[3h(z)^2 - \frac{4h(z)u(z)}{1+f(z)} + \frac{3u(z)^2}{(1+f(z))^2} + \frac{2(1+z)h(z)}{1+f(z)}\frac{du(z)}{dz}\right]. \tag{120}$$

The system of equations above has to be solved with initial conditions $f(0) = f_0$, $u(0) = u_0$, and $h(0) = 1$.

4.2. Methodology and Datasets

In this subsection, we provide a detailed explanation of the methodology used to estimate the posterior distributions of the model parameters, employing the Markov Chain Monte Carlo (MCMC) approach. This approach allows us to constrain model parameters by analyzing various observational datasets. The MCMC method explores the parameter space of theoretical models by sampling from the likelihood function and incorporating prior information, leading to a robust estimation of the posterior probability distribution [158]. Through this method, the model's parameter space is thoroughly explored. The MCMC algorithm relies on Bayes' theorem, which is given by

$$P(\theta|D) = \frac{L(D|\theta)P(\theta)}{P(D)}, \tag{121}$$

where $P(\theta|D)$ represents the probability of the parameters θ given the observational data D. The term $L(D|\theta)$ is the likelihood function, which measures how well the model fits the data. $P(\theta)$ is the prior distribution, incorporating any previous knowledge about the parameters, while $P(D)$ is the evidence, acting as a normalization factor [159].

One of the key advantages of the MCMC method is that it not only finds the most likely values for the parameters but also accounts for uncertainties in both the models. We define the likelihood function in such a way that it compares the theoretical predictions of the Hubble parameter ($H(z)$), luminosity distance ($\mu(z)$), transverse distance ($D_H(z)$), comoving angular diameter distance ($D_M(z)$), and comoving volume distance ($D_V(z)$) with observational data, taking into account uncertainties through covariance matrices.

The MCMC sampling is carried out using the `emcee` library[1] [160,161], which includes an affine-invariant ensemble sampler. This sampler is used to efficiently explore the parameter space. Specifically, we use 30 walkers, with a total of 100,000 steps for the chain. To ensure the removal of biases, the first 10,000 steps are discarded as burn-in, and the remaining steps are thinned by a factor of 30 to reduce autocorrelation between samples.

After running the MCMC chains, the posterior distributions of the parameters are analyzed, and the best-fit values are identified. To visualize and analyze the posterior distributions, we utilize the `GetDist` package[2] [162], which provides an extensive set of tools for generating 1D and 2D posterior distributions. In this work, we use three different observational datasets: Cosmic Chronometers, Type Ia Supernovae, and Baryon Acoustic Oscillations.

Below, we provide a detailed description of each dataset and how the likelihood has been formed based on them.

- **Cosmic Chronometers:** In this study, we use the Hubble measurements extracted based on the differential age approach. This technique leverages passively evolving massive galaxies, which formed at redshifts around $z \sim 2-3$, enabling a direct and model-independent determination of the Hubble parameter using the relationship $\Delta z/\Delta t$. This method significantly reduces the reliance on astrophysical assumptions [163]. For our analysis, we use 15 Hubble measurements selected from the 31 Hubble measurements, which cover a redshift range from $0.179 \leq z \leq 1.965$ [164–166]. We use the likelihood function provided by Moresco in his GitLab repository[3] which uses the full covariance matrix, accounting for both statistical and systematic uncertainties [167,168] .
- **Type Ia Supernova:** We also use the Pantheon$^+$ dataset without the SHOES calibration, which consists of 1701 light curves from 1550 Type Ia Supernovae (SNe Ia) covering a redshift range of $0 \leq z \leq 2.3$ [169]. To analyze this data, we adopt the likelihood function described in [170], which incorporates the total covariance matrix, $\mathbf{C}_{\text{total}}$, which includes both statistical ($\mathbf{C}_{\text{stat}}$) and systematic ($\mathbf{C}_{\text{sys}}$) uncertainties [171]. The likelihood function is given by $\mathcal{L}_{\text{SNe Ia}} = e^{\left(-\frac{1}{2}\mathbf{r}^T\mathbf{C}_{\text{total}}^{-1}\mathbf{r}\right)}$, where $\mathbf{r}$ represents the residual vector, defined as the difference between the observed and theoretical distance moduli: $r_i = \mu_{\text{obs}}(z_i) - \mu_{\text{th}}(z_i, \theta)$, where θ. Here, $\mathbf{C}_{\text{total}}^{-1}$ is the inverse of the total covariance matrix. The model-predicted distance modulus is calculated as $\mu_{\text{model}}(z_i) = 5\log_{10}\left(\frac{d_L(z)}{\text{Mpc}}\right) + \mathcal{M} + 25$, where the luminosity distance $d_L(z)$ in a flat FLRW Universe is given by $d_L(z) = c(1+z)\int_0^z \frac{dz'}{H(z')}$. Here, c is the speed of light, and $H(z)$ is the Hubble parameter. This formulation highlights the degeneracy between the nuisance parameter $\mathcal{M}$ and the Hubble constant H_0.
- **Baryon Acoustic Oscillation:** In our analysis, we also include the 13 recent Baryon Acoustic Oscillation (BAO) measurements from the Dark Energy Spectroscopic Instrument (DESI) Data Release 2 (DR2) [172]. These measurements cover a redshift range of $0.295 \leq z \leq 2.330$. They were obtained from observations of the Bright Galaxy Sample (BGS), Luminous Red Galaxies (LRG1, LRG2, LRG3), Emission Line Galaxies (ELG1 and ELG2), Quasars (QSO), and Lyman-α tracers[4]. The measurements are reported using three distance indicators: the Hubble distance $D_H(z)$, the comoving angular diameter distance $D_M(z)$, and the volume-averaged distance $D_V(z)$. To compare these with cosmological models, we compute the ratios $D_H(z)/r_d$, $D_M(z)/r_d$, and $D_V(z)/r_d$, where r_d represents the sound horizon at the drag epoch, occurring around redshift $z \sim 1060$. In a flat ΛCDM model, $r_d = 147.09 \pm 0.26$ Mpc [72]. However, in this study, we treat r_d as a free parameter, allowing late-time observations to constrain the corresponding model parameters [173–177]. The chi-squared statistic for the BAO measurements is given by $\chi_i^2 = (\vec{O}_i^{\text{th}} - \vec{O}_i^{\text{obs}})^T\,\mathbf{C}_i^{-1}\,(\vec{O}_i^{\text{th}} - \vec{O}_i^{\text{obs}})$, where $\vec{O}_i^{\text{th}}$ and $\vec{O}_i^{\text{obs}}$ denote the vectors of theoretical predictions and observed measurements, respectively, and $\mathbf{C}_i$ is the associated covariance matrix[5].

The posterior distribution of model parameters is obtained by maximizing the likelihood function. The total likelihood function is given by

$$\mathcal{L}_{\text{tot}} = \mathcal{L}_{\text{CC}} \times \mathcal{L}_{\text{SNe Ia}} \times \mathcal{L}_{\text{BAO}}. \tag{122}$$

First, we discuss how to extract the posterior distribution of the model parameters in the Barthel–Randers framework. To achieve these, we apply the standard approach outlined above. In these models, we treat the parameters H_0, Ω_{m0}, β, $\mathcal{M}$, and r_d as free parameters with the following priors: H_0 $(\text{km}\,\text{s}^{-1}\,\text{Mpc}^{-1}) \in [50, 100]$, $\Omega_{mo} \in [0, 1]$, $\mathcal{M} \in [-20, -18]$, r_d (Mpc) $\in [100, 200]$, and $\beta \in [0, 0.1]$, as the model is sensitive to this parameter. The Ω_{r0} is extracted using the following relation: $\Omega_{r0} = (4.183699 \times 10^{-5})h^{-2}$, where $h = \frac{H_0}{100}$. In the case of the linear and logarithmic models, $\Omega_{\Lambda 0}$ is extracted using the relation $\Omega_{\Lambda 0} = 1 - \beta + \frac{\beta^2}{4} - \Omega_{m0} - \Omega_{r0}$, while in the case of the exponential model, $\Omega_{\Lambda 0}$ is extracted using the relation $\Omega_{\Lambda 0} = 1 - 2\beta + \beta^2 - \Omega_{m0} - \Omega_{r0}$. This makes the explicit variation of Ω_{r0} and $\Omega_{\Lambda 0}$ redundant, as both are fully determined by the other parameters.

To constrain the Barthel–Kropina model, we begin by numerically solving the system of differential equations given in Equations (118)–(120). These equations are integrated using the *solve_ivp* function from *SciPy*, employing the Radau method with a fifth-order implicit Runge–Kutta scheme particularly effective for handling the stiff differential equations. For numerical stability and accuracy over the redshift range $z \in [0,3]$, we set the relative and absolute tolerances to 10^{-3} and 10^{-6}, respectively. After obtaining the numerical solutions, we use an MCMC approach to constrain the model parameters. For more information on how to handle these kinds of Hubble-like functions that appear within differential equations, please consult [178]. In the case of the Barthel–Kropina model, we take the H_0 (km s^{-1} Mpc^{-1}) $\in [50,100], u_o \in [0.0,3.0], \mathcal{M} \in [-20,-18]$, r_d (Mpc) $\in [100,200]$, and $f_0 \in [0,0.1]$. It is important to note that in both cases, we work with the normalized Hubble function. The final Hubble parameter $H(z)$ can be obtained by multiplying the normalized function $h(z)$ by the Hubble constant H_0, i.e., $H(z) = H_0 \cdot h(z)$.

Convergence Test

Gelman–Rubin Statistic $\left(\widehat{R}\right)$

To assess the convergence of multiple MCMC chains, we use the Gelman–Rubin criterion $\left(\widehat{R}\right)$, which compares the within-chain variance $\left(\widehat{W}\right)$ and the between-chain variance (B) to estimate the total variance $\left(\widehat{V}\right)$. These quantities are related by

$$\widehat{R} = \sqrt{\frac{\widehat{V}}{\widehat{W}}}. \tag{123}$$

If $\widehat{R} \approx 1$, it indicates convergence, suggesting that the chains are sampling from the same distribution. On the other hand, if $\widehat{R} > 1$, it suggests non-convergence, and values significantly above 1 (e.g., $\widehat{R} > 1.1$) indicate that more iterations are needed to ensure proper convergence.

Convergence of Chains and Trace Plots

In addition to the Gelman–Rubin statistic, trace plots are used to visualize whether the chains have converged. These plots show the evolution of each parameter over time, helping to verify if the chains are evolving in the convergence region or still exploring different regions of the parameter space.

5. Comparing Barthel–Randers, Barthel–Kropina and ΛCDM Models

After obtaining the mean parameter values from the MCMC simulations, we proceed to compare the predictions of the Barthel–Randers and Barthel–Kropina cosmological models with those of the standard ΛCDM model. This comparison is essential to assess how well the Barthel–Randers and Barthel–Kropina frameworks are consistent with the observed cosmic expansion history.

5.1. Evolution of the Hubble Parameter, Hubble Residual, and BAO Distance Scales

In this subsection, we compare the Hubble parameter $H(z)$ predicted by three Barthel–Randers models and the Barthel–Kropina model with those of the standard ΛCDM model. The ΛCDM model is given by the equation

$$H_{\Lambda\mathrm{CDM}}(z) = H_0\sqrt{\Omega_{m0}(1+z)^3 + \Omega_{\Lambda 0}},$$

where $H_0 = 67.8\,\mathrm{km\,s^{-1}\,Mpc^{-1}}$, $\Omega_{m0} = 0.3092$, and $\Omega_{\Lambda 0} = 0.608$. To quantify the deviation between the Barthel–Randers models and ΛCDM, we define the Hubble residual as

$$\Delta H(z) = H_{\text{Barthel-Randers Models}}(z) - H_{\Lambda\mathrm{CDM}}(z),$$

where $H_{\text{Barthel-Randers Models}}(z)$ denotes the Hubble parameter predicted by the respective Barthel–Randers model. Similarly, for the Barthel–Kropina model, we define

$$\Delta H(z) = H_{\text{Barthel-Kropina Model}}(z) - H_{\Lambda\text{CDM}}(z),$$

where $H_{\text{Barthel-Kropina Model}}(z)$ denotes the Hubble parameter predicted by the respective Barthel–Kropina model. We plot both $H(z)$ and $\Delta H(z)$ for the Barthel–Randers models and Barthel–Kropina model. A small or nearly constant residual indicates close agreement with the ΛCDM model, while significant deviations suggest potential extensions to the standard cosmological model. In addition to analyzing the Hubble parameter, we also consider the BAO distance scales, specifically the ratio between the angle-averaged distance $D_V(z) = (zD_M^2 D_H)^{1/3}$ and the ratio of transverse to line-of-sight comoving distances, $FAP(z) = D_M/D_H$, which are scaled by $z^{-2/3}$ and z^{-1}, respectively. By comparing these BAO distance ratios with the predictions of the best-fit flat ΛCDM model, we gain further insight into the behavior of the Barthel–Randers and Barthel–Kropina models in relation to the standard cosmological framework. The results of these comparisons are visualized by plotting $H(z)$, $\Delta H(z)$, and the BAO distance scales, providing a comprehensive view of how each model behaves relative to ΛCDM. We consider the CC data [164–166] in our analysis, and the BAO measurements are taken from [172].

5.2. *Cosmographic Analysis of Barthel–Randers and Barthel–Kropina Models*

Cosmography provides a model-independent framework to describe the kinematic behavior of the Universe by expanding cosmological quantities in terms of redshift [179–182]. In this work, we apply this approach to analyze the deceleration and jerk parameters within the Barthel–Randers and Barthel–Kropina cosmological models, which modify standard cosmic dynamics through different geometrical correction terms.

Deceleration Parameter $q(z)$ and Jerk Parameter $j(z)$

The deceleration parameter $q(z)$ describes the acceleration or deceleration of the cosmic expansion and is defined as

$$q(z) = -\frac{1}{H^2(z)}\frac{dH(z)}{dz} - 1.$$

A negative value of $q(z)$ indicates that the Universe is undergoing accelerated expansion. Two important phases derived from $q(z)$ are the present-day deceleration parameter, $q_0 = q(z = 0)$, which describes the current rate of acceleration, and the transition redshift, z_{tr}, defined by the condition $q(z_{\text{tr}}) = 0$, marking the point at which the Universe transitioned from decelerated to accelerated expansion. The jerk parameter $j(z)$ is a third-order cosmographic quantity defined as [183]

$$j(z) = \frac{1}{H^3(z)}\frac{d^2H(z)}{dz^2},$$

which characterizes the rate of change of the acceleration. Importantly, in the standard ΛCDM model, the jerk parameter remains constant at $j(z) = 1$. Any deviation from this value can signal dynamics beyond the standard cosmological model. For the three Barthel–Randers models and the Barthel–Kropina model, we evaluate both the deceleration parameter $q(z)$ and the jerk parameter $j(z)$ numerically and assess their behavior across the redshift range.

5.3. *Statistical Assessment of Barthel–Randers and Barthel–Kropina Models*

In this subsection, we apply a set of model selection criteria to evaluate the goodness of fit and model complexity of three variants of the Barthel–Randers cosmological models (linear, logarithmic and exponential) and Barthel–Kropina model. It is crucial to assess how well each model aligns with the data when compared to the ΛCDM model.

5.4. Goodness of Fit

The primary measure of model fit is the chi-squared statistic, χ^2, which quantifies the discrepancy between theoretical predictions and observational data. For a suitable comparison, as models generally have different numbers of parameters, we compute the reduced chi-square

$$\chi^2_{\rm red} = \frac{\chi^2_{\rm tot}}{\rm DOF}, \tag{124}$$

where $\chi^2_{\rm tot}$ is the total chi-squared value and DOF denotes the degrees of freedom, calculated as the number of data points minus the number of fitted parameters. A value of $\chi^2_{\rm red} \approx 1$ indicates a good statistical fit [184]. Lower values can suggest overfitting, while significantly higher values may reflect poor model performance.

5.4.1. Model Comparison Using AIC and BIC

In addition to chi-squared statistics, we apply information criteria that account for both fit quality and model complexity. These include the Akaike information criterion (AIC) and the Bayesian information criterion (BIC), defined respectively as [185–189]

$$\mathrm{AIC} = -2\log\mathcal{L}_{tot,max} + 2\mathcal{P}, \tag{125}$$
$$\mathrm{BIC} = -2\log\mathcal{L}_{tot,max} + \mathcal{P}\ln(\mathcal{N}), \tag{126}$$

where $\mathcal{P}$ is the number of free parameters, $\mathcal{N}$ is the total number of data points used in the analysis and $\mathcal{L}_{tot,max}$ is the total maximum likelihood. In our analysis, the total number of data points used is the sum of three different observational datasets: $\mathcal{N} = 1729$. For the model comparison, we note that the ΛCDM model has four free parameters, whereas the Barthel–Randers cosmological model has five free parameters, and the Barthel–Kropina model has six free parameters. These criteria penalize overly complex models to prevent overfitting. BIC generally applies a stricter penalty than AIC, especially in large datasets.

5.4.2. Relative Comparison: ΔAIC and ΔBIC

To directly compare the Barthel–Randers models with the ΛCDM model, we compute

$$\Delta\mathrm{AIC} = \mathrm{AIC}_{\text{Barthel-Randers Models}} - \mathrm{AIC}_{\Lambda\text{CDM Model}}, \tag{127}$$
$$\Delta\mathrm{BIC} = \mathrm{BIC}_{\text{Barthel-Randers Models}} - \mathrm{BIC}_{\Lambda\text{CDM Model}}. \tag{128}$$

Similarly, for the comparison between the Barthel–Kropina model and the ΛCDM model, we compute

$$\Delta\mathrm{AIC} = \mathrm{AIC}_{\text{Barthel-Kropina Model}} - \mathrm{AIC}_{\Lambda\text{CDM Model}}, \tag{129}$$
$$\Delta\mathrm{BIC} = \mathrm{BIC}_{\text{Barthel-Kropina Model}} - \mathrm{BIC}_{\Lambda\text{CDM Model}}. \tag{130}$$

According to the calibrated Jeffreys' scales [190], these differences offer insight into model preference:

- $|\Delta\mathrm{AIC}| \leq 2$: Comparable support.
- $4 \leq |\Delta\mathrm{AIC}| < 10$: Considerably less support.
- $|\Delta\mathrm{AIC}| \geq 10$: Strongly disfavored.
- $|\Delta\mathrm{BIC}| \leq 2$: Weak evidence against the model.
- $2 < |\Delta\mathrm{BIC}| \leq 6$: Moderate evidence against the model.
- $|\Delta\mathrm{BIC}| > 6$: Strong evidence against the model.

A negative value of ΔAIC or ΔBIC indicates that the Barthel–Randers and Barthel–Kropina models are preferred over ΛCDM, whereas positive values indicate that these models are less favored

5.4.3. p-Value Statistics

In our analysis, we also computed the p-value to assess the statistical significance of the fit for the studied cosmological models, calculated as follows:

$$p = 1 - \mathcal{F}_{\chi^2_{\rm min}}(\chi \mid \nu),$$

where $\mathcal{F}_{\chi^2_{\rm min}}(\chi \mid \nu)$ is the cumulative distribution function (CDF) of the chi-squared distribution with ν degrees of freedom, and ν represents the number of data points minus the number of free parameters [191].

The p-value quantifies the likelihood of observing results as extreme as those seen in the data, assuming the null hypothesis holds true. A p-value smaller than 0.05 ($p < 0.05$) indicates statistical significance, providing strong evidence against the null hypothesis and suggesting that the model is a good fit for the data. These corresponding statistical metrics ($\chi^2_{\rm min}$, $\chi^2_{\rm red}$, AIC, BIC, ΔAIC, ΔBIC, and p-value) are computed and compared with those of the ΛCDM model. These statistical metrics ensure that both the goodness of fit and the complexity of each model are thoroughly evaluated, enabling a comprehensive statistical assessment of the Barthel–Randers and Barthel–Kropina models in the context of cosmology.

6. Summary and Discussion of the Results

In the present section, we discuss the main results obtained by confronting the proposed cosmological models with observations, and by comparing their predictions and statistical significance with those of the ΛCDM model

6.1. MCMC and Convergence Results

Figure 2 shows the corner plot of the parameters of the Barthel–Randers and Barthel–Kropina models. The diagonal panels illustrate the one-dimensional marginalized posterior distributions for each parameter, highlighting their most probable values. The off-diagonal panels depict the two-dimensional marginalized contour plots, showing the 68% and 95% confidence intervals. These contour plots also reveal correlations or degeneracies between each model parameter. Figure 3 shows the corresponding trace plots, which depict the convergence behavior of the Markov chains for each parameter. As a convergence diagnostic, the Gelman–Rubin statistic was used for each model parameter. The results obtained are given by

$$\mathrm{BR}^{\text{Gelman-Rubin statistics}}_{\text{Linear Case}}\{H_0, \Omega_{mo}, b_1, \mathcal{M}, r_d\} = \{1.006, 1.001, 1.001, 1.006, 1.006\},$$

$$\mathrm{BR}^{\text{Gelman-Rubin statistics}}_{\text{Logarithmic Case}}\{H_0, \Omega_{mo}, b_1, \mathcal{M}, r_d\} = \{1.005, 1.002, 1.001, 1.005, 1.005\},$$

$$\mathrm{BR}^{\text{Gelman-Rubin statistics}}_{\text{Exponential Case}}\{H_0, \Omega_{mo}, b_1, \mathcal{M}, r_d\} = \{1.004, 1.002, 1.001, 1.004, 1.004\},$$

$$\mathrm{BK}^{\text{Gelman-Rubin statistics}}\{H_0, f_0, u_0, \omega, \mathcal{M}, r_d\} = \{1.007, 1.001, 1.002, 1.002, 1.0071.007\}.$$

In our case, the $\widehat{R}$ values for the Barthel–Randers (linear, logarithmic, and exponential) and Barthel–Kropina models are all very close to 1, indicating excellent convergence of the MCMC chains. However, it is important to note that in the BR models, $\Omega_{\Lambda 0}$ and Ω_{r0} are not independent parameters. Instead, they are determined algebraically once the other parameters constrained through the MCMC analysis are known. For this reason, we do not list $\Omega_{\Lambda 0}$ and Ω_{r0} among the constrained parameters, as they are derived quantities rather than directly fitted ones.

One can observe that the predicted values of H_0 from the ΛCDM and Barthel–Randers models (linear, logarithmic, and exponential cases), as well as from the osculating Barthel–Kropina dark energy model, are in close agreement with those obtained from Moresco's results. Specifically, Moresco's measurements, considering both the full covariance matrix and without systematics, are as follows: Without systematics: $H_0 = 66.2^{+3.8}_{-3.9}\,\mathrm{km\,s^{-1}\,Mpc^{-1}}$.

With systematics, we get $H_0 = 66.2^{+5.5}_{-5.6}\,\mathrm{km\,s^{-1}\,Mpc^{-1}}$. While the central value of our predicted H_0 is close to the estimates from Planck, DESI DR1, and DESI DR2, the associated uncertainty is notably larger.

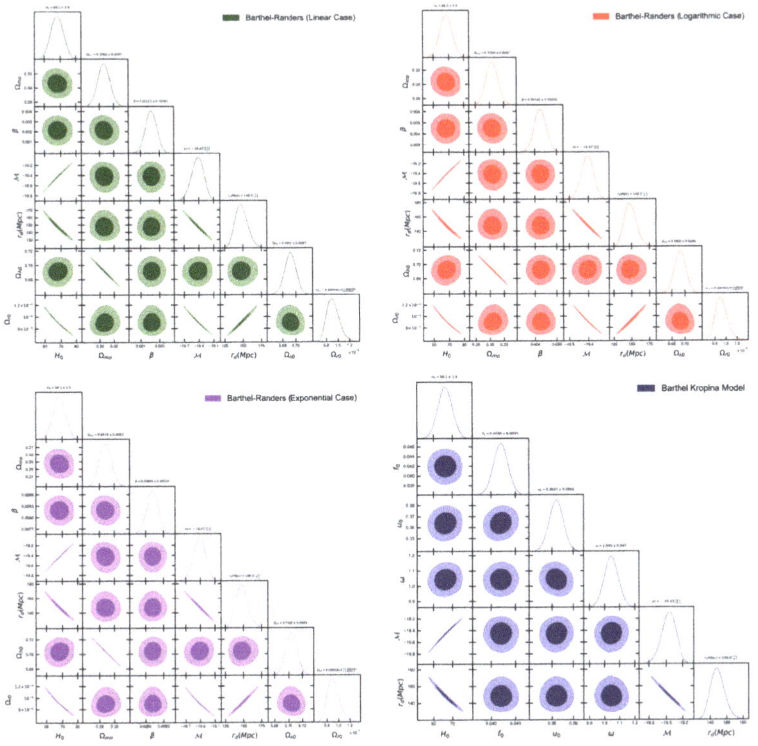

Figure 2. The figure shows the parameter constraints of the Barthel–Randers and Barthel–Kropina models using the CC + SNe Ia and BAO datasets together, and displaying both 1σ and 2σ confidence intervals. The contours show the correlations between these parameters, with marginalized probability distributions along the diagonal.

A similar trend is observed for r_d, where the predicted central values across all models are close to the Planck estimation, $r_d^{\mathrm{Planck}} = (147.09 \pm 0.26)$ Mpc, although the associated uncertainty is also noticeably larger. The primary reason for the higher uncertainty is the inclusion of the full cosmic covariance matrix, which accounts for uncertainties in several factors, including the estimation of stellar metallicity, star formation history, the adopted initial mass function (IMF), the stellar library used, the stellar population synthesis model, and potential contributions from any residual young components in the galaxy spectra.

Table 2 presents the mean values at corresponding 68% confidence intervals for the model parameters of the Barthel–Randers models (linear, logarithmic, and exponential cases), as well as for the osculating Barthel–Kropina dark energy model. The predicted value of the present-day matter density parameter in the case of the ΛCDM model, as well as the linear case of the Barthel–Randers model, is lower than the Planck estimation, $\Omega_{m0}^{\mathrm{Planck}} = 0.315 \pm 0.007$, and is closer to the values predicted by DESI DR1 and DESI DR2, $\Omega_{m0}^{\mathrm{DESI\ DR1}} = 0.295 \pm 0.015$ and $\Omega_{m0}^{\mathrm{DESI\ DR2}} = 0.2975 \pm 0.0086$.

Similarly, the predicted value of Ω_{m0} in the logarithmic case of the Barthel–Randers model is also in good agreement with the DESI DR1 and DR2 estimations. In contrast, the exponential case of the Barthel–Randers model predicts a matter density parameter that is lower than the values reported by both DESI DR1 and DR2. It is interesting to observe that the ΛCDM model, as well as the linear, logarithmic, and exponential cases of the Barthel–Randers model, predict a present-day dark energy density parameter, $\Omega_{\Lambda 0}$, that is greater than the value predicted by the Planck estimation: $\Omega_{\Lambda 0}^{\text{Planck}} = 0.685 \pm 0.007$.

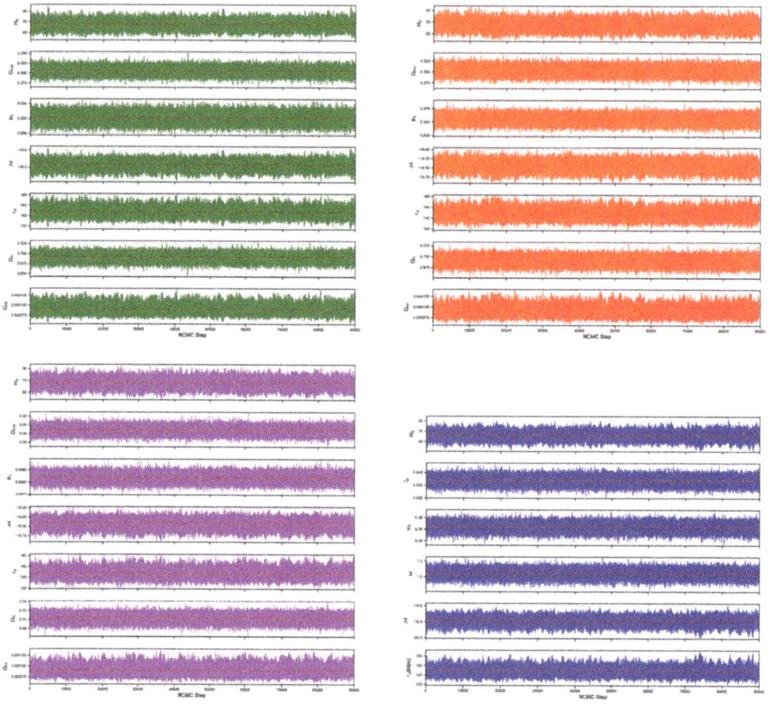

Figure 3. The figure illustrates the trace plots, which show the convergence behavior of the Markov chains for each parameter in the Barthel–Randers and Barthel–Kropina cosmological models.

Table 2. Summary of the mean values and 68% confidence intervals (1σ) for the parameters of the ΛCDM, Barthel–Randers, and Barthel–Kropina cosmological models.

Cosmological Models	Parameter	JOINT
ΛCDM Model	H_0	67.8 ± 3.7
	Ω_{m0}	0.3092 ± 0.0086
	$\mathcal{M}$	-19.43 ± 0.12
	r_d	148.5 ± 7.5
	$\Omega_{\Lambda 0}$	0.6908 ± 0.0086
BR (Linear Case)	H_0	68.1 ± 4.0
	Ω_{m0}	0.3064 ± 0.0087
	β	0.00213 ± 0.00061
	$\mathcal{M}$	-19.42 ± 0.12
	r_d	147.9 ± 7.6
	$\Omega_{\Lambda 0}$	0.6914 ± 0.0087
	Ω_{r0}	$(9.11 \pm 1.20) \times 10^{-5}$

Table 2. *Cont.*

Cosmological Models	Parameter	JOINT
BR (Logarithmic Case)	H_0	68.9 ± 3.9
	Ω_{m0}	0.3034 ± 0.0087
	β	0.00447 ± 0.00055
	$\mathcal{M}$	-19.42 ± 0.12
	r_d	148.1 ± 7.4
	$\Omega_{\Lambda 0}$	0.6921 ± 0.0086
	Ω_{r0}	$(9.13 \pm 1.20) \times 10^{-5}$
BR (Exponential Case)	H_0	68.2 ± 4.0
	Ω_{m0}	0.2878 ± 0.0083
	β	0.00820 ± 0.00016
	$\mathcal{M}$	-19.42 ± 0.12
	r_d	147.9 ± 7.4
	$\Omega_{\Lambda 0}$	0.7040 ± 0.0083
	Ω_{r0}	$(9.00 \pm 1.20) \times 10^{-5}$
Barthel–Kropina Model	H_0	66.2 ± 3.9
	f_0	0.0420 ± 0.0015
	u_0	0.3641 ± 0.0066
	ω	1.045 ± 0.047
	$\mathcal{M}$	-19.45 ± 0.12
	r_d	149.0 ± 7.5

6.2. Hubble Parameter, Hubble Residual, and BAO Distance Scale Results

Figure 4 shows the evolution of the Hubble function and the Hubble residuals for the linear, logarithmic, and exponential cases of the Barthel–Randers model, as well as the Barthel–Kropina model. These are compared with the predictions from the ΛCDM model and CC measurements. The left panel shows that at high redshift ($z > 0.7$), all the considered models deviate noticeably from the ΛCDM model. However, these deviations are not significant. At low redshift ($z < 0.7$), each model aligns closely with the ΛCDM predictions.

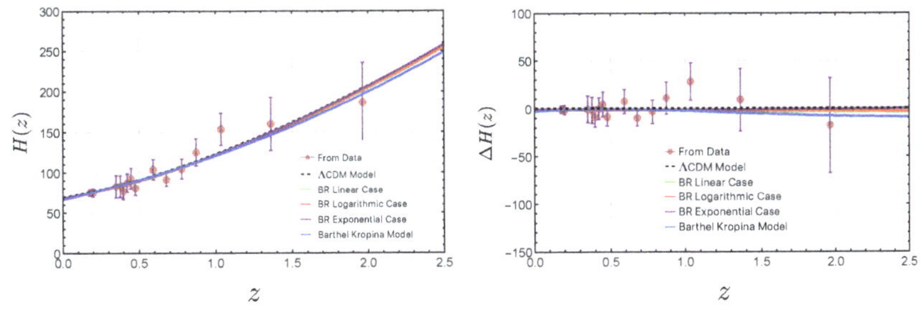

Figure 4. Comparison of the Barthel–Randers (linear, logarithmic, and exponential) and Barthel–Kropina models, with the ΛCDM model and CC measurements. The left panel shows the evolution of the Hubble parameter $H(z)$ using mean MCMC values. The right panel shows the Hubble residuals $\Delta H(z)$, indicating deviations from the standard model.

On the right plane, we can also observe similar behavior in the Hubble residual plots, where each model shows a slight deviation from the ΛCDM model at high redshift. However, at low redshift, all models exhibit behavior that closely matches the ΛCDM predictions. All considered models (Barthel–Randers variants and Barthel–Kropina) closely follow ΛCDM at low redshifts, indicating consistency with current observations. Their mild deviations at higher redshifts ($z > 0.7$) suggest a potentially different early Universe expansion history, offering alternative insights that may help address cosmological tensions such as the Hubble tension.

Figure 5 shows the evolution of different BAO distance scales. All three Barthel–Randers variants exhibit minimal deviation in both the volume-averaged distance $D_V/(r_d z^{2/3})$ and the transverse distance measure $D_M/(zD_H)$, indicating strong alignment with cosmological observations. The logarithmic and exponential BR models closely match ΛCDM at low to intermediate redshifts ($z < 1.5$), though some discrepancies appear beyond $z \approx 1.5$, particularly with the exponential model diverging in the D_V observable range and both models showing reduced accuracy for high-redshift QSO and Ly-α data. The Barthel–Kropina model displays a distinct profile, with the $D_V/(r_d z^{2/3})$ curve generally below the ΛCDM prediction, especially in the mid-redshift range ($z \sim 0.5 - 1.2$). Despite this, it remains within observational error for $z < 1$. In terms of $D_M/(zD_H)$, the Barthel–Kropina model performs similarly to ΛCDM but shows more pronounced deviations in residuals, particularly around $z \sim 1$, indicating less precise modeling in this region.

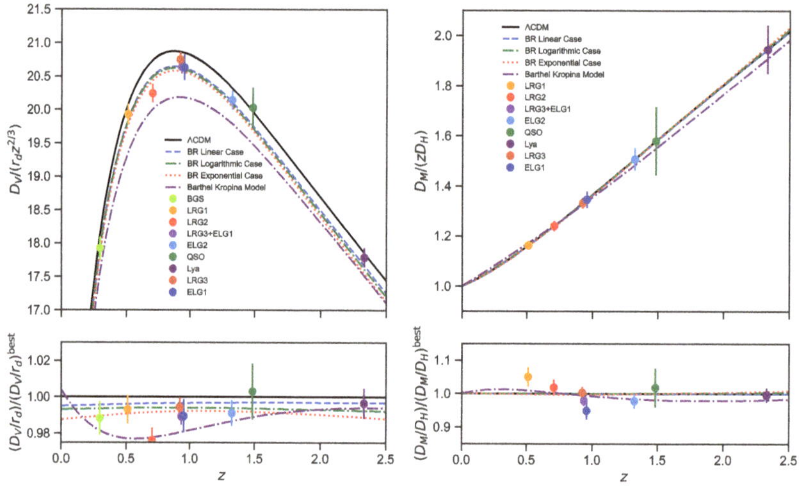

Figure 5. Comparison of the Barthel–Randers (linear, logarithmic, and exponential) and Barthel–Kropina models, with the ΛCDM model across various distance measures, with data from tracers such as LRG, QSO, and Lya. Measurement uncertainties are represented by error bars. In the left panel, the evolution of the ratio between the angle-averaged distance $D_V \equiv (zD_M^2 D_H)^{1/3}$ and the sound horizon at the baryon drag epoch, r_d, is shown. The right panel illustrates the ratio of transverse to line-of-sight comoving distances $FAP \equiv D_M/D_H$. To enhance clarity and compress the dynamic range, an arbitrary scaling of $z^{-2/3}$ has been applied to the left panel, and z^{-1} to the right. The bottom row presents the same models and data points as the top row, but now scaled as the ratio relative to the best-fit flat ΛCDM model predictions.

6.3. Cosmographic Results

Figure 6 shows the evolution of the cosmography parameters for the linear, logarithmic, and exponential cases of the Barthel–Randers model, along with the Barthel–Kropina model, compared to the ΛCDM model. The left panel shows the evolution of the deceleration parameter, $q(z)$.

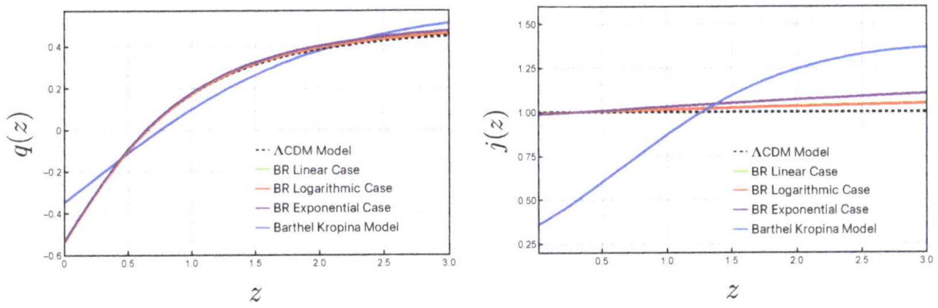

Figure 6. Evolution of the deceleration parameter $q(z)$ and jerk parameter $j(z)$ for the Barthel–Randers (linear, logarithmic, and exponential) and Barthel–Kropina models, and of the ΛCDM model. The left panel shows the evolution of $q(z)$ as a function of redshift, while the right panel presents $j(z)$, showing deviations from the ΛCDM value $j = 1$.

Notably, the Barthel–Kropina model exhibits a distinct evolution compared to both the standard ΛCDM model and the Barthel–Randers variants. At the present time ($z = 0$), all Barthel–Randers variants and ΛCDM predict a deceleration parameter of approximately $q_0 \approx -0.511$, while the Barthel–Kropina model predicts a less negative value of $q_0 \approx -0.366$. We also analyze an important cosmological feature: the transition redshift z_{tr}.

For the ΛCDM model, the transition from decelerated to accelerated expansion occurs at $z_{tr}^{\Lambda\text{CDM}} \approx 0.623$. The Barthel–Randers variants exhibit slightly later transitions: $z_{tr}^{\text{BR Linear}} \approx 0.610$, $z_{tr}^{\text{BR Logarithmic}} \approx 0.607$, and $z_{tr}^{\text{BR Exponential}} \approx 0.606$.

In contrast, the Barthel–Kropina model shows a significantly earlier transition at $z_{tr}^{\text{Barthel-Kropina}} \approx 0.723$. While at high redshift, all models show deviations from the ΛCDM model in their evolution, these deviations are not significant. The right panel shows the evolution of the jerk parameter for each model. At the present time ($z = 0$), all variants of the Barthel–Randers models exhibit good agreement with the ΛCDM model. Although each variant predicts a slightly different value for the jerk parameter, these deviations are not significant. Moreover, the Barthel–Kropina model shows a substantial deviation from the other models, predicting a present-day jerk value of $j_0^{\text{Barthel-Kropina}} \approx 0.448$.

At high redshift, all variants of the Barthel–Randers models, as well as the Barthel–Kropina model, deviate from the predictions of the ΛCDM model.

6.4. Statistical Results

Table 3 presents a comparative analysis of the linear case, logarithmic case, and exponential case of the Barthel–Randers model, as well as the Barthel–Kropina model, against the ΛCDM model, using several statistical metrics. The $\chi^2_{\text{tot,min}}$ value for the ΛCDM model is 1780.94. The Barthel–Randers models (linear, logarithmic, and exponential cases) show similar total $\chi^2_{\text{tot,min}}$ values ranging from 1781.22 to 1783.38, which are slightly higher than those of the ΛCDM model. The Barthel–Kropina model shows the lowest $\chi^2_{\text{tot,min}}$ value of 1762.37. This indicates that, in terms of overall goodness of fit, the Barthel–Kropina model provides the best fit, while the Barthel–Randers models perform slightly worse than the ΛCDM model.

The χ^2_{red} value provides insight into the goodness of fit adjusted for the number of data points and parameters in the model. The ΛCDM model has a χ^2_{red} of 1.032, which is very close to the Barthel–Randers models, all of which have values slightly above 1.033. The lowest χ^2_{red} value is from the Barthel–Kropina model (1.022), indicating that it is marginally the best in terms of adjusting for the complexity of the model and fitting the data well.

Table 3. Summary of χ^2_{tot}, χ^2_{red}, AIC, ΔAIC, BIC, ΔBIC, and p-value for the ΛCDM, Barthel–Randers, and Barthel–Kropina models.

Models	χ^2_{tot}	DoF	χ^2_{red}	AIC	ΔAIC	BIC	ΔBIC	p-Value
ΛCDM	1780.94	1725	1.032	1788.94	0	1810.76	0.170	0.167
BR (Linear Case)	1781.22	1724	1.033	1791.22	2.27	1818.49	7.74	0.164
BR (Logarithmic Case)	1781.60	1724	1.033	1791.60	2.65	1818.87	8.11	0.163
BR (Exponential Case)	1783.38	1724	1.034	1793.38	4.44	1820.65	9.89	0.155
Barthel–Kropina	1762.37	1723	1.022	1774.37	−14.57	1807.10	−3.65	0.249

The AIC is a measure of the relative quality of a model, accounting for both the goodness of fit and the number of parameters. The ΛCDM model has an AIC of 1788.94. The Barthel–Randers models (linear, logarithmic, and exponential) have AIC values around 1791.22, 1791.60, and 1793.38, which are higher by 2.27, 2.65, and 4.44, respectively, compared to the ΛCDM model, indicating a slightly worse balance between fit and model complexity. The Barthel–Kropina model has the lowest AIC value of 1774.37, which is 14.57 points lower than the ΛCDM model. According to the AIC, the Barthel–Kropina model provides the best balance between fit and complexity, while the Barthel–Randers models perform slightly worse than the ΛCDM model.

The BIC is another metric for model selection, similar to AIC but with a stronger penalty for the number of parameters. The ΛCDM model has a BIC of 1810.76, which is lower than the Barthel–Randers models but slightly higher than the Barthel–Kropina model. The Barthel–Randers models show BIC values between 1818.49 and 1820.65, with an increase of approximately 7.74 to 9.89 compared to the ΛCDM model. The Barthel–Kropina model has the lowest BIC of 1807.10, a decrease of 3.65 compared to the ΛCDM model. This suggests that the Barthel–Kropina model is the best according to the BIC criterion, while the Barthel–Randers models perform worse than the ΛCDM model.

The p-value indicates the statistical significance of the model in terms of explaining the data. The ΛCDM model has a p-value of 0.167, indicating that the model is statistically significant at a conventional level (i.e., it cannot be rejected at the 5% significance level). All the other models (Barthel–Randers linear, logarithmic, exponential, and Barthel–Kropina) have p-values ranging from 0.155 to 0.249, which are slightly higher or comparable, suggesting that none of the models can be rejected at the 5% significance level. The p-values imply that all models, including the base ΛCDM model, are statistically significant in explaining the data.

The Barthel–Kropina model outperforms the ΛCDM model in terms of goodness of fit, AIC, and BIC, offering a better balance between fit and model complexity. The Barthel–Randers models (linear, logarithmic, exponential), however, perform slightly worse than the ΛCDM model in this updated comparison.

Despite these improvements, all models, including the ΛCDM model, are statistically significant, with p-values indicating that none of them should be rejected. In terms of simplicity, however, the ΛCDM model remains the most straightforward and easiest to interpret. Therefore, while the Barthel–Kropina model appears to be the most favorable overall, the differences between the models are relatively small, and all offer comparable statistical significance.

7. Discussion and Final Remarks

One of the fundamental assumptions, and results, of present day physics is that the gravitational interaction can be successfully described only in geometric terms. However, which geometry can best describe gravity is still a matter of debate. The initial Riemannian framework of general relativity was extended to include geometries with non-metricity, torsion, or both. While non-metricity naturally arises in the Finslerian setting, the torsional equivalent is not as immediate. Hence, formulating teleparallel theories in the Finslerian setting could be an interesting line of research. A mathematical study of teleparallel connections in Finsler geometry is presented in [192], and it has been shown that general teleparallel gravity could be obtained from Finsler geometry [193]. This opens up new possibilities of research: the exploration of $f(T)$ theories in the Finslerian setting, which, to the best of our knowledge, has not yet been achieved. While the discussion above is entirely classical, it is widely believed that quantum effects should become significant at the Planck scale. A central focus of quantum gravity phenomenology—aimed at identifying observable signatures of quantum gravity—is the study of modified dispersion relations [194,195]. The most well-known examples arise from frameworks such as doubly special relativity (DSR) [196,197] and the κ-Poincaré algebra [198]. Interestingly, the symmetries of doubly special relativity can be realized within the framework of Finsler geometry [199]. Therefore, further investigation into the connection between quantum gravity and Finsler geometry represents a promising direction for research.

In the present work, we have briefly reviewed some recent advances in the application of another geometric framework that could provide some insight into the understanding of the gravitational force, namely Finsler geometry, and its various particular cases. In Finsler geometry, instead of the point x of Riemann geometry, the metric depends also on an internal degree of freedom y, which is often physically interpreted as velocity. Hence, the y dependence is the essential characteristic of Finsler spaces, and the presence of this internal degrees of freedom opens some new perspective on the physical interpretation of the Finsler geometric theories. Thus, Finsler geometries are generally anisotropic and nonlocal, and these properties significantly enlarge the possibilities of the physical applications.

On the other hand, from a physical point of view it is natural to assume that in some specific physical processes, the internal vector y becomes a function of x. Such specific situations may occur if we interpret y as describing spacetime fluctuations, and we perform an averaging over it, according to the rule $\langle\phi(x)\rangle = \int \phi(x,y)f(y)d^4y$, where $f(y)$ is a specific distribution function of y [57]. Hence, Finsler geometry with $\hat{g}(x) = g(x,y)|_{y=Y(x)}$ has a deep physical origin, which points towards interesting physical consequences. The averaging or the osculating process leads to a reduction of a Finsler geometry F to a corresponding Riemann geometry. The interpretation of the vector y as related to some quantum effects, like the vacuum fluctuation, is also supported by the fact that generally, in (α,β)-type gravitational theories, the matter energy–momentum tensor is not conserved. The most natural physical interpretation of this result is related to the possibility of particle production from the Finslerian vacuum in an expanding Universe, an effect that is also specific to quantum field theories in curved spacetimes [147–151]. Hence, Finslerian-type gravitational theories can offer a glimpse of the structure and classical limit of quantum field theories in Riemann geometries.

In our presentation, we have concentrated on a specific class of osculating geometries, the osculating (α,β) geometries, which have the remarkable property that their Barthel connection is the Levi–Civita connection. This property allows a significant simplification of the mathematical formalism, which becomes easily tractable by using the standard methods of Riemannian geometry. The osculating (α,β)-type theories can be interpreted as two metric theories. First, we have the Finsler metric function $F(\alpha,\beta)$, which generates its own metric, while α is a Riemannian metric given a priori. For cosmological applications, the FLRW metric can be adopted as the Riemann metric α, and we assume that this is the physical metric in which the real, observable gravitational processes take place. However, the Finsler metric $\hat{g}(x) = g(x,Y(x))$ also has an important imprint on gravitation, since its presence induces new terms in the gravitational field equations. We have investigated the effects of these new terms from a cosmological perspective only, by interpreting them as describing an effective dark energy that is responsible for the late acceleration of the Universe.

An important test of a cosmological model is represented by its consistency with the observations. We have used three observational datasets (Cosmic Chronometers, Type Ia Supernovae, Baryon Acoustic Oscillations) to perform a detailed analysis, and comparison of the Barthel–Randers, Barthel–Kropina and ΛCDM models. Generally, one can appreciate that all the considered models give a good description of the observational data, but important differences still appear between the various Finslerian approaches and the ΛCDM model. The differences appear, for example, in the prediction of the critical redshift z_{crit} at which the transition from deceleration to acceleration occurs, with the Barthel–Randers models predicting a value of z_{crit} of the order of $z_{crit} \approx 0.60$, while that in the Barthel–Kropina cosmology is $z_{crit} \approx 0.72$. Significant differences appear in the present values of the deceleration parameter $q(0)$, which in the Barthel–Kropina cosmology takes the value $q(0) \approx -0.37$, while in ΛCDM and Barthel–Randers models, it has the value $q(0) \approx -0.50$. Moreover, at redshifts $z > 3$, the predictions of all considered models show significant differences with respect to each other. Hence, cosmological observations at high redshift will give the possibility of discriminating between the various cosmological models based on modified gravity theory constructed with the help of Finsler (and other) geometries.

The main goal of the present review is to draw attention to the possibilities offered by Finsler geometry in the explanation of the cosmological phenomena, and to give a brief introduction to the existing results in a very specific and simple class of this very complex geometry. We have presented in some detail the generalized Friedmann cosmological equations, and the general theoretical framework in which cosmological models can be constructed. We would like to point out that from the point of view of calculations, the derivations of the field equations and of the evolution equations in these particular geometries are no more complicated than those in standard general relativity. This is related especially to the important fact that the connection in osculating-point Finsler geometries is, for the (α, β) metrics, the Barthel connection, which has the same mathematical form as the Levi–Civita connection. In our review, we have presented some basic theoretical tools that can be used for the further exploration of the applications of Finsler geometry in the description of gravitational phenomena.

Author Contributions: Conceptualization, S.V.S., R.H., S.S. and T.H.; methodology L.C., S.S. and S.V.S.; software A.B. and H.C.; review and editing T.H., L.C., A.B. and H.C. All authors have read and agreed to the published version of the manuscript.

Funding: The work of L.Cs. was supported by a grant of the Ministry of Research, Innovation and Digitization, CNCS/CCCDI-UEFISCDI, project number PN-IV-P8-8.1-PRE-HE-ORG-2023-0118, within PNCDI IV and Collegium Talentum Hungary. T. H. was supported by a grant from Collegium Talentum Hungary.

Data Availability Statement: The data underlying this article is already given with references during the analysis of this work.

Acknowledgments: We would like to thank the anonymous referees for comments and suggestions that helped to improve our manuscript.

Conflicts of Interest: The authors declare no conflicts of interest. The funders had no role in the design of the study; in the collection, analyses, or interpretation of data; in the writing of the manuscript; or in the decision to publish the results

Notes

1 https://emcee.readthedocs.io/en/stable/ (accessed on 14 June 2025)
2 https://getdist.readthedocs.io/en/latest/plot_gallery.html (accessed on 14 June 2025)
3 https://gitlab.com/mmoresco/CCcovariance (accessed on 14 June 2025)
4 https://github.com/CobayaSampler/bao_data (accessed on 14 June 2025)
5 https://github.com/CobayaSampler/bao_data/blob/master/desi_bao_dr2/desi_gaussian_bao_ALL_GCcomb_cov.txt (accessed on 14 June 2025)

References

1. Weyl, H. Gravitation und Elektrizität. In *Sitzungsberichte der Königlich Preussischen Akademie der Wissenschaften zu Berlin*, 1918th ed.; Königlich Preussische Akademie: Berlin, Germany, 1918; pp. 465–480.
2. Weyl, H. *Space, Time, Matter*; Dover Publications: New York, NY, USA, 1952.

3. Finsler, P. Über Kurven und Flächen in Allgemeinen Räumen. Ph.D. Thesis, University of Göttingen, Göttingen, Germany, 1918.
4. Einstein, A. Die Feldgleichungen der Gravitation. In *Sitzungsberichte der Königlich Preussischen Akademie der Wissenschaften zu Berlin*, 1915th ed.; Königlich Preussische Akademie: Berlin, Germany, 1915; pp. 844–847.
5. Einstein, A. Die Grundlage der allgemeinen Relativitätstheorie. *Ann. Phys.* **1916**, *49*, 769–822. [CrossRef]
6. Hilbert, D. Die Grundlagen der Physik. *Nachrichten Ges. Wiss. Göttingen Math.-Phys. KI.* **1915**, *1915*, 395–408.
7. Scholz, E. The Unexpected Resurgence of Weyl Geometry in late 20th-Century Physics. In *Beyond Einstein: Perspectives on Geometry, Gravitation, and Cosmology in the Twentieth Century*; Rowe, D.E., Sauer, T., Walter, S.A., Eds.; Springer: New York, NY, USA, 2018; pp. 261–360.
8. Riemann, B. Über die Hypothesen, welche der Geometrie zu Grunde liegen. *Abh. KöNiglichen Ges. Wiss. GöTtingen* **1868**, *13*, 133–150.
9. Chern, S.S. Finsler Geometry Is Just Riemannian Geometry without the Quadratic Restriction. *Not. Am. Math. Soc.* **1996**, *43*, 959–963.
10. Rund, H. *The Differential Geometry of Finsler Spaces*; Springer: Berlin, Germany, 1959.
11. Bejancu, A. *Finsler Geometry and Applications*; Ellis Horwood: New York, NY, USA, 1990.
12. Bao, D.; Chern, S.-S.; Shen, Z. *An Introduction to Riemann-Finsler Geometry*; Springer: New York, NY, USA, 2000.
13. Shen, Y.-B.; Shen, Z. *Introduction to Modern Finsler Geometry*; World Scientific: Singapore, 2016.
14. Will, C.M. The Confrontation between General Relativity and Experiment. *Living Rev. Relativ.* **2014**, *17*, 4. [CrossRef]
15. Abbott, B.P. et al. [LIGO Scientific Collaboration and Virgo Collaboration] Observation of Gravitational Waves from a Binary Black Hole Merger. *Phys. Rev. Lett.* **2016**, *116*, 061102.
16. Abbott, R. et al. [LIGO Scientific Collaboration and Virgo Collaboration] GW190814: Gravitational Waves from the Coalescence of a 23 Solar Mass Black Hole with a 2.6 Solar Mass Compact Object. *Astrophys. J. Lett.* **2020**, *896*, L44. [CrossRef]
17. Cartan, É. Sur les variétés à connexion affine et la théorie de la relativité généralisée. *Ann. L'École Norm. Supérieure* **1924**, *41*, 1–25. [CrossRef]
18. Cartan, É. Sur les variétés à connexion affine et la théorie de la relativité généralisée. *Ann. L'École Norm. Supérieure* **1925**, *42*, 17–88. [CrossRef]
19. Weitzenböck, R. *Invariantentheorie*; Noordhoff: Groningen, The Netherlands, 1923.
20. Ehlers, J.; Pirani, F.A.E.; Schild, A. The Geometry of Free Fall and Light Propagation. *Gen. Relativ. Gravit.* **2012**, *44*, 1587. [CrossRef]
21. Linnemann, N.; Read, J. Constructive Axiomatics in Spacetime Physics Part I: Walkthrough to the Ehlers-Pirani-Schild Axiomatisation. *arXiv* **2021**, arXiv:2112.14063.
22. Adlam, E.; Linnemann, N.; Read, J. Constructive Axiomatics in Spacetime Physics Part II: Constructive Axiomatics in Context. *arXiv* **2022**, arXiv:2211.05672.
23. Adlam, E.; Linnemann, N.; Read, J. Constructive Axiomatics in Spacetime Physics Part III: A Constructive Axiomatic Approach to Quantum Spacetime. *arXiv* **2022**, arXiv:2208.07249.
24. Pfeifer, C. Finsler Spacetime Geometry in Physics. *Int. J. Geom. Methods Mod. Phys.* **2019**, *16* (Suppl. S2), 1941004. [CrossRef]
25. Randers, G. On an Asymmetrical Metric in the Four-Space of General Relativity. *Phys. Rev.* **1941**, *59*, 195. [CrossRef]
26. Ingarden, R. On the Geometrically Absolute Optical Representation in the Electron Microscope. *Trav. Société Sci. Lettres WrocłAw Série B* **1957**, *45*, 3.
27. Miron, R. The Geometry of Ingarden Spaces. *Rep. Math. Phys.* **2004**, *54*, 131. [CrossRef]
28. Tavernelli, I. On the Geometrization of Quantum Mechanics. *Ann. Phys.* **2016**, *371*, 239. [CrossRef]
29. Tavernelli, I. On the Self-Interference in Electron Scattering: Copenhagen, Bohmian and Geometrical Interpretations of Quantum Mechanics. *Ann. Phys.* **2018**, *393*, 447. [CrossRef]
30. Liang, S.-D.; Sabau, S.V.; Harko, T. Finslerian Geometrization of Quantum Mechanics in the Hydrodynamical Representation. *Phys. Rev. D* **2019**, *100*, 105012. [CrossRef]
31. Tavernelli, I. Gravitational Quantum Dynamics: A Geometrical Perspective. *Found. Phys.* **2021**, *51*, 46. [CrossRef]
32. Hohmann, M.; Pfeifer, C.; Voicu, N. The Kinetic Gas Universe. *Eur. Phys. J. C* **2020**, *80*, 809. [CrossRef]
33. Pfeifer, C.; Voicu, N.; Friedl-Szász, A.; Popovici-Popescu, E. From Kinetic Gases to an Exponentially Expanding Universe. *arXiv* **2025**, arXiv:2504.08062.
34. Rätzel, D.; Rivera, S.; Schuller, F.P. Geometry of Physical Dispersion Relations. *Phys. Rev. D* **2011**, *83*, 044047. [CrossRef]
35. Duval, C. Finsler Spinoptics. *Commun. Math. Phys.* **2008**, *283*, 701–727. [CrossRef]
36. Schreck, M. Classical Lagrangians and Finsler Structures for the Nonminimal Fermion Sector of the Standard-Model Extension. *Phys. Rev. D* **2016**, *93*, 105017. [CrossRef]
37. Edwards, B.R.; Kostelecký, V.A. Riemann–Finsler Geometry and Lorentz-Violating Scalar Fields. *Phys. Lett. B* **2018**, *786*, 319–326. [CrossRef]
38. Horváth, J.I. A Geometrical Model for the Unified Theory of Physical Fields. *Phys. Rev.* **1950**, *80*, 901. [CrossRef]
39. Horváth, J.I.; Moór, A. Entwicklung einer einheitlichen Feldtheorie begründet auf die Finslersche Geometrie. *Z. Für Phys.* **1952**, *131*, 548. [CrossRef]
40. Takano, Y. Gravitational Field in Finsler Spaces. *Lett. Nuovo Cimento* **1974**, *10*, 747. [CrossRef]
41. Takano, Y. Variation Principle in Finsler Spaces. *Lett. Nuovo Cimento* **1974**, *11*, 486. [CrossRef]
42. Asanov, G.S. A Finslerian Extension of General Relativity. *Found. Phys.* **1981**, *11*, 137. [CrossRef]

43. Asanov, G.S. *Finsler Geometry, Relativity and Gauge Theories*; D. Reidel: Dordrecht, The Netherlands, 1985.
44. Asanov, G.S. Finslerian Solution for Static Spherically Symmetric Gravitational Field. *Fortschritte Phys.* **1991**, *39*, 185. [CrossRef]
45. Asanov, G.S. Finslerian Extension of Schwarzschild Metric. *Fortschritte Phys.* **1992**, *40*, 667. [CrossRef]
46. Malligawad, M.; Narasimhamurthy, S.K.; Nekouee, Z.; Yashwanth, B.R. Shadow analysis and light deflection in charged Finslerian Kiselev black holes under spherical accretion. *Ann. Phys.* **2025**, *477*, 70005. [CrossRef]
47. Nekouee, Z.; Narasimhamurthy, S.K.; Yashwanth, B.R.; Sanjay, T. Exploring null geodesic of Finslerian hairy black hole. *Class. Quantum Gravity* **2025**, *42*, 045002. [CrossRef]
48. Yashwanth, B.R.; Narasimhamurthy, S.K.; Nekouee, Z.; Malligawad, M. Geodesics of Finsler Hayward Black Hole Surrounded by Quintessence. *Eur. Phys. J. C* **2025**, *84*, 1276 [CrossRef]
49. Malligawad, M.; Narasimhamurthy, S.K.; Nekouee, Z. Exploring the quintessential influence on shadows of black holes in Finsler-Hayward geometry. *Phys. Lett. B* **2024**, *856*, 138963. [CrossRef]
50. Yashwanth, B.R.; Narasimhamurthy, S.K.; Nekouee, Z. Generalized Finslerian Wormhole Models in $f(R,T)$ Gravity. *Particles* **2024**, *7*, 747–767. [CrossRef]
51. Nekouee, Z.; Narasimhamurthy, S.K.; Pacif, S.K.J. Black Hole Solutions with Constant Ricci Scalar in a Model of Finsler Gravity. *J. Cosmol. Astropart. Phys.* **2024**, *2024*, 061. [CrossRef]
52. Malligawad, M.; Narasimhamurthy, S.K.; Nekouee, Z.; Kumbar, M.Y. Finslerian wormhole solution in the framework of modified gravity. *Phys. Scr.* **2024**, *99*, 045206. [CrossRef]
53. Sanjay, T.; Narasimhamurthy, S.K.; Nekouee, Z.; Manjunatha, H.M. Traversable wormhole model in Finslerian geometry. *Pramana—J. Phys.* **2024**, *98*, 16. [CrossRef]
54. Nekouee, Z.; Narasimhamurthy, S.K. Thermodynamic product formulae for Finslerian Kiselev black hole. *Eur. Phys. J. C* **2023**, *83*, 723. [CrossRef]
55. Manjunath, M.; Narasimhamurthy, S.K.; Nekouee, Z.; Yashwanth, B.R.; Mallikarjun, Y.K. Finsler geometry insights into wormhole traversability and physical properties. *Indian J. Phys.* **2025**. [CrossRef]
56. Miron, R.; Anastasiei, M. *Vector Bundles, Lagrange Spaces, and Applications to the Theory of Relativity*; Editura Academiei Române: Bucharest, Romania, 1987. (In Romanian)
57. Ikeda, S. *Advanced Studies in Applied Geometry*; Seizansha: Sagamihara, Japan, 1995.
58. Rutz, S.F. A Finsler Generalisation of Einstein's Vacuum Field Equations. *Gen. Relativ. Gravit.* **1993**, *25*, 1139. [CrossRef]
59. Li, X.; Chang, Z. Towards a gravitation theory in Berwald–Finsler space. *Chin. Phys. C* **2010**, *34*, 28.
60. Hohmann, M.; Pfeifer, C.; Voicu, N. Finsler Gravity Action from Variational Completion. *Phys. Rev. D* **2019**, *100*, 064035. [CrossRef]
61. Riess, A.G. et al. [Supernova Search Team] Observational Evidence from Supernovae for an Accelerating Universe and a Cosmological Constant. *Astron. J.* **1998**, *116*, 1009. [CrossRef]
62. Perlmutter, S. et al. [Supernova Cosmology Project] Measurements of Ω and Λ from 42 High-Redshift Supernovae. *Astrophys. J.* **1999**, *517*, 565. [CrossRef]
63. Knop, R.A.; Aldering, G.; Amanullah, R.; Astier, P.; Blanc, G.; Burns, M.S.; Conley, A.; Deustua, S.E.; Doi, M.; Ellis, R.; et al. New Constraints on Ω_M, Ω_Λ, and w from an Independent Set of Eleven High-Redshift Supernovae Observed with HST. *Astrophys. J.* **2003**, *598*, 102. [CrossRef]
64. Amanullah, R.; Lidman, C.; Rubin, D.; Aldering, G.; Astier, P.; Barbary, K.; Burns, M.S.; Conley, A.; Dawson, K.S.; Deustua, S.E.; et al. Spectra and Hubble space telescope Light Curves of Six Type Ia Supernovae at $0.511 < z < 1.12$ and the Union2 Compilation. *Astrophys. J.* **2010**, *716*, 712.
65. Weinberg, D.H.; Mortonson, M.J.; Eisenstein, D.J.; Hirata, C.; Riess, A.G.; Rozo, E. Observational Probes of Cosmic Acceleration. *Phys. Rep.* **2013**, *530*, 87.
66. Einstein, A. *Kosmologische Betrachtungen zur Allgemeinen Relativitätstheorie*; Sitzungsberichte der Königlich Preußischen Akademie der Wissenschaften: Berlin, Germany, 1917.
67. Weinberg, S. The Cosmological Constant Problem. *Rev. Mod. Phys.* **1989**, *61*, 1. [CrossRef]
68. Salucci, P.; Turini, N.; Di Paolo, C. Paradigms and Scenarios for the Dark Matter Phenomenon. *Universe* **2020**, *6*, 118. [CrossRef]
69. Alam, S. et al. [BOSS Collaboration] The Clustering of Galaxies in the Completed SDSS-III Baryon Oscillation Spectroscopic Survey: Cosmological Analysis of the DR12 Galaxy Sample. *Mon. Not. R. Astron. Soc.* **2017**, *470*, 2617. [CrossRef]
70. Abbott, T.M.C. et al. [DES Collaboration] Dark Energy Survey Year 1 Results: Cosmological Constraints from Galaxy Clustering and Weak Lensing. *Phys. Rev. D* **2018**, *98*, 043526. [CrossRef]
71. Tanabashi, M. et al. [Particle Data Group] Review of Particle Physics. *Phys. Rev. D* **2018**, *98*, 030001. [CrossRef]
72. Aghanim, N. et al. [Planck Collaboration] Planck 2018 Results. VI. Cosmological Parameters. *Astron. Astrophys.* **2020**, *641*, A6. [CrossRef]
73. Martel, H.; Shapiro, P.R.; Weinberg, S. Likely Values of the Cosmological Constant. *Astrophys. J.* **1998**, *492*, 29. [CrossRef]
74. Weinberg, S. The Cosmological Constant Problems. In *Sources and Detection of Dark Matter and Dark Energy in the Universe*; Cline, D.B., Ed.; Springer: Berlin, Germany, 2001; p. 18.
75. Lake, M.J. Why space could be quantised on a different scale to matter. *SciPost Phys. Proc.* **2021**, *4*, 014. [CrossRef]

76. Riess, A.G.; Casertano, S.; Yuan, W.; Macri, L.M.; Scolnic, D. Large Magellanic Cloud Cepheid Standards Provide Foundation for the Determination of the Hubble Constant and Stronger Evidence for Physics Beyond LCDM. *Astrophys. J.* **2019**, *876*, 85. [CrossRef]
77. Huang, C.D.; Riess, A.G.; Yuan, W.; Macri, L.M.; Zakamska, N.L.; Casertano, S.; Whitelock, P.A.; Hoffmann, S.L.; Filippenko, A.V.; Scolnic, D. Hubble Space Telescope Observations of Mira Variables in the SN Ia Host NGC 1559: An Alternative Candle to Measure the Hubble Constant. *Astrophys. J.* **2020**, *889*, 5. [CrossRef]
78. Pesce, D.W.; Braatz, J.A.; Reid, M.J.; Riess, A.G.; Scolnic, D.; Condon, J.J.; Gao, F.; Henkel, C.; Impellizzeri, C.M.V.; Kuo, C.Y.; et al. The Megamaser Cosmology Project. XIII. Combined Hubble Constant Constraints. *Astrophys. J. Lett.* **2020**, *891*, L1. [CrossRef]
79. Harko, T.; Lobo, F.S.N. Beyond Einstein's General Relativity: Hybrid Metric-Palatini Gravity and Curvature-Matter Couplings. *Int. J. Mod. Phys. D* **2020**, *29*, 2030008. [CrossRef]
80. Haghani, Z.; Harko, T.; Sepangi, H.R.; Shahidi, S. Weyl-Cartan-Weitzenböck Gravity as a Generalization of Teleparallel Gravity. *J. Cosmol. Astropart. Phys.* **2012**, *10*, 061. [CrossRef]
81. Haghani, Z.; Harko, T.; Sepangi, H.R.; Shahidi, S. Weyl-Cartan-Weitzenböck Gravity through Lagrange Multiplier. *Phys. Rev. D* **2013**, *88*, 044024. [CrossRef]
82. Nester, J.M.; Yo, H.-J. Symmetric Teleparallel General Relativity. *Chin. J. Phys.* **1999**, *37*, 113.
83. Beltrán Jiménez, J.; Heisenberg, L.; Koivisto, T. Coincident General Relativity. *Phys. Rev. D* **2018**, *98*, 044048. [CrossRef]
84. D'Agostino, R.; Luongo, O. Growth of Matter Perturbations in Nonminimal Teleparallel Dark Energy. *Phys. Rev. D* **2018**, *98*, 124013. [CrossRef]
85. Fontanini, M.; Huguet, E.; Le Delliou, M. Teleparallel gravity equivalent of general relativity as a gauge theory: Translation or Cartan connection? *Phys. Rev. D* **2019**, *99*, 064006. [CrossRef]
86. Koivisto, T.; Tsimperis, G. The Spectrum of Teleparallel Gravity. *Universe* **2019**, *5*, 80. [CrossRef]
87. Pereira, J.G.; Obukhov, Y.N. Gauge Structure of Teleparallel Gravity. *Universe* **2019**, *5*, 139. [CrossRef]
88. Blixt, D.; Hohmann, M.; Pfeifer, C. On the Gauge Fixing in the Hamiltonian Analysis of General Teleparallel Theories. *Universe* **2019**, *5*, 143. [CrossRef]
89. Coley, A.A.; van den Hoogen, R.J.; McNutt, D.D. Symmetry and Equivalence in Teleparallel Gravity. *J. Math. Phys.* **2020**, *61*, 072503. [CrossRef]
90. Haghani, Z.; Khosravi, N.; Shahidi, S. The Weyl–Cartan Gauss–Bonnet Gravity. *Class. Quantum Gravity* **2015**, *32*, 215016. [CrossRef]
91. Sotiriou, T.P.; Faraoni, V. f(R) Theories of Gravity. *Rev. Mod. Phys.* **2010**, *82*, 451. [CrossRef]
92. De Felice, A.; Tsujikawa, S. f(R) Theories. *Living Rev. Relativ.* **2010**, *13*, 3. [CrossRef]
93. Cai, Y.F.; Capozziello, S.; De Laurentis, M.; Saridakis, E.N. f(T) Teleparallel Gravity and Cosmology. *Rep. Prog. Phys.* **2016**, *79*, 106901. [CrossRef]
94. Nojiri, S.; Odintsov, S.D.; Oikonomou, V.K. Modified Gravity Theories on a Nutshell: Inflation, Bounce and Late-Time Evolution. *Phys. Rep.* **2017**, *692*, 1. [CrossRef]
95. Odintsov, S.D.; Oikonomou, V.K.; Giannakoudi, I.; Fronimos, F.P.; Lymperiadou, E.C. Recent Advances in Inflation. *Symmetry* **2023**, *15*, 1701. [CrossRef]
96. Basilakos, S.; Stavrinos, P. Cosmological Equivalence between the Finsler–Randers spacetime and the DGP Gravity Model. *Phys. Rev. D* **2013**, *87*, 043506. [CrossRef]
97. Exirifard, Q. Randers geometry as MOND/dark matter. *J. Cosmol. Astropart. Phys.* **2015**, *11*, 026. [CrossRef]
98. Silva, J.E.G.; Maluf, R.V.; Almeida, C.A.S. A nonlinear dynamics for the scalar field in Randers spacetime. *Phys. Lett. B* **2017**, *766*, 263. [CrossRef]
99. Papagiannopoulos, G.; Basilakos, S.; Paliathanasis, A.; Savvidou, S.; Stavrinos, P.C. Finsler–Randers cosmology: Dynamical analysis and growth of matter perturbations. *Class. Quantum Gravity* **2017**, *34*, 225008. [CrossRef]
100. Papagiannopoulos, G.; Basilakos, S.; Paliathanasis, A.; Pan, S.; Stavrinos, P. Dynamics in varying vacuum Finsler–Randers cosmology. *Eur. Phys. J. C* **2020**, *80*, 816. [CrossRef]
101. Raushan, R.; Chaubey, S. Finsler–Randers cosmology in the framework of a particle creation mechanism: A dynamical systems perspective. *Eur. Phys. J. Plus* **2020**, *135*, 228. [CrossRef]
102. Triantafyllopoulos, A.; Basilakos, S.; Kapsabelis, E.; Stavrinos, P.C. Schwarzschild-like solutions in Finsler-Randers gravity. *Eur. Phys. J. C* **2020**, *80*, 1200. [CrossRef]
103. Kapsabelis, E.; Triantafyllopoulos, A.; Basilakos, S.; Stavrinos, P.C. Applications of the Schwarzschild-Finsler-Randers model. *Eur. Phys. J. C* **2021**, *81*, 990. [CrossRef]
104. Kapsabelis, E.; Kevrekidis, P.G.; Stavrinos, P.C.; Triantafyllopoulos, A. Schwarzschild–Finsler–Randers spacetime: Geodesics, dynamical analysis and deflection angle. *Eur. Phys. J. C* **2022**, *82*, 1098. [CrossRef]
105. Nekouee, Z.; Narasimhamurthy, S.K.; Manjunatha, H.M.; Srivastava, S.K. Finsler–Randers model for anisotropic constant-roll inflation. *Eur. Phys. J. Plus* **2022**, *137*, 1388. [CrossRef]
106. Feng, W.; Yang, W.; Jiang, B.; Wang, Y.; Han, T.; Wu, Y. Theoretical analysis on the Barrow holographic dark energy in the Finsler–Randers cosmology. *Int. J. Mod. Phys. D* **2023**, *32*, 2350029. [CrossRef]
107. Das, P.D.; Debnath, U. Possible existence of traversable wormhole in Finsler–Randers geometry. *Eur. Phys. J. C* **2023**, *83*, 821. [CrossRef]

108. Triantafyllopoulos, A.; Kapsabelis, E.; Stavrinos, P.C. Raychaudhuri Equations, Tidal Forces, and the Weak-Field Limit in Schwarzshild–Finsler–Randers Spacetime. *Universe* **2024**, *10*, 26. [CrossRef]
109. Kapsabelis, E.; Saridakis, E.N.; Stavrinos, P.C. Finsler–Randers–Sasaki gravity and cosmology. *Eur. Phys. J. C* **2024**, *84*, 538. [CrossRef]
110. Praveen, J.; Narasimhamurthy, S.K.; Yashwanth, B.R. Exploring compact stellar structures in Finsler–Randers geometry with the Barthel connection. *Eur. Phys. J. C* **2024**, *84*, 597. [CrossRef]
111. r Liu, J.; Wang, R.; Gao, F. Nonlinear Dynamics in Variable-Vacuum Finsler–Randers Cosmology with Triple Interacting Fluids. *Universe* **2024**, *10*, 302. [CrossRef]
112. Nekouee, Z.; Narasimhamurthy, S.K.; Pourhassan, B.; Pacif, S.K.J. A phenomenological approach to the dark energy models in the Finsler–Randers framework. *Ann. Phys.* **2024**, *470*, 169787. [CrossRef]
113. Nekouee, Z.; Chaudhary, H.; Narasimhamurthy, S.K.; Pacif, S.K.J.; Malligawad, M. Cosmological tests of the dark energy models in Finsler-Randers Space-time. *J. High Energy Astrophys.* **2024**, *44*, 19. [CrossRef]
114. Praveen, J.; Narasimhamurthy, S.K. Matter bounce cosmology within Finsler-Randers geometry: A comprehensive study of anisotropic influences. *J. High Energy Astrophys.* **2024**, *44*, 300. [CrossRef]
115. Yashwanth, B.R.; Narasimhamurthy, S.K.; Praveen, J.; Malligawad, M. The influence of density models on wormhole formation in Finsler–Barthel–Randers geometry. *Eur. Phys. J. C* **2024**, *84*, 1272. [CrossRef]
116. Praveen, J.; Narasimhamurthy, S.K. The role of Finsler-Randers geometry in shaping anisotropic metrics and thermodynamic properties in black holes theory. *New Astron.* **2025**, *119*, 102404. [CrossRef]
117. Ikeda, S.; Saridakis, E.N.; Stavrinos, P.C.; Triantafyllopoulos, A. Cosmology of Lorentz Fiber-Bundle Induced Scalar-Tensor Theories. *Phys. Rev. D* **2019**, *100*, 124035. [CrossRef]
118. Fuster, A.; Pabst, C.; Pfeifer, C. Berwald Spacetimes and Very Special Relativity. *Phys. Rev. D* **2018**, *98*, 084062. [CrossRef]
119. Hohmann, M.; Pfeifer, C.; Voicu, N. Cosmological Finsler Spacetimes. *Universe* **2020**, *6*, 65. [CrossRef]
120. Hama, R.; Harko, T.; Sabau, S.V.; Shahidi, S. Cosmological Evolution and Dark Energy in Osculating Barthel–Randers Geometry. *Eur. Phys. J. C* **2021**, *81*, 742. [CrossRef]
121. Hama, R.; Rattanasak, R.; Harko, T.; Sabau, S.V. Dark Energy and Accelerating Cosmological Evolution from Osculating Barthel–Kropina Geometry. *Eur. Phys. J. C* **2022**, *82*, 385. [CrossRef]
122. Bouali, A.; Chaudhary, H.; Hama, R.; Harko, T.; Sabau, S.V.; San Martín, M. Cosmological Tests of the Osculating Barthel–Kropina Dark Energy Model. *Eur. Phys. J. C* **2023**, *83*, 121. [CrossRef]
123. Hama, R.; Harko, T.; Sabau, S.V. Conformal Gravitational Theories in Barthel–Kropina-Type Finslerian Geometry and Their Cosmological Implications. *Eur. Phys. J. C* **2023**, *83*, 1030. [CrossRef]
124. Tavakol, R.K.; Van den Bergh, N. Finsler Spaces and the Underlying Geometry of spacetime. *Phys. Lett. A* **1985**, *112*, 23. [CrossRef]
125. Tavakol, R.K. Geometry of Spacetime and Finsler Geometry. *Int. J. Mod. Phys. A* **2009**, *24*, 1678. [CrossRef]
126. Miron, R.; Hrimiuc, D.; Shimada, H.; Sabau, S.V. *The Geometry of Hamilton and Lagrange Spaces*; Kluwer Academic Publishers: Dordrecht, The Netherlands, 2001.
127. Javaloyes, M.A.; Sánchez, M. On the Definition and Examples of Finsler Metrics. *Ann. Sc. Norm. Super. Pisa Cl. Sci.* **2014**, *13*, 813. [CrossRef] [PubMed]
128. Yoshikawa, R.; Sabau, S.V. Kropina Metrics and Zermelo Navigation on Riemannian Manifolds. *Geom. Dedicata* **2014**, *171*, 119. [CrossRef]
129. Kropina, V.K. On Projective Finsler Spaces with a Metric of Special Form. *Nauchnye Dokl. Vyss. Shkoly Fiz.-Mat. Nauk.* **1959**, *2*, 38.
130. Kropina, V.K. On projective two-dimensional Finsler spaces with a special metric. *Trudy Sem. Vektor. Tenzor. Anal* **1961**, *11*, 277–292.
131. Matsumoto, M. On C-Reducible Finsler Spaces. *Tensor, New Ser.* **1972**, *24*, 29–37.
132. Matsumoto, M. Theory of Finsler Spaces with (α, β)-Metric. *Rep. Math. Phys.* **1992**, *31*, 43–83. [CrossRef]
133. Bácsó, S.; Cheng, X.; Shen, Z. Curvature Properties of (α, β)-Metrics. *Adv. Stud. Pure Math.* **2007**, *48*, 73–110.
134. Antonelli, P.L.; Ingarden, R.S.; Matsumoto, M. *The Theory of Sprays and Finsler Spaces with Applications in Physics and Biology*; Kluwer Academic Publishers: Dordrecht, The Netherlands, 1993.
135. Yoshikawa, R.; Sabau, S.V. Some remarks on the geometry of Kropina spaces. *Publ. Math. Debrecen* **2014**, *84*, 483–496. [CrossRef]
136. Sabau, S.V.; Shibuya, K.; Yoshikawa, R. Geodesics on Strong Kropina Manifolds. *Eur. J. Math.* **2017**, *3*, 1172–1224. [CrossRef]
137. Heefer, S.; Pfeifer, C.; van Voorthuizen, J.; Fuster, A. On the Metrizability of m-Kropina Spaces with Closed Null One-Form. *J. Math. Phys.* **2023**, *64*, 022502. [CrossRef]
138. Heefer, S. Berwald m-Kropina Spaces of Arbitrary Signature: Metrizability and Ricci-Flatness. *J. Math. Phys.* **2024**, *65*, 122502. [CrossRef]
139. Barthel, W. Zum Inhaltsbegriff in der Minkowskischen Geometrie. *Math. Z.* **1953**, *58*, 358. [CrossRef]
140. Barthel, W. Über eine Parallelverschiebung mit Längeninvarianz in lokal-Minkowskischen Räumen I, II. *Arch. Der Math.* **1953**, *4*, 346. [CrossRef]
141. Ingarden, R.S.; Tamássy, L. The Point Finsler Spaces and Their Physical Applications in Electron Optics and Thermodynamics. *Math. Comput. Model.* **1994**, *20*, 93. [CrossRef]
142. Nazim, A. *Über Finslersche Räume*; Dissertatio: München, Germany, 1936.
143. Varga, O. Zur Herleitung des Invarianten Differentials in Finslerschen Räumen. *Monatshefte Math. Phys.* **1941**, *50*, 165. [CrossRef]

144. Harko, T. Thermodynamic Interpretation of the Generalized Gravity Models with Geometry–Matter Coupling. *Phys. Rev. D* **2014**, *90*, 044067. [CrossRef]
145. Bertolami, O.; Böhmer, C.G.; Harko, T.; Lobo, F.S.N. Extra Force in $f(R)$ Modified Theories of Gravity. *Phys. Rev. D* **2007**, *75*, 104016. [CrossRef]
146. Harko, T.; Lobo, F.S.N.; Otalora, G.; Saridakis, E.N. Nonminimal Torsion–Matter Coupling Extension of $f(T)$ Gravity. *Phys. Rev. D* **2014**, *89*, 124036. [CrossRef]
147. Parker, L. Particle Creation in Expanding Universes. *Phys. Rev. Lett.* **1968**, *21*, 562. [CrossRef]
148. Parker, L. Quantized Fields and Particle Creation in Expanding Universes. I. *Phys. Rev.* **1969**, *183*, 1057. [CrossRef]
149. Zeldovich, Y.B.; Starobinsky, A.A. Particle Production and Vacuum Polarization in an Anisotropic Gravitational Field. *Zh. Eksp. Teor. Fiz.* **1971**, *61*, 2161, *Sov. Phys. JETP* **1972**, *34*, 1159.
150. Parker, L. Particle Creation in Isotropic Cosmologies. *Phys. Rev. Lett.* **1972**, *28*, 705, Erratum: *Phys. Rev. Lett.* **1972**, *28*, 1497. [CrossRef]
151. Fulling, S.A.; Parker, L.; Hu, B.L. Conformal Energy-Momentum Tensor in Curved Spacetime: Adiabatic Regularization and Renormalization. *Phys. Rev. D* **1974**, *10*, 3905. [CrossRef]
152. Parker, L. Particle Creation and Particle Number in an Expanding Universe. *J. Phys. A Math. Theor.* **2012**, *45*, 374023. [CrossRef]
153. Prigogine, I.; Geheniau, J.; Gunzig, E.; Nardone, P. Thermodynamics of Cosmological Matter Creation. *Proc. Natl. Acad. Sci. USA* **1988**, *85*, 7428. [CrossRef]
154. Calvão, M.O.; Lima, J.A.S.; Waga, I. On the Thermodynamics of Matter Creation in Cosmology. *Phys. Lett. A* **1992**, *162*, 223. [CrossRef]
155. Su, J.; Harko, T.; Liang, S.-D. Irreversible Thermodynamic Description of Dark Matter and Radiation Creation During Inflationary Reheating. *Adv. High Energy Phys.* **2017**, *2017*, 7650238. [CrossRef]
156. Lima, J.A.S.; Baranov, I.P. Gravitationally Induced Particle Production: Thermodynamics and Kinetic Theory. *Phys. Rev. D* **2014**, *90*, 043515. [CrossRef]
157. Saridakis, E.N.; González-Díaz, P.F.; Sigüenza, C.L. Unified Dark Energy Thermodynamics: Varying w and the -1-Crossing. *Class. Quantum Gravity* **2009**, *26*, 165003. [CrossRef]
158. Bernardo, J.M. Reference Prior Distributions for Bayesian Inference. *J. R. Stat. Soc. B* **1979**, *41*, 113–127. [CrossRef]
159. Joyce, J. Bayes' Theorem. In *The Stanford Encyclopedia of Philosophy*, Fall 2021 ed.; Zalta, E.N., Ed.; Stanford University: Stanford, CA, USA, 2021.
160. Foreman-Mackey, D.; Hogg, D.W.; Lang, D.; Goodman, J. emcee: The MCMC Hammer. *Publ. Astron. Soc. Pac.* **2013**, *125*, 306. [CrossRef]
161. Foreman-Mackey, D.; Hogg, D. W.; Lang, D.; Goodman, J. emcee v3: A Python Ensemble Sampling Toolkit for Affine-Invariant MCMC. *arXiv* **2019**, arXiv:1911.07688. [CrossRef]
162. Lewis, A. GetDist: A Python Package for Analysing Monte Carlo Samples. *arXiv* **2019**, arXiv:1910.13970.
163. Jimenez, R.; Loeb, A. Constraining Cosmological Parameters Based on Relative Galaxy Ages. *Astrophys. J.* **2002**, *573*, 37–51. [CrossRef]
164. Moresco, M.; Verde, L.; Pozzetti, L.; Jimenez, R.; Cimatti, A. Cosmological Constraints from Cosmic Chronometers: A New Approach. *J. Cosmol. Astropart. Phys.* **2012**, *2012*, 053. [CrossRef]
165. Moresco, M. Raising the Bar: New Constraints on the Hubble Parameter with Cosmic Chronometers at Z $\sim$2. *Mon. Not. R. Astron. Soc.* **2015**, *450*, L16–L20. [CrossRef]
166. Moresco, M.; Pozzetti, L.; Cimatti, A.; Jimenez, R.; Maraston, C.; Verde, L.; Thomas, D.; Citro, A.; Tojeiro, R.; Wilkinson, D. A 6% Measurement of the Hubble Parameter at Z $\sim$0.45: Direct Evid. Epoch Cosm. Re-Acceleration. *J. Cosmol. Astropart. Phys.* **2016**, *2016*, 014. [CrossRef]
167. Moresco, M.; Jimenez, R.; Verde, L.; Pozzetti, L.; Cimatti, A.; Citro, A. Cosmic Chronometers at Z $\sim$2: New Constraints Hubble Parameter. *Astrophys. J.* **2018**, *868*, 84. [CrossRef]
168. Moresco, M.; Jimenez, R.; Verde, L.; Cimatti, A.; Pozzetti, L. Setting the Stage for Cosmic Chronometers. II. Impact of Stellar Population Synthesis Models Systematics and Full Covariance Matrix. *Astrophys. J.* **2020**, *898*, 82. [CrossRef]
169. Brout, D.; Scolnic, D.; Popovic, B.; Riess, A.G.; Carr, A.; Zuntz, J.; Kessler, R.; Davis, T.M.; Hinton, S.; Jones, D.; et al. The Pantheon^{+} analysis: Cosmological constraints. *Astrophys. J.* **2022**, *938*, 110. [CrossRef]
170. Astier, P.; Guy, J.; Regnault, N.; Pain, R.; Aubourg, E.; Balam, D.; Basa, S.; Carlberg, R.; Fabbro, S.; Fouchez, D.; et al. The supernova legacy survey: Measurement of ω, ω and from the first year data set. *Astron. Astrophys.* **2006**, *447*, 31–48. [CrossRef]
171. Conley, A.; Guy, J.; Sullivan, M.; Regnault, N.; Astier, P.; Balland, C.; Basa, S.; Carlberg, R.; Fouchez, D.; Hardin, D.; et al. Supernova constraints and systematic uncertainties from the first three years of the supernova legacy survey. *Astrophys. J. Suppl. Ser.* **2010**, *192*, 1. [CrossRef]
172. Karim, M.A.; Aguilar, J.; Ahlen, S.; Alam, S.; Allen, L.; Prieto, C.A.; Alves, O.; Anand, A.; Andrade, U.; Armengaud, E.; et al. DESI DR2 Results II: Measurements of Baryon Acoustic Oscillations and Cosmological Constraints. *arXiv* **2025**, arXiv:2503.14738.
173. Pogosian, L.; Zhao, G.-B.; Jedamzik, K. Recombination-independent determination of the sound horizon and the Hubble constant from BAO. *Astrophys. J. Lett.* **2020**, *904*, L17. [CrossRef]
174. Jedamzik, K.; Pogosian, L.; Zhao, G.-B. Why reducing the cosmic sound horizon alone can not fully resolve the Hubble tension. *Commun. Phys.* **2021**, *4*, 123. [CrossRef]

175. Pogosian, L.; Zhao, G.-B.; Jedamzik, K. A consistency test of the cosmological model at the epoch of recombination using DESI BAO and Planck measurements. *Astrophys. J. Lett.* **2024**, *973*, L13. [CrossRef]
176. Lin, W.; Chen, X.; Mack, K.J. Early universe physics insensitive and uncalibrated cosmic standards: Constraints on Ω_m and implications for the Hubble tension. *Astrophys. J.* **2021**, *920*, 159. [CrossRef]
177. Vagnozzi, S. Seven hints that early-time new physics alone is not sufficient to solve the Hubble tension. *Universe* **2023**, *9*, 393. [CrossRef]
178. Chaudhary, H.; Csillag, L.; Harko, T. Semi-Symmetric Metric Gravity: A Brief Overview. *Universe* **2024**, *10*, 419. [CrossRef]
179. Cattoën, C.; Visser, M. Cosmographic Hubble fits to the supernova data. *Phys. Rev. D* **2008**, *78*, 063501. [CrossRef]
180. Visser, M.; Cattoën, C. Cosmographic analysis of dark energy. In *Dark Matter in Astrophysics and Particle Physics*; Klapdor-Kleingrothaus, H.V., Krivosheina, I.V., Eds.; World Scientific Publishing Co. Pte. Ltd.: Singapore, 2009; pp. 287–300.
181. Visser, M. Cosmography: Cosmology without the Einstein equations. *Gen. Relativ. Gravit.* **2005**, *37*, 1541. [CrossRef]
182. Luongo, O. Cosmography with the Hubble parameter. *Mod. Phys. Lett. A* **2011**, *26*, 1459. [CrossRef]
183. Visser, M. Jerk, snap and the cosmological equation of state. *Class. Quantum Gravity* **2004**, *21*, 2603. [CrossRef]
184. Andrae, R.; Schulze-Hartung, T.; Melchior, P. Dos and don'ts of reduced chi-squared. *arXiv* **2010**, arXiv:1012.3754.
185. Liddle, A.R. Information criteria for astrophysical model selection. *Mon. Not. R. Astron. Soc. Lett.* **2007**, *377*, L74. [CrossRef]
186. Vrieze, S.I. Model selection and psychological theory: A discussion of the differences between the Akaike information criterion (AIC) and the Bayesian information criterion (BIC). *Psychol. Methods* **2012**, *17*, 228. [CrossRef]
187. Tan, M.Y.J.; Biswas, R. The reliability of the Akaike information criterion method in cosmological model selection. *Mon. Not. R. Astron. Soc.* **2012**, *419*, 3292. [CrossRef]
188. Arevalo, F.; Cid, A.; Moya, J. AIC and BIC for cosmological interacting scenarios. *Eur. Phys. J. C* **2017**, *77*, 1. [CrossRef]
189. Burnham, K.P.; Anderson, D.R. *Model Selection and Multimodel Inference*, 2nd ed.; Springer: New York, NY, USA, 2010.
190. Jeffreys, H. *The Theory of Probability*; Oxford University Press: Oxford, UK, 1998.
191. Andrade, C. The P value and statistical significance: Misunderstandings, explanations, challenges, and alternatives. *Indian J. Psychol. Med.* **2019**, *41*, 210–215. [CrossRef]
192. Vargas, J.G.; Torr, D.G. The construction of teleparallel Finsler connections and the emergence of an alternative concept of metric compatibility. *Found. Phys.* **1997**, *27*, 825–843.
193. Tong, Y. General teleparallel gravity from Finsler geometry. *Commun. Theor. Phys.* **2023**, *75*, 095403. [CrossRef]
194. Amelino-Camelia, G. Quantum-Spacetime Phenomenology. *Living Rev. Relativ.* **2013**, *16*, 5. [CrossRef]
195. Caroff, S.; Pfeifer, C.; Bolmont, J.; Terzić, T.; Campoy-Ordaz, A.; Kerszberg, D.; Martinez, M.; Pensec, U.; Plaid, C.; Strišković, J.; et al. Discriminating between different modified dispersion relations from gamma-ray observations. *Phys. Rev. D* **2025**, *111*, 083021. [CrossRef]
196. Amelino-Camelia, G. Doubly special relativity. *Nature* **2002**, *418*, 34–35. [CrossRef]
197. Letizia, M.; Liberati, S. Deformed relativity symmetries and the local structure of spacetime. *Phys. Rev. D* **2017**, *95*, 046007. [CrossRef]
198. Majid, S.; Ruegg, H. Bicrossproduct structure of κ-Poincaré group and non-commutative geometry. *Phys. Lett. B* **1994**, *334*, 348–354. [CrossRef]
199. Amelino-Camelia, G.; Barcaroli, L.; Gubitosi, G.; Liberati, S.; Loret, N. Realization of doubly special relativistic symmetries in Finsler geometries. *Phys. Rev. D* **2014**, *90*, 125030. [CrossRef]

universe

MDPI

Review

Cosmological Models in Lovelock Gravity: An Overview of Recent Progress

Sergey Pavluchenko

Department of Bioengineering, University of Illinois Urbana-Champaign, Urbana, IL 61801, USA; sergey.pavluchenko@gmail.com

Abstract: In the current review, we provide a summary of the recent progress made in the cosmological aspect of extra-dimensional Lovelock gravity. Our review covers a wide variety of particular model/matter source combinations: Einstein–Gauss–Bonnet as well as cubic Lovelock gravities with vacuum, cosmological constant, perfect fluid, spatial curvature, and some of their combinations. Our analysis suggests that it is possible to set constraints on the parameters of the above-mentioned models from the simple requirement of the existence of a smooth transition from the initial singularity to a realistic low-energy regime. Initially, anisotropic space naturally evolves into a configuration with two isotropic subspaces, and if one of these subspaces is three-dimensional and is expanding while another is contracting, we call it realistic compactification. Of course, the process is not devoid of obstacles, and in our paper, we review the results of the compactification occurrence investigation for the above-mentioned models. In particular, for vacuum and Λ-term EGB models, compactification is not suppressed (but is not the only possible outcome either) if the number of extra dimensions is $D \geqslant 2$; for vacuum cubic Lovelock gravities it is always present (however, cubic Lovelock gravity is defined only for $D \geqslant 3$ number of extra dimensions); for the EGB model with perfect fluid it is present for $D = 2$ (we have not considered this model in higher dimensions yet), and in the presence of spatial curvature, the realistic stabilization of extra dimensions is always present (however, such a model is well-defined only in $D \geqslant 4$ number of extra dimensions).

Keywords: modified gravity; Einstein–Gauss–Bonnet gravity; Lovelock gravity; extra-dimensional theory; cosmology

PACS: 04.50.-h; 11.25.Mj; 98.80.Cq

1. Introduction

By now, mathematical modeling has become an integral part of any quantitative or qualitative study in natural or life sciences and beyond. The very first models in the history of humankind were what we could call "cosmological models", as they described the whole world surrounding the ancient scientists—the planets revolving around the Earth. The development of differential calculus by Newton and Leibniz became an important milestone, which boosted both physics and mathematics and, through this, the modeling. However, it took three more centuries and the development of even more powerful tensor calculus for cosmological modeling to become a full-fledged discipline. Nowadays, modeling is an essential part of cosmology and we rely on models for all of its aspects.

The typical workflow in cosmology is the same as in other natural sciences—we formulate a hypothesis, build a model based on this hypothesis, calculate predictions, and confront them with experimental data to decide if the model is viable, and if so, to find a realistic range of parameters. However, there are exceptions from this workflow and one of them is the Very Early Universe. Indeed, there is not much experimental and/or observational evidence from the early stages of the Universe's evolution and most of them are indirect; the earlier the stage, the less evidence we have. In this case, we can rely only on models and use only the most general assumptions of what to expect.

A good example of such a situation is the topic of the current review—Lovelock gravity in the Early Universe. Lovelock gravity is the generalization of General Relativity (GR) in the following way: we know [1–3] that the Einstein tensor is, in any dimension, the only symmetric and conserved tensor depending only on the metric and its first and second derivatives (with a linear dependence on second derivatives). If we drop the requirement of linear dependence of the second derivatives and use a similar procedure as in GR, we naturally obtain the Lovelock tensor [4]. Lovelock gravity manifests itself as a curvature correction to GR: the zeroth (constant) term is just the boundary term associated with the cosmological constant, the first (linear) term is GR, the second (quadratic) term is known as the Gauss–Bonnet (GB) term, and so on. Due to the Lovelock tensor construction procedure, nontrivial—higher-than-GR—contributions from Lovelock gravity manifest themselves only in a higher-than-three number of spatial dimensions, making it extra-dimensional theory. For the Gauss–Bonnet contribution to become nontrivial, we need at least four spatial dimensions, and for the cubic Lovelock term, at least six, and so on. As we just mentioned, Lovelock gravity manifests itself as a curvature correction, so this theory is distinguishable from GR only in the high-curvature regime, which is naturally achieved during the early stages of the Universe's evolution. That is why it is mostly an Early Universe theory.

A nonlinear theory, Lovelock gravity shares the same curse with other nonlinear theories—unwanted singularities, which we call "nonstandard singularities", in contrast with standard cosmological singularities. These nonstandard singularities are referred to as a situation when some of the dynamical variables diverge while others are regular at some finite time; they are considered "weak" by Tipler's classification [5], and "type II" in tje classifications by Kitaura and Wheeler [6,7]. Studies demonstrate that they naturally appear during evolution and, being physical singularities (they cause curvature invariants to diverge), cease the Universe's evolution.

On the other hand, Lovelock gravity is formulated in a greater-than-three number of spatial dimensions, while we clearly live in a three-dimensional world.[1] So, the natural question here is how to connect these two within the same theory. The widely accepted answer is that extra dimensions are compact—their scale is much smaller than that of three dimensions, so we do not "sense" them. This explanation, however, raises another question—how did this compactification occur? In the most elegant way, this compactification should happen naturally during the Universe's evolution. In other words, compactification should be a natural attractor of the system.

So, we have Lovelock gravity, which is formulated in a higher number of spatial dimensions, and we want it to describe our Universe at late times. However, there are next to no observational data from the very early times of the Universe's evolution, so, heeding in mind the previously mentioned potential problems, we want the following natural assumptions to be held: (1) the late-time regime should be low-curvature (GR); (2) the late-time regime should have three expanding dimensions while the remaining should be compact (contracting or constant with the scale much smaller than "our" three dimensions); (3) the late-time regime should be achieved smoothly (i.e., without any nonstandard singularities). These three assumptions sound reasonable but we will see that they are sufficient to set constraints on the theory under consideration.

As we already mentioned, the models under consideration have two features: (a) they are multidimensional and (b) we consider them within Lovelock gravity. Each of these features has mechanisms that may prevent the model from reaching the realistic regime. Since the model is multidimensional, it may compactify into wrong spatial splitting—it could be that there is no three-dimensional subspace after compactification and the resulting regime is unrealistic—findings on this feature are summarized in Section 9. Another feature is that we consider Lovelock gravity in action, and it brings additional possibilities for interfering with the evolution—this includes unrealistic late-time asymptotes (such as power-law regimes) or even finite-time nonstandard singularities; the rest of the sections are dedicated to the study of these possibilities. And since we are living in three-dimensional space, which experiences accelerated expansion, that is what we are aiming for as a late-time asymptote.

Let us emphasize that we focus on cosmological behavior in the very early Universe—indeed, Lovelock gravity contains terms that are higher-order in curvature and these terms manifest themselves only when the curvature itself is high, and for cosmological context (especially since we are working mostly with spatially flat models) it is usually near the initial singularity (or recollapse, but this is an undesired outcome). This way, we ensure that if realistic compactification is reached, the future asymptote is GR and its behavior is considered the same as in standard GR, as will its manifestations—direct and indirect observables, qualitative behavior, and so on (BBN predictions, cosmic expansion history, structure formation, etc.). So, to put it simply, in the current manuscript, we are summarizing findings on constraints on the parameters of the theory, which allows realistic compactification for Lovelock gravity.

The manuscript is organized as follows: first, we shall provide a brief historical outline of the research in cosmological aspects of Lovelock gravity, and after that we will set up a model. With the model being set up, we will consider different Lovelock corrections and matter sources—vacuum and cosmological term in Einstein–Gauss–Bonnet (EGB) gravity (Sections 4 and 5, respectively), vacuum in cubic Lovelock gravity (Section 6), perfect fluid in EGB gravity (Section 7), spatial curvature in EGB gravity (Section 8), and finally, general anisotropic case in EGB gravity (Section 9). After that, we discuss the results and draw some conclusions.

2. Historical Outline

Surprisingly, the very idea of extra dimensions precedes GR—the first extra-dimensional theory was formulated by Nordström in 1914 [8] and it represents the unification of Nordström's second gravity theory [9] with Maxwell's electromagnetism. However, soon after that, Albert Einstein introduced GR [10], which competed with Nordström's scalar gravity for being the correct theory. This competence was resolved in 1919 in favor of GR: Nordström's scalar gravity, similarly to most other scalar gravities, predicts no light bending near massive bodies, while observations made during the solar eclipse in 1919 clearly demonstrated that the deflection angle is in agreement with GR predictions.

With that, Nordström's scalar gravity was abandoned but his brilliant idea survived. Kaluza proposed [11] a similar model but based on GR: five-dimensional Einstein equations could be decomposed into four-dimensional Einstein equations and Maxwell's electromagnetism. His model has one extra dimension and it should be "curled" (or compactified) into a circle of a very small size and "cylindrical conditions" should be imposed in order to perform dimensional decomposition. That was followed by Klein who proposed [12,13] an interesting quantum mechanical interpretation of this extra dimension and the resulting theory was named Kaluza–Klein, after its founders. It is worth mentioning that the Kaluza–Klein theory united all interactions known at that time. As time passes, new interactions were discovered and it became clear that, to unite all of them, one needs a larger number of extra dimensions. As of now, a few models unite all interactions and M/string theory is one of the most promising of them.

In recent times, theories similar to (or, rather, generalizing) the Kaluza–Klein theory[2] are used to address a number of problems in modern physics, with the hierarchy problem being one of the most important of them. In particular, Arkani-Hamed, Dimopoulos, and Dvali [15,16] proposed a model that involves large extra dimensions with the effective Planck mass at about 1 TeV, while Randall and Sundrum considered a couple of models with warped extra dimensions [17,18], addressing the same hierarchy problem. Observations do not rule out such models but severely constrain them (see, e.g., [19] for a constraint coming from neutron stars observations by Fermi-LAT telescope), but it is worth mentioning that the models with really large extra dimensions (at least about the length of gravitational waves ($\sim$100 km)) are literally undetectable (see, e.g., [20] for constraints coming from the analysis of the gravitational waves damping). Extra-dimensional theories, as well as the manifestations of the background theories (M/string), could be detected in particle experiments, but so far, results are discouraging [21,22].

From the gravitational standpoint, one of the distinguishing features of M/string theories is the presence of the curvature-squared corrections in the Lagrangian. Scherk and Schwarz [23] noted the presence of the R^2 and $R_{\mu\nu}R^{\mu\nu}$ terms in the Lagrangian of the Virasoro–Shapiro model [24,25]; Candelas et al. [26] demonstrated the presence of the curvature-squared term of the $R^{\mu\nu\lambda\rho}R_{\mu\nu\lambda\rho}$ type in the low-energy limit of the $E_8 \times E_8$ heterotic superstring theory [27] to compensate the kinetic term of the Yang–Mills field. Later on, Zwiebach showed [28] that in order to keep the theory ghost-free, one has to use a specific combination of quadratic terms known as the Gauss–Bonnet (GB) term, as follows:

$$L_{GB} = R_{\mu\nu\lambda\rho}R^{\mu\nu\lambda\rho} - 4R_{\mu\nu}R^{\mu\nu} + R^2.$$

This term, originally discovered by Lanczos [29,30] (and thus sometimes called the Lanczos term) is an Euler topological invariant in (3 + 1)-dimensional spacetime, but starting from (4 + 1), and in higher dimensions, it gives a nontrivial contribution to the equations of motion. Zumino [31] extended Zwiebach's analysis with the higher-than-squared curvature terms taken into account; his results suggested that the low-energy limit of the unified theory might have a Lagrangian density represented by a sum of different powers of curvature—then, the earlier-mentioned Lovelock gravity would be a natural candidate for such a theory.

There are two approaches to the compactification problem, with the first of them being "spontaneous compactification" [32,33] (see also [34] for more cosmology-relevant solutions) and the other is "dynamical compactification". Roughly speaking, the difference between these two is as follows: in the former, you make extra dimensions compact "by hand" and see if the resulting setup is viable within the gravity theory under consideration, while in the latter, you do not impose small extra dimensions in the beginning, but they become small in the process of the evolution. As "dynamical compactification" proposes a more elegant way, it involves different approaches [35,36] and setups [37,38]. Apart from cosmology, studies of extra dimensions include the investigation of black holes in Gauss–Bonnet [39–46] and Lovelock [47–51] gravities, features of the gravitational collapse in these theories [52–54], general peculiarities of spherical-symmetric solutions [55], the formation of shock due to nonlinearity [56], and many others.

The structure of the equations of motion in EGB and more general Lovelock gravity is more complicated than that in GR, as the former are nonlinear theories, unlike the latter. Thus, finding exact solutions within these theories is a very non-trivial task, and to overcome it, one usually applies metric *ansatz* of some sort. For cosmology, the usual *ansätzen* are power-law and exponential; the former of them resembles the Friedmann stage while the latter, the accelerated expansion nowadays or inflationary stage in the Early Universe—the de Sitter stage. The cosmological power-law solution within Lovelock and EGB gravities was studied in [33,57] and more recently in [58–63], resulting in some understanding of their dynamics. One of the first considerations of the exponential solutions within the considered theories could be found in [64], while recent works include [65–75]; the separate description of the exponential solutions with variable and constant volume was performed in [76] and [77], respectively; we could also refer to [78] for the discussion about the link between the existence of power-law and exponential solutions and for the discussion about the physical manifestations of different branches of the solutions within these theories. We also offer the full description of the general scheme for finding all possible exponential solutions in arbitrary dimensions and with arbitrary Lovelock contributions taken into account in [79]. A deeper investigation of the exponential solutions reveals that not all of the solutions found in [79] are stable [80]; a more general approach to the stability of exponential solutions in EGB gravity could be found in [81], while in [82–85], some particular cases are described more closely.

The above-mentioned exponential and power-law *ansätzen* are considered as dynamical late-time attractors for the Universe's evolution; however, there could be static attractors as well. As we demonstrated in [86,87] (and further investigated in [88]) if we consider extra dimensions to have negative spatial curvature, for some parameter combinations, there exist asymptotic regimes with expanding three and constant-size extra dimensions. Generally, in GR, positive spatial curvature is considered; say, positive spatial curvature could change inflationary asymptotic [89,90]. However, in EGB gravity, it is negative curvature which gives rise to the stabilization of extra dimensions; positive curvature could stabilize as well but for a very narrow range of parameters [91]; see also [92,93] for some more details on the difference between the cases with positive and negative spatial curvature.

This finalizes our review of preliminary studies dedicated to cosmological solutions in EGB and more general Lovelock gravity. With power-law and exponential solutions being well-described, it is time to find out if they could be realized during the natural evolution of the model, which is exactly what the rest of the paper is dedicated to.

3. The Model

As we already mentioned, we work with extra-dimensional theory and we are particularly interested in obtaining compactification in a natural way, meaning that the Universe ends up with three expanding dimensions while extra dimensions remain compact. Then, it is natural to consider the metric *ansatz* for such a model as spatially flat anisotropic (Bianchi-I-type), as follows:

$$ds^2 = \mathrm{diag}(-1, a_1^2(t), a_2^2(t), \dots, a_{\tilde{D}}^2(t)), \qquad (1)$$

where $a_i(t)$ is the scale factor corresponding to ith spatial dimension and $\tilde{D}$ is the total number of spatial dimensions. To obtain equations of motion, we remind the reader how Lovelock gravity is structured: Lovelock invariants have the form [4]

$$L_n = \frac{1}{2^n}\delta^{i_1 i_2 \dots i_{2n}}_{j_1 j_2 \dots j_{2n}} R^{j_1 j_2}_{i_1 i_2} \dots R^{j_{2n-1} j_{2n}}_{i_{2n-1} i_{2n}},$$

where $\delta^{i_1 i_2 \dots i_{2n}}_{j_1 j_2 \dots j_{2n}}$ is the generalized Kronecker delta of the order $2n$. Then, the Lagrangian density has a form

$$\mathcal{L} = \sqrt{-g}\sum_n \alpha_n L_n, \qquad (2)$$

with g being the determinant of the metric tensor, α_n being the coupling constant, and the summation over all n in consideration is assumed. The rest of the derivation is quite straightforward (see, e.g., [58]), and resulting equations, rewritten in terms of Hubble parameters ($H_i = \dot{a}_i/a_i$), read as follows:

$$(2n-1) \sum_{k_1 > k_2 > \dots k_{2n}} H_{k_1} H_{k_2} \dots H_{k_{2n}} = 0 \qquad (3)$$

for constraint equation and

$$\sum_{\substack{m=1 \\ m \neq i}}^{D} \left[(\dot{H}_m + H_m^2) \sum_{\substack{\{j_1, j_2, \dots j_{2n-2}\} \neq \{i, m\} \\ j_1 > j_2 > \dots j_{2n-2}}} H_{j_1} H_{j_2} \dots H_{j_{2n-2}} \right] + \\ +(2n-1) \sum_{\substack{\{k_1, k_2, \dots k_{2n}\} \neq i \\ k_1 > k_2 > \dots k_{2n}}} H_{k_1} H_{k_2} \dots H_{k_{2n}} = 0 \qquad (4)$$

for dynamical equation corresponding to ith spatial coordinate. Please mind that the equations above are obtained for the vacuum case; in the presence of a nontrivial source, the right-hand sides of (3) and (4) are modified appropriately by adding density and pressure, respectively.

Our studies demonstrated that $(3+D)$ spatial splitting[3] could be realized under some conditions (see Section 9 for details), so that we can use this splitting to study the dynamics of the corresponding model. Then, the $(3+D)$ spatial splitting (D now stands for number of extra dimensions) metric has the form

$$ds^2 = \mathrm{diag}(-1, a^2(t), a^2(t), a^2(t), b^2(t), \ldots, b^2(t)), \tag{5}$$

and the resulting equations of motion could be obtained from (3) and (4) by applying metric *ansatz* (5). Then, if we limit ourselves with the most cubic Lovelock contribution ($n = 1, 2, 3$ in (2); that is, the highest Lovelock order we are dealing with in this review) and normalize Gauss–Bonnet and cubic Lovelock couplings to GR coupling ($\alpha_1 \equiv 1$, $\alpha_2 \equiv \alpha$, $\alpha_3 \equiv \beta$), and use notations $H \equiv \dot{a}/a$, $h \equiv \dot{b}/b$, dynamical equations could be rewritten as

$$\begin{aligned}
&2\left[2\dot{H} + 3H^2 + D\dot{h} + \frac{D(D+1)}{2}h^2 + 2DHh\right] + 8\alpha\left[2\dot{H}\left(DHh + \frac{D(D-1)}{2}h^2\right) + \right.\\
&+D\dot{h}\left(H^2 + 2(D-1)Hh + \frac{(D-1)(D-2)}{2}h^2\right) + 2DH^3h + \frac{D(5D-3)}{2}H^2h^2 + \\
&\left. +D^2(D-1)Hh^3 + \frac{(D+1)D(D-1)(D-2)}{8}h^4\right] + \\
&+144\beta\left[\dot{H}\left(Hh^3\frac{D(D-1)(D-2)}{3} + h^4\frac{D(D-1)(D-2)(D-3)}{12}\right) + \right.\\
&+D\dot{h}\left(H^2h^2\frac{(D-1)(D-2)}{2} + Hh^3\frac{(D-1)(D-2)(D-3)}{3} + \right.\\
&\left. + h^4\frac{(D-1)(D-2)(D-3)(D-4)}{24}\right) + H^3h^3\frac{D(D-1)(D-2)}{3} + \\
&+ H^2h^4\frac{D(D-1)(D-2)(7D-9)}{24} + Hh^5\frac{D^2(D-1)(D-2)(D-3)}{12} + \\
&\left. + h^6\frac{(D+1)D(D-1)(D-2)(D-3)(D-4)}{144}\right] - \Lambda = 0
\end{aligned} \tag{6}$$

for a dynamical equation that corresponds to H

$$\begin{aligned}
&2\left[3H^2 + 3DHh + \frac{D(D-1)}{2}h^2\right] + 24\alpha\left[DH^3h + \frac{3D(D-1)}{2}H^2h^2 + \right.\\
&\left. + \frac{D(D-1)(D-2)}{2}Hh^3 + \frac{D(D-1)(D-2)(D-3)}{24}h^4\right] + 720\beta\left[H^3h^3\frac{D(D-1)(D-2)}{6} + \right.\\
&+H^2h^4\frac{D(D-1)(D-2)(D-3)}{8} + Hh^5\frac{D(D-1)(D-2)(D-3)(D-4)}{40} + \\
&\left. +h^6\frac{D(D-1)(D-2)(D-3)(D-4)(D-5)}{720}\right] = \Lambda
\end{aligned} \tag{7}$$

for a constraint equation and

$$
\begin{aligned}
&2\left[3\dot{H}+6H^2+(D-1)\dot{h}+\frac{D(D-1)}{2}h^2+3(D-1)Hh\right]+8\alpha\left[3\dot{H}\left(H^2+\right.\right.\\
&+2(D-1)Hh+\left.\frac{(D-1)(D-2)}{2}h^2\right)+(D-1)\dot{h}\left(3H^2+3(D-2)Hh+\right.\\
&+\left.\frac{(D-2)(D-3)}{2}h^2\right)+3H^4+9(D-1)H^3h+3(D-1)(2D-3)H^2h^2+\\
&+\left.\frac{3(D-1)^2(D-2)}{2}Hh^3+\frac{D(D-1)(D-2)(D-3)}{8}h^4\right]+\\
&+144\beta\left[\dot{H}\left(H^2h^2\frac{3(D-1)(D-2)}{2}+Hh^3(D-1)(D-2)(D-3)+\right.\right.\\
&+\left.h^4\frac{(D-1)(D-2)(D-3)(D-4)}{8}\right)+(D-1)\dot{h}\left(H^3h(D-2)+\right.\\
&+H^2h^2\frac{3(D-2)(D-3)}{2}+Hh^3\frac{(D-2)(D-3)(D-4)}{2}+\\
&+\left.h^4\frac{(D-2)(D-3)(D-4)(D-5)}{24}\right)+H^4h^2\frac{3(D-1)(D-2)}{2}+\\
&+H^3h^3\frac{(D-1)(D-2)(11D-27)}{6}+H^2h^4\frac{3(D-1)(D-2)^2(D-3)}{4}+\\
&+Hh^5\frac{(D+1)(D-1)(D-2)(D-3)(D-4)}{12}+\\
&+\left.h^6\frac{D(D-1)(D-2)(D-3)(D-4)(D-5)}{144}\right]-\Lambda=0
\end{aligned}
\tag{8}
$$

for a dynamical equation that corresponds to h.

Finally, the last metric configuration we are going to consider in this review is the case with the spatial curvature. In this case, the metric takes a form

$$ds^2 = -dt^2 + a(t)^2 d\Sigma^2_{(3)} + b(t)^2 d\Sigma^2_{(D)}, \tag{9}$$

where $d\Sigma^2_{(3)}$ and $d\Sigma^2_{(D)}$ stand for the metric of two constant curvature manifolds $\Sigma_{(3)}$ and $\Sigma_{(D)}$[4]. It is worth pointing out that even a negative constant curvature space can be compactified by making the quotient of the space from a freely acting discrete subgroup of $O(D,1)$ [94].

The complete derivation of the equations of motion is similar to the previous cases and could be found in [86,87]; if we use the following rescaling of the coupling constants

$$\alpha = \frac{(D+3)(D+2)(D+1)}{6}c_0, \quad \beta = \frac{(D+1)D(D-1)}{6}c_1, \quad \gamma = \frac{(D-1)(D-2)(D-3)}{6}c_2, \tag{10}$$

and the following notations

$$
\begin{aligned}
A_{(1)} &= \frac{\ddot{a}}{a}, \quad C = \frac{\dot{a}\dot{b}}{ab}, \quad B_{(1)} = \frac{\ddot{b}}{b}, \\
A_{(2)} &= \frac{\left[\gamma_{(3)} + (\dot{a})^2\right]}{a^2}, \quad B_{(2)} = \frac{\left[\gamma_{(\mathbf{D})} + \left(\dot{b}\right)^2\right]}{b^2}
\end{aligned} \tag{11}
$$

the equations of motion could be written in the following form:

$$
\begin{aligned}
&\mathcal{E}_0 = 0 \Leftrightarrow 0 = \alpha + \beta\left(B_{(2)} + \frac{6}{D-1}C + \frac{6}{D(D-1)}A_{(2)}\right) + \gamma\left(B_{(2)}^2 + \frac{12A_{(2)}B_{(2)}}{(D-2)(D-3)} + \right. \\
&\left. + \frac{24C^2}{(D-2)(D-3)} + \frac{12B_{(2)}C}{(D-3)} + \frac{24A_{(2)}C}{(D-1)(D-2)(D-3)}\right),
\end{aligned} \tag{12}
$$

$$
\begin{aligned}
&\mathcal{E}_i = 0 \Leftrightarrow 0 = \alpha + \beta\left(B_{(2)} + \frac{4A_{(1)}}{D(D-1)} + \frac{2B_{(1)}}{D-1} + \frac{2A_{(2)}}{D(D-1)} + \frac{4C}{(D-1)}\right) + \gamma\left(B_{(2)}^2 + \right. \\
&+ \frac{16A_{(1)}C}{(D-1)(D-2)(D-3)} + \frac{8B_{(2)}C}{D-3} + + \frac{8A_{(1)}B_{(2)}}{(D-2)(D-3)} + \frac{8A_{(2)}B_{(1)}}{(D-1)(D-2)(D-3)} + \\
&\left. + \frac{16B_{(1)}C}{(D-2)(D-3)} + \frac{4B_{(1)}B_{(2)}}{(D-3)} + \frac{4A_{(2)}B_{(2)}}{(D-2)(D-3)} + \frac{8C^2}{(D-2)(D-3)}\right),
\end{aligned} \tag{13}
$$

while the equation $\mathcal{E}_a = 0$ reads

$$
\begin{aligned}
&\mathcal{E}_a = 0 \Leftrightarrow 0 = \frac{D}{(D-4)}\alpha + \frac{(D-2)}{(D-4)}\beta\left(B_{(2)} + \frac{6A_{(1)}}{(D-1)(D-2)} + \frac{2B_{(1)}}{D-2} + \frac{6A_{(2)}}{(D-1)(D-2)} + \right. \\
&\left. + \frac{6C}{(D-2)}\right) + \gamma\left(B_{(2)}^2 + \frac{48A_{(1)}C}{(D-2)(D-3)(D-4)} + \frac{12B_{(2)}C}{D-4} + \frac{24C^2}{(D-3)(D-4)} + \right. \\
&+ \frac{12A_{(1)}B_{(2)}}{(D-3)(D-4)} + \frac{24A_{(2)}B_{(1)}}{(D-2)(D-3)(D-4)} + \frac{24B_{(1)}C}{(D-3)(D-4)} + \frac{4B_{(1)}B_{(2)}}{(D-4)} + \\
&\left. + \frac{12A_{(2)}B_{(2)}}{(D-3)(D-4)} + \frac{24A_{(2)}C}{(D-2)(D-3)(D-4)} + \frac{24A_{(1)}A_{(2)}}{(D-1)(D-2)(D-3)(D-4)}\right).
\end{aligned} \tag{14}
$$

This finalizes the list of the models we will use throughout this review. The equations of motion are provided here for the general case (for the arbitrary number of spatial dimensions) and for a vacuum; however, particular cases we will be considering have specified a number of dimensions, so terms of a different nature but the same structure will be summarized and the final equations of motion will look simpler. As for the source, we will consider vacuum, cosmological constant (Λ-term), and perfect fluid; for non-vacuum cases, these equations need to be appropriately modified—density added to the constraint equation and pressure to dynamical equations corresponding to spatial dimensions.

4. EGB: Vacuum Case

In this section, we review the results for EGB gravity (i.e., maximally with quadratic contribution taken into account) and with no matter source (vacuum model). In this case, the equations of motion are exactly as in (6) and (7) with $\Lambda = 0$ and $\beta = 0$. We shall use $D = 1$ as an example of the method and of the features and consider it in more detail. The equations of motion for the $D = 1$ vacuum case reads (H-equation, h-equation, and constraint, respectively)

$$4\dot{H} + 6H^2 + 2\dot{h} + 2h^2 + 4Hh + 8\alpha\left(2(\dot{H} + H^2)Hh + (\dot{h} + h^2)H^2\right) = 0, \tag{15}$$

$$6\dot{H} + 12H^2 + 24\alpha(\dot{H} + H^2)H^2 = 0, \tag{16}$$

$$6H^2 + 6Hh + 24\alpha H^3 h = 0, \tag{17}$$

where H is the Hubble parameter corresponding to the three-dimensional world, h is the Hubble parameter corresponding to extra dimensions, and α is Gauss–Bonnet coupling.

From (17), we can easily see that

$$h = -\frac{H}{1 + 4\alpha H^2}, \tag{18}$$

and so H and h have opposite signs for $\alpha > 0$, but could have the same sign in the $\alpha < 0$ case. Also, we can resolve (16) with respect to $\dot{H}$ to obtain

$$\dot{H} = -\frac{2H^2(1 + 2\alpha H^2)}{1 + 4\alpha H^2}; \tag{19}$$

and after that, with the use of (19), one can solve (15) to obtain

$$\dot{h} = -\frac{2H^2(8\alpha^2 H^4 + 2\alpha H^2 - 1)}{(1 + 4\alpha H^2)(16\alpha^2 H^4 + 8\alpha H^2 + 1)}. \tag{20}$$

One can see $(1 + 4\alpha H^2)$ in the denominator of (18)–(20); it becomes zero at $H^2 = -1/(4\alpha)$ and it corresponds to what we called nonstandard singularity in the Introduction: H is finite while h and curvature become infinite. Later on, for more complicated cases, we shall see that nonstandard singularities could happen for both H and h being finite, and nonstandard singularities form smooth lines on the (H, h) plane.

For given α, Equations (19) and (20) form a system of equations that govern the evolution of the Universe. In [95], we performed a full analysis of the system with all technical details (see also [96] for another review) while here we just provide the idea: for a given α, (19) and (20) could be analyzed to see the asymptotic regimes and stable points and then combined together. The result of this analysis is presented in Figure 1. On the (a) panel we provided the evolutionary behavior on the (H, h) plane for $\alpha > 0$, while on the (b) panel, for $\alpha < 0$, arrows correspond to the direction of evolution.

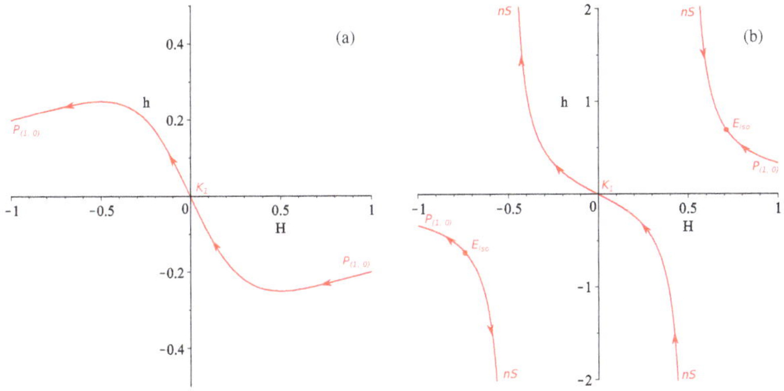

Figure 1. Resulting regimes for the $D = 1$ vacuum case: $\alpha > 0$ on (**a**) and $\alpha < 0$ on (**b**) (see the text for more details).

Let us have a closer look at Figure 1a: within the $(H > 0, h < 0)$ quadrant we have $P_{(1,0)} \to K_1$ regime transition. These regimes stand for the following: $P_{(1,0)}$ stands for the power-law regime with $p_H = 1$ and $p_h = 0$, where p_H and p_h are Kasner exponents: $p_i = -H_i^2/\dot{H}_i$ (they characterize the power of power-law behavior $a(t) \propto t^p$). Then, K_1 is another power-law regime, namely, the Kasner regime in GR [97] (that is why "1" index—in GR for Kasner regime $\sum\limits_p = 1$). Then, for $\alpha > 0$, we have a regime with $H > 0$ (our three-dimensional Universe is expanding) and $h < 0$, where extra dimensions are contracting with the Kasner regime as a future asymptote, meaning that the mentioned expansion and contractions are power-law. However, the observed expansion of our Universe is accelerated, which corresponds to the exponential solution. That derived regime could explain the Friedmann stage of the Universe's evolution, but the exponents are different from what is expected in standard cosmology (see [95] for details). Thus, the resulting regime cannot describe our current state of evolution. The other branch, which exists in the $(H < 0, h > 0)$ quadrant, is just the time-reversal of the just-described transition: $K_1 \to P_{(1,0)}$.

Now, let us have a closer look at Figure 1b and focus on the $H > 0$ half-plane (another $H < 0$ half-plane is just a time-reversal of the $H > 0$ one). This particular example is a plot for $\alpha = -1$ so that there is a nonstandard singularity (designed as nS) at $H^2 = 1/4$ and we clearly see it at $H = 1/2$: h diverges both from the left and the right. Then, for small H, we have the $nS \to K_1$ transition, and for larger H, there are two: $nS \to E_{iso}$ and $P_{(1,0)} \to E_{iso}$. The E_{iso} is the isotropic exponential solution (isotropic in an extra-dimensional sense so that all four spatial dimensions are expanding and equal to each other) and obviously it contradicts observations. Then, we are left with no viable regimes in this case.

Summarizing the $D = 1$ vacuum case, for $\alpha > 0$ there is a "would-be" viable transition $P_{(1,0)} \to K_1$, but the late-time asymptote is the Kasner solution, which is not favorable according to observations; for $\alpha < 0$, there are no viable transitions at all.

Quite similarly, we perform an analysis for the vacuum model with another D: $D = 2$, $D = 3$, and general $D \geqslant 4$ cases (equations of motion (6) and (7) have different structures for cases $D = 1, 2$, and 3, while for all $D \geqslant 4$, they remain the same—that is why we consider these cases separately). The difference appears in the number of branches—for $D = 1$, the constraint Equation (17) has only one root for h—Equation (17); however, for $D = 2$, there will be two, and for $D \geqslant 3$, there will be three; then, the number of branches of solutions will also increase accordingly. The results for the analysis, similar to those presented for $D = 1$, are presented in Figure 2. There, on the (a) and (b) panels, we presented the results for $D = 2$ ($\alpha > 0$ on (a) and $\alpha < 0$ on (b)), on the (c) and (d) panels—for $D = 3$ ($\alpha > 0$ on (c) and $\alpha < 0$ on (d)), and on (e) and (f)—for general $D \geqslant 4$ case ($\alpha > 0$ on (e) and $\alpha < 0$ on (f)); different colors correspond to different branches.

Browsing $\alpha < 0$ cases (panels (b), (d), and (f)), we can conclude that there are no viable regimes—there is a $K_3 \to K_1$ regime (where K_3 is GB Kasner with $\sum\limits_p = 3$ (see, e.g., [58] for the generalization of Kasner regimes on higher Lovelock orders)), but the late-time asymptote is K_1. On the contrary, for $\alpha > 0$ and $D \geqslant 2$, there is always a viable transition $K_3 \to E_{3+D}$. There, E_{3+D} stands for an exponential solution with different exponents for three and extra dimensions. Since the solution in question is found in the $(H > 0, h < 0)$ quadrant, it has expanding three and compacting extra dimensions, which is exactly what we are looking for. Let us note that for $D = 3$, since both subspaces are three-dimensional, it does not matter which of them is expanding and which is contracting, as there are two different exponential solutions E_{3+3} that satisfy our conditions.

Concluding EGB vacuum cases, for $\alpha > 0$ and for all $D \geqslant 2$, there always exists the $K_3 \to E_{3+D}$ transition, which is viable as the late-time regime is exponential with expanding three and contracting extra dimensions.

Before moving to the next case, let us support our choice of discarding power-law but not exponential regimes: the widely accepted parameter, which quantifies the acceleration of the Universe, is the deceleration parameter defined as

$$q = -\frac{\ddot{a}a}{\dot{a}^2},$$

where a is the scale factor. For the exponential *ansatz*, the scale factor reads $a(t) = \exp(Ht)$, so that $\dot{a} = H\exp(Ht)$ and $\ddot{a} = H^2\exp(Ht)$, leading to $q_{exp} \equiv -1$, meaning that the exponential solution always expands with acceleration. On the other hand, for the power-law *ansatz*, the scale factor reads $a(t) = a_0 t^p$, so $\dot{a} = pt^{p-1}$ and $\ddot{a} = p(p-1)t^{p-2}$, resulting in

$$q_{power} = -\frac{p(p-1)t^{p-2+p}}{p^2 t^{2(p-1)}} = -1 + \frac{1}{p}.$$

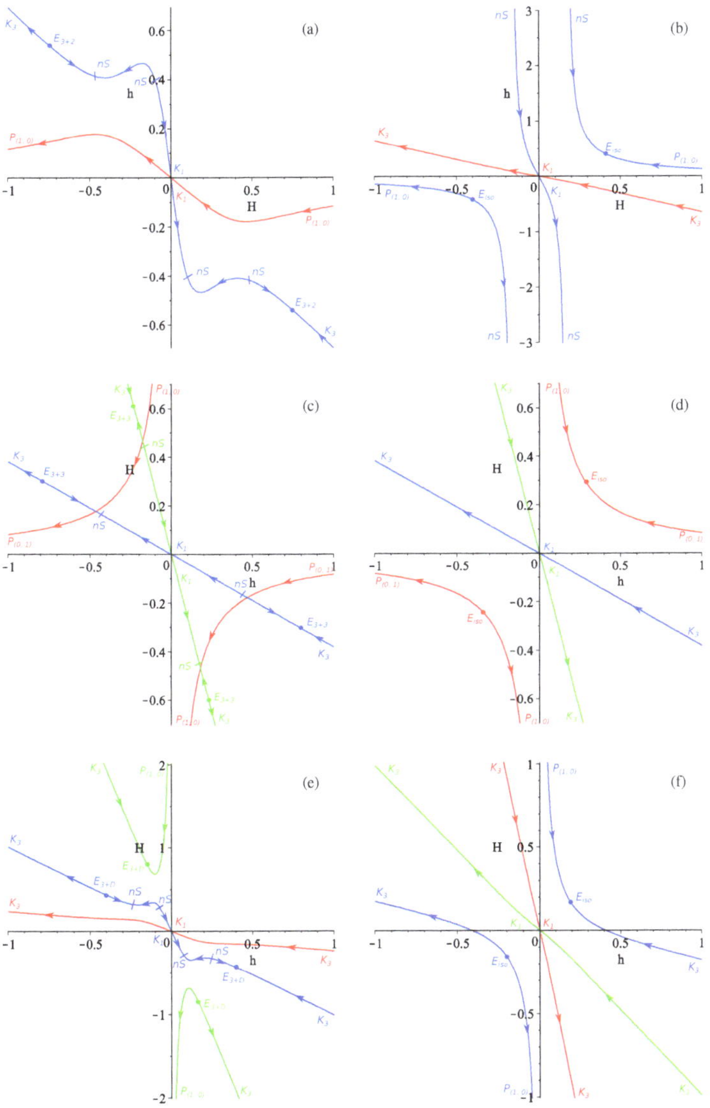

Figure 2. Resulting regimes for EGB vacuum cases: $D = 2$ (panel (**a**) for $\alpha > 0$ and panel (**b**) for $\alpha < 0$), $D = 3$ (panel (**c**) for $\alpha > 0$ and panel (**d**) for $\alpha < 0$), and general $D \geqslant 4$ case (panel (**e**) for $\alpha > 0$, and panel (**f**) for $\alpha < 0$) (see the text for more details).

For accelerated expansion, we need $q < 0$, and for that, $p > 1$ is required. Apparently, all values for p_H—power-law exponents corresponding to three-dimensional subspace—originally found in [95], are $0 < p_H < 1$, so all asymptotic K_1 regimes cannot be called realistic. The same is true for $P_{(1,0)}$, as it has $p_H = 1$, which leads to $q_{power} \equiv 0$, which is still not accelerated expansion. Concluding, exponential solutions are always viable, while power-law solutions could be viable, if $p_H > 1$ which, however, is not the case according to [95], they are $0 < p_H < 1$. So, hereafter, we consider E_{3+D} realistic compactification regimes while K_1 is not viable.

5. EGB: Λ-Term Case

Our next step is to consider EGB gravity but with Λ-term as a source. The equations of motion are (6)–(7) but with $\beta = 0$. The equations of motion seem to be slightly different—we only added Λ to the right-hand side—but the resulting dynamics change drastically. First of all, there is no more low-energy Kasner regime K_1—Kasner regime is a vacuum regime—it approaches $(H = 0, h = 0)$ in a power-law way, and for vacuum, $(H = 0, h = 0)$ is a solution (though a trivial solution), as seen from (7). For the Λ-term case, however, r.h.s. of (7) is replaced with nonzero Λ and so $(H = 0, h = 0)$ is no longer a solution.

An analysis of all possible transitions also becomes more complicated—instead of only one parameter α in the vacuum case, now we have two— α and Λ—and transition availability is now a function of both.

The procedure for the Λ-term case is the same as for the vacuum case, so we immediately turn to the results; the detailed analysis with all technical details could be found in [98] for $D = 1, 2$ and in [99] for $D = 3$ and general $D \geqslant 4$ cases; see also [96] for the shorter review version. Analysis suggests that there are no viable regimes for the $D = 1$ case, but for $D \geqslant 2$ there are. In particular, for $D = 2$, there is a $K_3 \to E_{3+2}$ transition, which exists for $\alpha > 0$ and $\alpha\Lambda < 1/2$ (including the entire $\Lambda < 0$ region). In addition, again for $\alpha > 0$ and $\alpha\Lambda < -1/6$ there is another transition $P_{(1,0)} \to E_{3+2}$ to the same exponential solution, so for $\alpha > 0$ and $\alpha\Lambda < -1/6$ we have one viable exponential solution E_{3+2}, which is a future asymptote, and there could be two different past asymptotes—K_3—Gauss–Bonnet Kasner solution and $P_{(1,0)}$—another power-law solution with $p_H = 1$ and $p_h = 0$; hereafter, we shall denote such a "double" transition as $K_3 \to E_{3+2} \leftarrow P_{(1,0)}$.

For $D = 3$, the situation is slightly different—again, similar to the vacuum case, since both subspaces are three-dimensional, it does not matter which of them is expanding and which is contracting, which "doubles" the number of viable regimes and, correspondingly, transitions. Thus, for $\alpha < 0$, $\Lambda > 0$, $\alpha\Lambda \leqslant -3/2$, there are two "double transitions" $K_3^1 \to E_{3+3}^1 \leftarrow P_{(1,0)}$ and $K_3^2 \to E_{3+3}^2 \leftarrow P_{(0,1)}$; please mind that K_3^1 and K_3^2 are two different GB Kasner solutions and E_{3+3}^1 and E_{3+3}^2 are two different viable exponential solutions. The fact that there are two of them is coming from the fact that both subspaces are three-dimensional. The same two double transitions ($K_3^1 \to E_{3+3}^1 \leftarrow P_{(1,0)}$ and $K_3^2 \to E_{3+3}^2 \leftarrow P_{(0,1)}$) also exist for the entire $\alpha > 0$, $\Lambda < 0$, while for $\alpha > 0$, $\Lambda > 0$, $\alpha\Lambda \leqslant 1/2$, they reduce to a couple of transitions $K_3^{1,2} \to E_{3+3}^{1,2}$.

Finally, the general $D \geqslant 4$ case has an additional regime—$K_3 \to K_3$—transition between two different GB Kasner regimes. Similar to the $K_3 \to K_1$ transition in the vacuum case, we cannot call it entirely viable, as the future regime is power-law, but the existence of K_3 as a future asymptote is interesting; the transition is demonstrated in Figure 3a.

Similarly to the $D = 3$ case, realistic transitions exist in three domains on the (α, Λ) space. The first of them is $\alpha < 0$, $\Lambda > 0$ with the additional constraint $\alpha\Lambda \leqslant \zeta_1$; let us note that within this domain, there were no realistic transitions for $D = 2$. Realistic transitions here are $K_3 \to E_{3+D}^1 \leftarrow P_{(1,0)}$—the same realistic exponential solution could be reached from two different high-curvature power-law past asymptotes; these transitions are demonstrated in Figure 3b.

Mentioned above, ζ_1 was derived in [99] (though denoted as ζ_3 there) and it reads

$$\zeta_1 = -\frac{D(D-1)}{4(D-2)(D-3)}; \tag{21}$$

this limit replaces the corresponding $\alpha\Lambda = -3/2$ limit from the $D = 3$ case.

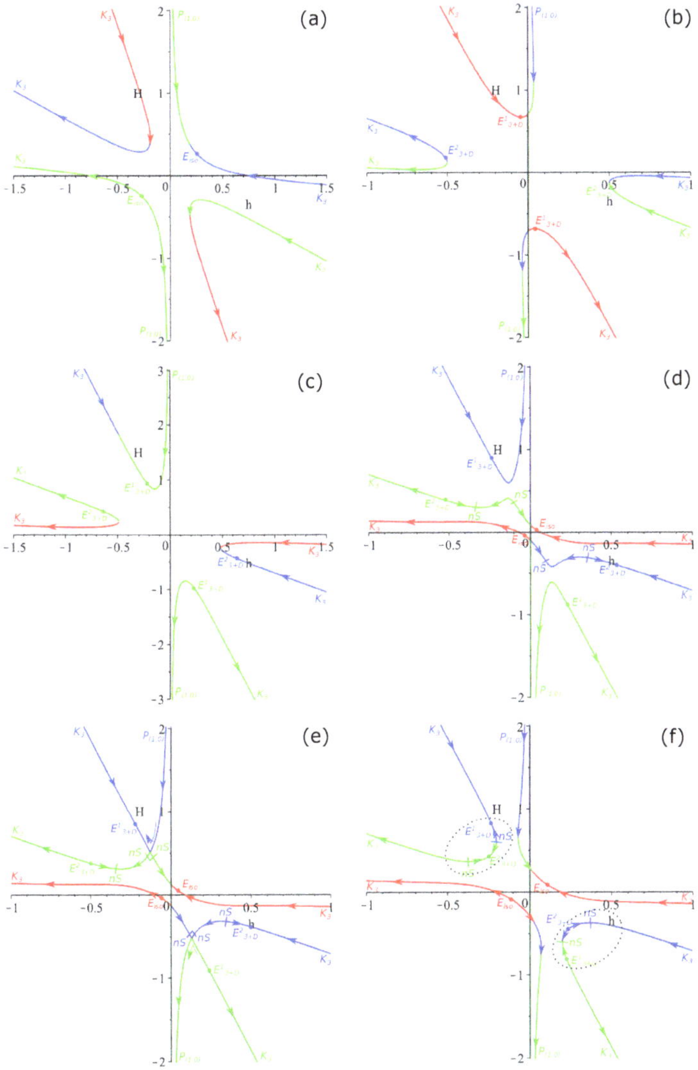

Figure 3. Viable regimes for $D \geqslant 4$ EGB cosmology with Λ-term: (**a**) panel: $\alpha < 0$, $\Lambda < 0$ featuring $K_3 \to K_3$ regime; (**b**) panel: $\alpha < 0$, $\Lambda > 0$, $\alpha\Lambda \leqslant \zeta_1$ featuring $K_3 \to E^1_{3+D} \leftarrow P_{(1,0)}$ transitions; (**c**) panel: $\alpha > 0$, $\Lambda < 0$ featuring the same $K_3 \to E^1_{3+D} \leftarrow P_{(1,0)}$ transitions; (**d**) panel: $\alpha > 0$, $\Lambda > 0$, $\alpha\Lambda < \zeta_2$ still featuring the same $K_3 \to E^1_{3+D} \leftarrow P_{(1,0)}$ transitions; (**e**) panel: $\alpha > 0$, $\Lambda > 0$, $\alpha\Lambda = \zeta_2$ with $K_3 \to E^1_{3+D} \leftarrow P_{(1,0)}$ transitions; (**f**) panel: $\alpha > 0$, $\Lambda > 0$, $\zeta_2 < \alpha\Lambda < \zeta_3$ (see the text for more details).

The same transitions exist in another domain as well—in $\alpha > 0$, $\Lambda < 0$, where they exist within the entire domain; this situation is demonstrated in Figure 3c.

Finally, the same transitions also exist in the $\alpha > 0$, $\Lambda > 0$ domain, but there, its location and surroundings depend on $\alpha\Lambda$: for $\alpha\Lambda < \zeta_2$, the situation is depicted in Figure 3d, while for $\alpha\Lambda = \zeta_2$, it is presented in Figure 3e. For $\alpha\Lambda > \zeta_2$, the situation is presented in Figure 3f and one can see that the double transition $K_3 \to E^1_{3+D} \leftarrow P_{(1,0)}$ degrades to a single $K_3 \to E^1_{3+D}$ for $\alpha\Lambda > \zeta_2$. Further, for $\alpha\Lambda > \zeta_3$, E^1_{3+D} is replaced with nonstandard singularity and so no realistic transitions exist for $\alpha\Lambda > \zeta_3$.

Mentioned above, ζ_2 and ζ_3 are also derived from [99], where they are denoted as ζ_2 and ζ_6, respectively:

$$\zeta_2 = \frac{\sqrt[3]{\mathcal{D}_2(D-1)^2}}{12(D-2)(D-1)D(D+1)} +$$

$$+\frac{(D^6-6D^5+10D^4-20D^2+24D+36)(D-1)}{3D(D-2)(D+1)\sqrt[3]{\mathcal{D}_2(D-1)^2}} + \frac{D^3-9D^2+8D+24}{12D(D-2)(D+1)}, \text{ where}$$

$$\mathcal{D}_2 = 10D^{10} + 6D^9\mathcal{D}_1 - 100D^9 - 30D^8\mathcal{D}_1 + 330D^8 + 30D^7\mathcal{D}_1 - 240D^7 + 54D^6\mathcal{D}_1 - \\ -600D^6 - 84D^5\mathcal{D}_1 + 240D^5 - 24D^4\mathcal{D}_1 + 1520D^4 + 48D^3\mathcal{D}_1 + 640D^3 - 2880D^2 + 1728 \tag{22}$$

and

$$\mathcal{D}_1 = \frac{(D-4)(D-3)(D+2)}{(D-1)(D+1)}\sqrt{\frac{(D-4)(D+2)}{D(D-2)}};$$

$$\zeta_3 = \frac{1}{4}\frac{3D^2-7D+6}{D(D-1)}; \tag{23}$$

let us note that ζ_3 replaces the corresponding limit $\alpha\Lambda = 1/2$ from $D = 2, 3$ cases.

So, concluding with realistic transitions existing in EGB cosmologies with Λ-term as a source, starting from $D \geqslant 2$, we always have realistic transitions—either double $K_3 \to E_{3+D} \leftarrow P_{(1,0)}$ or single $K_3 \to E_{3+D}$; the area of parameters on the (α, Λ) space where they exist depends on D, but it is always an open region.

6. Cubic Lovelock: Vacuum Case

For this case, we use the same set of Equations (6) and (7) but with $\Lambda = 0$. The original study is performed in [100] for $D = 3, 4$ and in [101] for $D = 5$ and general $D \geqslant 6$ cases, and here, we summarize our findings. Similar to the EGB case, depending on D, different terms will be nonzero, so we consider all possible D concluding with a general D case where all terms are present. This way, we start with $D = 3$, as it is the lowest dimension where cubic Lovelock terms are nontrivial; analysis suggests [100] that in that case, there are two sets of regimes, $P_{(1,0)} \to E^1_{3+3}$ and $P_{(0,1)} \to E^2_{3+3}$, which exist for $\alpha > 0$, $\beta < 0$ and $P_{(1,0)} \to K^1_1$, $P_{(0,1)} \to K^2_1$, in turn which exist for $\alpha > 0$, $\beta > 0$. We have two sets for the same reason we have the "doubling" of regimes in $D = 3$ EGB cases because both subspaces are three-dimensional. Let us note that the second regime is Kasner and cannot be called viable for the same reason as for EGB gravity cases. Let us also note that both regimes exist for $\alpha > 0$.

The next case to consider is $D = 4$ (see [100] for a full analysis); similar to the $D = 3$ case, we have two transitions. The first of them is $P_{(1,0)} \to E_{3+4}$, which exists for $\alpha > 0$, $\mu \leqslant \mu_1$, where $\mu \equiv \beta/\alpha^2$ and $\mu_1 = -4\sqrt[3]{98}/135 + 38/135 \approx 0.1449$ were found in [100]. So, the entire $\beta < 0$ is included in the existence region for this transition. Another transition is $P_{(1,0)} \to K_1$, which exists for $\alpha > 0$, $\mu > \mu_1$. Again, since the low-curvature regime is Kasner, we cannot truly call it realistic. However, it is interesting to note that, similar to the $D = 3$ case, both regimes exist only for $\alpha > 0$.

Surprisingly, for $D = 5$, we have exactly the same regimes with exactly the same locations but with another μ_1 (see [101] for details):

$$\mu_1 = -\frac{\sqrt[3]{2150+210\sqrt{105}}}{210} + \frac{2}{21\sqrt[3]{2150+210\sqrt{105}}} + \frac{11}{42} \approx 0.1903.$$

Finally, the case $D \geqslant 6$ is supposed to be general—indeed, for $D \geqslant 6$, all possible terms are present in the equations of motion (6) and (7), and so equations of motion for $D \geqslant 6$ contain all possible terms. However, analysis of these equations reveals [101] that the resulting regimes are different in $D = 6, 7$ and $D \geqslant 8$. Realistic regimes in $D = 6, 7$ are the same as in $D = 3, 4, 5$ but with different values for μ_1: for $D = 6$ we have $\mu_1 \approx 0.2151$

while for $D = 7$ we have $\mu_1 \approx 0.2324$ (please mind that in the original research [101] they are denoted differently due to different numbering of special points in different D cases).

The mentioned above regimes for $D = 3 \div 7$ are demonstrated in Figure 4a—$P_{(1,0)} \to E_{3+D}$—and in Figure 4b—$P_{(1,0)} \to K_1$.

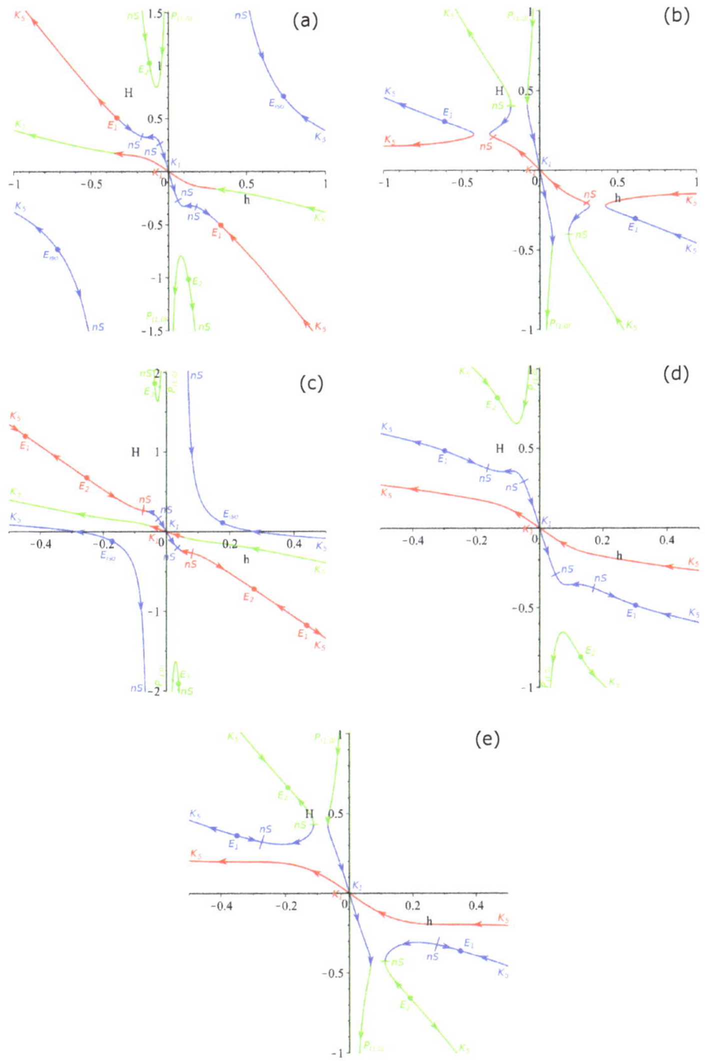

Figure 4. Viable regimes for vacuum cubic Lovelock cosmology (all viable regimes are located in the second quadrant): (**a**) panel: $P_{(1,0)} \to E_{3+D}$ transition on the green branch for $D = 3 \div 7$ and $\alpha > 0$, $\mu \leqslant \mu_1$; (**b**) panel: $P_{(1,0)} \to K_1$ transition on the green-blue branch for $D = 3 \div 7$ and $\alpha > 0$, $\beta > 0$, $\mu > \mu_1$; (**c**) panel: $P_{(1,0)} \to E_{3+D}$ transition on the green branch as well as $K_5 \to E_{3+D}$ transition on the red branch for $D \geqslant 8$ and $\alpha > 0$, $\beta < 0$, $\mu \leqslant \mu_2$; (**d**) panel: $P_{(1,0)} \to E_{3+D} \leftarrow K_5$ double transition on the green branch for $D \geqslant 8$ and $\alpha > 0$, $\beta > 0$, $\mu \leqslant \mu_3$; (**e**) panel: $P_{(1,0)} \to K_1$ transition on the right green-blue branch and $K_5 \to E_{3+D}$ transition on the left green-blue branch for $D \geqslant 8$ and $\alpha > 0$, $\beta > 0$, $\mu > \mu_3$ (see the text for more details).

At last, for $D \geqslant 8$, the abundance and structure of the viable solutions changes drastically: for $\alpha > 0$ and $\beta < 0$ there are two realistic compactifications, $P_{(1,0)} \to E^1_{3+D}$, which exist everywhere in $\beta < 0$, demonstrated in Figure 4c, and $K_5 \to E^2_{3+D}$, which exists for $\mu \leqslant \mu_2$, where

$$\mu_2 = -\frac{D^4 + 30D^3 + 189D^2 - 540D + 324}{D^4 - 6D^3 - 25D^2 + 102D - 72},$$

again found in [101]; this regime is demonstrated in Figure 4d. So, for $\mu \leqslant \mu_2$, there exist two different compactification schemes for two different exponential solutions E^1_{3+D} and E^1_{3+D}. For $\alpha > 0$, $\beta > 0$, $\mu \leqslant \mu_3$, we have double transition $K_5 \to E^3_{3+D} \leftarrow P_{(1,0)}$, demonstrated in Figure 4d, which splits into $K_5 \to E^3_{3+D}$ and $P_{(1,0)} \to K_1$ for $\mu > \mu_3$, demonstrated in Figure 4e. The quoted value μ_3 is the lesser root of a certain sixth-order polynomial mentioned in [101].

Concluding with the realistic compactification regimes in vacuum models in cubic Lovelock gravity, we can notice that they exist in all $D \geqslant 3$, but their structure and abundance differ for $D = 3 \div 7$ and $D \geqslant 8$: for $D = 3 \div 7$, we need $\mu \leqslant \mu_1$, while for $D \geqslant 8$, realistic compactifications exist for all β; another interesting feature is that all of them exist only for $\alpha > 0$.

7. EGB: Perfect Fluid Case

To study EGB cosmology with perfect fluid as a source, we use Equations (6) and (7) with $\beta = 0$ and replace Λ with $(-p)$ in (6) and (8) and Λ with ρ in (7) where p is the pressure and ρ is the density of the perfect fluid. We use ω for the equation of state which links pressure and density as $p = \omega\rho$. Please mind that ω, being an equation of state, is constant, while ρ is a dynamical variable. Formally, since we have matter in the form of a perfect fluid, the system should be supplemented with a continuity equation, but the full system (6)–(7) plus continuity equation is overdetermined, so we drop the continuity equation; similar techniques are used for the analysis of Friedmann equations with perfect fluid. Originally, research was performed in [102] and covers only low-dimensional cases ($D = 1, 2$), which we are going to review here.

The $D = 1$ case demonstrates the existence of three nonsingular regimes, but neither of them has realistic compactification. In contrast, in $D = 2$, there is a number of nonsingular regimes, but only one of them is realistic—$K_3 \to E_{3+2}$—and it happens for $\alpha > 0$ and $\omega < 1/3$; the measure of the initial conditions leading to this regime increases with the growth of ω, reaching the maximum at $\omega = 1/3 - 0$.

These regimes are demonstrated in Figure 5: (a) and (b) panels correspond to the $\omega < 0$ case (the (a) panel shows large-scale structure of the $H > 0, h < 0$ quadrant while the (b) panel focuses on the vicinity of E_{3+2} exponential solution), the (c) and (d) panels—to $0 < \omega < 1/3$ (again, the (c) and (d) panels represent large-scale and fine structures, respectively) and the (e) panel—to $\omega > 1/3$ and we can see that $K_3 \to E_{3+2}$ is no more; we have $nS \to E_{3+2}$ instead.

As Figure 5 differs significantly from the previous cases, it requires additional explanation. First of all, evolution curves: for vacuum and Λ-term cases, evolution curves are determined by nonzero parameters of the system; α, β, and Λ: in this case, we have variable parameter—density ρ—so that instead of predetermined evolution curves, we have a field, which is represented by red arrows. However, this is a physical parameter, and it has to be strictly positive, but not all (H, h) combinations correspond to $\rho > 0$. We denoted such "forbidden" $\rho < 0$ regions in dark blue. To simplify the analysis of the regimes, we also added $\dot{H} = 0$ lines in green, and $\dot{h} = 0$ lines in black so that their intersections have $\dot{H} = \dot{h} = 0$, which corresponds to exponential solutions (which are also fixed points of the system). The last piece is the nonstandard singularity; their locations are marked with dashed blue lines.

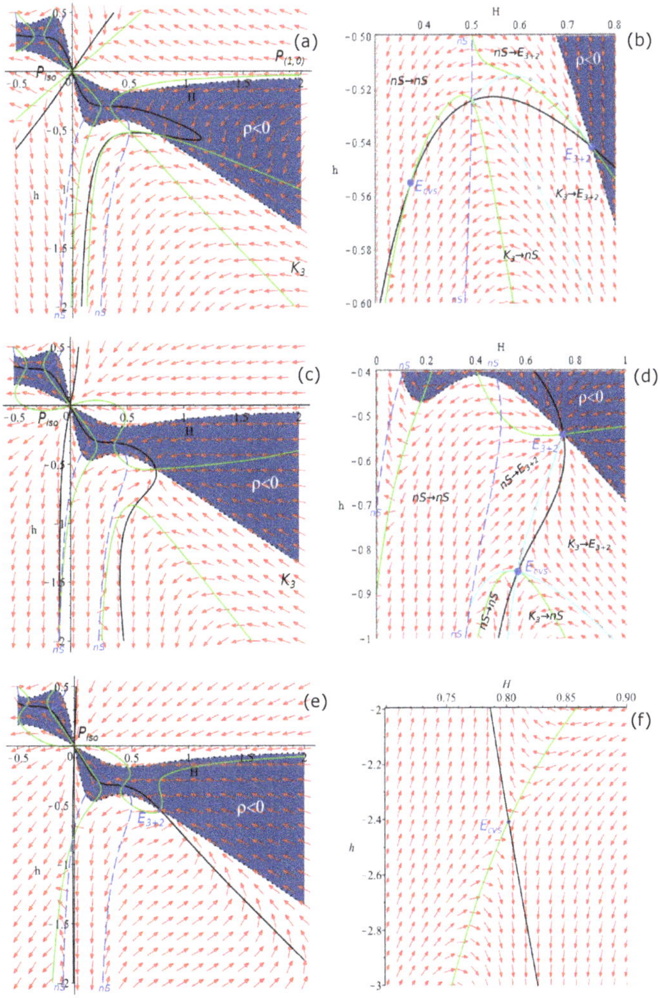

Figure 5. Viable regimes for EGB model with a perfect fluid as a source: (**a**) panel: large-scale structure of the $H > 0$, $h < 0$ quadrant for $D = 2$, $\alpha > 0$, $\omega < 0$; (**b**) panel: vicinity of E_{3+2} stable point for $D = 2$, $\alpha > 0$, $\omega < 0$, initial conditions leading to $K_3 \to E_{3+2}$ transition are bounded by light-blue lines; (**c**) panel: large-scale structure of the $H > 0$, $h < 0$ quadrant for $D = 2$, $\alpha > 0$, $1/3 > \omega > 0$; (**d**) panel: vicinity of E_{3+2} stable point for $D = 2$, $\alpha > 0$, $1/3 > \omega > 0$, initial conditions leading to $K_3 \to E_{3+2}$ transition are bounded by light-blue lines; (**e**) panel: large-scale structure of the $H > 0$, $h < 0$ quadrant for $D = 2$, $\alpha > 0$, $\omega > 1/3$, only $nS \to E_{3+2}$ transition remains; (**f**) vicinity of the exponential constant volume solution (E_{CVS}) (see the text for more details).

These notations are enough for analyzing the resulting regimes. Additional numerical simulations reveal particular past or future asymptotes and the vector field demonstrates the general flow from and to these asymptotes. For instance, on the (a) panel, in the bottom part we can see that the past asymptote is K_3 (found numerically), and the vector field shows the direction to nS, making the transition $K_3 \to nS$; we perform a similar analysis for all other regimes.

The light-blue lines on Figure 5b,d are bound initial conditions, which lead to E_{3+2} from K_3, so that initial conditions form the only realistic compactification regime $K_3 \to E_{3+2}$. Comparing Figure 5b with Figure 5d, one can notice an increase in this area, so that the measure of the realistic compactification regime is increasing as ω is growing until it reaches $1/3$; for $\omega > 1/3$, we still have E_{3+2}, but K_3 is no longer reachable as the field changes direction (compare Figure 5c and Figure 5e) and it is nS as a past asymptote, resulting in an $nS \to E_{3+2}$ transition, which is not realistic compactification.

After considering the EGB model with perfect fluid as a source, we can conclude that, similar to previously considered EGB models, realistic compactification exists in $D = 2$; we have not considered this model in higher dimensions yet. It is interesting that realistic compactification exists only for $\omega \leqslant 1/3$; we are going to return to this question in the Discussion section.

8. EGB: Spatial Curvature Case

Spatial curvature plays an important role in GR: negative curvature typically leads to expansion, while positive could lead to recollapse; particularly, positive curvature could change the measure of the trajectories reaching inflationary asymptotic [89,90]. In the context of extra dimensions, the effects of spatial curvature were not investigated much until we found a particular solution where extra dimensions could be stabilized if the curvature of extra dimensions is negative and the parameters are within some non-vanishing range; this situation is illustrated in Figure 6a, where we presented behavior for Hubble parameters and one can see that the 3-dimensional subspace is expanding ($H > 0$) while the size of extra dimensions is stabilized (as $h \to 0$, $b(t) \to$ const). Initially, this regime was described in [86] by numerically solving (12)–(14) for a variety of parameters; with the follow-up investigation performed in [87], where we described all other regimes that could appear within the same model, and [88], where the influence of matter in the form of perfect fluid on regime availability was studied. In [92], we additionally investigated the influence of the spatial curvature of three-dimensional subspace on the dynamics of the regimes.

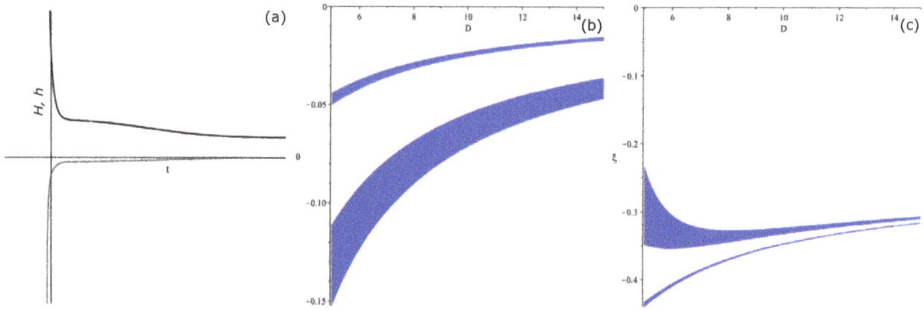

Figure 6. Illustrations for the dynamics of the EGB case with spatial curvature: regime with stabilization of extra dimensions ($H > 0$, $h \to 0$) on (**a**) panel, areas on the parameters space where stabilization of extra dimensions with positive spatial curvature is possible and stable, as a function of D (**b**,**c**) panels (see the text for more details).

Other notable contributions to the study of this regime and its availability and abundance include [103], where it was demonstrated that in the presence of the cubic Lovelock term, the stabilization of extra dimensions could coexist with isotropization—it is not so in the EGB case. It is also worth mentioning that we observed a similar effect for the vacuum model with a cubic Lovelock contribution—successful compactification schemes could coexist even with two different isotropizations—a situation we do not have in vacuum EGB (see Sections 4 and 6). In addition to this contribution, we can also mention [104], where a certain case is investigated numerically, and [105], where the $(3 + 7)$-dimensional case in the presence of the cubic Lovelock contribution is investigated numerically.

Finally, the stability of the extra dimensions stabilization regime is investigated in [91]. There, we build a system of the perturbation equations based on (12)–(14) and perturbed around the exact solution $H \equiv \dot{a}(t)/a(t) = \text{const} = H_0$, $b(t) = \text{const} = b_0$. The results of the investigation suggest that for negative curvature of extra dimensions, the resulting regime with stabilization of extra dimensions always exists and is stable. However, for many reasons, the case of positive curvature is more attractive—it is more obvious how to make extra dimensions compact in the case of positive spatial curvature—so we kept looking for the same regime but with positive curvature.

The results of our investigations are as follows: for $D = 2$, there formally exist solutions with the stabilization of extra dimensions, but they are unstable. On the contrary, starting from $D = 3$, there are stable solutions of the desired type for both positive and negative curvatures of extra dimensions. This way, for $D = 3$, for positive curvature, stable solutions exist for $\alpha < 0$ and $\xi \in (-0.5448, -0.5) \cup (0, +\infty)$ (where $\xi = \alpha\Lambda$), and thus for entire $\Lambda < 0$ and the small positive range of Λ. For negative curvature of extra dimensions in $D = 3$, stable solutions exist for $\alpha < 0$, $\xi < -3/2$; it is interesting to note that both of them exist only for $\alpha < 0$. Next, for $D = 4$, stable solutions for positive curvature of extra dimensions exist for $\alpha < 0$, $\xi \in (-27/54, -15/32) \cup (-0.3, 3/8)$—again, for both signs of Λ, but now both ranges are finite. For negative curvature, there are two domains with different conditions: one is the same as for $D = 3$: $\alpha < 0$, $\xi < -3/2$ and the other is $\alpha > 0$, $\Lambda < 0$. Finally, for the general $D \geqslant 5$ case, the situation is as follows: for positive curvature of extra dimensions, stable solutions exist for $\alpha < 0$, $\xi \in (\xi_1, \xi_2) \cup (\xi_3, \xi_4)$ (see Figure 6b,c where we provide this range in $\theta = \alpha H_0^2$ and ξ variables). For negative curvature, one of the domains is the same as before, $\alpha < 0$, $\xi < -3/2$, while another is $\alpha > 0$, $\xi < -D(D-1)/(4(D-2)(D-3))$.

It is interesting to note that for positive curvature, we require $\alpha < 0$, while for negative α, both could be signs for $D \geqslant 4$.

Overall, the stabilization of extra dimensions could take place in all $D \geqslant 3$: for the negative curvature of extra dimensions, it is a natural process, while for positive curvature, it requires fine-tuning the parameters, and this fine-tuning becomes more and more tight with the growth of D.

9. EGB: General Anisotropic Case

Last but not least, the case to report here is the case where all spatial dimensions are initially totally anisotropic, making it a Bianchi-I-type configuration. In this case, the equations of motion are (3)–(4) with $n = 2$ maximum. Then, we start with some initial conditions and integrate the equations to see the past and future asymptotic regimes. For the case with four spatial dimensions, the investigation was performed in [63] for vacuum, Λ-term, and perfect fluid cases. However, since there are no realistic compactifications for $D = 1$ neither in vacuum nor in Λ-term or perfect fluid cases, the only nonsingular regime reported there is isotropization. By singular regimes, we mean the situation when evolution is interrupted by nonstandard singularity, defined the same as before, but for totally anisotropic cases, nonstandard singularities are not points on the evolutional curve but are surphases in the phase space and are quite abundant (see, e.g., [63] for the simplest case with four spatial dimensions). The analysis was further continued in [92] for five spatial dimensions, where we know compactification for $D = 2$ exists, presented in Figure 7a, while isotropization for the same case is presented in Figure 7b. Since there are no exponential solutions with other spatial splittings in five spatial dimensions, these two are the only possible future asymptotes. However, in six spatial dimensions, the situation is different, as there could be two different principal spatial splittings: $[4+2]$, with expanding four and contracting two dimensions, and $[3+3]$ with expanding three and contracting another three dimensions. Of these two splittings, only the latter is realistic, but we can end up in both starting from the initially anisotropic case, which is demonstrated in Figure 7c. There, we plot a two-dimensional slice of the six-dimensional Hubble parameter space with the distribution of the initial conditions, leading to either $[3+3]$ splitting or $[4+2]$. One can see that the area for $[4+2]$ is larger; however, it is a two-dimensional slice and to judge the real measure, one needs to take into account the distribution across all six dimensions.

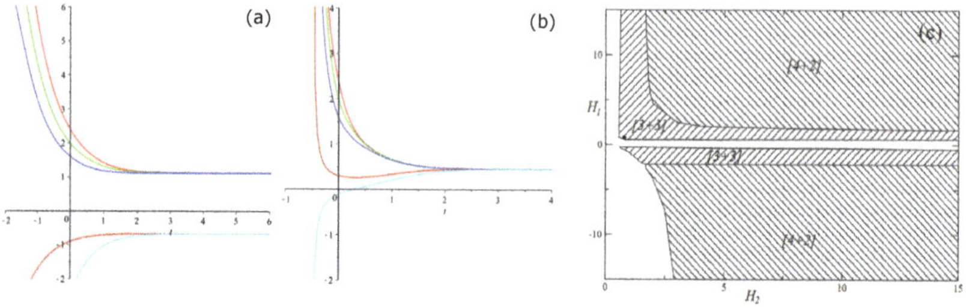

Figure 7. Illustrations for the dynamics of the EGB case initial total anisotropy (Bianchi-I-type): different initial conditions for the model with 5 spatial dimensions could lead to either $[3+2]$ spatial splitting ((**a**) panel) or isotropization ((**b**) panel); distribution of the initial conditions for the model with 6 spatial dimensions leading to either $[4+2]$ or $[3+3]$ spatial splittings ((**c**) panel) (see the text for more details).

In addition to the mentioned references in [106], the additional numerical analysis of the five and six spatial dimensional cases was performed. As in [92], we performed an analysis of both the anisotropic model and the curvature, and we proposed a scheme that would allow us to address the question of reaching the stabilization of the extra dimensions regime from an initially anisotropic state. It cannot be performed directly (as the initial state is Bianchi-I and it is spatially flat while the stabilization is achieved if there is a spatial curvature) so we developed an indirect scheme: we first address the possibility of reaching $[3+D]$ spatial splitting from the initial anisotropy for some parameters, then address the possibility of reaching compactification from the initial conditions with initially very small but nonzero curvature. So, in [105], this scheme was tested for seven spatial dimensions: initially, the possibility of forming $[3+4]$ spatial splitting was addressed, and after that, the possibility of building the stabilization of extra dimensions from $[3+4]$ spatial splitting with initially small negative curvature.

Summarizing, an initially anisotropic Universe in the course of evolution would end up in one of the spatial splittings, available for a given number of spatial dimensions, or encounter nonstandard singularity. For the former case, we always have an isotropic solution as a possibility (though it exists not for all values of the parameters) and, starting with five spatial dimensions, we have nontrivial spatial splitting into two (or more) isotropic subspaces, again, subject to existence and stability. Among them, we are interested in configurations with an expanding three-dimensional subspace and contracting other subspaces (all other subspaces, if more than one; however, due to high computational difficulty we have not investigated cases where more than one extra-dimensional subspace exists). So, our investigation reveals that in five spatial dimensions, everything is good, while in higher dimensions, other spatial splitting exists and it could be that the Universe ends up with "wrong" spatial splitting. It will not be three-dimensional subspace which would expand, resulting in the conclusion that the higher the number of spatial dimensions, the less likely that we will end up with "correct" compactification.

10. Summary and Discussion

Now, let us summarize and discuss the results reported in this review. The vacuum EGB model is the simplest to consider and it has realistic regimes. For $\alpha < 0$ (negative Gauss–Bonnet coupling), the $K_3 \to K_1$ transition exists, from high-curvature (Gauss–Bonnet) Kasner regime to low-curvature standard (GR) Kasner. Despite having the desired behavior of the scale factors, (expansion of the three- and contraction of extra-dimensional subspaces) the resulting regime is power-law and the resulting exponents do not fit the Friedmann stage, so that, despite being potentially viable, it still does not fit the observations. For $\alpha > 0$, the counterpart of this regime is $P_{(1,0)} \to K_1$—the same future but different past asymptotes. For this case, it is another power-law regime with $p_H = 1$ and $p_h = 0$. Since it also has $\sum p_i = 3$, it could be mistaken for GB Kasner, but it is a different regime. In a way, it is closer to the Taub solution [107], but the exact relationship between these two is not established.

Apart from these regimes, there is another: for $\alpha > 0$ and for all $D \geqslant 2$, the $K_3 \to E_{3+D}$ transition always exists, which is realistic—three dimensions expand exponentially while extra dimensions exponentially contract; exponential expansion of three-dimensional subspace agrees with the observed accelerated expansion of the Universe.

The next case to consider is the EGB model with the Λ-term. This model is more complicated, both in dynamics and regime/transition abundance, compared to the vacuum model. We again have a power-law transition there; this time it is the $K_3 \to K_3$ transition between two different high-curvature (GB) Kasner regimes. Again, they are power-law and cannot be realistic.

Realistic transitions for the model with Λ-term include either double $K_3 \to E_{3+D} \leftarrow P_{(1,0)}$ or single $K_3 \to E_{3+D}$—in both cases, the late-time asymptote is an exponential solution with expanding three and contracting extra dimensions. These transitions are always present for $D \geqslant 2$, but exact ranges of the parameters depend on D:

- For $D = 2$, they exist only for $\alpha > 0$: for $\alpha\Lambda < 1/2$ (including the entire $\Lambda < 0$) the $K_3 \to E_{3+2}$ transition exists, and for $\alpha\Lambda < -1/6$, additional $P_{(1,0)} \to E_{3+2}$ transitions to the same exponential solution exist, making it a double transition $K_3 \to E_{3+2} \leftarrow P_{(1,0)}$;
- For $D = 3$, we additionally have transitions for $\alpha < 0$, $\Lambda > 0$, $\alpha\Lambda \leqslant -3/2$. There, we have two double transitions $K_3^1 \to E_{3+3}^1 \leftarrow P_{(1,0)}$ and $K_3^2 \to E_{3+3}^2 \leftarrow P_{(0,1)}$; the same double transitions exist for the entire $\alpha > 0$, $\Lambda < 0$, while for $\alpha > 0$, $\Lambda > 0$, $\alpha\Lambda \leqslant 1/2$, they reduce to a couple of single transitions $K_3^{1,2} \to E_{3+3}^{1,2}$;
- The general $D \geqslant 4$ case has the same regimes as $D = 3$, but the limits of their existence are different and D-dependent. Double transitions in the negative α domain exist for $\alpha < 0$, $\Lambda > 0$, $\alpha\Lambda \leqslant \zeta_1$, and for positive α single $K_3 \to E_{3+D}$, transitions exist for $\alpha\Lambda < \zeta_3$, while for $\alpha\Lambda \leqslant \zeta_2$, they are supplemented with $P_{(1,0)} \to E_{3+D}$ transitions and form double transitions.

So, this analysis demonstrates how the usage of simple requirements for the existence of realistic Universe evolution can help us impose constraints on the parameters of the theory. However, Gauss–Bonnet gravity is used not only in cosmology. There are various aspects of Gauss–Bonnet gravity in AdS spaces, such as shear viscosity to entropy ratio, causality violations, or CFTs in dual gravity description, which allow us to put constraints on the parameters as well; these constraints could be summarized as follows [108–115]:

$$-\frac{(D+2)(D+3)(D^2+5D+12)}{8(D^2+3D+6)^2} \equiv \eta_2 \leqslant \alpha\Lambda \leqslant \eta_1 \equiv \frac{(D+2)(D+3)(3D+11)}{8D(D+5)^2}. \quad (24)$$

Limits for dS ($\Lambda > 0$) are less numerous and are based on different aspects (causality violations, perturbation propagation, and so on) of black hole physics in dS spaces. The most stringent constraint coming from these considerations is [44,48,116]

$$\alpha\Lambda \geqslant \eta_3 \equiv -\frac{D^2+7D+4}{8(D-1)(D+2)}. \quad (25)$$

Then, we can confront our limits with those obtained from other considerations of GB: we put our constraints in Figure 8a (with $\zeta_{1,2,3}$ being defined in (21)–(23)) and mentioned constraints from the other literature in Figure 8b ($\alpha > 0$, $\Lambda > 0$, which is not shaded but is included) and their intersection is presented in Figure 8c. From there, one can see that $\alpha > 0$ is favored by the combination of constraints and the resulting constraint on $\alpha\Lambda$ reads

$$\frac{3D^2-7D+6}{4D(D-1)} \equiv \zeta_3 \geqslant \alpha\Lambda \geqslant \eta_2 \equiv -\frac{(D+2)(D+3)(D^2+5D+12)}{8(D^2+3D+6)^2}. \quad (26)$$

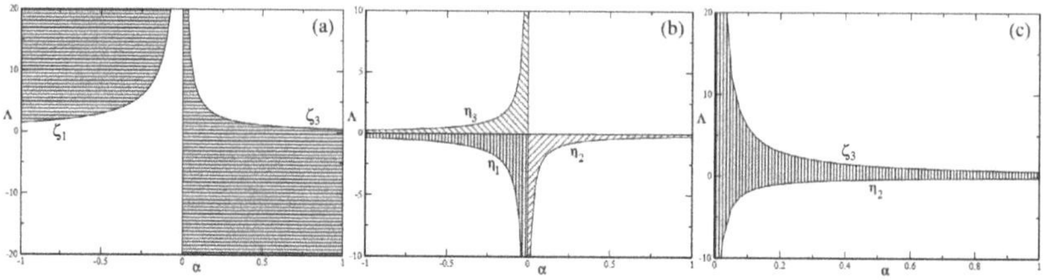

Figure 8. Summary of the bounds on (α, Λ) from this paper alone on (**a**) panel; from other considerations found in the literature on (**b**) panel; and the intersection between them on (**c**) panel (see the text for more details).

It is worth mentioning that initially this result was obtained in [99], but as it is an important milestone, we decided to reproduce it here.

The next model we considered is the vacuum cubic Lovelock gravity. Analysis suggests that for all $D \geqslant 3$ cases, there are realistic compactification schemes, though the regime abundance and structure is different in $D = 3 \div 7$ and in $D \geqslant 8$. For $D = 3 \div 7$, there is one realistic transition $P_{(1,0)} \to E_{3+D}$, which exists for $\alpha > 0$, $\mu \leqslant \mu_1$ (including entire $\beta < 0$); the other "would-be" viable transition is $P_{(1,0)} \to K_1$, but the corresponding late-time asymptote K_1 is Kasner regime. It is power-law and its expansion rate is insufficient to describe the current expansion of the Universe; these regimes are demonstrated in Figure 4a,b. For $D \geqslant 8$, we have the following realistic compactifications: for $\alpha > 0$, $\beta < 0$, we have $P_{(1,0)} \to E_{3+D}$, which exists everywhere in $\beta < 0$ and $K_5 \to E_{3+D}$, which exists for $\mu \leqslant \mu_2$; for $\alpha > 0$, $\beta > 0$, there is double transition $K_5 \to E_{3+D} \leftarrow P_{(1,0)}$, for $\mu \leqslant \mu_3$ which "decouples" into $K_5 \to E_{3+D}$ and $P_{(1,0)} \to K_1$ (the latter not being quite viable) for $\mu > \mu_3$; these regimes are demonstrated in Figure 4c–e. This way, realistic compactification exists only for $\alpha > 0$, but for all β, there is at least one viable compactification regime.

Another model we consider is the EGB model, with perfect fluid as a source. Due to complexity, so far we have considered only low-dimensional cases ($D = 1, 2$), and these are the cases we are reporting here. So, in $D = 1$, there are three nonsingular transitions, but none of them have realistic late-time regimes. On the contrary, for $D = 2$, there is realistic compactification $K_3 \to E_{3+2}$, and it is achieved for $\alpha > 0$, $\omega < 1/3$ with the measure of the trajectories experiencing this transition, increasing as $\omega \to 1/3 - 0$; the situation is demonstrated in Figure 5.

The model with perfect fluid has two interesting features we want to highlight for our readers. The first of them is what we call "constant volume solution"—exponential solution with constant volume, defined as follows in the continuity equation:

$$\dot{\rho} + \rho(1+\omega)\sum_i H_i = 0, \tag{27}$$

where we used $p = \omega\rho$ equation of state; for exponential solutions $H_i \equiv \text{const}$, so that l.h.s. of the equations of motion are constants, and so the density $\rho \equiv \text{const}$, then the continuity Equation (27) reduces to

$$(\rho + p)\sum_i H_i = 0 \iff \left[\begin{array}{c} \rho = 0 \quad (\mathbf{a}) \\ p = -\rho \quad (\mathbf{b}) \\ \sum_k H_k = 0 \quad (\mathbf{c}) \end{array}\right. \tag{28}$$

Then, we can clearly see all three possibilities when exponential solutions could exist—it is either vacuum (case (a)), Λ-term (case (b)), or the third possibility with $\sum_k H_k = 0$ (case (c)). One can see that in the third possibility, the sum of Hubble parameters is zero, so the comoving volume is constant, that is why we call these exponential solutions "constant volume solutions". We have a separate paper dedicated to the study of their properties [77] (see also [76] for the general scheme and [79] for non-constant-volume solutions) and here we notice their appearance on the actual evolution curves. Well, "appearance on the evolution curves" is not quite correct—they appear, but not on the evolution curves—they are "avoided" by evolution flux. In Figure 5f, we presented the vicinity of the constant-volume solution and one can see that it is a saddle point. Since it has $\sum_k H_k = 0$, it is not formally stable in either $t \to +\infty$ or $t \to -\infty$ (that would require either $\sum_k H_k > 0$ or $\sum_k H_k < 0$; see [81]), so it is natural that it is a saddle point. The important point is that, despite the fact that we described these solutions in [77], their actual appearance and nature remained unknown, so the fact that we discovered their presence in a model with perfect fluid as a saddle point is an important result.

The second feature we want to highlight for the readers is the fact that for $\rho \to 0$, we obtain vacuum EGB behavior, reported in Section 4. Indeed, from Figure 5b,d, one can notice that E_{3+2} is located on the $\rho = 0$ boundary, and it is quite clear why. From (28), we can see that there are only three possibilities for building an exponential solution—option (c) gives us CVS, and (b) is not an option since we do not have the Λ-term here, so the only option remaining is (a), and E_{3+2} is located exactly on $\rho = 0$. Moreover, taking the $\rho = 0$ slice from the evolution vector field would reconstruct vacuum EGB behavior, so that the vacuum EGB model is the $\rho \to 0$ limit of the perfect fluid model. While this is somewhat expected, it is still an interesting fact.

The less expected and equally interesting fact is the boundary on the equation of state—the $K_3 \to E_{3+2}$ transition takes place only for $\omega < 1/3$, while for $\omega > 1/3$, it is replaced by $nS \to E_{3+2}$. The explanation lies in the difference in scalings for matter and curvature and $\omega = 1/3$ is a critical value for the GB term (see [62] for details). For contraction (it defines the stability of the past asymptote, and K_3 here is a past asymptote), if $\omega < 1/3$, curvature dominates, and GB Kasner is preserved, for $\omega > 1/3$, matter dominates and GB Kasner is no more. That is what is happening with K_3 as a past asymptote.

The next case to consider is the case with the spatial curvature in EGB gravity. After the discovery of the regime, where the scale factor of the extra dimensions could be stabilized [86], in EGB gravity, if the spatial curvature of the extra-dimensional section is negative, we started to investigate this regime and the models with nonzero spatial curvature in more detail. Indeed, the regime is interesting, but negative curvature is hard to handle, while positive curvature is more associated with compact manifolds. So, if there was a similar regime but with positive curvature, this would be ideal. Indeed, we found this regime in the case of positive curvature, but its abundance is much less than that of the negative curvature. What is more, its abundance is decreasing with an increase in the number of extra dimensions. Thus, it exists and is stable, but to end up in this regime, we require some sort of fine-tuning.

And finally, the totally anisotropic case—the case where initially we had all dimensions equally distributed (Bianchi-I-type). All previous cases assumed that all spatial dimensions are divided into subspaces—isotropic three-dimensional which represent our Universe, and isotropic extra-dimensional. However, in the most natural scenario, we should start with totally anisotropic space, and it should naturally evolve into a configuration with two isotropic subspaces—this evolution is what we hope to find. The results suggest that for the cases where stable exponential solutions exist, such transitions are possible. For instance, for four spatial dimensions, the only stable exponential solution is isotropic, so only it could be reached. However, in five spatial dimensions, there are two, isotropic and $[3+2]$ splitting—with expanding three and contracting two dimensions—so that the final configurations could be reached (see Figure 7a,b). Similarly, in higher numbers of dimensions, more different spatial splittings are allowed, which makes it less probable to end up with the "correct" spatial splitting.

Finally, let us put together all cases where realistic compactification to the exponential solution is happening, and so, ΛCDM could be recovered in a three-dimensional subspace. All such cases are summarized in Table 1. There, we indicate the type of model and the source of matter in the first column, the number of extra dimensions in the second column, and the conditions for the parameters under which compactification occurs.

Table 1. All cases with realistic compactification/ΛCDM recovery.

Model	Number of Extra Dimensions	Parameters
EGB, vac	$D \geqslant 2$	$\alpha > 0$
EGB, Λ	$D = 2$	$\alpha\Lambda < 1/2$ (incl. $\Lambda < 0$)
	$D = 3$	$\alpha < 0, \alpha\Lambda \leqslant -3/2$ $\alpha > 0, \alpha\Lambda < 1/2$ (incl. $\Lambda < 0$)
	$D \geqslant 4$	$\alpha < 0, \alpha\Lambda \leqslant \zeta_1$ $\alpha > 0, \alpha\Lambda < \zeta_3$ (incl. $\Lambda < 0$)
cubic Lovelock, vac	$D = 3 \div 7$ $D \geqslant 8$	$\alpha > 0, \mu < \mu_1$ $\alpha > 0$
EGB, perfect fluid	$D = 2$	$\alpha > 0, \omega < 1/3$
EGB, curvature	$D \geqslant 3$	$\gamma_D < 0, \alpha < 0, \alpha\Lambda < -3/2$
	$D = 3$	$\gamma_D > 0, \alpha < 0, \alpha\Lambda \in (-0.5448, -0.5) \cup (0, +\infty)$
	$D = 4$	$\gamma_D > 0, \alpha < 0, \alpha\Lambda \in (-27/54, -15/32) \cup (-0.3, 3/8)$
		$\gamma_D > 0, \alpha > 0, \Lambda > 0$
	$D \geqslant 5$	$\gamma_D > 0, \alpha < 0, \alpha\Lambda \in (\xi_1, \xi_2) \cup (\xi_3, \xi_4)$ (see Figure 6c)
		$\gamma_D < 0, \alpha > 0, \alpha\Lambda < -D(D-1)/(4(D-2)(D-3))$

11. Conclusions

To conclude, we have demonstrated that it is possible to set constraints on the parameters of Lovelock gravity using generic requirements, such as the existence of a nonsingular transition from a high-curvature to a realistic low-curvature regime. Since the theory is multidimensional, "realistic" here stands for expanding three- and contracting or static extra-dimensional subspaces. These results are obtained under the condition that the total space is split into two isotropic subspaces—three- and extra-dimensional with the former representing our Universe while the latter is the "hidden" extra dimensions. Our results also suggest that the transition from a totally anisotropic (Bianchi-I-type) Universe to the configuration with two isotropic subspaces is quite natural and happening for quite a wide range of parameters. However, in a high number of spatial dimensions, compactification could happen to a different spatial splitting, e.g., six spatial dimensions could end up both as $(3+3)$—three contracting and three expanding—which favors observations, or as $(4+2)$—four expanding and two contracting—which obviously contradicts; the exact outcome depends on the parameters and initial conditions.

This review reflects the current status of the Lovelock cosmology exploration. From it, one can see that despite some progress, we still are quite far from fully understanding the dynamics of Lovelock cosmology. Most of the progress is conducted within EGB gravity, which is the lowest non-GR contribution of Lovelock gravity. The results reported for cubic Lovelock gravity demonstrate that we cannot extrapolate results for EGB on higher-order Lovelock gravity, we need to investigate at least cubic Lovelock gravity to see if higher orders could be extrapolated from EGB and cubic contributions. So far we investigated only vacuum regimes in cubic Lovelock gravity, and some progress was made on exponential solutions and solutions with spatial curvature, but a full investigation of the regimes in cubic Lovelock gravity with Λ-term is still ongoing. It is more complicated than the cases considered in this review and it has more parameters, so both the structure of the regimes' abundance and their distribution over parameter space is much more complicated—that slows the investigation, but we are still hoping to complete it in the future.

However, even within EGB gravity, we lack a full understanding and full description of all regimes and cases—vacuum and Λ-term transitions are described while the model with the perfect fluid as a source still lacks full description. We described low-D cases, which demonstrate a promise in explaining successful compactification in the presence of ordinary matter but are still struggling with the general high-D case. Still, we are planning to complete it in the future, which will broaden our understanding of the dynamics of EGB cosmologies.

Overall, as just stated, the results reported here demonstrate promise in describing our observed Universe as a three-dimensional subspace of the higher-dimensional theory utilizing Lovelock gravity, which exists only in higher dimensions. As the results demonstrated promise, we are going to continue our investigation of the model and the pursuit of new exciting features and regimes that this beautiful theory could provide.

Funding: This research received no external funding.

Conflicts of Interest: The author declares no conflicts of interest.

Notes

1. Good proof comes from Newtonian gravity: if defined in more than three spatial dimensions, it has no stable orbits, while the Earth is rotating around the Sun for billions of years.
2. Apparently, the original Kaluza–Klein theory is considered incorrect for a number of reasons and follow-up investigations generalize it—see, e.g., [14] for review.
3. We use this term to describe the situation when an initially anisotropic space evolves into a configuration with two isotropic subspaces; for $(3+D)$ spatial splitting it would be $(a_1^2, a_2^2, \ldots, a_{3+D}^2) \to (a^2, a^2, a^2, b^2, \ldots, b^2)$.
4. We consider *ansatz* for spacetime in form of a warped product $M_4 \times b(t)M_D$, where M_4 is a Friedmann–Robertson–Walker manifold with scale factor $a(t)$ whereas M_D is a D-dimensional Euclidean compact and constant curvature manifold with scale factor $b(t)$.

References

1. Vermeil, H. Nachrichten von der Königlichen Gesellschaft der Wissenschaften zu Göttingen. *Math.-Phys. Klasse* **1917**, *21*, 334–344.
2. Weyl, H. *Raum, Zeit, Materie*, 4th ed.; Springer: Berlin, Germany, 1921.
3. Cartan, E. Sur les équations de la gravitation d'Einstein. *J. Math. Pure Appl.* **1922**, *1*, 141–203.
4. Lovelock, D. The Einstein Tensor and Its Generalizations. *J. Math. Phys.* **1971**, *12*, 498–501. [CrossRef]
5. Tipler, F.J. Singularities in conformally flat spacetimes. *Phys. Lett. A* **1977**, *64*, 8–10. [CrossRef]
6. Kitaura, T.; Wheeler, J.T. Anisotropic, time-dependent solutions in maximally Gauss-Bonnet extended gravity. *Nucl. Phys. B* **1991**, *355*, 250–277. [CrossRef]
7. Kitaura, T.; Wheeler, J.T. New singularity in anisotropic, time-dependent, maximally Gauss-Bonnet extended gravity. *Phys. Rev. D* **1993**, *48*, 667–672. [CrossRef]
8. Nordström, G. Über die Möglichkeit, das Elektromagnetische Feld und das Gravitationsfeld zu vereiningen. *Phys. Z.* **1914**, *15*, 504–506.
9. Nordström, G. Zur Theorie der Gravitation vom Standpunkt des Relativitätsprinzips. *Ann. Phys.* **1913**, *347*, 533–554. [CrossRef]
10. Einstein, A. The Foundation of the General Theory of Relativity. *Ann. Phys.* **1916**, **49**, 769–822. [CrossRef]
11. Kaluza, T. Zum Unitätsproblem der Physik. *Sit. Preuss. Akad. Wiss.* **1921**, *K1*, 966.
12. Klein, O. Quantentheorie und fünfdimensionale Relativitätstheorie. *Z. Phys.* **1926**, *37*, 895–906. [CrossRef]
13. Klein, O. The Atomicity of Electricity as a Quantum Theory Law. *Nature* **1926**, *118*, 516. [CrossRef]
14. Wesson, P.S. *Five-Dimensional Physics: Classical and Quantum Consequences of Kaluza-Klein Cosmology*; World Scientific: Hackensack, NJ, USA, 2006.
15. Arkani-Hamed, N.; Dimopoulos, S.; Dvali, G. The Hierarchy problem and new dimensions at a millimeter. *Phys. Lett. B* **1998**, *429*, 263–272. [CrossRef]
16. Arkani-Hamed, N.; Dimopoulos, S.; Dvali, G. Phenomenology, astrophysics and cosmology of theories with submillimeter dimensions and TeV scale quantum gravity. *Phys. Rev. D* **1999**, *59*, 086004. [CrossRef]
17. Randall, L.; Sundrum, R. Large Mass Hierarchy from a Small Extra Dimension. *Phys. Rev. Lett.* **1999**, *83*, 3370–3373. [CrossRef]
18. Randall, L.; Sundrum, R. An Alternative to Compactification. *Phys. Rev. Lett.* **1999**, *83*, 4690–4693. [CrossRef]
19. Fermi-LAT Collaboration. Limits on large extra dimensions based on observations of neutron stars with the Fermi-LAT. *JCAP* **2012**, *1202*, 012.
20. Pardo, K.; Fishbach, M.; Holz, D.E.; Spergel, D.N. Limits on the number of spacetime dimensions from GW170817. *JCAP* **2018**, *1807*, 048. [CrossRef]
21. ATLAS Collaboration. Search for microscopic black holes and string balls in final states with leptons and jets with the ATLAS detector at $\sqrt{s}$ = 8 TeV. *JHEP* **2014**, *1408*, 103.
22. ATLAS Collaboration. Search for strong gravity in multijet final states produced in pp collisions at $\sqrt{s}$ = 13 TeV using the ATLAS detector at the LHC. *JHEP* **2016**, *1603*, 026.
23. Scherk, J.; Schwarz, J.H. Dual Models for Nonhadrons. *Nucl. Phys. B* **1974**, *81*, 118–144. [CrossRef]

24. Virasoro, M.A. Alternative Constructions of Crossing-Symmetric Amplitudes with Regge Behavior. *Phys. Rev.* **1969**, *177*, 2309. [CrossRef]
25. Shapiro, J.A. Electrostatic analog for the Virasoro model. *Phys. Lett. B* **1970**, *33*, 361–362. [CrossRef]
26. Candelas, P.; Horowitz, G.T.; Strominger, A.; Witten, E. Vacuum configurations for superstrings. *Nucl. Phys. B* **1985**, *258*, 46–74. [CrossRef]
27. Gross, D.J.; Harvey, J.A.; Martinec, E.; Rohm, R. Heterotic String. *Phys. Rev. Lett.* **1985**, *54*, 502. [CrossRef] [PubMed]
28. Zwiebach, B. Curvature squared terms and string theories. *Phys. Lett. B* **1985**, *156*, 315–317. [CrossRef]
29. Lanczos, C. A Hadrons and Nuclei. *Z. Phys.* **1932**, *73*, 147–168. [CrossRef]
30. Lanczos, C. A Remarkable Property of the Riemann-Christoffel Tensor in Four Dimensions. *Ann. Math.* **1938**, *39*, 842–850. [CrossRef]
31. Zumino, B. Gravity theories in more than four dimensions. *Phys. Rep.* **1986**, *137*, 109–114. [CrossRef]
32. Müller-Hoissen, F. Spontaneous compactification with quadratic and cubic curvature terms. *Phys. Lett. B* **1985**, *163*, 106–110. [CrossRef]
33. Deruelle, N.; Fariña-Busto, L. Lovelock gravitational field equations in cosmology. *Phys. Rev. D* **1990**, *41*, 3696–3708. [CrossRef] [PubMed]
34. Müller-Hoissen, F. Dimensionally continued Euler forms: Kaluza-Klein cosmology and dimensional reduction. *Class. Quant. Gravity* **1986**, *3*, 665–677. [CrossRef]
35. Mena Marugán, G.A. Dynamically generated four-dimensional models in Lovelock cosmology. *Phys. Rev. D* **1992**, *46*, 4340. [CrossRef] [PubMed]
36. Elizalde, E.; Makarenko, A.N.; Obukhov, V.V.; Osetrin, K.E.; Filippov, A.E. Stationary vs. singular points in an accelerating FRW cosmology derived from six-dimensional Einstein–Gauss–Bonnet gravity. *Phys. Lett. B* **2007**, *644*, 1–6. [CrossRef]
37. Maeda, K.I.; Ohta, N. Inflation from superstring and M-theory compactification with higher order corrections. *Phys. Rev. D* **2005**, *71*, 063520. [CrossRef]
38. Maeda, K.I.; Ohta, N. Cosmic acceleration with a negative cosmological constant in higher dimensions. *JHEP* **2014**, *1406*, 095. [CrossRef]
39. Boulware, D.G.; Deser, S. String-generated gravity models. *Phys. Rev. Lett.* **1985**, *55*, 2656–2660. [CrossRef]
40. Wheeler, J.T. Symmetric Solutions to the Gauss-Bonnet Extended Einstein Equations. *Nucl. Phys. B* **1986**, *268*, 737–746. [CrossRef]
41. Nojiri, S.; Odintsov, S.D. Anti-de-Sitter black hole thermodynamics in higher derivative gravity and new confining–deconfining phases in dual CFT. *Phys. Lett. B* **2001**, *521*, 87–95; Erratum in *Phys. Lett. B* **2002**, *542*, 301. [CrossRef]
42. Nojiri, S.; Odintsov, S.D.; Ogushi, S. Friedmann-Robertson-Walker brane cosmological equations from the five-dimensional bulk (A)dS black hole. *Int. J. Mod. Phys. A* **2002**, *17*, 4809–4870. [CrossRef]
43. Cvetic, M.; Nojiri, S.; Odintsov, S.D. Black Hole Thermodynamics and Negative Entropy in deSitter and Anti-deSitter Einstein-Gauss-Bonnet gravity. *Nucl. Phys. B* **2002**, *628*, 295–330. [CrossRef]
44. Cai, R.G. Gauss-Bonnet black holes in AdS spaces. *Phys. Rev. D* **2002**, *65*, 084014. [CrossRef]
45. Torii, T.; Maeda, H. Spacetime structure of static solutions in Gauss-Bonnet gravity: Neutral case. *Phys. Rev. D* **2005**, *71*, 124002. [CrossRef]
46. Torii, T.; Maeda, H. Spacetime structure of static solutions in Gauss-Bonnet gravity: Charged case. *Phys. Rev. D* **2005**, *72*, 064007. [CrossRef]
47. Wilshire, D.L. Spherically symmetric solutions of Einstein-Maxwell theory with a Gauss-Bonnet term. *Phys. Lett. B* **1986**, *169*, 36–40. [CrossRef]
48. Cai, R.G. A Note on Thermodynamics of Black Holes in Lovelock Gravity. *Phys. Lett. B* **2004**, *582*, 237–242. [CrossRef]
49. Grain, J.; Barrau, A.; Kanti, P. Exact Results for Evaporating Black Holes in Curvature-Squared Lovelock Gravity: Gauss-Bonnet Greybody Factors. *Phys. Rev. D* **2005**, *72*, 104016. [CrossRef]
50. Cai, R.G.; Ohta, N. Black holes in pure Lovelock gravities. *Phys. Rev. D* **2006**, *74*, 064001. [CrossRef]
51. Camanho, X.O.; Edelstein, J.D. A Lovelock black hole bestiary. *Class. Quant. Gravity* **2013**, *30*, 035009. [CrossRef]
52. Maeda, H. Final fate of spherically symmetric gravitational collapse of a dust cloud in Einstein-Gauss-Bonnet gravity. *Phys. Rev. D* **2006**, *73*, 104004. [CrossRef]
53. Nozawa, M.; Maeda, H. Effects of Lovelock terms on the final fate of gravitational collapse: Analysis in dimensionally continued gravity. *Class. Quant. Gravity* **2006**, *23*, 1779–1800. [CrossRef]
54. Maeda, H. Effects of Gauss–Bonnet term on the final fate of gravitational collapse. *Class. Quant. Gravity* **2006**, *23*, 2155–2170. [CrossRef]
55. Dehghani, M.H.; Farhangkhah, N. Asymptotically flat radiating solutions in third order Lovelock gravity. *Phys. Rev. D* **2008**, *78*, 064015. [CrossRef]
56. Reall, H.S.; Tanahashi, N.; Way, B. Shock formation in Lovelock theories. *Phys. Rev. D* **2015**, *91*, 044013. [CrossRef]
57. Deruelle, N. On the approach to the cosmological singularity in quadratic theories of gravity: The Kasner regimes. *Nucl. Phys. B* **1989**, *327*, 253–266. [CrossRef]
58. Pavluchenko, S.A. General features of Bianchi-I cosmological models in Lovelock gravity. *Phys. Rev. D* **2009**, *80*, 107501. [CrossRef]
59. Pavluchenko, S.A.; Toporensky, A.V. A note on differences between (4 + 1)- and (5 + 1)-dimensional anisotropic cosmology in the presence of the Gauss-Bonnet term. *Mod. Phys. Lett. A* **2009**, *24*, 513–521. [CrossRef]

60. Ivashchuk, V.D. On anisotropic Gauss-Bonnet cosmologies in $(n+1)$ dimensions, governed by an n-dimensional Finslerian 4-metric. *Gravitat. Cosmol.* **2010**, *16*, 118–125. [CrossRef]
61. Ivashchuk, V. On cosmological-type solutions in multi-dimensional model with Gauss-Bonnet term. *Int. J. Geom. Meth. Mod. Phys.* **2010**, *7*, 797–819. [CrossRef]
62. Kirnos, I.V.; Makarenko, A.N.; Pavluchenko, S.A.; Toporensky, A.V. The nature of singularity in multidimensional anisotropic Gauss-Bonnet cosmology with a perfect fluid. *Gen. Relativ. Gravit.* **2010**, *42*, 2633–2641. [CrossRef]
63. Pavluchenko, S.A. The dynamics of the flat anisotropic models in the Lovelock gravity. I: The even-dimensional case. *Phys. Rev. D* **2010**, *82*, 104021. [CrossRef]
64. Ishihara, H. Cosmological solutions of the extended Einstein gravity with the Gauss-Bonnet term. *Phys. Lett. B* **1986**, *179*, 217–222. [CrossRef]
65. Kirnos, I.V.; Pavluchenko, S.A.; Toporensky, A.V. New features of flat (4+1)-dimensional cosmological model with a perfect fluid in Gauss-Bonnet gravity. *Gravitat. Cosmol.* **2010**, *16*, 274–282. [CrossRef]
66. Ivashchuk, V.D.; Kobtsev, A.A. On exponential cosmological type solutions in the model with Gauss-Bonnet term and variation of gravitational constant. *Eur. Phys. J. C* **2015**, *75*, 177; Erratum in *Eur. Phys. J. C* **2016**, *76*, 584. [CrossRef]
67. Ernazarov, K.K.; Ivashchuk, V.D.; Kobtsev, A.A. On exponential solutions in the Einstein–Gauss–Bonnet cosmology, stability and variation of *G*. *Gravit. Cosmol.* **2016**, *22*, 245–250. [CrossRef]
68. Ivashchuk, V.D. On stable exponential solutions in Einstein–Gauss–Bonnet cosmology with zero variation of *G*. *Gravit. Cosmol.* **2016**, *22*, 329–332; Erratum in *Gravit. Cosmol.* **2017**, *23*, 401. [CrossRef]
69. Ernazarov, K.K.; Ivashchuk, V.D. Stable exponential cosmological solutions with zero variation of G in the Einstein–Gauss–Bonnet model with a Λ-term. *Eur. Phys. J. C* **2017**, *77*, 89. [CrossRef]
70. Ivashchuk, V.D.; Kobtsev, A.A. Stable exponential cosmological solutions with two factor spaces in the Einstein–Gauss–Bonnet model with a Λ-term. *Gen. Relativ. Gravit.* **2018**, *50*, 119. [CrossRef]
71. Ivashchuk, V.D.; Kobtsev, A.A. Exact exponential cosmological solutions with two factor spaces of dimension m in EGB model with a Λ-term. *Int. J. Geom. Meth. Mod. Phys.* **2018**, *16*, 1950025. [CrossRef]
72. Ernazarov, K.K.; Ivashchuk, V.D. Examples of Stable Exponential Cosmological Solutions with Three Factor Spaces in EGB Model with a Λ-Term. *Gravit. Cosmol.* **2019**, *25*, 164–168. [CrossRef]
73. Ivashchuk, V.D.; Kobtsev, A.A. Exponential cosmological solutions with two factor spaces in EGB model with $\Lambda = 0$ revisited. *Eur. Phys. J. C* **2019**, *79*, 824. [CrossRef]
74. Ernazarov, K.K.; Ivashchuk, V.D. Stable exponential cosmological solutions with three different Hubble-like parameters in EGB model with a Λ-term. *Eur. Phys. J. C* **2020**, *80*, 543. [CrossRef]
75. Pavluchenko, S. Exponential cosmological solutions in Einstein-Gauss-Bonnet gravity with two subspaces: General approach. *arXiv* **2021**, arXiv:2104.10423.
76. Chirkov, D.; Pavluchenko, S.; Toporensky, A. Exact exponential solutions in Einstein-Gauss-Bonnet flat anisotropic cosmology. *Mod. Phys. Lett. A* **2014**, *29*, 1450093. [CrossRef]
77. Chirkov, D.; Pavluchenko, S.; Toporensky, A. Constant volume exponential solutions in Einstein-Gauss-Bonnet flat anisotropic cosmology with a perfect fluid. *Gen. Relativ. Gravit.* **2014**, *46*, 1799. [CrossRef]
78. Pavluchenko, S.A.; Toporensky, A.V. Note on properties of exact cosmological solutions in Lovelock gravity. *Gravitat. Cosmol.* **2014**, *20*, 127–131. [CrossRef]
79. Chirkov, D.; Pavluchenko, S.; Toporensky, A. Non-constant volume exponential solutions in higher-dimensional Lovelock cosmologies. *Gen. Relativ. Gravit.* **2015**, *47*, 137. [CrossRef]
80. Pavluchenko, S.A. Stability analysis of the exponential solutions in Lovelock cosmologies. *Phys. Rev. D* **2015**, *92*, 104017. [CrossRef]
81. Ivashchuk, V.D. On stability of exponential cosmological solutions with nonstatic volume factor in the Einstein-Gauss-Bonnet model. *Eur. Phys. J. C* **2016**, *76*, 431. [CrossRef]
82. Chirkov, D.M.; Toporensky, A.V. On stable exponential cosmological solutions in the EGB model with a cosmological constant in dimensions D = 5, 6, 7, 8. *Gravit. Cosmol.* **2017**, *23*, 359–366. [CrossRef]
83. Ernazarov, K.K.; Ivashchuk, V.D. Stable exponential cosmological solutions with zero variation of GG and three different Hubble-like parameters in the Einstein-Gauss-Bonnet model with a Λ-term. *Eur. Phys. J. C* **2017**, *77*, 402. [CrossRef]
84. Ivashchuk, V.D.; Kobtsev, A.A. Stable exponential cosmological solutions with 3- and l-dimensional factor spaces in the Einstein-Gauss-Bonnet model with a Λ-term. *Eur. Phys. J. C* **2018**, *78*, 100. [CrossRef]
85. Ivashchuk, V.D.; Kobtsev, A.A. On stable exponential cosmological solutions with two factor spaces in (1 + m + 2)-dimensional Einstein–Gauss–Bonnet model with Λ-term. *Phil. Trans. A Math. Phys. Eng. Sci.* **2022**, *380*, 20210177. [CrossRef] [PubMed]
86. Canfora, F.; Giacomini, A.; Pavluchenko, S.A. Dynamical compactification in Einstein-Gauss-Bonnet gravity from geometric frustration. *Phys. Rev. D* **2013**, *88*, 064044. [CrossRef]
87. Canfora, F.; Giacomini, A.; Pavluchenko, S.A. Cosmological dynamics in higher-dimensional Einstein-Gauss-Bonnet gravity. *Gen. Relativ. Gravit.* **2014**, *46*, 1805. [CrossRef]
88. Canfora, F.; Giacomini, A.; Pavluchenko, S.A.; Toporensky, A. Friedmann dynamics recovered from compactified Einstein-Gauss-Bonnet cosmology. *Gravit. Cosmol.* **2018**, *24*, 28–38. [CrossRef]

89. Pavluchenko, S.A. The generality of inflation in closed cosmological models with some quintessence potentials. *Phys. Rev. D* **2003**, *67*, 103518. [CrossRef]
90. Pavluchenko, S.A. Constraints on inflation in closed universe. *Phys. Rev. D* **2004**, *69*, 021301. [CrossRef]
91. Chirkov, D.; Giacomini, A.; Pavluchenko, S.A.; Toporensky, A. Cosmological solutions in Einstein-Gauss-Bonnet gravity with static curved extra dimensions. *Eur. Phys. J. C* **2021**, *81*, 136. [CrossRef]
92. Pavluchenko, S.; Toporensky, A. Effects of spatial curvature and anisotropy on the asymptotic regimes in Einstein-Gauss-Bonnet gravity. *Eur. Phys. J. C* **2018**, *78*, 373. [CrossRef]
93. Chirkov, D.; Pavluchenko, S.A. Some aspects of the cosmological dynamics in Einstein-Gauss-Bonnet gravity. *Mod. Phys. Lett. A* **2021**, *36*, 2150092. [CrossRef]
94. Wolf, J.A. *Spaces of Constant Curvature*, 4th ed.; Publish or Perish: Wilmington, DE, USA, 1984; p. 69.
95. Pavluchenko, S.A. Cosmological dynamics of spatially flat Einstein-Gauss-Bonnet models in various dimensions. Vacuum case. *Phys. Rev. D* **2016**, *94*, 024046. [CrossRef]
96. Pavluchenko, S.A. Realistic Compactification Models in Einstein–Gauss–Bonnet Gravity. *Particles* **2018**, *1*, 36–55. [CrossRef]
97. Kasner, E. Geometrical Theorems on Einstein's Cosmological Equations. *Am. J. Math.* **1921**, *43*, 217–221. [CrossRef]
98. Pavluchenko, S.A. Cosmological dynamics of spatially flat Einstein-Gauss-Bonnet models in various dimensions: Low-dimensional Λ-term case. *Phys. Rev. D* **2016** *94*, 084019. [CrossRef]
99. Pavluchenko, S.A. Cosmological dynamics of spatially flat Einstein-Gauss-Bonnet models in various dimensions: High-dimensional Λ-term case. *Eur. Phys. J. C* **2017**, *77*, 503. [CrossRef]
100. Pavluchenko, S.A. Realistic compactification in spatially flat vacuum cosmological models in cubic Lovelock gravity: Low-dimensional case. *Eur. Phys. J. C* **2018**, *78*, 551. [CrossRef]
101. Pavluchenko, S.A. Realistic compactification in spatially flat vacuum cosmological models in cubic Lovelock gravity: High-dimensional case. *Eur. Phys. J. C* **2018**, *78*, 611. [CrossRef]
102. Pavluchenko, S.A. Dynamics of the cosmological models with perfect fluid in Einstein–Gauss–Bonnet gravity: Low-dimensional case. *Eur. Phys. J. C* **2019**, *79*, 111. [CrossRef]
103. Chirkov, D.; Giacomini, A.; Toporensky, A. Dynamic compactification with stabilized extra dimensions in cubic Lovelock gravity. *Gen. Relativ. Gravit.* **2018**, *50*, 98. [CrossRef]
104. Chirkov, D.; Giacomini, A.; Toporensky, A. Anisotropic cosmological dynamics in Einstein–Gauss–Bonnet gravity: An example of dynamical compactification in 7+1 dimensions. *Gen. Relativ. Gravit.* **2020**, *52*, 30. [CrossRef]
105. Chirkov, D.; Toporensky, A. Stability Analysis of Compactification in 3rd Order Lovelock Gravity. *Gravit. Cosmol.* **2023**, *29*, 262–268. [CrossRef]
106. Chirkov, D.; Toporensky, A. Splitting into two isotropic subspaces as a result of cosmological evolution in Einstein-Gauss-Bonnet gravity. *Gravit. Cosmol.* **2019**, *25*, 243–249. [CrossRef]
107. Taub, A.H. Empty Space-Times Admitting a Three Parameter Group of Motions. *Ann. Math.* **1951**, *53*, 472–490. [CrossRef]
108. Brigante, M.; Liu, H.; Myers, R.C.; Shenker, S.; Yaida, S. Viscosity Bound Violation in Higher Derivative Gravity. *Phys. Rev. D* **2008**, *77*, 126006. [CrossRef]
109. Brigante, M.; Liu, H.; Myers, R.C.; Shenker, S.; Yaida, S. Viscosity Bound and Causality Violation. *Phys. Rev. Lett.* **2008**, *100*, 191601. [CrossRef]
110. Buchel, A.; Myers, R.C. Causality of Holographic Hydrodynamics. *JHEP* **2009**, *0908*, 016. [CrossRef]
111. Hofman, D.M. Higher Derivative Gravity, Causality and Positivity of Energy in a UV complete QFT. *Nucl. Phys. B* **2009**, *823*, 174–194. [CrossRef]
112. de Boer, J.; Kulaxizi, M.; Parnachev, A. AdS_7/CFT_6, Gauss-Bonnet Gravity, and Viscosity Bound. *JHEP* **2010**, *1003*, 087. [CrossRef]
113. Camanho, X.O.; Edelstein, J.D. Causality constraints in AdS/CFT from conformal collider physics and Gauss-Bonnet gravity. *JHEP* **2010**, *1004*, 007. [CrossRef]
114. Buchel, A.; Escobedo, J.; Myers, R.C.; Paulos, M.F.; Sinha, A.; Smolkin, M. Holographic GB gravity in arbitrary dimensions. *JHEP* **2010**, *1003*, 111. [CrossRef]
115. Ge, X.-H.; Sin, S.-J. Shear viscosity, instability and the upper bound of the Gauss-Bonnet coupling constant. *JHEP* **2009**, *0905*, 051. [CrossRef]
116. Cai, R.G.; Guo, Q. Gauss-Bonnet black holes in dS spaces. *Phys. Rev. D* **2004**, *69*, 104025. [CrossRef]

MDPI AG
Grosspeteranlage 5
4052 Basel
Switzerland
Tel.: +41 61 683 77 34

Universe Editorial Office
E-mail: universe@mdpi.com
www.mdpi.com/journal/universe

www.ingramcontent.com/pod-product-compliance
Lightning Source LLC
LaVergne TN
LVHW070540071126
851858LV00011BA/249